Handbook of Nutraceuticals

Volume II

Scale-Up, Processing and Automation

Handbook of Nutraceuticals
Volume II
Scale-Up, Processing and Automation

Edited by
Yashwant Pathak

CRC Press
Taylor & Francis Group
Boca Raton London New York

CRC Press is an imprint of the
Taylor & Francis Group, an **informa** business

Reprint, 2018

CRC Press
Taylor & Francis Group
6000 Broken Sound Parkway NW, Suite 300
Boca Raton, FL 33487-2742

International Standard Book Number: 978-1-498-78314-9

Visit the Taylor & Francis Web site at
http://www.taylorandfrancis.com

and the CRC Press Web site at
http://www.crcpress.com

Printed and bound in India by Bhavish Print Solutions Pvt.Ltd.,

Dedicated to all the Rushis, sages, shamans, medicine men and women, and people of ancient traditions and cultures who contributed to the development of drugs and nutraceuticals worldwide and kept the science of health alive for the last several millennia

Contents

Foreword

> I enjoy talking with very old people. They have gone before us on a road by which we, too, may have to travel, and I think we do well to learn from them what it is like.
>
> **Socrates**

Progress in the past 150 years or so in the field of medicine has served us well, particularly in the treatment of communicable diseases, trauma, and surgically correctable conditions. However, the current health care system has become extremely technocratic, depersonalizing, expensive, and unable to deal with the chronic disease crises and unsustainable health care costs the United States is facing. The hundreds of billions of dollars poured into biomedical or Cartesian–Newtonian model-based research in the hope of finding "magic bullet" solutions have been largely unsuccessful.

For example, as shown in four recent large trials from the Nateglinide and Valsartan in Impaired Glucose Tolerance Outcomes Research (NAVIGATOR) and Action to Control Cardiovascular Risk in Diabetes (ACCORD) groups, the push by the pharmaceutical industry of drugs designed to treat risk factor(s), at the cost of billions of dollars, for primary prevention of chronic disease has not only been ineffective in preventing cardiac events, diabetes, and mortality, but also causes harm by aggressively treating risk factors (*N Engl J Med* 2010, 362(18): 1746–48).

Health care economics, science of epidemiology of chronic diseases and diet and lifestyle, public demand, and recognition by some health care providers and politicians alike of the inadequacy of the biomedical model to deal with health and health care cost have forced the system to grudgingly accept the wisdom of yesteryears that health is more than the absence of diseases. Health is the state of equilibrium of physiological system functions and sensorial, mental, and spiritual well-being. This reorganization brought the issue of disease prevention through proper nutrition and lifestyle changes into focus during the 2010 health care reform debate. These newly enacted health care reforms emphasize prevention in dealing with the escalating health care costs and chronic disease crises, and the expanding nutraceutical and functional food industry may have the opportunity to derive great economic benefit in the development of health-promoting products that people in the United States are seeking.

The term *nutraceutical* was coined recently by Stephen L. DeFelice; the concept that a food or parts of a food have health benefits including the prevention and treatment of disease has been known in the comprehensive medical system of India, known as Ayurveda (science of life), for several millennia as *Rasayana*.

रसायनं च तत् ज्ञेयं यत् जरा- व्याधि-नाशनम्

Rasāyanam ca tajjnyeyam yajjaravyādhināśanam.
The therapy that helps to retard aging and disease is Rasayana.
Susrut Samhita, Part 1, 14:13 (600 BCE, *Chikitsa Sthana*, Chapters 26–30 on Rasayana, ed. and trans. into English by G. D. Singhal, P. Mamgain, S. N. Tripathi, L. V. Guru, and R. H. Singh. Surbharathi, India: Choukhamba, 2001).

The science of Ayurvedic Rasayana is based on the philosophy that the optimum health of a given individual can be achieved only when all aspects of human dimensions are integrated. That means that healthfulness is an outcome of a dynamic interaction among our genetic (or physiological), mental (or psychological), emotional, spiritual, social, and environmental factors. The Rasayanas consist of an amplification and synergy of the natural health-promoting potential of certain plants and food products and related measures that are supposed to retard aging and to impart longevity, improve immunity and body resistance against disease, improve mental faculties, and add vitality and luster to the body. For more information on Rasayana therapy, please refer to Chapter 1, Volume I of this series.

The *Handbook of Nutraceuticals* series, edited by the most dynamic and integrative thinker in the field of nutraceuticals, Dr. Yashwant Pathak, has brought together international experts to shed light on the much needed processing and standardization of information. There is no doubt that the chapters in this Volume II will prove valuable for the industry and scientists alike.

Vimal Patel, PhD, retired professor
Indiana University School of Medicine
Indianapolis, Indiana

Preface

Improving the Credibility of Nutraceuticals and Functional Foods

Humanity survived many millennia on herbs, minerals, and products developed out of these for their health care needs. Nature has been the greatest guru for humans in this regard. Nature has taught and provided us with many herbs and food products to help us live longer. Most of the ancient traditions and cultures that have existed for thousands of years believed that the human life needs to be at least 100 years. In Sanskrit, one of our oldest languages, the blessing given to young children says "*Shatayushi Bhava*," meaning "Live longer than 100 years." In Sanskrit there is also a saying "*Nastimulum Vanushadhi*," which means that every plant has medicinal value—the key is "One needs to know how to use these *Yojakas Tatra Durlabhaha*."

The nutraceutical market is at the crossroads. The momentum is continuously building. The numbers are promising, with more than a $170 billion market for both nutraceuticals and functional foods expected by end of 2013. Some recent developments are encouraging for the growth of the nutraceutical market.

As reported in a recent *Nutraceuticals World* feature entitled "Bone Health Ingredients" (see complete reference information below), research suggests that vitamins are deficient in the diets of millions of children in the United States. The Albert Einstein College of Medicine conducted an assessment of more than 6,000 children to measure their intake of vitamins, particularly vitamin D. Several small studies found a high prevalence of vitamin D deficiency in specific populations of children in United States. This leads to the threat of bone and heart disease. Vitamin D was listed in the "Bone Health Ingredients" feature, which explained that the vitamin combines with calcium in bone formation, as part of the normal cycle of creation and destruction that takes place in the bones over the course of a lifetime (Joanna Cosgrave, "Bone Health Ingredients," *Nutraceuticals World*, 2009, http://www.nutraceuticalsworld.com/contents/view/14130).

Functional ingredients—particularly omega 3—have been associated with a "major" impact on the mood of pregnant women. Research in a 2009 article in the journal *Epidemiology* (see complete reference information below) shows a higher prevalence of depression in pregnant women who eat little or no seafood. The study reveals that in Western countries, depression is not an unusual characteristic of pregnancy—but depression is virtually nonexistent in the East, where seafood intake is higher. Researchers add that depression can have an adverse effect not only on the mother but also on her unborn child. A test of nearly 10,000 women whose omega-3 intake was calculated based on their seafood consumption found that participants with little of the fatty acid in their diet were 50% more likely to experience depression than those who ate three portions of fish per week (Jean Golding, Colin Steer, Pauline Emmett, John M. Davis, and Joseph R. Hibbeln, "High Levels of Depressive Symptoms in Pregnancy with Low Omega-3 Fatty Acid Intake from Fish," *Epidemiology*, 2009, 20:598–603).

Executive Director of the Global Organization for EPA/DHA Omega 3s, Adam Ismail, recently wrote in *Nutraceuticals World* that the fatty acids are still in their infancy as functional ingredients, even though the first products specifically containing them were launched 21 years ago (Adam Ismail, "Low Omega 3 Dose Fails to Help Heart Attack Survivors," *Nutraceuticals World*, 2010, http://www.nutraceuticalsworld.com/contents/view/29371).

Phytosterols are derived from plants and are similar in many ways to cholesterol; however, their consumption has been linked with lower cholesterol levels in the body. They have been seen to act in the membrane of cancer cells, reducing cholesterol there while activating caspase, an enzyme with action on cell death. With these dual nutraceutical benefits in mind, researchers from the Department of Animal Science and the University of Manitoba's Richardson Centre for Functional Foods and Nutraceuticals suggest that the molecules could help to prevent cancer when consumed. "This combined evidence strongly supports an anti-carcinogenic action of phytosterols and hence advocates their dietary inclusion" (T. C. Rideout and P. J. H. Jones, "Plant Sterols: An Essential Aspect of Cardiovascular Preventative Medicine," *Pulse*, 2010, 29: 1) Previous research from Wageningen University and Unilever found that the action of phytosterols on lowering cholesterol is consistent whether they are in fatty or nonfatty foods.

Protein plaques formed during the progression of Alzheimer's disease could be targeted using a combination of functional ingredients. Researchers at the University of California, Los Angeles looked into curcumin, which is derived from turmeric, and vitamin D3 as two functional ingredients used together. They found that the pair can help to ease the presence of beta-amyloid deposits in the brains of persons with Alzheimer's disease. "We hope that vitamin D3 and curcumin, both naturally occurring nutrients, may offer new preventive and treatment possibilities for Alzheimer's disease" (Barbi Trejo, Vitamin D3 and Curcumin Offer Hope to Alheimer's Patients," *Natural News*, 2009, http://www.naturalnews.com/026861_curcumin_vitamin_D3_disease.html#ixzz1C5KV2vU6). Giving HIV drugs to babies as functional ingredients in their milk—and offering breastfeeding mothers highly active treatment—is safe and can help prevent transmission of the virus, says a new study. Scientists at the University of North Carolina School of Medicine found that feeding breastfeeding infants daily with syrup containing antiretroviral ingredients and/or treating their mothers with antiretroviral drugs both helped limit infections (see http://www.med.unc.edu/infdis/research/hiv-stds). The death or infection rate among those infants who received no treatment was roughly double that of those whose mothers received treatment. Those who were fed the syrup directly as functional ingredients in their milk had an even higher survival rate. The study's lead investigator, Dr. Charles van der Horst, said that the treatment could spare mothers in the developing world a "horrible choice" of choosing to continue breastfeeding or relying on an unsafe water supply (M. Braun, M. M. Kabue, E. D. McCollum, et al. "Inadequate Coordination of Maternal and Infant HIV Services Detrimentally Affects Early Infant Diagnosis Outcomes in Lilongwe, Malawi," *Journal of Acquired Immune Deficiency Syndromes*, 2011, http://www.ncbi.nlm.nih.gov/pubmed). Of the 420,000 infants infected with HIV every year, approximately half were infected through breast milk, says the study. Many such studies show the benefits of nutraceuticals and functional foods as providing solutions for health care problems globally, and perhaps also helping to reduce health care expenditure worldwide.

Some of the challenges nutraceuticals and functional foods face are absence of standard testing methods accepted universally, regulatory ignorance, marketing incompetence, and ethical impunity. Even though many researchers believe that there is a connection between nutraceuticals and functional foods and reduced health care expenses, as well as disease prevention, these challenges still remain. Consumers like to have access to these products, but they also demand credibility and improved quality control. There is a need for rigorous quality control programs for these products. The industry needs more standardized testing, reliable reproducible clinical studies to prove the efficacies of these products in vivo, and possible in-process controls for these products to maintain the qualities universally. There is a need for providing adequate scientific evidence to the claims written on the label of these products. I very much understand that the nutraceuticals and functional foods market is a different ball game and need not be compared with the stringent pharmaceutical industry. It would not be advisable to apply the rules of football to basketball; however, the nutraceutical industry needs to develop their own rules for playing an even game. There is a need for improving the credibility of these products in the global market.

In our first volume, I recommended a new definition for nutraceuticals, which covers the regulatory aspect, as well as insisted on current good manufacturing practices and standard operating procedures for nutraceutical manufacturing. In the second volume of the *Handbook of Nutraceuticals*, there is a collection of 22 chapters covering various aspects of nutraceutical scaling up, processing, and automation. I believe the nutraceutical industry is moving in the right direction. Chapters 2 through 4 cover various extraction methods and advances in this area. Chapters 5 through 14 address various unit operations involved in scaling up and processing of the nutraceuticals, including the rheological and flow properties of nutraceutical products as characterization parameters. Chapters 15 through 20 address various aspects of fortification and value enhancement using nanotechnology, green concepts in food industry, flavoring of nutraceuticals, and clinical batch nutraceutical manufacturing. The last two chapters are on new technologies to prevent counterfeiting and automation in the nutraceutical industry. These chapters are written by leading authorities in these fields and have brought together hundreds of years of combined experience.

I am extremely thankful to all the chapter authors for their contributions and pains they have taken in completing this book. The contributions through the various chapters will guide nutraceutical research and lead to increased credibility of the products if their suggestions are applied by the industry and researchers.

I am thankful to Stephen Zollo, Patricia Roberson, and Kathryn Younce from Taylor and Francis/CRC press for their kind help and patience.

I am thankful to Allison Koch, Event Coordinator at Sullivan University College of Pharmacy and in charge of special projects, for her kind help in communications, maintaining the files, and word processing and formatting this book.

Any such project always takes time from my family, my wife Seema and son Sarvadaman, and from my colleagues in the nonprofit organizations ICCS USA Inc., SEWA International, and HSS USA Inc., in which I have been involved for many years. I am highly indebted to all of them.

Yashwant Pathak

Contributors

Samiul Amin
Malvern Instruments Limited
Malvern, Worcestershire, United Kingdom

Roger Barnum
Jenike & Johanson, Inc.
Tyngsboro, Massachusetts

Eduardo A. C. Batista
Laboratory of Extraction, Applied Thermodynamics and Equilibrium (ExTrAE)
Faculty of Food Engineering
University of Campinas (UNICAMP)
Campinas, São Paulo, Brazil

Thomas Baxter
Jenike & Johanson, Inc.
Tyngsboro, Massachusetts

Weiyuan Chang
Department of Environmental and Occupational Health
School of Public Health University of Louisville
Louisville, Kentucky

Donald Chiou
School of Chemical and Biomolecular Engineering
The University of Sydney
Sydney, New South Wales, Australia

Jacqueline Concha
Departamento de Bioquímica
Facultad de Farmacia
Universidad de Valparaíso and
Escuela de Ingeniería Bioquímica
Pontificia Universidad Católica de Valparaíso
Valparaíso, Chile

Mariana C. Costa
Laboratory of Extraction, Applied Thermodynamics and Equilibrium (ExTrAE)
Faculty of Food Engineering
University of Campinas (UNICAMP)
Campinas, São Paulo, Brazil

Shantanu Das
Riddet Institute
Massey University
Palmerston North, New Zealand

John Eatough
School of Pharmacy & Pharmaceutical Sciences
Manchester, United Kingdom

Dilip Ghosh
Neptune Bio-Innovations Pty. Ltd.,
Sydney, New South Wales, Australia

Qingrong Huang
Department of Food Science
Rutgers University
New Brunswick, New Jersey

Yuping Huang
Department of Food Science
Rutgers University
New Brunswick, New Jersey

William Igou, Jr.
Weber Flavors
Wheeling, Illinois

Aparna Keskar
Imbibe
Wilmette, Illinois

Jayant Khambekar
Jenike & Johanson, Inc.
Tyngsboro, Massachusetts

Se-Kwon Kim
Department of Chemistry and Marine Bioprocess Research Center
Pukyong National University
Busan, South Korea

Girish J. Kotwal
Sullivan University College of Pharmacy
Louisville, Kentucky

Timothy A. G. Langrish
School of Chemical and Bimolecular Engineering
The University of Sydney
Sydney, New South Wales, Australia

Marcelo Lanza
Laboratory of Extraction, Applied Thermodynamics and Equilibrium (ExTrAE)
Faculty of Food Engineering
University of Campinas (UNICAMP)
Campinas, São Paulo, Brazil

Uyen Le
Sullivan University College of Pharmacy
Louisville, Kentucky

Brian Lockwood
School of Pharmacy & Pharmaceutical Sciences
Manchester, United Kingdom

Jayant Lokhande
Naturomic LLC
Anaheim, California

Antonio J. A. Meirelles
Laboratory of Extraction, Applied Thermodynamics and Equilibrium (ExTrAE)
Faculty of Food Engineering
University of Campinas (UNICAMP)
Campinas, São Paulo, Brazil

Roland Meylan
AlpVision
Vevey, Switzerland

Dai-Hung Ngo
Department of Chemistry
Pukyong National University
Busan, South Korea

Melvin A. Pascall
Department of Food Science and Technology
The Ohio State University
Columbus, Ohio

Yashwant Pathak
University of South Florida School of Pharmacy
Tampa, Florida

Eduardo Pérez
Centro Regional de Estudios en Alimentos Saludables
Valparaíso, Chile

James Prescott
Jenike & Johanson, Inc.
Tyngsboro, Massachusetts

Vijay Kumar Puli
Sullivan University College of Pharmacy
Louisville, Kentucky

Naresh Rajanna
Center of Excellence – Integrated Supply Chain
PepsiCo Chicago
Aurora, Illinois

Sri Hyndhavi Ramadugu
Sullivan University College of Pharmacy
Louisville, Kentucky

Don Rosendale
Vector Corporation
Marion, Iowa

Anwesha Sarkar
Riddet Institute
Massey University
Palmerston North, New Zealand

Harjinder Singh
Riddet Institute
Massey University
Palmerston North, New Zealand

Carmen Soto
Centro Regional de Estudios en Alimentos Saludables and
Escuela de Ingeniería Bioquímica
Pontificia Universidad Católica de Valparaíso
Valparaíso, Chile

Krassimir P. Velikov
Unilever Research and Development Vlaardingen
Vlaardingen, the Netherlands

Lijun Wang
Biological Engineering Program
North Carolina Agricultural and Technical State University
Greensboro, North Carolina

Isuru Wijesekara
Department of Chemistry
Pukyong National University
Busan, South Korea

Mike Witt
Louis Pharma
Louisville, Kentucky

Xiaoqing Yang
Department of Food Science
Rutgers University
New Brunswick, New Jersey

María Elvira Zúñiga
Centro Regional de Estudios en Alimentos Saludables and
Escuela de Ingeniería Bioquímica
Pontificia Universidad Católica de Valparaíso
Valparaíso, Chile

CHAPTER 1

Nutraceuticals

Advancing in a Right Direction

Uyen Le and Yashwant Pathak

CONTENTS

1.1 CHANGING LIFESTYLES OF THE PEOPLE

After the industrial and communication revolutions, the lifestyles of humans are changing significantly. Everyone feels the scarcity of time, whether a young child or a retired adult. Previously it was normal for a person to regularly walk a few miles; now lifestyles have changed, and there are fewer opportunities for people to walk or take regular exercise. Everyone is looking for alternative ways to keep healthy and trying to change their current lifestyles, leading to much healthier lifestyles.

Recent trends in the pharmaceutical industry show that the experiment of "solo treatment" is leading to potent drugs with severe side and toxic effects. The pharmaceutical industry tried to purify and modify the structure of the chemical entity leading to potent drugs but with significant side effects. Even though new drugs are introduced in the market every year, people are still looking for alternatives.

As a result of the explosion of electronic data information systems, people now have access to nutritional information and are aware of recent advances in the various sciences. Recent discoveries in genetic sciences have led the average person to think more about the alternatives. As the sciences have developed, people are looking more for preventive measures as well as for natural remedies for their health conditions or the health conditions they are expecting as a result of their genetic backgrounds. Young people who have parents with diabetes or arthritis are looking at preventive measures that may delay the onset of these diseases for them, or are trying to take precautions to make the disease less severe.

1.2 GROWING NUTRACEUTICALS MARKET

The huge wave of health consciousness that started in the early 1990s has drastically changed the factors dictating nutritional choices. The interest in exploring diet regimens beyond the recommended dietary allowances, with the reason that it may reduce the risk of chronic diseases, increased significantly. Dietary supplements started becoming part of the mainstream food served in American homes. The belief that these products will prevent disease, enhance performance, and may delay the onset of aging has attracted a large number of US citizens, leading to a voluptuous market of nutraceuticals growing to more than $80 billion, almost competing with the pharmaceutical industry (Pandey and Saraf 2010).

At the conference of the Foundation for Innovation in Medicine (FIM), Dr. Stephen DeFelice announced that the nutraceuticals market will grow threefold larger than the pharmaceutical market in the near future. FIM is claimed to be the "eyes and ears" of the nutraceuticals industry (FIM 2011).

Besides the aging population and increasing number of people, attaching importance to health in the United States and Europe, Japan, India, China, and many other countries will affect the nutraceuticals market worldwide. Functional foods also are contributing significantly to the numbers. It is expected that sales of functional foods are also growing in the West and may reach $25 billion by 2010.

This is very well reflected in the growth of the nutraceutical and functional food market in past decade. Growth of the nutraceutical food market from 2007 to 2013 is predicted to be almost 142%, while that of nutraceutical supplements is estimated to increase 125%, and nutraceutical beverages may increase a whopping 185%.

1.3 PREVENTION VERSUS CURE

The huge wave that started during the 1990s in the United States has drastically changed the market trends for nutraceuticals. Consumers' awareness is ever-increasing, and more scientific information is available for making better choices; finally, the number of well-informed customers is also ever-increasing. Consumers have started subscribing to diet regimens that reduce the risk of chronic diseases. People are using nutraceuticals and functional foods more for the prevention than for the cure of diseases. People have started using functional foods and dietary supplements that will have prevention values as well as be a part of the recommended dietary allowances. It is interesting to observe that people prefer more natural substances than purely synthetic ones and opt for natural and organic foods, functional beverages, and natural supplements. Interestingly, in the past two decades, the media have played an important role and drawn people's attention to scientific developments in the health and nutrition field. The market for nutraceuticals has tremendously benefited from this media attention.

According to a new technical market research report, the nutraceuticals global market was reported to be worth $117.3 billion in 2007 (Technical Market Research 2007). It was projected that this will increase to $123.9 billion in 2008 and reach $176.7 billion by 2013.

The report by Global Industry Analyst, Inc. suggested that the increasing consumer desire for leading a healthy life and increasing scientific evidence supporting health foods continue to drive the nutraceuticals market. They projected a healthy growth in the market to cross $187 billion by 2010 (Global Industry Analyst 2008). They predicted that the United States, Europe, and Japan will dominate the global market and will constitute 86% of the total global nutraceuticals market. They also suggested that the expanding elderly population, enhanced awareness, high income levels of consumers, widespread preferences for specialty nutritional and herbal products, increasing trends promoting preventive medicines, and self-treatment are responsible for the phenomenal growth prediction of the nutraceuticals market (Global Industry Analyst 2008).

The other emerging market worldwide (Heller 2008) includes India, with $540 million in sales in 2006, and an expected increase of almost 38% per year in the coming years, crossing the billion mark by the end of 2009 (Research Studies 2002). China is another market that is growing significantly, followed by Brazil ($881 million in 2006), Turkey ($200 million in 2006), and Australia, New Zealand, and Middle East and African countries ($300 million in 2006).

1.4 NUTRACEUTICAL DEFINITIONS

"Nutraceuticals are in their formative years. But make no mistake, the Nutraceutical boom is coming and it will be worth billions to the companies who define it," predicted Jim Wagner, editor of *Nutritional Outlook* in June 2002 (Wagner 2002, p. 1).

Since then, the nutraceuticals market, a multibillion dollar industry, has been growing significantly, leading to its impact on not only health foods but also on the pharmaceutical markets. Recent numbers show that it has almost reached the dollar value market currently earned by the pharmaceutical industry worldwide.

There is no clear definition of nutraceuticals that is accepted universally by all the stakeholders; hence it is difficult to forecast the size of the nutraceuticals market. For ages, every country had its own system of products, which had both nutritional and medicinal values. Looking at the traditional medicinal system of African tribes or Native American nations, Australian aboriginal people, Chinese, Japanese, Hindus, Maoris, and many other indigenous tribes and cultures worldwide, all of them developed a functional, indigenous system of medicine, in which they extensively used the local herbs and natural minerals for their nutritional and medicinal values. Today in the allopathic medical world, most of the drugs marketed find their origin in the form of a lead compound based on information acquired from these traditional systems. Nutraceuticals in Indian traditional systems are defined as extracts of foods claimed to have medical effects on human health and are usually contained in a nonfood matrix such as capsule or tablets. We authored a chapter in Volume I describing the importance of nutraceuticals in the Indian Ayurvedic system of medicines, which is thousands of years old.

In recent years, the word *nutraceutical* was first coined by Dr. Stephen DeFelice (2002), who defined it as "a substance that is a food or a part of food and provides medical and health benefits, including prevention and treatment of disease." It was coined as a hybrid of the two terms *nutrition* and *pharmaceuticals*. Such products may range from isolated nutrients, dietary supplements, and specific diets to genetically engineered designer foods, herbal products, and processed food such as cereals, soups, vegetable juices, and beverages. It is important to note that this definition applies to all categories of food and parts of food, including folic acid, antioxidant food substances, stimulant functional food, and pharma food.

Another definition was suggested by Lockwood (2007) in his book, stating that "'Nutraceutical' is the term used to describe a medicinal or nutritional component that includes a food, plant, or naturally occurring material which may have been purified or concentrated, and that is used for the improvement of health, by preventing or treating a disease" (p. 1). Zeisal (1999) suggested a definition, claiming that "Nutraceuticals is a diet supplement that delivers a concentrated form of a presumed bioactive agent from a food, presented in a non food matrix and used to enhance health in dosages that exceed those that could be obtained from normal food" (p. 1813).

It is interesting to know that many countries have adopted different terminologies for nutraceutical and have defined the word in different ways. Most countries are trying to establish a regulatory framework for these products. Interestingly, the nutraceuticals revolution is becoming part of the mainstream medical discovery process. Medical practitioners worldwide are now accepting nutraceuticals as part of their mainstream clinical practice as a result of more and more scientific medical research (Abedowale, Liang, and Eddington 2000).

The functional food concept was first introduced in Japan; until 1998 Japan was the only country that legally defined foods for specified health use. A functional food is natural or formulated food that has enhanced physiological performance or prevents or treats a particular disease (Hardy 2000; Lockwood 2007). In Canada, functional foods are defined as similar in appearance to conventional food and consumed as a regular part of the diet; whereas nutraceuticals are defined as a product produced from food but sold as pills, tablets, capsules, and other medicinal forms not generally associated with food. In the United Kingdom, the Department of Environment, Food, and Rural Affairs defines functional food as a food that has a component incorporated into it to give it a specific medical or physiological benefit other than a purely nutritional one.

In the United States, a new product may be introduced as a food or dietary supplement or as a medical food (Litov 1998) under the US Dietary Supplement Health and Education Act (DSHEA). According to DSHEA, the dietary supplement can state that these products can offer nutritional support to those with nutrient-deficient diseases, but the companies are expected to write that this statement has not been evaluated by the US Food and Drug Administration (FDA). Further, they must state that this product is not intended to diagnose, treat, cure, or prevent any disease. The US FDA expects that nutraceuticals are manufactured under current good manufacturing practices (cGMPs), and gradually the FDA is moving toward stricter regulations for nutraceuticals.

1.5 MODIFICATION IN THE DEFINITION OF NUTRACEUTICALS

In consideration of these regulations, we would like to put forward a modification in the present nutraceuticals definition as follows (Pathak 2009, p. 17):

> Nutraceuticals are the products developed from either food or dietary substance, or from traditional herbal or mineral substance or their synthetic derivatives or forms thereof, which are delivered in the pharmaceutical dosage forms such as pills, tablets, capsules, liquid orals, lotions, delivery systems, or other dermal preparations, and are manufactured under strict cGMP procedures. These are developed according to the pharmaceutical principles and evaluated using one or several parameters and in process controls to ensure the reproducibility and therapeutic efficacy of the product.

1.6 CONCERT VERSUS SOLO PERFORMANCE

Galenicals and nutraceuticals are products that offer therapeutic value by concert performance. There are combinations of several ingredients present in these products, which offer the medicinal effect through combination of their actions. Hence we refer to their actions as *concert performances*. It is observed that there are several ingredients present in the herbal product or other nutraceuticals that have either similar therapeutic action or sometimes antagonistic action, and they control the effect of each other, which thereby reduces the side and toxic effects of the products. To clarify, reserpine is an alkaloid present in the plant *Rauwolfia serpentina*, which is used for its antihypertensive action in heart problems and blood pressure. Interestingly, it also has an ingredient that can be used to increase blood pressure. Therefore, both ingredients work in synchronization, providing a positive effect. All major antihypertensive drugs have their lead obtained from this plant, and further drug development has led to several solo-performing drugs, which are either derivatives or modified versions of the reserpine molecule. Similar relationships can be seen in many solo drugs that have been developed based on the lead compound from their herbal analog. When a single drug is used for the treatment in some suitable drug dosage form or delivery system, we refer to it as a *solo performance.* Most allopathic drugs have solo performances; recently, there are increasingly more and more combinations used with many side effects, creating more complications after the treatments. It is observed that all the statins used for lowering cholesterol found their origin in Chinese red rice yeast; unfortunately, most statins have shown reasonable side effects to question their utility as solo drugs. It might be good to use the concert-performing Chinese red rice yeast in place of solo statins, and then followed with the application of nutraceuticals more for prevention. One can quote several such examples, and that is one of the reasons why the nutraceuticals market is growing significantly—as people are more aware of these facts, they are opting for natural products for their preventive care, rather than looking for a solo performance cure.

1.7 CHANGING STATE OF COMPLIANCE

It is observed that over time the regulatory compliance requirements are changing fast for nutraceuticals products. The US FDA passed DSHEA in 1994, addressing the sudden boom in the nutraceuticals market. DSHEA allowed manufacturers to make claims about the applications of nutraceuticals products, which they were not allowed before the act was passed. Thus, DSHEA also helped to educate consumers about the marketed products. Under DSEHA, the dietary supplement manufacturer is responsible for ensuring that a dietary supplement is safe before it is marketed and that the product label information is truthful and not misleading.

Since then, the US FDA was made responsible for taking action against any unsafe dietary supplement product after it reaches the market. Generally manufacturers need not register with the FDA or get FDA approval before producing or selling dietary supplements (http://www.fda.gov). In addition to these rules, FDA regulations related to cGMP manufacturing are also applicable to nutraceutical manufacturing, ensuring the quality, purity, safety, and reproducibility of the nutraceuticals products.

1.8 FUTURE OF FDA REGULATIONS

Looking at the trends worldwide and especially in the United States, it appears that the FDA will develop more stringent regulations for nutraceuticals and functional food products to be marketed. The rest of the world will also follow in their footsteps. FDA regulations and cGMP guidelines will be implemented similarly as in the pharmaceuticals industry. This trend will prove to be costly to the nutraceuticals market. Many are welcoming the changes in the regulations. The problem that the nutraceuticals market will face is the profitability and intellectual protection of the products. As for patenting nutraceuticals or functional food products, this profitability protection is not available to their manufacturers. Hence, as regulations become stringent, manufacturing will become costlier, and the innovators and small companies in the market will not be able to afford these changes, resulting in the market being left to the big players.

1.9 SCALING UP OF NUTRACEUTICALS

New processes of a product are usually developed from a laboratory at a small or pilot scale with small volumes. The products addressed to the public and consumed by people need mass production or scaling up at a plant or industrial scale with large volumes. Under the latter conditions, the processes can lead to different reactions occurring than in the original lab scale. Scaling up and technology transfer for the products involve many activities that relate to adaptation of lab and pilot scale to the commercial production. The goal of scaling up is to maintain equivalent product quality between laboratory and commercial processes. To achieve this goal, understanding strategies and issues that may occur during the processes is essential. Unfortunately, dissemination of technologies and practice in nutraceutical scaling up are problematic as a result of complex properties of ingredients and various types of dosage forms. The scale-up process may require multiple attempts and trials. Every nutraceutical and delivery system needs an appropriate modification applied to the scale-up process to ensure that the products resemble purity and potency compared with those present during clinical trials. Significant research efforts and process technology on nutraceuticals provide a substantial portfolio of nutraceutical products.

1.10 IMPORTANCE OF SCALING UP IN DEVELOPING THE PROCESSES

When scaling up a process, the vessel volumes cannot be assumed to be increased by orders of magnitude and homogenous reaction conditions. In fact, different regions in reaction vessels may have different process conditions, and that issue can be significantly important in exothermic reactions. Process scale up is critical in confirming process conditions and validating process models. It should be borne in mind that successfully scaled-up volumes do not occur by accident but rather require a reliable scale-up procedure developed from focused attention, strategic planning, management, and resource allocation. A focus on scaling up is required while determining needs and designing pilot interventions. All these are based on sound engineering fundamentals and thorough process understanding. As a part of process development, scaling up should be placed closely with other key factors to facilitate a smooth technology transfer from development to high-volume manufacturing.

1.11 ROLE OF SCALING UP IN PROCESS DEVELOPMENT AND FACTORS INVOLVED IN DECIDING THE PROCESSES

Process development is an important part of research and development in developing a new process concept to commercial exploitation. It helps to determine process conditions, define technical and economic problems, eliminate possible practical obstacles, and collect necessary data for the implementation of the commercial process (Sie and Krishna 1998). In general, process development involves "reducing the difficulty in complex practical world into a problem in the relevant technical area for which solutions can be sought with the appropriate scientific approach in a laboratory" (Sie and Krishna 1998, p. 47).

When scaling up a nutraceutical process, it is important to consider factors such as choice of raw materials, sourcing, supply chain, purity, and security. DePalma (2008) recommended "tips for successful process development and scale-up" (para. 29) for bioproducts that can be selectively applied for nutraceuticals.

- Define the process.
- Adopt platform technologies.
- Use statistical methods.
- Invest in process development.
- Develop relationships with vendors.
- Qualify the analytical methods.

The aim of process development is to raise process productivity and robustness, reduce raw material cost, and improve scalability (Junker et al. 2008). During process development, the following factors need to be considered (Kramer et al. 2009).

- Baseline process
- Characterization of the methodology
- Properties of each step in the process
- Scale-up procedure
- Scale-up assessment

The key issue to ensure consistent product quality is to keep constant appropriate strategies, detailed process characterization, relevant process parameters, and scale-up parameters. A "simplified" formulation facilitates the scaling up and helps to reduce the cost of manufacturing. The success of scaling up generally depends on separate process development and optimization on each scale rather than conclusive experimental strategy (Schmidt 2005).

1.12 PROCESSING OF NUTRACEUTICALS

Unlike other products, nutraceuticals are a complex mix of interacting chemical and physical systems. With a multistep process of nutraceuticals (Figure 1.1), the process controls are critical to ensure the bioactivity without changing the structure of the compounds. A large percentage of nutraceuticals are herbal-based products. The best growth market of nutraceuticals was expected in the soy protein nutrients, herbal extracts of green tea and garlic, functional food, beverage additives, lutein, lycopene, and coenzyme Q10 (Little 2000). The compounds are commonly unstable under various conditions. Nutraceutical manufacturing involves several major steps, starting with preprocessing (e.g., raw material harvesting and storage) through processing. The preprocessing ensures that the active ingredients are maintained before processing. The processing is a critical aspect of nutraceutical production, especially for low yield of extracts (Aziz, Sarmidi, and Kumaresan 2008). The common extraction methods are high pressure water extraction, supercritical fluid extraction, pressurized low polarity water extraction, membrane separation, distillation, dehydration, and bioprocessing technologies. Nutraceutical products can be formulated in a variety of forms, such as tablets, capsules, bags, extracts, and essential oils. Generally, the processing of nutraceuticals is different from that of other products for several reasons.

1. *Selection of technologies and equipment*: This is highly dependent on not only the properties of active ingredients but also dosage forms and package matrix. Different specific processes, such as thermo process (Brady et al. 2005), extrusion process (Freedonia 2006), and freezing–drying process (Martin and Zhao 2006), can also play a key role in changing the product profiles, and thus may influence sensory or pharmacological properties of nutraceuticals. The most common scale-up processing methodology is combining traditional batch processing and continuous processing, in which the products are batched in large containers and continued to move through the system. The method helps to control discrete units of production and reduce the hold time of the products. Another important factor

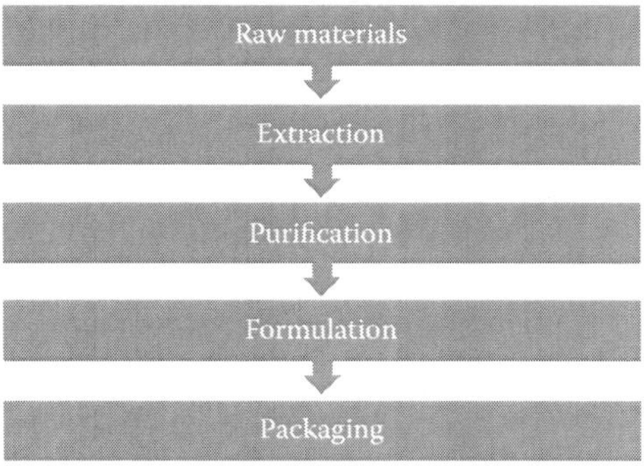

Figure 1.1 Processing of nutraceuticals.

in nutraceutical processing is the product transfer. For each type of dosage forms (e.g., liquid, solid, or semisolid), different equipment (e.g., pumps, valves, piping, conveyor, and tubing) is required to handle materials. Care must be taken when opening and positioning this equipment to prevent contamination of products from outside sources.

2. *Process controls*: Developing nutraceutical process control is hard and problematic. Nutraceuticals need producing under controlled conditions to ensure batch uniformity and integrity of the products. Typical processes found in their manufacturing are extraction, purification, filtering, freeze drying, and centrifuging. Each process requires various measurements with accurate and flexible instruments. Measurements are commonly focused on temperatures, pressures, pump control, valve control, flows, nutraceutical composition, tank level, agitator control, and product weight. Unfortunately, as mentioned previously, nutraceuticals are mixed compounds; therefore, adoption of manufacturing control, especially standardization, seems to be complex. In addition, manufacturers may struggle with process validation as a result of a lack of studies on the compatibility of active ingredients and recipients, of final products and packaging materials, and of stability, etc.
3. *Reproducibility of the product*: The reproducibility of the product is challenging. Nutraceuticals, especially herbs, are not composed of single chemicals but many ingredients that are different in proportions from source to source (Bratman and Girman 2003). Large variation in effectiveness is possible from one batch of the product to another. To overcome this difficulty, nutraceutical "standardization" was suggested as a part of the process (Schulz, Hansel, and Tyler 1998). In this part, an extract of the raw materials is dried until reaching a certain concentration of some ingredients. The extract is then formulated into an appropriate dosage form (e.g., tablets, capsules, or bottles) with the listed constituents and considered as a "tag" of the standardization. Nevertheless, this method still demonstrates weaknesses because of the lack of unlisted or unidentified active ingredients. Another procedure recommended is to use the actual products that were tested in double-blind studies as standardized extracts. However, not all tested extracts are available, and this method also can increase remarkably the cost of the process.

Process analytical technology (PAT) applied in nutraceutical processing is still limited. PAT has been described as "a system for designing and controlling manufacturing through timely measurement (i.e., during processing) of critical quality and performance attribute for raw and in-process materials and also processes with the goal of ensuring final product quality" (Arbindakshya et al. 2008, para. 1).

Unlike quality assurance, which examines products only at the end stage, PAT acts as a "streaming video" of monitoring and detecting the products from early to end point. PAT helps to improve product quality, enhance production efficiency, reduce costs, and shorten time to market (Arbindakshya et al. 2008). For nutraceuticals, PAT is applied from preprocessing (e.g., harvesting) until final packaging. The aim of the PAT-based approach is to ensure the safety, effectiveness, and affordability of the products. Nutraceuticals have been produced based on existing experimental methods and established manufacturing processes. Improved technologies are constantly applied in nutraceutical production, but this makes it difficult to determine the extending process capabilities of new technologies.

The benefits that PAT can bring to nutraceutical production can be briefly listed as follows.

- Good selection of raw materials from the beginning
- High homogenization in material quality
- Enhanced understanding and process control
- Reduced variation and production waste
- More stable products
- Decreased costs of production in the long run

1.13 AUTOMATION IN NUTRACEUTICAL MANUFACTURING

The global nutraceuticals market is still hot, approximately $120 billion in 2007 with a compound growth rate of more than 7% annually (Mirasol 2009). Large consumption demands mass production. Manufacturing done with automated machines improves productivity, increases quality of products, facilitates product quality control, and reduces labor costs. Nutraceuticals are not exceptional. Process automation may be feasibly applied in all stages of production from extraction up to final packaging.

In the stage of extraction, material loading, water draining, discharging temperature and timer of extraction, filtering, extruding, vacuum suction conveying, and cleaning are automatically controlled from the central department of the agitating equipment. The process is fully closed to ensure a sanitary environment. In the industrial processes, the condense systems are commonly preferred to provide the atmosphere pressure that helps to maintain the flavor and essence of oil. Various automated equipment of extraction are on the market for specific active ingredients, such as the supercritical fluid extraction machine, supercritical CO_2 extraction machine, cassava extracting machine, sell extraction machine, bitumen automatic extraction machine, hydraulic olive oil extraction machine, starch extraction machine, oil extractor, and so on.

Formulations of nutraceuticals can be classified as liquid or solid dosage forms. Nutraceutical liquids are formulated in stirring vessels or filled in containers. Solids are developed through a multistep process, such as weighing, granulating, tableting, capsuling, and filling into bottles. The whole process used with automatic equipment provides reliable, precise, and repetitive controls. Automatic equipment for these products is similar to those for other pharmaceuticals.

1.14 NUTRACEUTICAL DISTRIBUTION

Nutraceuticals cross a wide range of various products, market sections, and economic conditions, making it difficult to draw generalizations about a typical distribution. Online distribution plays an important role in introducing the products to consumers. Food and dietary supplement companies have had the "inside track" on the market for years, but pharmaceutical companies are not far behind (Nutraceuticals Trend 1997). The nonpharmaceutical companies are well positioned in both expertise and marketing for the products. Popular distributors for nutraceuticals include Nutraceutical, Mouton Logistics, and Ware-Pak Nutrix. Some distribute the products through one or more middlemen (e.g., prewholesaler, wholesalers, stores, customers). Others use logistics service providers or direct distribution and have advanced marketing and management for the warehousing, inventory control, and fulfillment of nutraceutical products.

One obstacle of nutraceutical distribution is how to ensure the shelf-life of products in different storage conditions because of their unstable ingredients. In addition, the usual bulky sizes of the containers contribute to limited storage space. The major challenge is deregulation of nutraceuticals, which results in difficult distribution alliances with pharmaceutical companies for consumers. Limitation in proprietary value and medical values of nutraceuticals hinders investments from both manufacturing and distribution.

1.15 EDUCATING THE RETAILERS AND CONSUMERS

Food and *functional foods* are common words to people. *Nutraceutical*, a word marrying *nutrition* and *pharmaceutical* with various definitions, however, is still unfamiliar to most consumers. Many scientists call nutraceuticals food or part of the food that provides medical or health benefits. Food companies may refer to food as nutraceuticals when appropriate. Federal regulators will not consider nutraceuticals as products in a regulatory category. Others might have no idea about what nutraceuticals are. Nevertheless, nutraceuticals is a beloved term used by consultants and marketing pros as a way of capturing the healthy eating trend (Shell 1998).

Based on either a conventional or modified definition, nutraceuticals are products that consumers can purchase freely on the market. In the modern society, the trend of using functional food, dietary supplements, and other natural products, especially for stress release and size control, are increasing. A survey in the United States demonstrated that people rely not only on their physicians but also on the media for medical advice (Katz 2009). For this appealing self-usage habit, the following important factors can be used to educate consumers:

- Labeling
- Branding
- Packaging
- Marketing and advertising

Katz (1999) presented two purposes of labels: "enticing the consumer via pictures and words" (para. 3) and "providing information regarding directions and ingredients" (para. 3). Whether consumers are sufficiently informed about the function and medicinal effects of the products is still a big question for most nutraceuticals on the market. Many publications freely distributed without the FDA's review can be great marketing tools for distributors and manufacturers.

1.16 EXCITING TIME AHEAD FOR NUTRACEUTICALS

The popular "Let food be thy medicine and medicine be thy food" saying of the Greek physician Hippocrates 2,500 years ago has been recognized thoroughly. When preventive care is as important as medical care, nutrition is an integral part of health care management. Obviously, nutraceutical consumers continue to cogitate on the natural products that make their lifestyles healthier during stressful times. Nutraceutical companies keep investing in manufacturing and distribution to allow for their future growth and expansion. The nutraceutical industry is one of the largest in the United States, and the world (Freedonia 2006). With the trend of a strong international consumer base, combined with advances in medical science and pharmaceuticals, the business of nutraceuticals is supported now more than ever.

REFERENCES

Adebowale, A. O., Z. Liang, and N. D. Eddington. 2000. Nutraceuticals, a Call for Quality Control of Delivery Systems: A Case Study with Chondroitin Sulfate and Glucosamine. *Journal of Nutraceuticals, Functional & Medical Foods* 2: 15–30.

Arbindakshya, M., B. Saurabh, B. Nikhil, M. Prakash, and K. Pranjali. 2008. Process Analytical Technology (PAT): Boon to Pharmaceutical Industry. *Pharmaceutical Reviews* 6(6). http://www.pharmainfo.net/reviews/process-analytical-technology-pat-boon-pharmaceutical-industry.

Aziz, R.A., M. R. Sarmidi, and S. Kumaresan. 2008. Phytochemical Processing: The Next Emerging Field in Chemical Engineering—Aspects and Opportunities. *Journal Kejuruteraan Kimia Malaysia* 3: 45–60.

BCC Research. 2008. *Nutraceuticals: Global Markets and Processing Technologies*. http://www.bccresearch.com/report/FOD013C.html.

Brady, K., C. T. Ho, R. T. Rosen, S. Sang, and M. V. Karwe. 2005. Effects of Processing on the Nutraceutical Profile of Quinoa. *Food Chemistry* 100: 1209–16.

Bratman, S., and A. M. Girman. 2003. *Mosby's Handbook of Herbs and Supplements and Their Therapeutic Uses*. St. Louis: Mosby, Inc.

DeFelice, S. L. 2002. *FIM Rationale and Proposed Guidelines for the Nutraceutical Research & Education Act—NREA. Foundation for Innovation in Medicine*. http://www.fimdefelice.org/archives/arc.researchact.html.

DePalma, A. 2008. Platform Technologies Ease Scale-Up Pain. *Genetic Engineering & Biotechnology News*. http://www.genengnews.com/gen-articles/platform-b-technologies-ease-scale-b-b-up-pain-b/2558/.

Dogan, H., Z. Nanqun, C. T. Ho, R. T. Rosen, and M. W. Karwe. 2001. Effect of Processing on Nutraceutical Potential of Quinoa. Paper presented at the Annual Meeting of the Institute of Food Technologists, New Orleans, LA, June 2001.

Freedonia. 2006. *World Nutraceuticals 2010—Market Research, Market Share, Market Size, Sales, Demand Forecast, Market Leaders, Company Profiles, Industry Trend*, 486. http://www.freedoniagroup.com/World-Nutraceuticals.html.

Foundation for Innovation in Medicine. 2011. *About the Foundation*. http://www.fimdefelice.org.

Global Industry Analyst Inc. 2008. *Global Nutraceutical Market to Cross US$187 Billion by 2010*. http://www.strategyr.com/nutraceuticals_market_report.asp.

Hardy, G. 2000. Nutraceuticals and Functional Foods: Introduction and Meaning. *Nutrition* 16: 688–89.

Heller, L. 1998. Emerging Nutraceuticals Market Report. http://www.nutraingredients-usa.com/Consumer-Trends/Emerging-nutraceutical-markets.

Junker, B., A. Walker, M. Hesse, D. Vesey, J. Christensen, B. Burgess, and N. Connors. 2008. Pilot-Scale Process Development and Scale Up for Antifungal Production. *Bioprocess and Biosystems Engineering* 32: 443–58.

Katz, S. 2009. Nutraceuticals Labeling. http://www.labelandnarrowweb.com/articles/2009/05/nutraceuticals-labeling.

Kramer, T., D. M. Kremer, M. J. Pikal, W. J. Petre, E. Y. Shalaev, and L. A. Gatlin. 2009. A Procedure to Optimize Scale-up for the Primary Drying Phase of Lyophillization. *Journal of Pharmaceutical Science* 98: 307–18.

Little, W. R. 2000. Herbal Products: A Retail Pharmacists Perspective. *Nutraceuticals World.* http://www.nutraceuticalsworld.com/contents/view/12820.

Litov, R. E. 1998. Developing Claims for New Phytochemical Products. *Phytochemicals: A New Paradigm.* Edited by W. R. Bidlack, S. T. Omaye, M. S. Meskin, and D. Jahner. Lancaster, PA: Technomic Publishing, 173–78.

Lockwood, B. 2007. *Nutraceuticals*, 2nd ed. London: Pharmaceutical Press, 1.

Martin, R. R., and Y. Zhao. 2006. Effect of Food Processing Technologies on the Nutraceutical Content of Raspberries, Strawberries, and Blackberries. Project of Oregon State University. http://www.ars.usda.gov/research/projects/projects.htm?accn_no=409160&showpars=true&fy=2007.

Mirasol, F. 2009. Nutraceuticals Find a Niche in Beauty Market. http://www.icis.com/articles/2009/11/09/9260304/nutraceuticals-find-a-niche-in-beauty-market.html.

Nutraceuticals Trend Takes Root Despite Definition Challenges. 1997. *Nutrition Business Journal* 7. http://www.fimdefelice.org/clippings/clip.business97.html.

Pandey, R. K. V., and Saraf S. A. 2010. Nutraceuticals: New Era of Medicine and Health. *Asian Journal of Pharmaceutical and Clinical Research* 3: 11–15.

Pathak, Y. 2009. *Handbook of Nutraceuticals: Ingredients, Formulations, and Applications.* Boca Raton, FL, CRC Press Publishing, vol. 1, 15–26.

Research Studies–Freedonia Group. 2002. *World Nutraceutical Chemicals Demand to Grow Over 6% Annually Through the Year 2006.* http://www.allbusiness.com/specialty-businesses/355577-1.html.

Schmidt, F. R. 2005. Optimization and Scale Up of Industrial Fermentation Processes. *Applied Microbiology and Biotechnology* 68: 425–35.

Schulz, V., R. Hansel R., and V. Tyler. 1998. *Rational Phytotherapy. A Physician's Guide to Herbal Medicine*. Berlin and Heidelberg, Germany: Springer-Verlag.

Shell, E. R. 1998. The Hippocractic Wars. *The New York Times Magazine,* 34.

Sie, S. T., and R. Krishna. 1998. Process development and scale up. *Reviews in Chemical Engineering* 14: 47–88.

Stauffer, J. E. 1999. Nutraceuticals. *Cereal Foods World* 44: 115–17.

Technical Market Research Report. 2007. *Nutraceuticals: Global Markets and Processing Technologies*. http://www.bccresearch.com/report/FOD013C.html.

Wagner, J. 2002. The Future of Nutraceuticals. *Nutritional Outlook* June/July: 1–2. http://www.fimdefelice.org/clippings/clip.future.html.

Zeisal, S. H. 1999. Regulation of Nutraceuticals. *Science* 285: 1853–55.

CHAPTER 2

Advances in Extraction of Plant Products in Nutraceutical Processing

Lijun Wang

CONTENTS

2.1 INTRODUCTION

Plants produce a large amount of primary metabolites of lipids, protein, and carbohydrates. Plants also produce a broad range of bioactive compounds of secondary metabolites, which can be classified into three main groups of phenolics, terpenoids, and alkaloids. Extraction techniques have been widely investigated to obtain such valuable natural compounds from plants to be used in the food, pharmaceutical, and cosmetics industries (Wang and Weller 2006). Many of those phytochemicals, such as flavonoids and carotenoids, have been determined to be beneficial to the human body in preventing or treating one or more diseases or improving physiological performance (Wildman 2002).

Solvent extraction of solid materials, which is commonly known as solid–solvent extraction or leaching, has been used for many decades. The solid–solvent extraction of natural products from plant matrices is based on the choice of solvents such as hexane and water, coupled with the use of heating and/or agitation. Traditional solid–solvent extraction methods, such as Soxhlet extraction, are time-consuming and require relatively large quantities of solvents (Luque de Castro and Garcia-Ayuso 1998). There is an increasing demand for new extraction techniques with shortened extraction time, reduced organic solvent consumption, increased pollution prevention, and increased extract yield and quality. Novel methods including ultrasound (Vinatoru 2001), pulsed electric field (Toepfl 2006), enzyme digestion (Gaur et al. 2007), extrusion (Lusas and Watkins 1988), microwave heating (Kaufmann and Christen 2002), ohmic heating (Lakkakula, Lima, and Walker 2004), supercritical fluids (Marr and Gamse 2000; Lang and Wai 2001; Meireles and Angela 2003; Wang et al. 2008), and accelerated solvents (Kaufmann and Christen 2002; Smith 2002) have been investigated to enhance the solid–solvent extraction of phytochemicals from plant materials. These emerging extraction techniques can use environment-friendly solvents such as water and CO_2 and have the possibility to greatly decrease the time of extraction and improve the yield and quality of extracts.

The novel extraction techniques have become relatively mature, and some potential applications for the extraction of nutraceuticals from solid plant matrices have been reported. This chapter provides theoretical background on the solid–solvent extraction and several emerging extraction techniques. Potential applications of those extraction methods are reviewed. The practical issues for each extraction method, such as matrix characteristics, solvent choice, liquid–solid ratio, temperature, pressure, and extraction time, are discussed.

2.2 FUNDAMENTALS OF SOLVENT–SOLID EXTRACTION

2.2.1 Principle and Mechanism

Classical solid–solvent extraction techniques for the extraction of nutraceuticals from plant matrices are based on the choice of solvent coupled with the use of heat and/or agitation. Existing classical solid–solvent techniques used to obtain nutraceuticals from plants include maceration, Soxhlet extraction, and hydrodistillation. Maceration extraction is a simple and cheap solid–solvent extraction method. During maceration, the plant materials are soaked in a solvent such as hot oil to rupture their cell membrane and release compounds into the solvent. The solid residue is then filtered from the liquid phase of solvent and extract after extraction. The extract is usually separated from the solvent by vaporization (Zhang et al. 2005).

Soxhlet extraction, which has been used for a long time, is a standard technique and the main reference for evaluating the performance of other solid–liquid extraction methods. Soxhlet extraction is a general and well-established technique, which surpasses in performance other conventional extraction techniques except for, in limited field of applications, the extraction of thermolabile compounds (Luque de Castro and Garcia-Ayuso 1998). An overview of Soxhlet extraction of solid materials was given by Luque de Castro and Garcia-Ayuso (1998). In a conventional Soxhlet system as shown in Figure 2.1, plant material is placed in a thimble holder and filled with condensed fresh solvent from a distillation flask. When the liquid reaches the overflow level, a siphon aspirates the solution of the thimble holder and unloads it

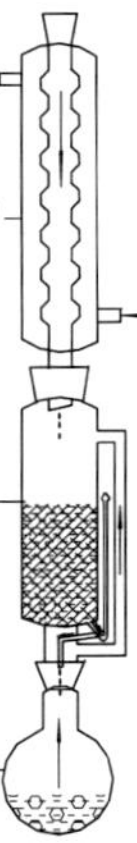

Figure 2.1 Experimental Soxhlet extraction apparatus. (From Wang, L. J. and C. L. Weller, *Trends Food Sci. Technol.*, 17, 300, 2006. With permission.)

back into the distillation flask, carrying extracted solutes into the bulk liquid. In the solvent flask, solute is separated from the solvent using distillation. Solute is left in the flask, and fresh solvent passes back into the plant solid bed. The operation is repeated until complete extraction is achieved. The advantages of conventional Soxhlet extraction over maceration extraction include (1) the displacement of transfer equilibrium by repeatedly bringing fresh solvent into contact with the solid matrix, (2) maintaining a relatively high extraction temperature with heat from the distillation flask, and (3) no filtration requirement after leaching (Luque de Castro and Garcia-Ayuso 1998).

Hydrodistillation is a traditional method for removal of essential oils. Hydrodistillation using the Clevenger apparatus is the standard Association of Analytical Communities International method for the analysis of volatile oils from spices. Hydrodistillation does not involve organic solvents and can be performed before dehydration of the plant materials. There are three types of hydrodistillation: water immersion, combined water immersion and vapor injection, and direct vapor injection. During hydrodistillation, the wet plant materials are distilled, and the volatile oils are extracted into the vapor. When the vapor is condensed, the water and volatile oils are separated (Silva et al. 2005). Some volatile components may be lost at a high extraction temperature during hydrodistillation. Vacuum may be used to decrease the distillation temperature.

2.2.2 Solvent Choice

A suitable extracting solvent should be selected for the extraction of targeted nutraceuticals during solid–solvent extraction. Different solvents will yield different extracts and extract compositions (Zarnowski and Suzuki 2004). The most widely used solvent to extract edible oils from plant sources is hexane. Hexane has a fairly narrow boiling point range of approximately 63°C–69°C, and it is an excellent solvent for the extraction of oil in terms of oil solubility and ease of recovery. However, *n*-hexane, the main component of commercial hexane, is listed number 1 on the list of 189 hazardous air pollutants by the US Environmental Protection Agency (Mamidipally and Liu 2004). In recent years, the US Environmental Protection Agency has enacted emission standards for the vegetable oil extraction industry, imposing stringent restrictions and financial penalties on hexane losses to environment.

The use of alternative solvents, such as isopropanol, ethanol, hydrocarbons, and even water, has increased as a result of environmental, health, and safety concerns. Mamidipally and Liu (2004) used *d*-limonene and hexane in the extraction of oil from rice bran. It was observed that *d*-limonene extracted a significantly higher amount of oil than hexane under any given set of conditions. Water has been extensively investigated as an alternative to hexane for the extraction of oil from plant materials (Hanmoungjai, Pyle, and Niranjan 2000).

However, alternative solvents often generate less recovery as a result of a decreased molecular affinity between solvents and solutes. The costs of alternative solvents could be higher. A cosolvent is sometimes added to adjust the polarity of the liquid phase. A mixture of solvents such as isopropanol and hexane has been reported to increase the extraction yield and kinetics of oil from soybeans (Li, Pordesimo, and Weiss 2004).

2.2.3 Operating Conditions

Traditional solid–solvent extraction such as Soxhlet strongly depends on matrix characteristics and particle size as the internal diffusion may be the limiting step during extraction. For the Soxhlet extraction of total fat from oleaginous seeds, a 2-hour extraction obtained 99% extraction efficiency if the particle size was 0.4 mm, whereas a 12-hour extraction was necessary to obtain similar efficiency if the particle size was 2.0 mm (Luque-Garcia and Luque de Castro 2004).

Solid–solvent extraction may occur at a high temperature. For example, the hydrodistillation extraction usually occurs at the boiling point of water (i.e., 100°C), and the Soxhlet extraction with hexane occurs at the boiling points of hexane (i.e., ~69°C). During Soxhlet extraction, the solvent is usually recovered by evaporation. The extraction and evaporation temperatures have a significant effect on the quality of final products. Mamidipally and Liu (2004) found that *d*-limonene extracted rice bran oil was slightly darker compared with hexane extracted oil, probably as a result of higher extraction and evaporation temperatures used during the *d*-limonene solvent extraction. The high boiling temperature for solvent recovery can be decreased by using vacuum or membrane separation to recover the solvent.

2.2.4 Challenges in Existing Solid–Solvent Extraction Processes

The long time requirement and the requirement of large amounts of solvent lead to wide criticism of the conventional solid–solvent extraction methods such as Soxhlet extraction. The main challenges of conventional solid–solvent extraction include (1) the extraction time is long; (2) a large amount of solvent is used; (3) agitation may be needed but cannot be provided in the Soxhlet device to accelerate the process; (4) the large amount of solvent used requires an evaporation–concentration procedure for solvent recovery; and (5) the possibility of thermal decomposition of the target compounds cannot be ignored as the extraction usually occurs at the boiling point of the solvent for a long time (Luque de Castro and Garcia-Ayuso 1998).

2.3 EMERGING EXTRACTION PROCESSES

2.3.1 Ultrasound-Assisted Extraction

2.3.1.1 Principles and Mechanism

Sound waves, which have frequencies higher than 20 kHz, are mechanical vibrations in a solid, liquid, and gas. Unlike electromagnetic waves, sound waves must travel through matter, and they involve expansion and compression cycles during travel in the medium. Expansion pulls molecules apart, and compression pushes them together. The expansion can create bubbles in a liquid and produce negative pressure. The bubbles form, grow, and finally collapse. Close to a solid boundary, cavity collapse is asymmetric and produces high-speed jets of liquid. The liquid jets have strong impact on the solid surface (Luque-Garcia and Luque de Castro 2003).

Two general designs of ultrasound-assisted extractors are ultrasonic baths or closed extractors fitted with an ultrasonic horn transducer. The mechanical effects of ultrasound induce a greater penetration of solvent into cellular materials and improve mass transfer. Ultrasound in extraction can also disrupt biological cell walls, facilitating the release of contents. Therefore, efficient cell disruption and effective mass transfer are cited as two major factors leading to the enhancement of extraction with ultrasonic power (Mason, Paniwnyk, and Lorimer 1996). Scanning electron micrographs (SEMs) have provided evidence of the mechanical effects of ultrasound, mainly shown by the destruction of cell walls and release of cell contents. In contrast to conventional extractions, plant extracts diffuse across cell walls as a result of ultrasound, causing cell rupture over a shorter period (Vinatoru, Toma, and Mason 1999; Toma et al. 2001; Chemat et al. 2004; Li, Pordesimo, and Weiss 2004).

It is necessary to take into account plant characteristics, such as moisture content and particle size, and solvent used for the extraction to obtain an efficient and effective ultrasound-assisted extraction. Furthermore, many factors govern the action of ultrasound, including frequency, pressure, temperature, and sonication time. Ultrasound frequency has great effects on extraction yield and kinetics. However, the effects of ultrasound on extraction yield and kinetics differ depending on the nature of the plant material to be extracted. A small change in frequency can increase the yield of extract approximately 32% for ultrasound-assisted solid–hexane extraction of pyrethrins from *pyrethrum* flowers. However, ultrasound has weak effects on both yield and kinetics for the extraction of oil from *woad* seeds (Romdhane and Gourdon 2002).

The ultrasonic wave distribution inside an extractor is also a key parameter in the design of an ultrasonic extractor. The maximum ultrasound power is observed in the vicinity of the radiating surface of the ultrasonic horn. Ultrasonic intensity decreases rather abruptly as the distance from the radiating surface increases. Also, ultrasound intensity is attenuated with the increase of the presence of solid particles (Romdhane, Gourdon, and Casamatta 1995). To avoid standing waves or the formation of solid-free regions for the preferential passage of the ultrasonic waves, additional agitation or shaking is usually used (Vinatoru et al. 1997).

2.3.1.2 Advantages and Disadvantages

Ultrasound-assisted extraction is an inexpensive, simple, and efficient alternative to conventional extraction techniques. The main benefits of the use of ultrasound in solid–liquid extraction include the increase of extraction yield and faster kinetics. Ultrasound can also reduce the operating temperature, allowing the extraction of thermolabile compounds. Compared with other novel extraction techniques such as microwave-assisted extraction, the ultrasound apparatus is cheaper and its operation is easier. Furthermore, the ultrasound-assisted extraction, like Soxhlet extraction, can be used with any solvent for extracting a wide variety of natural compounds.

The use of ultrasound allows changes in the processing condition such as a decrease of temperature and pressure from those used in extractions without ultrasound (Wu, Lin, and Chau 2001; Romdhane and Gourdon 2002). For solid–hexane extraction of pyrethrins from *pyrethrum* flowers without ultrasound, extraction yield increases with the extraction temperature, and maximum yield is achieved at 66°C. With ultrasound, the effect of temperature in the range of 40°C–66°C on the yield is negligible, such that optimal extraction occurs across the range of temperature from 40°C–66°C. Therefore, use of ultrasound-assisted extraction is advisable for thermolabile compounds, which may be altered under Soxhlet operating conditions as a resulet of the high extraction temperature (Romdhane and Gourdon 2002). However, it should be noted that because ultrasound generates heat, it is important to accurately control the extraction temperature (Salisova, Toma, and Mason 1997). The sonication time should also be considered carefully because excess of sonication can damage the quality of extracts.

The effects of ultrasound on extraction yield and kinetics may be linked to the nature of the plant matrix. The presence of a dispersed phase contributes to the ultrasound wave attenuation, and the active part of ultrasound inside the extractor is restricted to a zone located in the vicinity of the ultrasonic emitter. Therefore, those two factors must be considered carefully in the design of ultrasound-assisted extractors.

2.3.1.3 Applications of Ultrasound-Assisted Extraction

Ultrasound-assisted extraction has been used to extract nutraceuticals from plants such as essential oils and lipids (Chemat et al. 2004; Cravotto et al. 2004; Li, Pordesimo, and Weiss 2004; Luque-Garcia and Luque de Castro 2004; Sharma and

Gupta 2004), dietary supplements (Hui, Etsuzo, and Masao 1994; Salisova, Toma, and Mason 1997; Wu, Lin, and Chau 2001; Bruni et al. 2002; Melecchi et al. 2002; Albu et al. 2004). An overview of the uses of ultrasound in food technology was prepared by Mason, Paniwnyk, and Lorimer (1996). An overview of ultrasound-assisted extraction of bioactive compounds from herbs was drafted by Vinatoru (2001).

Ultrasound can increase extraction yield. Sharma and Gupta (2004) found that ultrasonication was a critical pretreatment to obtain high yields of oils from almond, apricot, and rice bran. The yield of oil extracted from soybeans also increased significantly using ultrasound (Li, Pordesimo, and Weiss 2004). For ultrasound-assisted extraction of saponin from ginseng, the observed total yield and saponin yield increased by 15% and 30%, respectively (Hui, Etsuzo, and Masao 1994).

Ultrasound can increase extraction kinetics and even improve the quality of extracts. Cravotto et al. (2004) found that rice bran oil extraction can be efficiently performed in 30 minutes under high-intensity ultrasound either using hexane or a basic aqueous solution. Extraction rates of carvone and limonene by ultrasound-assisted extraction with hexane were 1.3–2 times more rapid than those by the conventional extraction depending on temperature (Chemat et al. 2004). Furthermore, the yield and quality of carvone obtained by the ultrasound-assisted extraction were better than those by a conventional method. The ultrasound was also applied to the cartridge of a Soxhlet extraction for the extraction of total fat from oleaginous seeds such as sunflower, rape, and soybean seeds. The use of ultrasound reduced the extraction by at least half of the time needed by conventional extraction methods without any change in the composition of extracted oils (Luque-Garcia and Luque de Castro 2004). Wu, Lin, and Chau (2001) found the ultrasound-assisted extraction of ginseng saponins occurred about 3 times faster than Soxhlet extraction.

Ultrasound-assisted extraction was considered an efficient method for extracting bioactive compounds from *Salvia officinalis* (Salisova, Toma, and Mason 1997) and *Hibiscus tiliaceus L.* flowers (Melecchi et al. 2002), antioxidants from *Rosmarinus officinalis* (Albu et al. 2004), and steroids and triterpenoids from *Chresta* spp. (Schinor et al. 2004). The use of ultrasound as an adjunct to conventional extraction provides qualitatively acceptable tocols from *Amaranthus caudatus* seeds but much more quickly, more economically, and using equipment commonly available (Bruni et al. 2002).

2.3.2 Pulsed Electric Field-Assisted Extraction

2.3.2.1 *Principles and Mechanism*

Pulsed electric fields (PEFs) can be used to destroy the cell membrane structure for enhancing extraction. When a living cell is suspended in an electrical field, an electric potential across the membrane of the cell is induced. The electric potential causes an electrostatic charge separation in the cell membrane based on the dipole nature of the membrane molecules. When the transmembrane potential exceeds a critical value of approximately 1 V, the repulsion between charge-carrying molecules initiates the formation of pores in weak areas of the membrane. This results in a drastic increase of permeability (Bryant and Wolfe 1987). PEF treatment has been used to inactivate enzyme activity and microorganisms in liquid foods such as juices, milk, liquid egg, model beer, and nutrient broth (Wang 2008). PEF has also been investigated to improve the extraction of intracellular compounds of plant and animal tissues by rupturing or permeabilizing cell membrane (Toepfl et al. 2006).

A simple circuit for a PEF treatment with exponential decay pulses is given in Figure 2.2 (Wang 2008). A treatment chamber consists of two electrodes that are held in position by insulation materials. A PEF process can operate in either a batch or continuous mode depending on the treatment chamber design. The high PEF is generated by high-voltage electrodes. In practice, the time between pulses is much longer than the pulse width. Therefore, the generation of pulses involves slow charging and fast discharging of an electrical energy storage device such as a capacitor. A high pulsed-shaped voltage power supply is needed to charge the capacitor for the PEF treatment. The PEF treatment can be described by its:

- Form of pulse
- Voltage of electric field
- Distance between two electrodes
- Pulse duration time
- Total number of pulses

Electric field strength is determined by the voltage across two electrodes and the distance between two electrodes. For a certain electric field strength, the required voltage will increase with the increase of the gap between two electrodes. A large flux of electrical current must flow through the plant material in a treatment chamber for a short period (i.e., microseconds) (Zhang, Barbosa-Canovas, and Swanson 1995).

The electric field strength of PEF treatment for food pasteurization is usually 20–40 kV/cm. However, the electric field strength of PEF treatment for extraction is much lower (e.g., 2.5 kV/cm) (Eshtiaghi and Knorr 2002). The pulse width is

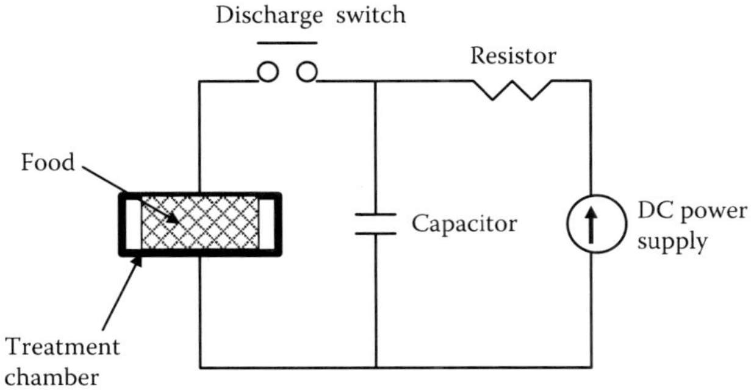

Figure 2.2 A simple circuit for a pulsed electric field treatment of plant materials with exponential decay pulses. (From Wang, L. J., *Energy Efficiency and Management in Food Processing Facilities*, CRC Press, Boca Raton, FL, 2008. With permission.)

short, several microseconds. A product is usually treated for several to a couple hundred pulses. Thus, the total treatment time is less than 1 second (Gongora-Nieto et al. 2003).

2.3.2.2 Advantages and Disadvantages

The PEF treatment can increase the mass transfer during extraction by rupturing cell membrane, thus increasing extract yield and decreasing extraction time. PEF treatment of plant tissues to enhance extraction requires low energy intensities to achieve maximum permeabilization. The temperature increase during PEF treatment at such low energy intensity is almost negligible (Knorr and Angersbach 1998). The low-heat PEF treatment can minimize the degradation of heat-sensitive ingredients such as flavors and proteins (Ade-Omowaye et al. 2001). The PEF treatment of plant materials before extraction allows the application of milder processing conditions during extraction (López et al. 2009). It is also possible to develop a continuous PEF process to treat plant materials (Puértolas et al. 2010).

The effectiveness of PEF treatment strongly depends on the process parameters, including field strength, specific energy input, pulse number, treatment temperature, and the properties of materials to be treated. PEF treatment can be used for processing liquid and semiliquid materials. Application of PEF processing is restricted to materials with no air bubbles and with low electrical conductivity. The maximum particle size in the liquid must be smaller than the gap between two electrodes of the treatment chamber to ensure proper treatment.

2.3.2.3 Applications of Pulsed Electric Field-Assisted Extraction

PEF treatment has been used to enhance the extraction of juice from fruits, vegetables, and sugar beet (Bazhal and Vorobiev 2000; Eshtiaghi and Knorr 2002). PEF can increase the yield of juice by 40% at a pressure of 0.2–0.3 MPa and 12% at a pressure of 3 MPa compared with untreated apples (Bazhal and Vorobiev 2000). Schilling et al. (2007) found the juice yield increased with increasing electric field intensities during PEF-assisted extraction of apple mash.

PEF treatment can significantly reduce the required extraction time and temperature. It was found the extraction time for the sugar beet treated with PEF was only 30 minutes compared with 70 minutes for traditional thermal extraction (Eshtiaghi and Knorr 2002). The red beetroots pretreated with PEF at 7 kV/cm for 5 pulses released approximately 90% of total betaine to extracting water at a pH value of 3.5 and temperature of 30°C within 300 minutes, which is fivefold quicker than the samples without treatment. Pressing the pretreated samples at a pressure of approximately 1 MPa during extraction further reduced the extraction time to 35 minutes (López et al. 2009).

PEF-assisted extraction is also an energy efficient extraction method. An almost total permeabilization of apple and potato tissue can be achieved with an electric energy input in the range of 1–5 kJ/kg of product compared with 20–40 kJ/kg for mechanical, 60–100 kJ/kg for enzymatic, and more than 100 kJ/kg for thermal degradation of plant tissue (Toepfl et al. 2006). PEF treatment at 1 kV/cm was shown to be an effective method of permeabilization for the extraction of pigment from beetroots, with a low energy consumption of approximately 7 kJ/kg (Fincan et al. 2004).

2.3.3 Enzyme-Assisted Extraction

2.3.3.1 Principles and Mechanism

Plant tissues contain many high-value phytochemical compounds such as phenolics. Some phytochemicals in plant matrices are dispersed in the cytoplasm of cells, which is not easily accessible in the extracting process with a solvent. Some compounds are retained in the polysaccharide–lignin network in the plant cells by hydrogen or hydrophobic bonding. They may also be associated with plant cell wall polymers via ether and/or ester linkage. Bound phytochemicals is unavailable for solvent extraction. Application of cell wall degrading enzymes such as pectinase, cellulase, protease, and phospholipase before extraction can hydrolyze the cell wall and the bound phytochemicals to enhance the release of those high-value compounds.

The effectiveness of enzyme-assisted extraction depends on the type of plant materials and the enzymes used. Gaur et al. (2007) found that the yields of oil during enzyme-assisted aqueous extraction of soybean, rice bran, and mango kernel were 98%, 86%, and 79%, respectively. The enzyme-assisted extracted oil was found to contain higher free fatty acids and phosphorus contents than the oil from traditional hexane-extracted oil (Dominguez et al. 1995). Lamsal, Murphy, and Johnson (2006) found that the protease hydrolysis of proteins in soybeans can recover more oil from flaked and extruded soybeans. Treating extruded flakes with cellulase, however, did not enhance oil extraction either alone or in combination with protease.

2.3.3.2 Advantages and Disadvantages

Enzyme treatment can make it possible to use water as an alternative solvent to hexane for extracting oily compounds. The nonpolar solvent of hexane, which can dissolve oil, is usually used to extract edible oily compounds from plant materials such as oilseeds. Because of the environmental concerns with hexane, there has been an increasing interest in aqueous extraction because it is regarded as green processing with little environmental impact, and there may be opportunities to add more value to coproducts. Once the ground seed is placed in water during aqueous extraction, the oil in the seed is released and floats as an emulsified cream in the water. Because oil is insoluble in water, however, the recovery of oil during aqueous extraction is only 60% compared with 95% for hexane extraction. Enzymatic pretreatment of plant materials before aqueous extraction has been used to improve the recovery of oil such as garlic volatile oil (Sowbhagya et al. 2009).

Enzyme-assisted aqueous extraction can simultaneously extract oil and protein factions from plant materials. Protein molecules are denaturized at a high temperature, pressure, and mechanical shear force during traditional extraction with alkaline solutions. Protease can be used to enhance the aqueous extraction of protein from plant materials under a mild condition (Shen et al. 2008). Some phytochemicals such as borage seed oil should be extracted under mild process conditions to reduce the oil oxidation. However, cold extraction such as cold pressing can produce better product quality but lower extract yield than traditional high-temperature solid–solvent extraction. Enzyme treatment can be used to enhance the cold extraction (Soto, Chamy, and Zuniga 2007).

The enzyme treatment can increase the recovery of phytochemicals from plant materials. The addition of 0.5% protease and phospholipase (grams of enzyme per grams of soybean) could recover 97.5% of total oil from extruded soybean flakes at a solid concentration of 10% and 40°C during 1-hour aqueous extraction, compared with approximately 60% for the samples with enzyme treatment (Jung, Maurer, and Johnson 2009).

Najafian et al. (2009) found that the treatment of olive fruit with 0.02%–0.04% pectinase (v/w) could not only increase the oil recovery from olive but also improve the oil quality by increasing the contents of high-value compounds such as phenolics.

However, the activity of enzymes is affected by moisture content and operating conditions such as pH value, temperature, and agitation. The efficiency of enzymatic hydrolysis is very low if plant materials have low moisture content (Dominguez et al. 1995). The operating conditions should be optimized and precisely controlled to maximize the performance of enzymes. Enzyme treatment is usually a slow process, and it may take from several hours to several days (Soto, Chamy, and Zuniga 2007).

2.3.3.3 Applications of Enzyme-Assisted Extraction

Fruit peels such as apple and citrus, tea, and olive oil contain many phenolics (Li, Smith, and Hossain 2006; Pinelo, Zornoza, and Meyer 2008). Li et al. (2006) found that the cellulase enzyme-assisted aqueous extraction process could recover 65.5% of the phenolics from citrus peels at an enzyme concentration of 1.5% (grams of enzyme per grams of plant materials) and 50°C, which was approximately 87.9% of the yield obtained with the traditional solid–solvent extraction with organic solvents. The main parameters that affected the yield of phenolics include the species of the peels, condition of the peels, extraction temperature, types of enzymes, and enzyme concentration (Li, Smith, and Hossain 2006). The release of different types of phenolic compounds differs under the same extraction conditions. Thus, the phenols can be selectively extracted by varying the extraction conditions and by adding cell wall degradation enzymes (Pinelo, Zornoza, and Meyer 2008).

Garlic oil and celery oil are widely used in the perfumery and pharmaceutical industries because they have several nutraceutical attributes such as anticoagulation activity of blood plasma and prevention of cardiovascular diseases. Garlic oil and celery oil are usually extracted by steam distillation or hydrodistillation. Sowbhagya et al. (2009, 2010) found that the yields of garlic and celery oil, after cellulase, pectinase, protease, and viscozyme pretreatment, were in the range of 0.39%–0.51% and

2.2%–2.3% for enzyme-assisted steam distillation compared with 0.28% and 1.8% for traditional steam distillation, respectively. Their study showed that the enzymes had little change in either flavor profile or physicochemical properties of the oil.

Lycopene is another high value nutraceutical compound. Lycopene is a bright red carotene and carotenoid pigment found in red fruits and vegetables such as red carrots, tomato, and watermelons. Cellulase and pectinase enzymes were used to extract lycopene from tomato tissues. For whole tomatoes, Choudhari and Ananthanarayan (2007) found that the pectinase was more effective than cellulase. The pectinase treatment increased the lycopene yield by 224% or 108 μg/g for the whole tomato and 206% or 1104 μg/g for tomato peels, compared with the yield obtained from the samples without enzymatic treatment.

2.3.4 Extrusion-Assisted Extraction

2.3.4.1 Principles and Mechanism

Extrusion, which is a combined unit operation of fluid flow, heat transfer, compression, and reaction, has found increasing number of applications in the manufacture of food, pharmaceuticals, feed products, plastics, and polymer. The single-screw extruder as shown in Figure 2.3, which has a simple geometry, is a popular design (Wang et al. 2004). Twin-screw extruders consist of two screws, having a "figure-eight" cross section, mounted in a barrel as shown in Figure 2.4 (Wang et al. 2006). Therefore, the twin-screw channel has two regions: the C-shaped region and the intermeshing region. There are a large number of types of twin-screw extruders. The corotating intermeshing twin-screw extruder is one of the most frequently used designs. The main difference between the screw channels of a corotating intermeshing twin-screw extruder and a single-screw extruder is the existence of an intermeshing region in the twin-screw channel. Twin-screw extruders have several advantages such as excellent feeding and mixing capacities compared with a single-screw extruder.

Extrusion processing begins with conveying raw granular materials down the screw channel. The free-flowing granular particles of the material are conveyed

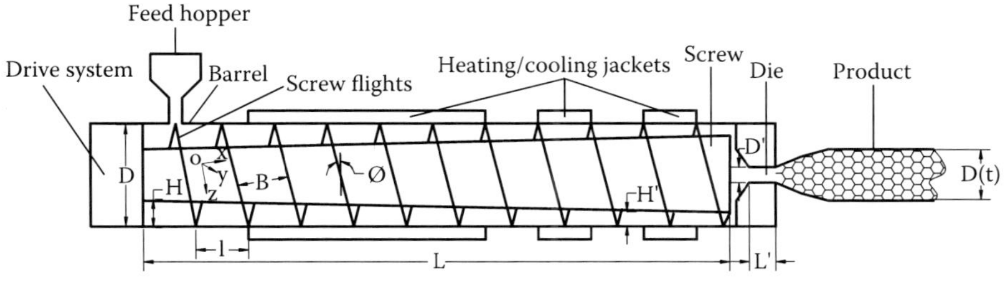

Figure 2.3 Cross section of a typical single-screw extruder. (From Wang, L. J., G. M. Ganjyal, D. D. Jones, C. L. Weller, and M. A. Hanna, *J. Food Sci.*, 69, 212, 2004. With permission.)

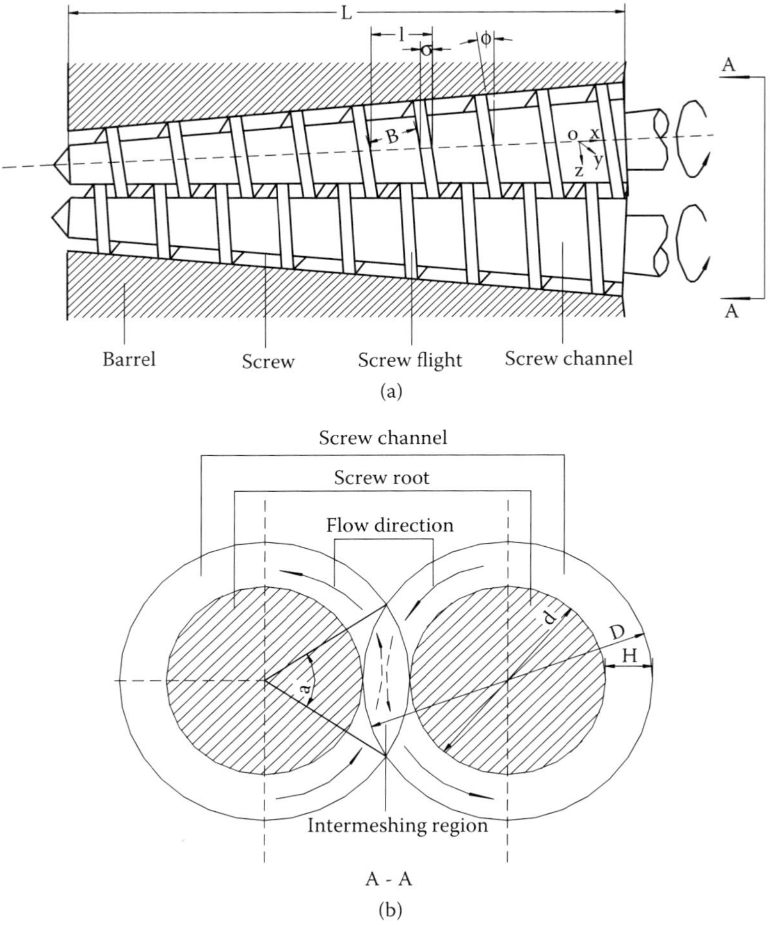

Figure 2.4 Cross sections and flow diagram of a typical intermeshing, corotating twin-screw extruder. (a) Intermeshing corotating twin-screw elements. (b) A cross section of the extruder. (From Wang, L. J., D. D. Jones, C. L. Weller, and M. A. Hanna, *Adv. Polym. Technol.*, 25, 22, 2006. With permission.)

through the feed zone in a manner similar to the action of a screw conveyor. Through heating and compression, the loose granular materials are gradually melted to form a dense plastic-like dough in the transition zone. The plasticized melt dough is further exposed to heat, pressure, and shear effects through the metering zone and finally puffed on exit from a die channel. Processing conditions, including screw speed, flow rate, moisture content, barrel temperature, and screw and die configuration, determine the final product quality through system variables such as pressure gradient, temperature profile, shear rate, and residence time. The extruder generates considerable frictional heat. The barrel can also be equipped with one or more jackets for contact heating with steam or steam injection valves for direct steam injection.

However, it is sometimes desirable to cool the jacket section closest to the feed end to prevent blowback of steam (Wang et al. 2004, 2006).

The extrusion of plant materials such as oilseeds before solid–solvent extraction is another method to enhance the extraction. Ground materials such as flours are usually used in solvent extraction. However, grinding alone does not completely rupture cell membrane, which is a key barrier to recovering oil by solvent extraction, particularly aqueous extraction. Fine grinding may also make a stable emulsion cream phase of oil, protein, and fiber particles that must be broken to further recover oil. Extrusion provides another form of mechanically and thermally treating plant materials that do not require extensive grinding while achieving greater cell distortion. Extruding flakes provides vigorous thermal and mechanic treatment to rupture cell walls for expelling the free oil inside cells.

2.3.4.2 Advantages and Disadvantages

Extrusion of plant materials before solvent extraction can (1) increase the recovery of oil and the capacity of solvent extractors as a result of the increased density of extruded materials, (2) decrease the holdup of solvent in the material bed and the energy requirement for solvent recovery, (3) improve percolation of the material bed, (4) deactivate undesirable enzymes such as lipase in rice bran, and phospholipase in soybeans and other oilseeds, and (5) make it possible to conduct other chemical reactions and product formation while processing materials (Lusas and Watkins 1988).

The equipment cost may be a drawback for the extrusion-assisted extraction process, but it could be a promising alternative for the production of high-value nutraceutical products. Extrusion alone such as extrusion expelling and screw pressing can be used to squeeze oil out of plant matrices. However, these processes leave 7%–10% residual oil contents in the meal. Both of these mechanical procedures, if operated at a high temperature, also denature protein, rendering it deficient in functional properties for food use (Lamsal, Murphy, and Johnson 2006).

2.3.4.3 Applications of Extrusion-Assisted Extraction

It is estimated that approximately 60% of the current domestic soybean crush and 50% of the cottonseed crush are now processed with an extruder or expander (Lusas and Watkins 1988). Continuous extrusion expelling and screw pressing to extract oil from soybeans have become viable alternatives to hexane extraction (Lansal et al. 2006).

Alginates are natural polysaccharides that are usually extracted from brown seaweeds in a batch process with an alkaline solution. Vauchel, Kaas, and Arhaliass (2008) used a twin-screw extruder at a rotational speed of 400 rpm to extract alginates from seaweeds. The seaweed was fed into the extruder at a feed rate of 1 kg/h while a 4% (w/w) Na_2CO_3 solution was cofed into the extruder at a feed rate of 1 L/h. As alginate starts to degrade at 40°C, the barrel temperature was maintained at 20°C. Their results showed that the extraction time for the continuous extrusion-assisted extraction process was only several minutes compared with 1 hour for the traditional batch process. The extraction yield was more than 15% higher than that obtained using the batch process. Both water and alkaline consumption was significantly reduced to less than half for the batch process. Furthermore, alginates extracted by reactive extrusion clearly had a superior quality.

2.3.5 Microwave-Assisted Extraction

2.3.5.1 Principles and Mechanism

Microwaves are electromagnetic radiations with a frequency from 0.3 to 300 GHz. Domestic and industrial microwaves generally operate at 2.45 GHz, and occasionally at 0.915 GHz, in the United States and at 0.896 GHz in Europe. Microwaves are transmitted as waves, which can penetrate biomaterials and interact with polar molecules such as water in the biomaterials to create heat. Consequently, microwaves can heat a whole material to penetration depth simultaneously.

Microwave-assisted extraction (MAE) offers a rapid delivery of energy to a total volume of solvent and solid plant matrix with subsequent heating of the solvent and solid matrix, efficiently and homogeneously. Because water within the plant matrix absorbs microwave energy, cell disruption is promoted by internal superheating, which facilitates desorption of chemicals from the matrix, improving the recovery of nutraceuticals (Kaufmann, Christen, and Veuthey 2001a). Kratchanova, Pavlova, and Panchev (2004) observed using SEMs that microwave pretreatment of fresh orange peels led to destructive changes in the plant tissue. These changes in the plant tissue caused by microwave heating gave a considerable increase in the yield of extractable pectin. Furthermore, the migration of dissolved ions increased solvent penetration into the matrix and thus facilitated the release of the chemicals. Therefore, the effect of microwave energy is strongly dependent on the dielectric susceptibility of both the solvent and the solid plant matrix.

There are two types of commercially available MAE systems: closed extraction vessels under controlled pressure and temperature, and focused microwave ovens at atmospheric pressure (Kaufmann and Christen 2002). The closed MAE system is generally used for extraction under drastic conditions such as high extraction temperature. The pressure in the vessel essentially depends on the volume and the boiling point of the solvents. The focused MAE system can be operated at a maximum temperature determined by the boiling point of the solvents at atmospheric pressure. Ericsson and Colmsjo (2000) introduced a dynamic MAE system, which was demonstrated to yield extract equivalent to yield of extract from Soxhlet extraction, but in a much shorter time.

Because MAE depends on the dielectric susceptibility of solvent and matrix, better recoveries can be obtained by moistening samples with a substance that possesses a relatively high dielectric constant such as water. If a dry biomaterial is rehydrated before extraction, the matrix itself can thus interact with microwaves and hence facilitate the heating process. The microwave heating leads to the expansion and rupture of cell walls and is followed by the liberation of chemicals into the solvent (Spar Eskilsson and Bjorklund 2000). In this case, the surrounding solvent can have a low dielectric constant and thus remains cold during extraction. This method can be used to extract thermosensitive compounds such as essential oils (Brachet et al. 2002). However, it was found that it was impossible to perform a good MAE for completely dry as well as for wet samples when a nonpolar solvent such as hexane was used as the extraction solvent (Molins et al. 1997).

Plant particle size and size distribution usually have a significant influence on the efficiency of MAE. The particle sizes of the extracted materials are usually in the range of 100 μm to 2 mm (Spar Eskilsson and Bjorklund 2000). Fine powder can enhance the extraction because the limiting step of the extraction is often the diffusion of chemicals out of the plant matrix, and the larger surface area of a fine powder provides contact between the plant matrix and the solvent. For example, for MAE of cocaine, finely ground coca powder was more easily extracted than large particles (Brachet et al. 2002).

Solvent choice for MAE is dictated by the solubility of the extracts of interest, by the interaction between solvent and plant matrix, and finally by the microwave-absorbing properties of the solvent determined by its dielectric constant. Csiktusnadi Kiss et al. (2000) investigated the efficiency and selectivity of MAE for the extraction of color pigments from *paprika* powders using 30 extracting solvent mixtures. Their results showed that efficacy and selectivity of MAE depend significantly on the dielectric constant of the extraction solvent mixture. Usually the chosen solvent should possess a high dielectric constant and strongly absorb microwave energy. Solvents such as ethanol, methanol, and water are sufficiently polar to be heated by microwave energy (Brachet et al. 2002). Nonpolar solvents with low dielectric constants such as hexane and toluene are not potential solvents for MAE. The extracting selectivity and the ability of the solvent to interact with microwaves can be modulated by using mixtures of solvents (Brachet et al. 2002). One of the most commonly used mixtures is hexane–acetone (Spar Eskilsson and Bjorklund 2000). A small amount of water (e.g., 10%) can also be incorporated in nonpolar solvents such as hexane, xylene, or toluene to improve the heating rate (Spar Eskilsson and Bjorklund 2000).

During extraction, the solvent volume must be sufficient to ensure that the solid matrix is entirely immersed. Generally, a higher ratio of solvent volume to solid matrix mass in conventional extraction techniques can increase the recovery. However, in the MAE, a higher ratio may give lower recoveries. This is probably the result of inadequate stirring of the solvent by microwaves (Spar Eskilsson et al. 1999).

Temperature is another important factor contributing to the recovery yield. Generally, elevated temperatures result in improved extraction efficiencies. However, for the extraction of thermolabile compounds, high temperatures may cause the degradation of extracts. In this case, the chosen power during MAE has to be set correctly to avoid excess temperatures, leading to possible solute degradation (Font et al. 1998).

2.3.5.2 Advantages and Disadvantages

MAE has been considered a potential alternative to traditional solid–liquid extraction for the extraction of metabolites from plants. It has been used to extract nutraceuticals for several reasons: (1) reduced extraction time, (2) reduced solvent usage, and (3) improved extraction yield. MAE is also comparable with other modern extraction techniques such as supercritical fluid extraction (SFE) because of its process simplicity and low cost. By considering economical and practical aspects, MAE is a strong novel extraction technique for the extraction of nutraceuticals.

However, compared with SFE, an additional filtration or centrifugation is necessary to remove the solid residue during MAE. Furthermore, the efficiency of microwaves can be poor when either the target compounds or the solvents are nonpolar, or when they are volatile.

2.3.5.3 Applications of Extrusion-Assisted Extraction

Although MAE of organic compounds from environmental matrices and microwave-assisted leaching in process metallurgy has widely been investigated (Ganzler, Salgo, and Valko 1986; Tomaniova et al. 1998; Lorenzo et al. 1999; Spar Eskilsson and Bjorklund 2000; Barriada-Pereira et al. 2003; Al-Harahsheh and Kingman 2004), few applications have been published in the nutraceutical area. An overview of publications on MAE of natural products was prepared by Kaufmann and Christen (2002).

MAE can extract nutraceutical products from plant sources in a faster manner than conventional solid–liquid extractions. MAE of the puerarin from the herb *Radix puerariae* could be completed within 1 minute (Guo et al. 2001). MAE (80% methanol) could dramatically reduce the extraction time of *ginseng* saponins from 12 hours using conventional extraction methods to a few seconds (Kwon et al. 2003). It took only 30 seconds to extract cocaine from leaves with the assistance of microwave energy quantitatively similar to those obtained by conventional solid–liquid extraction for several hours (Brachet et al. 2002). For extracting an equivalent amount and quality of tanshinones from *Salvia miltiorrhiza Bunge*, MAE only needed 2 minutes, whereas extraction at room temperature, Soxhlet extraction, ultrasonic extraction, and heat reflux extraction needed 24 hours, 90 minutes, 75 minutes, and 45 minutes, respectively (Pan, Niu, and Liu 2002). Williams et al. (2004) found MAE was efficient in recovering approximately 95% of the total capsaicinoid fraction from *capsicum* fruit in 15 minutes compared with 2 hours for reflux extraction and 24 hours for maceration.

A higher extraction yield can be achieved in a shorter extraction time using MAE. A 12-minute MAE could recover 92.1% of artemisinin from *Artemisia annua L*, whereas several-hour Soxhlet extraction could only achieve approximately 60% recovery (Hao et al. 2002). A 4- to 5-minute MAE (ethanol–water) of glycyrrhizic acid from *licorice* root achieved a higher extraction yield than extraction (ethanol–water) at room temperature for 20–24 hours (Pan et al. 2000). For the extraction of tea polyphenols and caffeine from green tea leaves, a 4-minute MAE achieved a higher extraction yield than an extraction at room temperature for 20 hours, ultrasonic extraction for 90 minutes, and heat reflux extraction for 45 minutes, respectively (Pan, Niu, and Liu 2003). Shu et al. (2003) reported that the extraction yield of ginsenosides from *ginseng* root obtained by a 15-minute MAE (ethanol–water) was higher than that obtained by 10-hour conventional solvent extraction (ethanol–water).

MAE can also reduce solvent consumption. Focused MAE was applied to the extraction of withanolides from air-dried leaves of *Iochroma gesnerioides* (Kaufmann, Christen, and Veuthey 2001a). The main advantages of MAE over Soxhlet extraction are associated with the drastic reduction in organic solvent consumption (5 mL vs. 100 mL) and extraction time (40 seconds vs. 6 hours). It was also found that the presence of water in the solvent of methanol had a beneficial effect and allowed faster extractions than with organic solvent alone.

2.3.6 Ohmic Heating-Assisted Extraction

2.3.6.1 Principles and Mechanism

Ohmic heating, like microwave heating, is another advanced thermal processing method. During ohmic heating, a plant material serves as an electrical resistor. When electricity is passed through it, electrical energy is dissipated into heat, which results in a rapid and uniform temperature increase. Ohmic heating is also called electrical resistance heating, Joule heating, or electroheating. Ohmic heating can be used for heating liquid or semiliquid materials, usually with high moisture content such as soups, fruit slices in syrups, and sauces. Potential applications of ohmic heating include its use in blanching, evaporation, dehydration, fermentation, and extraction.

2.3.6.2 Advantages and Disadvantages

Plant materials have poor thermal conductivity. Conventional thermal processing based on heat conduction has a slow heating rate and causes a big temperature gradient in the materials, which decreases the efficiency and damages the product quality during extraction. Ohmic heating, like microwave heating, can rapidly and uniformly heat the entire mass of a material. Liquid egg, for example, can be uniformly heated in a fraction of a second without coagulating using the ohmic heating method.

Additionally, ohmic heating can reduce the formation of fouling on the contact surface of a heater. Ohmic heating is useful for the treatment of materials with a high protein content, which tend to denature and coagulate during traditional thermal processing. Ohmic heating can also be used to inactivate enzymes in juices without affecting their flavor. Ohmic heating has been shown to increase the extraction yield of sucrose from sugar beets (Katrokha et al. 1984), apple juice from apples (Lima and Sastry 1999), soymilk from soybeans (Kim and Pyun 1995), and oil from rice bran (Lakkakula, Lima, and Walker 2004).

Ohmic heating like microwave heating cannot work for dry materials because of their low electrical conductivity. An economic analysis conducted at the University of Minnesota in the early 1990s indicated that ohmic heating would be economically viable for premium quality foods. However, costs of ohmic systems have decreased greatly since then, and the range of products for which ohmic heating is economical has expanded considerably.

2.3.6.3 Applications of Ohmic Heating-Assisted Extraction

Lakkakula et al. (2004) used ohmic heating to enhance the oil extraction from rice bran. Their results showed that ohmic heating is not an effective method for dry rice bran because of its low electrical conductivity. The temperature of rice bran at a moisture content of 10.5% on a wet basis was increased less than 1°C after 15-minute ohmic heating. However, ohmic heating was an effective method to stabilize rice bran oil and increase the oil yield after the moisture content was adjusted to 21% by adding deionized water. Ohmic heating increased the total percentage of lipids extracted from rice bran to a maximum of 92%, from 53% of total lipids extracted from the samples without ohmic heating. They also found that lowering the frequency of alternating current significantly increased the amount of oil extracted.

2.3.7 Supercritical Fluid Extraction

2.3.7.1 Principles and Mechanism

Supercritical state is achieved when the temperature and the pressure of a substance are raised over its critical value. The supercritical fluid has characteristics of both gases and liquids. Compared with liquid solvents, supercritical fluids have several major advantages: (1) the dissolving power of a supercritical fluid solvent depends on its density, which is highly adjustable by changing the pressure or/and temperature; and (2) the supercritical fluid has a higher diffusion coefficient and lower viscosity and surface tension than a liquid solvent, leading to more favorable mass transfer.

A SFE system is shown in Figure 2.5. During SFE, raw plant material is loaded into an extraction vessel, which is equipped with temperature controllers and pressure valves at both inlet and outlet to keep desired extraction conditions. The extraction vessel is pressurized with the fluid by a pump. The fluid and the dissolved compounds are transported to separators, where the salvation power of the fluid is decreased by decreasing the pressure or increasing the temperature of the fluid. The product is then collected via a valve located in the lower part of the separators. The fluid is further regenerated and cycled (Sihvonen et al. 1999).

To develop a successful SFE, several factors must be taken into consideration. These factors include the selection of supercritical fluids, plant material preparation, modifiers, and extraction conditions.

Selection of supercritical fluids is critical for the development of an SFE process. With a reduction in the price of CO_2 and restrictions in the use of other organic solvents, CO_2 has begun to move from some marginal applications to being the

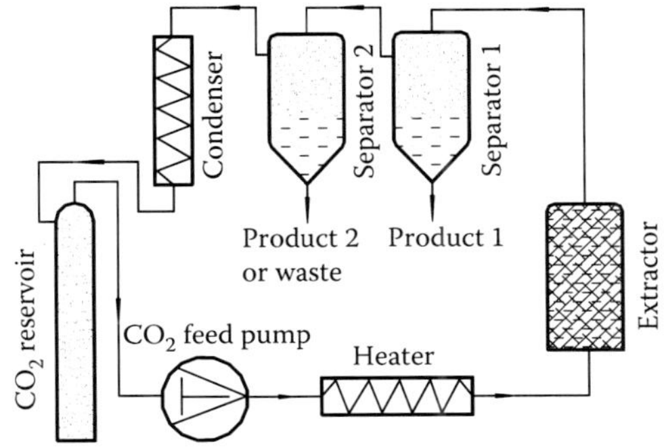

Figure 2.5 Schematic diagram of a process-scale supercritical fluid extraction system. (From Wang, L. J. and C. L. Weller, *Trends Food Sci. Technol.*, 17, 300, 2006. With permission.)

major solvent for SFE (Hurren 1999). The critical state of CO_2 fluid is at a temperature of only 304 K and pressure of 7.3 MPa. Also, CO_2 is nonflammable and nontoxic. Supercritical CO_2 is a good solvent for the extraction of nonpolar compounds such as hydrocarbons (Vilegas, de Marchi, and Lancas 1997). To extract polar compounds, some polar supercritical fluids such as Freon 22 (DuPont, Wilmington, DE), nitrous oxide, and hexane have been considered. However, their applications are limited because of their unfavorable properties with respect to safety and environmental considerations (Lang and Wai 2001). Although supercritical water and superheated water have certain advantages, such as higher extraction ability for polar compounds, they are not suitable for thermally labile compounds (Lang and Wai 2001).

Many nutraceuticals such as phenolics, alkaloids, and glycosidic compounds are poorly soluble in CO_2 and hence not extractable. Techniques aimed at overcoming the limited solubility of polar substances in supercritical CO_2 have been sought. Addition of polar cosolvents (modifiers) to the supercritical CO_2 is known to significantly increase the solubility of polar compounds. Among all the modifiers, including methanol, ethanol, acetonitrile, acetone, water, ethyl ether, and dichloromethane, methanol is the most commonly used because it is an effective polar modifier and is up to 20% miscible with CO_2. However, ethanol may be a better choice in SFE of nutraceuticals because of its lower toxicity (Lang and Wai 2001; Hamburger, Baumann, and Adler 2004). Furthermore, the use of methanol as a modifier requires a slightly higher temperature to reach the supercritical state, and this could be disadvantageous for thermolabile compounds. A mixture of modifiers can also be used in SFE. The best modifier usually can be determined based on preliminary experiments. One disadvantage of using a modifier is that it can cause poor selectivity.

Preparation of plant materials is another critical factor for SFE of nutraceuticals. Fresh plant materials are frequently used in the SFE of nutraceuticals. When fresh plant materials are extracted, the high moisture content can cause mechanical difficulties such as restrictor clogging as a result of ice formation. Although water is only approximately 0.3% soluble in supercritical CO_2, highly water-soluble solutes would prefer to partition into the aqueous phase, resulting in low efficiency of SFE. Thus, some chemicals such as Na_2SO_4 and silica gel are mixed with the plant materials to retain the moisture for SFE of fresh materials (Lang and Wai 2001).

Plant particle size is also important for a good SFE process. Large particles may result in a long extraction process because the process may be controlled by internal diffusion. However, fine powder can speed up the extraction but may also cause difficulty in maintaining a proper flow rate. Coelho et al. (2003) used supercritical CO_2 to extract *Foeniculum vulgare* volatile oil from fennel fruits with difference in mean particle sizes of 0.35, 0.55, and 0.75 mm. They found that the decrease of particle size had a small decrease in the total yield of the extracted oil. Therefore, in this case, some rigid inert materials such as glass beads and sea sand are packed with the fine plant powder to maintain a desired permissibility of the particle bed.

The solubility of a target compound in a supercritical fluid is a major factor in determining its extraction efficiency. The temperature and density of the fluid control the solubility. The choice of a proper density of a supercritical fluid such as CO_2 is the crucial point influencing solvent power and selectivity, and the main factor determining the extract composition (Cherchi et al. 2001). It is often desirable to extract the compound right above the point where the desired compounds become soluble in the fluid so that the extraction of other compounds can be minimized. For supercritical CO_2 extraction of Chilean hop (*Humulus lupulus*) ecotypes, del Valle et al. (2003) found that limited increases in extraction rate were observed when applying pressure greater than 20 MPa at the temperature of 40°C; rather, the increase in pressure increased the coextraction of undesirable compounds.

By controlling the fluid density and temperature, fractionation of the extracts could also be achieved. For supercritical CO_2 extraction of squalene and stigmasterol from the entire plant of *Spirodela polyrhiza*, Choi et al. (1997) found the relative extraction yield of squalene was much higher than that of stigmasterol at 10 MPa and 50°C or 60°C. The extraction of squalene was comparable with *n*-hexane extraction, but stigmasterol was not detected under these conditions. Their result confirmed that SFE could selectively extract substances from the plant materials by controlling conditions such as temperature and pressure (density).

The extraction time has been proven to be another parameter that determines extract composition. Lower molecular weight and less polar compounds are more readily extracted during supercritical CO_2 extraction because the extraction mechanism is usually controlled by internal diffusion (Poiana, Fresa, and Mincione 1999; Cherchi et al. 2001). Therefore, the extract composition varies with the extraction time. However, Coelho et al. (2003) reported that the increase of CO_2 flow rate did not seem to influence the composition for the supercritical CO_2 extraction of *F. vulgare* volatile oil, although it increased the extraction rate.

2.3.7.2 Advantages and Disadvantages

SFE offers unusual possibilities for selective extractions and fractionations because the solubility of a chemical in a supercritical fluid can be manipulated by changing the pressure and/or temperature of the fluid. Furthermore, supercritical fluids have a density of a liquid and can solubilize a solid like a liquid solvent. The solubility of a solid in a supercritical fluid increases with the density of the fluid, which can be achieved at high pressures.

The dissolved nutraceutical compounds can be recovered from the fluid by the reduction of the density of the supercritical fluid, which can usually be reduced by decreasing pressure (Poiana, Sicari, and Mincione 1998). Therefore, SFE can eliminate the concentration process, which usually is time-consuming. Furthermore, the solutes can be separated from a supercritical solvent without a loss of volatiles as a result of the extreme volatility of the supercritical fluid.

Additionally, the diffusivity of a supercritical fluid is one to two orders of magnitude higher than that of other liquids, which permits rapid mass transfer, resulting in a larger extraction rate than that obtained by conventional solvent extractions (Roy et al. 1996). Supercritical CO_2 extraction uses a moderate extraction temperature as low as 30°C. The low supercritical temperature of CO_2 makes it attractive for the extraction of heat-sensible compounds. Because SFE uses no or only minimal organic solvent (organic modifiers) in extraction, it is a more environmentally friendly extraction process than conventional solvent–solid extraction. SFE can be directly coupled with a chromatographic method for simultaneously extracting and quantifying highly volatile extracted compounds.

However, the economics and onerous operating conditions of the SFE processes have restricted the applications to some specialized fields such as essential oil extraction and coffee decaffeination, as well as to university research.

2.3.7.3 Applications of Supercritical Fluid Extraction

SFE is a potential alternative to conventional extraction methods using organic solvents for extracting biologically active components from plants (Modey, Mulholland, and Raynor 1996; Choi et al. 1997; Dean et al. 1998; Dean and Liu 2000; Szentmihalyi et al. 2002; Ellington et al. 2003; Hamburger, Baumann, and Adler 2004; Andras et al. 2005; Wang et al. 2007, 2008). It has been used to extract plant materials, especially lipids (Bernardo-Gil et al. 2002), essential oils (Berna et al. 2000; Coelho et al. 2003; Marongiu et al. 2003), and flavors (Sass-Kiss et al. 1998; Giannuzzo et al. 2003). Overviews of fundamentals and applications of supercritical fluids in different processes have been prepared by Sihvonen et al. (1999), Marr and Gamse (2000), and Hauthal (2001). Turner, King, and Mathiasson (2001) gave a review about supercritical fluids in the extraction and chromatographic separation of fat-soluble vitamins. An overview of published data for the SFE of different materials was given by Marr and Gamse (2000), Lang and Wai (2001), and Meireles and Angela (2003).

SFE can prevent the oxidation of lipids. Bernardo-Gil et al. (2002) found that the contents of free fatty acids, sterols, triacylglycerols, and tocopherols in the hazelnut oil extracted by SFE were comparable with those obtained with *n*-hexane extraction. However, the SFE-extracted oil was more protected against oxidation of the unstable polyunsaturated fatty acids than the *n*-hexane-extracted oil. Oil extracted with supercritical CO_2 was clearer than the one extracted by *n*-hexane.

SFE can achieve higher yield and quality of essential oils, flavors, and natural aromas than conventional steam distillation. The mean percentage yields of cedarwood oil for supercritical CO_2 extraction and steam distillation were 4.4% and 1.3%, respectively (Eller and King 2002). The yield of supercritical CO_2 extraction of essential oil from *juniper* wood at 50°C and 10 MPa was 14.7% (w/w), whereas hydrodistillation gave a yield of 11% (w/w) (Marongiu et al. 2003). Coelho et al. (2003) found that moderate supercritical CO_2 conditions (9 MPa and 40°C) could achieve an efficient extraction of *F. vulgare* volatile oil, enabling approximately 94% of the oil to be extracted within 150 minutes. Compared with hydrodistillation, SFE (20 MPa and 50°C) led to higher concentrations of light oxygenated compounds in the oil extracted from Egyptian marjoram leaves, which gave the oil a superior aroma. The antioxidant property of the SFE extract was also markedly higher than that of the hydrodistillation extract (El-Ghorab et al. 2004). However, it should be addressed that the composition of the essential oil was determined by two important parameters: CO_2 density (pressure) and extraction time (Roy et al. 1996; Cherchi et al. 2001).

The number of industrial-scale applications of SFE in plant extraction has remained small because of the lipophilic nature of supercritical CO_2. Comparison of different extraction methods for the extraction of oleoresin from dried onion showed that the yield after supercritical CO_2 extraction was 22 times higher than that after steam distillation, but it was only 7% of the yield achieved by the extraction using a polar solvent of alcohol at 25°C (Sass-Kiss et al. 1998). However, the flavor and biological activity of onion are attributed mainly to its sulfur-containing compounds. The concentration of sulfur was the highest in steam-distilled onion oil, whereas it was the lowest in the extract of alcohol at 25°C. The oleoresin produced by supercritical CO_2 extraction had the best sensory quality. Many active substances in plants such as phenolics, alkaloids, and glycosidic compounds are poorly soluble in CO_2 and hence not extractable (Hamburger, Baumann, and Adler 2004).

Modifiers such as methanol and ethanol are widely used in the supercritical CO_2 extraction of polar substances. Supercritical CO_2 modified with 15% ethanol gave higher yields than pure supercritical CO_2 to extract naringin (a glycosylated flavonoid) from *citrus paradise* at 9.5 MPa and 58.6°C (Giannuzzo et al. 2003). The use of a 10% ethanol cosolvent resulted in a much higher yield of *epicatechin* (13 mg/100 g seed coat) than that with pure CO_2 (22 μg/100 g seed coat) from sweet Thai *tamarind* seed coat (Luengthanaphol et al. 2004). Supercritical CO_2 extraction with methanol in the range of 3%–7% as modifier was proven to be an efficient and fast method to recover higher than 98% of colchicines and colchicoside and 97% of 3-demethylcolchicine from seeds of *Colchicum autumnale* (Ellington et al. 2003).

2.3.8 Accelerated Solvent Extraction

2.3.8.1 Principles and Mechanism

Accelerated solvent extraction (ASE) is a solid–liquid extraction process performed at elevated temperatures, usually between 50°C and 200°C and at pressures between 10 MPa and 15 MPa. Therefore, ASE is a form of pressurized solvent

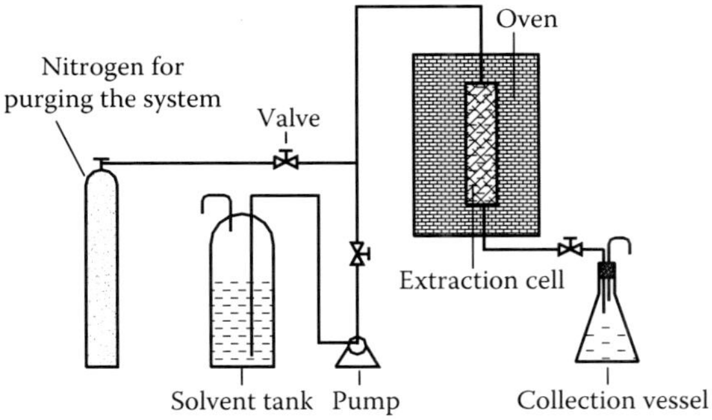

Figure 2.6 Schematic diagram of an accelerated solvent extraction system. (From Wang, L. J. and C. L. Weller, *Trends Food Sci. Technol.*, 17, 300, 2006. With permission.)

extraction that is similar to SFE. Extraction is carried out under pressure to maintain the solvent in its liquid state at high temperature. The solvent is still below its critical condition during ASE. Increased temperature accelerates the extraction kinetics, and elevated pressure keeps the solvent in the liquid state, thus achieving safe and rapid extraction. Also, pressure allows the extraction cell to be filled faster and helps to force liquid into the solid matrix. Elevated temperatures enhance diffusivity of the solvent, resulting in increased extraction kinetics (Richter et al. 1996; Brachet et al. 2001; Kaufmann and Christen 2002). A typical schematic diagram of an ASE system is given in Figure 2.6.

Although the solvent used in ASE is usually an organic solvent, pressurized hot water or subcritical water can also be used in an ASE apparatus, which is usually called pressurized hot water extraction or subcritical water extraction (Eskilsson et al. 2004).

2.3.8.2 Advantages and Disadvantages

Use of nontoxic extracting solvents such as CO_2 and water has economic and environmental benefits. Supercritical CO_2 extraction has been reported to be a valuable novel extraction technique for the extraction of nutraceuticals. However, a considerable quantity of polar modifier has to be added to CO_2 to extract polar compounds. ASE is considered a potential alternative technique to SFE for the extraction of polar compounds (Brachet et al. 2001). Compared with traditional Soxhlet extraction, there is a dramatic decrease in the amount of solvent and the extraction time for ASE (Richter et al. 1996). Particular attention should be paid to the ASE performed at a high extraction temperature, which may lead to degradation of thermolabile compounds.

2.3.8.3 Applications of Accelerated Solvent Extraction

ASE is usually used for the extraction of high-temperature stable organic pollutants from environmental matrices. Few applications of ASE have been published in the field of nutraceuticals. Kaufmann and Christen (2002) reviewed recent developments in ASE for natural products. A review of fundamentals and practical use of pressurized hot water extraction was given by Smith (2002).

ASE was developed for the rapid extraction of cocaine and benzoylecgonine from *coca* leaves using methanol as solvent. The optimal pressure, temperature, extraction time, and particle size were found to be 20 MPa, 80°C, 10 minutes, and 90–150 µm (Brachet et al. 2001). Their result showed there was a substantial reduction in extraction time, compared with several-hour conventional extraction methods. Kaufmann, Christen, and Veuthey (2001b) compared the performance of ASE with traditional Soxhlet extraction for the recovery of steroids from the leaves of *I. gesnerioides*. They found ASE produced similar results to Soxhlet extraction in terms of recovery, repeatability, and selectivity. However, both extraction time and solvent consumption were dramatically reduced with ASE. More applications of ASE for natural products can be found in the literature (Kaufmann and Christen 2002). Denery et al. (2004) compared the extraction of carotenoids and kavalactones from algae using pressurized solvent extraction and traditional solid–solvent extraction. They found that the pressurized solvent at a pressure of approximately 10 MPa required only half the amount of the solvent as traditional extraction and approximately 20 minutes compared with 90 minutes for traditional extraction.

2.4 APPLICATIONS OF EMERGING EXTRACTION METHODS FOR NUTRACEUTICAL PRODUCT DEVELOPMENT

2.4.1 Plant-Based Nutraceutical Products

Nutraceuticals were defined as a wide range of foods and food components with claimed medical or health benefits by Stephen DeFelici in the 1980s. The benefits of these components range from the supply of essential minerals, vitamins, or other dietary supplements for human health to protection against several infectious diseases such as cancer, diabetes, and cardiovascular disease (Hugenholtz and Smid 2002). There is increasing interest to use phytochemicals, which are nutritional plant sources of supplements such as genistein from soybeans and flavonoids from garlic, citrus fruit, and berries. Plant cells produce several primary metabolites, mainly including lipids, proteins, and carbohydrates. Plant cells also produce many high-value organic compounds of secondary metabolites. Most of those plant-based natural compounds can be classified into three major groups: phenolics, terpenoids, and alkaloids. Those high-value compounds are retained in the polysaccharide–lignin network in the plant cells by hydrogen or hydrophobic bonding. They may also be associated with plant cell wall polymers via ether and/or ester linkage (Wildman 2001).

Phenolic compounds such as flavonoids are a group of secondary metabolites in plant materials. Phenolic compounds, which affect the color, flavor, and tastes of plant materials, have high sensory, nutritional, and medical values. Flavonoids, for example, are most commonly known for their antioxidant activity and potential role in the prevention of cancers and cardiovascular disease. Terpenes and terpenoids are the primary constituents of the essential oils of many types of plants and flowers. Essential oils are used widely as natural flavor additives for food, as fragrances in perfumery, and traditionally as alternative medicines such as aromatherapy. Carotenoids, which are pigments to absorb visible light, are part of terpenoids. The pigments are attached to proteins in photosynthetic structures in plant chloroplast membrane. Carotenoids can reduce the risk of certain types of cancers. Alkaloids are naturally occurring chemical compounds containing basic nitrogen atoms. Plant alkaloids are produced as toxins to grazing or browsing animals. Many alkaloids such as cocaine, morphine, caffeine, and nicotine can be isolated and purified from plant materials by acid–base extraction.

2.4.2 Selection of Extraction Methods for Nutraceutical Development

Traditional solid–solvent extraction methods, including maceration, hydrodistillation, and Soxhlet, are generally well-established techniques. Wide industrial applications, better reproducibility and efficiency, and less extract manipulation are the advantages of Soxhlet extraction over emerging extraction methods, such as ultrasound, PEFs, enzyme, extrusion, microwave heating, ohmic heating, supercritical fluids, and ASEs. However, compared with the novel fast extraction techniques, Soxhlet extraction is an old-fashioned and time- and solvent-consuming technique. Some solvents used in conventional Soxhlet extraction have recently been questioned because of their toxicity. For Soxhlet extraction, the advantages such as sample–fresh solvent contact during the whole, no-filtration procedure and simple manipulation should be retained. Meanwhile, auxiliary features such as a vacuum pump, a membrane separation unit, a source of ultrasound and microwave, and supercritical fluids can be incorporated into the conventional Soxhlet method to improve its performance.

Plant-based nutraceuticals are retained inside cells or associated with the cell membrane wall. Mechanical and/or thermal processing is usually used to denaturize the cell membrane for releasing compounds inside the cell into an extraction liquid. However, a high temperature and nonuniform temperature distribution during thermal processing may degrade the extracted compounds. Ultrasound, PEF, enzymatic treatment, and extrusion are promising methods to rupture cell membrane for enhancing the release of nutraceuticals from plant cells. Temperature is an important parameter affecting the extraction yield, extraction kinetics, and product quality. Because of low thermal conductivity of plant materials in nature, the traditional heating process based on heat conduction is slow and nonuniform. Microwave and ohmic heating can provide rapid and uniform heating of plant materials during extraction. Organic solvents such as hexane are widely used to extract phytochemicals from plant matrix. Because of the environmental and safety concerns about the use of organic solvents, several environmentally friendly solvents such as supercritical CO_2 and water have been investigated as alternatives to organic solvents. Furthermore, the supercritical state of solvents and accelerated pressure can improve the mass transfer between solid plant materials and solvents.

Table 2.1 Comparison of Different Extraction Methods for Selected Nutraceuticals

Phytochemicals	Plants	Solvent	Extraction Method	Extraction Time (min)	Yield (mg/kg matrix)	Reference
Tocols	*Amaranthus caudatus*	Methanol	CSE (R: 1/20, 25°C)	1440	76.32 (mg/kg)	Bruni et al. 2002
		Methanol	UAE (R: 1/20, 25°C)	60	63.7 (mg/kg)	
		CO_2	SFE (R: 1/30, 25°C, 400 atm)	15	129.27 (mg/kg)	
β-Sitosterol/ α-tocopherol/ γ-tocopherol	Okra seed	*n*-Hexane	Sox	—	2010/127/380 (mg/kg)	Andras et al. 2005
		Ethanol	Sox	—	2680/129/494 (mg/kg)	
		CO_2	SFE (R: 1/24–1/80, 50°C, 450 bar)	240–800	2390/148/407 (mg/kg)	
Saponins	Ginseng	80% Methanol	CSE (R: 1/10, 75°C)	180	5.24 (g/100 g)	Kwon et al. 2003
		80% Methanol	MAE(R: 1/10, 75°C)	0.5	5.31 (g/100g)	
Sulfur/oleoresin	Onion	Alcohol	Sox (R: 1/20)	240	3.78/350 (g/kg)	Sass-Kiss et al. 1998
		Steam	Distillation (R: 7/120)	300	0.167/0.4 (g/kg)	
		n-Hexane	CSE (R: 1/20, 25°C)	120	0.087/11 (g/kg)	
		Alcohol	CSE (R: 1/20, 25°C)	120	0.895/126 (g/kg)	
		CO_2	SFE (R: 1/14, 65°C, 300 bar)	180	0.208/9 (g/kg)	
Naringin	*Citrus* × *paradisi*	Ethanol/water (70:30)	Sox (R: 1/10)	480	15.2 (g/kg)	Giannuzzo et al. 2003
		Ethanol/water (70:30)	CSE (R:1/5, 22–25°C)	180	13.5 (g/kg)	
		CO_2/Ethanol (85:15)	SFE (R: —, 58.6°C, 95 bar)	45	14.4 (g/kg)	
Carvone/limone	Caraway seeds	*n*-Hexane	Sox (R: 1/20)	300	16.28/15.15 (mg/g)	Chemat et al. 2004
		n-Hexane	CSE (R: 1/20, 69°C)	60	13.38/12.63 (mg/g)	
		n-Hexane	UAE (R: 1/20, 69°C)	60	14.45/14.27 (mg/g)	
		n-Hexane	UAE (R: 1/20, 20–38°C)	60	17.16/16.16 (mg/g)	
Oil	Rose hip seeds	*n*-Hexane	Sox (R: 1/25)	180	48.5 (g/kg)	Szentmihalyi et al. 2002
		n-Hexane	UAE (R: 1/25, 69°C)	60	32.5 (g/kg)	
		n-Hexane	MAE (R: 1/3.5; 40°C)	30	52.6 (g/kg)	
		CO_2	SFE (35°C, 250 bar)	80	57.2 (g/kg)	
		CO_2/propane	SFE (28°C, 100 bar)	35	66.8 (g/kg)	

Source: Wang, L. J. and C. L. Weller, *Trends Food Sci. Technol.*, 17, 300, 2006. With permission.
CSE, conventional solvent extraction; Sox, Soxhlet extraction; UAE, ultrasound-assisted extraction; MAE, microwave-assisted extraction; SFE, supercritical fluid extraction; R, ratio of solid to solvent (g/mL).

These extraction enhancement techniques can increase the recovery, extraction kinetics, and product quality, and make it possible to decrease extraction temperature, solvent consumption, and use environmentally friendly solvents such as water and CO_2. To obtain the most effective and potential extract, it is necessary to take into account the characteristics of plant materials and nutraceutical products, the nature of solvents used for extraction, and the techniques used to enhance solvent extraction by enhancing the rupturing of cells and the heating of the materials. Sometimes, the high yield of extract will not ensure a high yield of bioactive components in the extract. Some bioactive components such as free fatty acids and tocopherols are sensitive to oxygen and heat. In this case, more care should be taken to prevent the oxidation and thermal degradation of those components. Therefore, the yield and quality of bioactive components should also be considered when an extraction method is selected. A comparison of different extraction methods for selected nutraceuticals is given in Table 2.1.

2.5 SEVERAL PRACTICAL ISSUES FOR APPLICATIONS OF EMERGING EXTRACTION METHODS

2.5.1 Technical Barriers of Novel Extraction Techniques

Efficient cell disruption, effective mass and heat transfer, and use of environmentally friendly solvents are three major factors leading to the development of the techniques for enhancing the existing solid–solvent extraction. However, each of these individual techniques has some technical and/or economical challenges.

The presence of a dispersed phase contributes to ultrasound wave attenuation, and the active part of ultrasound inside the extractor is restricted to a zone located in the vicinity of the emitter. Therefore, these two factors must be carefully considered in the design of ultrasound-assisted extractors. PEF treatment can be used for processing liquid and semiliquid materials with no air bubbles and low electrical conductivity. The activity of enzymes is affected by moisture content and operating conditions such as pH value, temperature, and agitation. Furthermore, enzyme treatment is usually a slow process. The high temperature and homogeneous temperature distribution over penetration depth reached by microwave heating reduces dramatically the extraction time and solvent consumption. The efficiency of microwaves can be poor when either the target compounds or the solvents are nonpolar, or when they are volatile. Like microwave heating, ohmic heating cannot work for dry materials. Many nutraceuticals are thermally unstable and may degrade during microwave or ohmic heating-assisted extraction. More research is needed to investigate the interaction between microwaves or electric field, and plant materials and solvents. SFE is one of the most successful and recent contributions to extraction techniques of nutraceuticals from plants. However, SFE, which exclusively uses CO_2 as the extracting solvent, is also restricted to nonpolar extracts. ASE under elevated temperature and high pressure can be supplementary to SFE for extraction of polar compounds.

High cost is another barrier to commercialize the emerging extraction technologies. More research is needed to reduce the capital and operating costs of the emerging extraction technologies such as PEF-assisted extraction, extrusion-assisted extraction, SFE, and ASE.

2.5.2 Improvement in Design of Novel Extraction Systems

High-pressure and microwave energy can be combined with conventional Soxhlet extraction, leading to a new design of high-pressure and microwave-assisted Soxhlet extractors. Supercritical CO_2 solvent also can be used in Soxhlet extraction. These combined techniques retain the advantages of conventional Soxhlet extraction, while overcoming the limitations of the novel extraction techniques such as ultrasound-assisted, microwave-assisted, supercritical fluid, and accelerated solvent extractions. Ultrasound can be used to enhance the mass transfer in SFE (Riera et al. 2004). SFEs can be coupled with silica gels to improve the overall extraction selectivity and on-line fractionations. Enzyme treatment and extrusion have also been combined to enhance the extraction of oil from oilseeds. Ultrasound, PEFs, and microwave and ohmic heating can be easily applied in continuous extraction processes.

2.5.3 Scaling Up of Novel Extraction Techniques

Emerging solid–solvent extraction technologies, such as ultrasound, PEFs, enzyme, extrusion, microwave heating, ohmic heating, supercritical fluids, and accelerated solvents, are promising for the extraction of nutraceuticals from plants. However, most of these novel extraction techniques are still done successfully at the laboratory or bench-scale testing, although several industrial applications of SFE and extrusion-assisted extraction can be found. More research is needed to exploit industrial applications of the novel extraction techniques.

Novel extraction processes are complex thermodynamic systems with higher capital costs. The engineering design of novel extraction systems requires knowledge of the thermodynamic constraints of solubility and selectivity, and kinetic constraints of mass transfer rate. Modeling of novel extraction processes can provide a better understanding of the extraction mechanisms and be used to quickly optimize extraction conditions and scale-up any design (Alonso et al. 2002; Mezzomo, Martinez, and Ferreira 2009).

2.6 CONCLUSIONS

The need to extract nutraceuticals from plant materials prompts continued searching for economically and ecologically feasible extraction technologies. Traditional solid–liquid extraction methods, including maceration, Soxhlet extraction, and hydrodistillation, require a large quantity of solvent and are time-consuming. The large amount of solvent used not only increases operating costs but also causes additional environmental problems. Several novel extraction techniques have been developed as an alternative to conventional extraction methods, offering advantages with respect to extraction time, solvent consumption, extraction yields, and reproducibility. However, novel extraction techniques have only been found in a limited field of applications. More research is needed to improve the understanding of extraction mechanism, remove technical barriers, improve the design, and decrease the costs and scale-up of the novel extraction systems for their industrial applications.

REFERENCES

Ade-Omowaye, B. I. O., A. Angersbach, N. M. Eshtiaghi, and D. Knorr. 2001. Impact of High Intensity Electric Field Pulses on Cell Permeabilisation and as Pre-Processing Step in Coconut Processing. *Innovative Food Science and Emerging Technologies* 1: 203–09.

Albu, S., E. Joyce, L. Paniwnyk, J. P. Lorimer, and T. J. Mason. 2004. Potential for the Use of Ultrasound in the Extraction of Antioxidants from *Rosmarinus officinalis* for the Food and Pharmaceutical Industry. *Ultrasonics Sonochemistry* 11: 261–65.

Al-Harahsheh, M., and S. W. Kingman. 2004. Microwave-Assisted Leaching—A Review. *Hydrometallurgy* 73: 189–203.

Alonso, E., F. J. Cantero, J. Garcia, and M. J. Cocero. 2002. Scale-Up for a Process of Supercritical Extraction with Adsorption of Solute onto Active Carbon. Application to Soil Remediation. *Journal of Supercritical Fluids* 24: 123–35.

Andras, C. D., B. Simandi, F. Orsi, C. Lambrou, D. Missopolinou-Tatala, C. Panayiotou, J. Domokos, and F. Doleschall. 2005. Supercritical Carbon Dioxide Extraction of Okra (*Hibiscus esculentus L*) Seeds. *Journal of the Science of Food and Agriculture* 85: 1415–19.

Barriada-Pereira, M., E. Concha-Grana, M. J. Gonzalez-Castro, S. Muniategui-Lorenzo, P. Lopez-Mahia, D. Prada-Rodriguez, and E. Fernandez-Fernandez. 2003. Microwave-Assisted Extraction versus Soxhlet Extraction in the Analysis of 21 Organochlorine Pesticides in Plants. *Journal of Chromatography A* 1008: 115–22.

Bazhal, M., and E. Vorobiev. 2000. Electrical Treatment of Apple Cossettes for Intensifying Juice Pressing. *Journal of the Science of Food and Agriculture* 80: 1668–74.

Berna, A., A. Tarrega, M. Blasco, and S. Subirats. 2000. Supercritical CO_2 Extraction of Essential Oil from Orange Peel: Effect of the Height of the Bed. *Journal of Supercritical Fluids* 18: 227–37.

Bernardo-Gil, M. G., J. Grenha, J. Santos, and P. Cardoso. 2002. Supercritical Fluid Extraction and Characterization of Oil from Hazelnut. *European Journal of Lipid Science and Technology* 104: 402–09.

Brachet, A., P. Christen, and J. L. Veuthey. 2002. Focused Microwave-Assisted Extraction of Cocaine and Benzoylecgonine from *Coca* Leaves. *Phytochemical Analysis* 13: 162–69.

Brachet, A., S. Rudaz, L. Mateus, P. Christen, and J. Veuthey. 2001. Optimisation of Accelerated Solvent Extraction of Cocaine and Benzoylecgonine from *Coca* Leaves. *Journal of Separation Science* 24: 865–73.

Bruni, R., A. Guerrini, S. Scalia, C. Romagnoli, and G. Sacchetti. 2002. Rapid Techniques for the Extraction of Vitamin E Isomers from *Amaranthus caudatus* Seeds: Ultrasonic and Supercritical Fluid Extraction. *Phytochemical Analysis* 13: 257–61.

Bryant, G., and J. Wolfe. 1987. Electromechanical Stress Produced in the Plasma Membranes of Suspended Cells by Applied Electrical Fields. *Journal of Membrane Biology* 96: 129–39.

Chemat, S., A. Lagha, H. AitAmar, P. V. Bartels, and F. Chemat. 2004. Comparison of Conventional and Ultrasound-Assisted Extraction of Carvone and Limonene from *Caraway* Seeds. *Flavour and Fragrance Journal* 19: 188–95.

Cherchi, G., D. Deidda, B. de Gioannis, B. Marongiu, R. Pompei, and S. Porcedda. 2001. Extraction of *Santolina insularis* Essential Oil by Supercritical Carbon Dioxide: Influence of Some Process Parameters and Biological Activity. *Flavour and Fragrance Journal* 16: 35–43.

Choi, Y. H., J. Kim, M. J. Noh, E. S. Choi, and K. P. Yoo. 1997. Comparison of Supercritical Carbon Dioxide Extraction with Solvent of Nonacosan-10-ol, α-Amyrin Acetate, Squalene, and Stigmasterol from Medicinal Plants. *Phytochemical Analysis* 8: 233–37.

Choudhari, S. M., and L. Ananthanarayan. 2007. Enzyme Aided Extraction of Lycopene from Tomato Tissues. *Food Chemistry* 102: 77–81.

Coelho, J. A. P., A. P. Pereira, R. L. Mendes, and A. M. F. Palavra. 2003. Supercritical Carbon Dioxide Extraction of *Foeniculum vulgare* Volatile Oil. *Flavour and Fragrance Journal* 18: 316–19.

Cravotto, G., A. Binello, G. Merizzi, and M. Avogadro. 2004. Improving Solvent-Free Extraction of Policosanol from Rice Bran by High-Intensity Ultrasound Treatment. *European Journal of Lipid Science and Technology* 106: 147–51.

Csiktusnadi Kiss, G. A., E. F. Forgacs, T. Cserhati, T. Mota, H. Morais, and A. Ramos. 2000. Optimisation of the Microwave-Assisted Extraction of Pigments from *Paprika (Capsicum annuum L.)* Powders. *Journal of Chromatography A* 889: 41–49.

Dean, J. R., and B. Liu. 2000. Supercritical Fluid Extraction of Chinese Herbal Medicines: Investigation of Extraction Kinetics. *Phytochemical Analysis* 11: 1–6.

Dean, J. R., B. Liu, and R. Price. 1998. Extraction of Magnolol from *Magnolia officinalis* Using Supercritical Fluid Extraction and Phytosol Solvent Extraction. *Phytochemical Analysis* 9: 248–52.

del Valle, J. M., O. Rivera, O. Teuber, and M. Teresa Palma. 2003. Supercritical CO_2 Extraction of Chilean Hop (*Humulus lupulus*) Ecotypes. *Journal of the Science of Food and Agriculture* 83: 1349–56.

Denery, J. R., K. Dragull, C. S. Tang, and Q. X. Li. 2004. Pressurized Fluid Extraction of Carotenoids from *Haematococcus pluvialis* and *Dunaliella salina* and Kavalactones from *Piper methysticum*. *Analytica Chimica Acta* 501: 175–81.

Dominguez, H., M. J. Ntiiiez, and J. M. Lema. 1995. Enzyme-Assisted Hexane Extraction of Soybean Oil. *Food Chemistry* 54: 223–31.

El-Ghorab, A., A. F. Mansour, and K. F. El-massry. 2004. Effect of Extraction Methods on the Chemical Composition and Antioxidant Activity of Egyptian Marjoram (*Majorana hortensis Moench*). *Flavour and Fragrance Journal* 19: 54–61.

Eller, F. J., and J. W. King. 2000. Supercritical Carbon Dioxide Extraction of Cedarwood Oil: A Study of Extraction Parameters and Oil Characteristics. *Phytochemical Analysis* 11: 226–31.

Ellington, E., J. Bastida, F. Viladomat, and C. Codina. 2003. Supercritical Carbon Dioxide Extraction of Colchicines and Related Alkaloids from Seeds of *Colchicum autumnale L.* *Phytochemical Analysis* 14: 164–69.

Ericsson, M., and A. Colmsjo. 2000. Dynamic Microwave-Assisted Extraction. *Journal of Chromatography A* 877: 141–51.

Eshtiaghi, M. N., and D. Knorr. 2002. High Electric Field Pulse Pretreatment: Potential for Sugar Beet Processing. *Journal of Food Engineering* 52: 265–72.

Eskilsson, C. S., K. Hartonen, L. Mathiasson, and M. L. Riekkola. 2004. Pressurized Hot Water Extraction of Insecticides from Process Dust—Comparison with Supercritical Fluid Extraction. *Journal of Separation Science* 27: 59–64.

Fincan, M., F. deVito, and P. Dejmek. 2004. Pulsed Electric Field Treatment for Solid–Liquid Extraction of Red Beetroot Pigment. *Journal of Food Engineering* 64: 381–88.

Font, N., F. Hernandez, E. A. Hogendoorn, R. A. Baumann, and P. van Zoonen. 1998. Microwave-Assisted Solvent Extraction and Reversed-Phase Liquid Chromatography-UV Detection for Screening Soils for Sulfonylurea Herbicides. *Journal of Chromatography A* 798: 179–86.

Ganzler, K., A. Salgo, and K. Valko. 1986. Microwave Extraction. A Novel Sample Preparation Method for Chromatography. *Journal of Chromatography A* 371: 299–306.

Gaur, R., A. Sharma, S. K. Khare, and M. N. Gupta. 2007. A Novel Process for Extraction of Edible Oils: Enzyme Assisted Three Phase Partitioning (EATPP). *Bioresource Technology* 98: 696–99.

Giannuzzo, A. N., H. J. Boggetti, M. A. Nazareno, and H. T. Mishima. 2003. Supercritical Fluid Extraction of *Naringin* from the Peel of Citrus Paradise. *Phytochemical Analysis* 14: 221–23.

Gongora-Nieto, M. M., P. D. Pedrow, B. G. Swanson, and G. V. Barbosa-Ganovas. 2003. Energy Analysis of Liquid Whole Egg Pasteurized by Pulsed Electric Fields. *Journal of Food Engineering* 57: 209–16.

Guo, Z., Q. Jin, G. Fan, Y. Duan, C. Qin, and M. Wen. 2001. Microwave-Assisted Extraction of Effective Constituents from a Chinese Herbal Medicine *Radix puerariae*. *Analytica Chimica Acta* 436: 41–47.

Hamburger, M., D. Baumann, and S. Adler. 2004. Supercritical Carbon Dioxide Extraction of Selected Medicinal Plants—Effects of High Pressure and Added Ethanol on Yield of Extracted Substances. *Phytochemical Analysis* 15: 46–54.

Hanmoungjai, P., L. Pyle, and K. Niranjan. 2000. Extraction of Rice Bran Oil Using Aqueous Media. *Journal of Chemical Technology and Biotechnology* 75: 348–52.

Hao, J. Y., W. Han, S. D. Huang, B. Y. Xue, and X. Deng. 2002. Microwave-Assisted Extraction of Artemisinin from *Artemisia annua L. Separation and Purification Technology* 28: 191–96.

Hauthal, W. H. 2001. Advances with Supercritical Fluids [Review]. *Chemosphere* 43: 123–35.

Hugenholtz, J., and E. J. Smid. 2002. Nutraceutical Production with Food-Grade Microorganisms. *Food Biotechnology* 13: 497–507.

Hui, L., O. Etsuzo, and I. Masao. 1994. Effects of Ultrasound on the Extraction of Saponin from Ginseng. *Japanese Journal of Applied Physics* 33(5B): 3085–87.

Hurren, D. 1999. Supercritical Fluid Extraction with CO_2. *Filtration and Separation* 36: 25–27.

Jung, S., D. Maurer, and L. A. Johnson. 2009. Factors Affecting Emulsion Stability and Quality of Oil Recovered from Enzyme-Assisted Aqueous Extraction of Soybeans. *Bioresource Technology* 100: 5340–47.

Katrokha, I., A. Matvienko, L. Vorona, M. Kupchik, and V. Zaets. 1984. Intensification of Sugar Extraction from Sweet Sugar Beet Cossettes in an Electric Field. *Sakharnaya Promyshlennost* 7: 28–31.

Kaufmann, B., and P. Christen. 2002. Recent Extraction Techniques for Natural Products: Microwave-Assisted Extraction and Pressurized Solvent Extraction. *Phytochemical Analysis* 13: 105–13.

Kaufmann, B., P. Christen, and J. L. Veuthey. 2001a. Parameters Affecting Microwave-Assisted Extraction of *Withanolides*. *Phytochemical Analysis* 12: 327–31.

Kaufmann, B., P. Christen, and J. L. Veuthey. 2001b. Study of Factors Influencing Pressurized Solvent Extraction of Polar Steroids from Plant Material. *Chromatographia* 54: 394–98.

Kim, J., and Y. Pyun. 1995. Extraction of Soy Milk Using Ohmic Heating. Abstract of the Ninth Congress of Food Science and Technology, Budapest, Hungary, July–August 1995.

Knorr, D., and A. Angersbach. 1998. Impact of High-Intensity Electric Field Pulses on Plant Membrane Permeabilization. *Trends in Food Science and Technology* 9: 185–91.

Kratchanova, M., E. Pavlova, and I. Panchev. 2004. The Effect of Microwave Heating of Fresh Orange Peels on the Fruit Tissue and Quality of Extracted Pectin. *Carbohydrate Polymers* 56: 181–86.

Kwon, J. H., J. M. R. Belanger, J. R. Jocelyn Pare, and V. A. Yaylayan. 2003. Application of Microwave-Assisted Process (MAP TM) to the Fast Extraction of *Ginseng saponins*. *Food Research International* 36: 491–98.

Lakkakula, N. R., M. Lima, and T. Walker. 2004. Rice Bran Stabilization and Rice Bran Oil Extraction Using Ohmic Heating. *Bioresource Technology* 92: 157–61.

Lamsal, B. P., P. A. Murphy, and L. A. Johnson. 2006. Flaking and Extrusion as Mechanical Treatments for Enzyme-Assisted Aqueous Extraction of Oil from Soybeans. *Journal of the American Oil Chemists' Society* 83: 973–79.

Lang, Q., and C. M. Wai. 2001. Supercritical Fluid Extraction in Herbal and Natural Product Studies—A Practical Review. *Talanta* 53: 771–82.

Li, B. B., B. Smith, and M. M. Hossain. 2006. Extraction of Phenolics from Citrus Peels. II. Enzyme-Assisted Extraction Method. *Separation and Purification Technology* 48: 189–96.

Li, H., L. Pordesimo, and J. Weiss. 2004. High Intensity Ultrasound-Assisted Extraction of Oil from Soybeans. *Food Research International* 37: 731–38.

Lima, M., and S. K. Sastry. 1999. The Effect of Ohmic Heating on Hot-Air Drying Rate and Juice Yield. *Journal of Food Engineering* 41: 115–19.

López, N., E. Puértolas, S. Condón, J. Raso, and I. Álvarez. 2009. Enhancement of the Extraction of Betanine from Red Beetroot by Pulsed Electric Fields. *Journal of Food Engineering* 90: 60–67.

Lorenzo, R. A., M. J. Vazquez, A. M. Carro, and R. Cela. 1999. Methylmercury Extraction from Aquatic Sediments. *Trends in Analytical Chemistry* 18: 410–16.

Luengthanaphol, S., D. Mongkholkhajornsilp, S. Douglas, P. L. Douglas, L. Pengsopa, and S. Pongamphai. 2004. Extraction of Antioxidants from Sweet *Thai tamarind* Seed Coat—Preliminary Experiments. *Journal of Food Engineering* 63: 247–52.

Luque de Castro, M. D., and L. E. Garcia-Ayuso. 1998. Soxhlet Extraction of Solid Materials: An Outdated Technique with a Promising Innovative Future. *Analytica Chimica Acta* 369: 1–10.

Luque-Garcia, J. L., and M. D. Luque de Castro. 2003. Ultrasound: A Powerful Tool for Leaching. *Trends in Analytical Chemistry* 22: 41–47.

Luque-Garcia, J. L., and M. D. Luque de Castro. 2004. Ultrasound-Assisted Soxhlet Extraction: An Expeditive Approach for Solid Sample Treatment—Application to the Extraction of Total Fat from *Oleaginous* Seeds. *Journal of Chromatography A* 1034: 237–42.

Lusas, E. W., and L. R. Watkins. 1988. Oilseeds: Extrusion for Solvent Extraction. *Journal of the American Oil Chemists' Society* 65: 1109–14.

Mamidipally, P. K., and S. X. Liu. 2004. First Approach on Rice Bran Oil Extraction Using Limonene. *European Journal of Lipid Science and Technology* 106: 122–25.

Marongiu, B., S. Porcedda, A. Caredda, B. De Gioannis, L. Vargiu, and P. La Colla. 2003. Extraction of *Juniperus oxycedrus* spp. *Oxycedrus* Essential Oil by Supercritical Carbon Dioxide: Influence of Some Process Parameters and Biological Activity. *Flavour and Fragrance Journal* 18: 390–97.

Marr, R., and T. Gamse. 2000. Use of Supercritical Fluids for Different Processes Including New Developments—A Review. *Chemical Engineering and Processing* 39: 19–28.

Mason, T. J., L. Paniwnyk, and J. P. Lorimer. 1996. The Uses of Ultrasound in Food Technology. *Ultrasonics Sonochemistry* 3: 253–60.

Meireles, A., and M. Angela. 2003. Supercritical Extraction from Solid: Process Design Data (2001–2003). *Current Opinion in Solid State and Materials Science* 7: 321–30.

Melecchi, M. I. S., M. M. Martinez, F. C. Abad, P. P. Zini, I. N. Filho, and E. B. Caramao. 2002. Chemical Composition of *Hibiscus tiliaceus L.* Flowers: A Study of Extraction Methods. *Journal of Separation Science* 25: 86–90.

Mezzomo, N., J. Martinez, and S. R. S. Ferreira. 2009. Supercritical Fluid Extraction of Peach (*Prunus persica*) Almond Oil: Kinetics, Mathematical Modeling and Scale-Up. *Journal of Supercritical Fluids* 51: 10–16.

Modey, W. K., D. A. Mulholland, and M. W. Raynor. 1996. Analytical Supercritical Fluid Extraction of Natural Products. *Phytochemical Analysis* 7: 1–15.

Molins, C., E. A. Hogendoorn, H. A. G. Heusinkveld, P. van Zoonen, and R. A. Baumann. 1997. Microwave Assisted Solvent Extraction (MASE) of Organochlorine Pesticides from Soil Samples. *International Journal of Environmental Analytical Chemistry* 68: 155–69.

Najafian, L., A. Ghodsvali, M. H. Haddad Khodaparast, and L. L. Diosady. 2009. Aqueous Extraction of Virgin Olive Oil Using Industrial Enzymes. *Food Research International* 42: 171–75.

Pan, X., H. Liu, G. Jia, and Y. Y. Shu. 2000. Microwave-Assisted Extraction of Glycyrrhizic Acid from *Licorice* Root. *Biochemical Engineering Journal* 5: 173–77.

Pan, X., G. Niu, and H. Liu. 2002. Comparison of Microwave-Assisted Extraction and Conventional Extraction Techniques for the Extraction of Tanshinones from *Salvia miltiorrhiza Bunge*. *Biochemical Engineering Journal* 12: 71–77.

Pan, X., G. Niu, and H. Liu. 2003. Microwave-Assisted Extraction of Tea Polyphenols and Tea Caffeine from Green Tea Leaves. *Chemical Engineering and Processing* 42: 129–33.

Pinelo, M., B. Zornoza, A. S. Meyer. 2008. Selective Release of Phenols from Apple Skin: Mass Transfer Kinetics During Solvent and Enzyme-Assisted Extraction. *Separation and Purification Technology* 63: 620–27.

Poiana, M., R. Fresa, and B. Mincione. 1999. Supercritical Carbon Dioxide Extraction of *Bergamot* Peels. Extraction Kinetics of Oil and Its Components. *Flavour and Fragrance Journal* 14: 358–66.

Poiana, M., V. Sicari, and B. Mincione. 1998. Supercritical Carbon Dioxide (SC-CO_2) Extraction of Grape Fruit *Flavedo*. *Flavour and Fragrance Journal* 13: 125–30.

Puértolas, E., N. López, G. Saldaña, I. Álvarez, and J. Raso. 2010. Evaluation of Phenolic Extraction During Fermentation of Red Grapes Treated by a Continuous Pulsed Electric Fields Process at Pilot-Plant Scale. *Journal of Food Engineering* 119(3): 1063–70.

Richter, B. E., B. A. Jones, J. L. Ezzell, N. L. Porter, N. Avdalovic, and C. Pohl. 1996. Accelerated Solvent Extraction: A Technology for Sample Preparation. *Analytical Chemistry* 68: 1033–39.

Riera, E., Y. Golas, A. Blanco, J. A. Gallego, M. Blasco, and A. Mulet. 2004. Mass Transfer Enhancement in Supercritical Fluids Extraction by Means of Power Ultrasound. *Ultrasonics Sonochemistry* 11: 241–44.

Romdhane, M., and C. Gourdon. 2002. Investigation in Solid-Liquid Extraction: Influence of Ultrasound. *Chemical Engineering Journal* 87: 11–19.

Romdhane, M., C. Gourdon, and G. Casamatta. 1995. Local Investigation of Some Ultrasonic Devices by Means of a Thermal Sensor. *Ultrasononics* 33: 221–27.

Roy, B. C., M. Goto, A. Kodama, and T. Hirose. 1996. Supercritical CO_2 Extraction of Essential Oils and Cuticular Waxes from *Peppermint* Leaves. *Journal of Chemical Technology and Biotechnology* 67: 21–26.

Salisova, M., S. Toma, and T. J. Mason. 1997. Comparison of Conventional and Ultrasonically Assisted Extractions of Pharmaceutically Active Compounds from *Salvia officinalis*. *Ultrasonics Sonochemistry* 4: 131–34.

Sass-Kiss, A., B. Simandi, Y. Gao, F. Boross, and Z. Vamos-Falusi. 1998. Study on the Pilot-Scale Extraction of Onion Oleoresin Using Supercritical CO_2. *Journal of the Science of Food and Agriculture* 76: 320–26.

Schilling, S., T. Alber, S. Toepfl, S. Neidhart, D. Knorr, A. Schieber, and R. Carle. 2007. Effects of Pulsed Electric Field Treatment of Apple Mash on Juice Yield and Quality Attributes of Apple Juices. *Innovative Food Science and Emerging Technologies* 8: 127–34.

Schinor, E. C., M. J. Salvador, I. C. C. Turatti, O. L. A. D. Zucchi, and D. A. Dias. 2004. Comparison of Classical and Ultrasound-Assisted Extractions of Steroids and Triterpenoids from Three *Chresta* spp. *Ultrasonics Sonochemistry* 11: 415–21.

Sharma, A., and M. N. Gupta. 2004. Oil Extraction from Almond, Apricot and Rice Bran by Three-Phase Partitioning after Ultrasonication. *European Journal of Lipid Science and Technology* 106: 183–86.

Shen, L., X. Wang, Z. Wang, Y. Wu, and J. Chen. 2008. Studies on Tea Protein Extraction Using Alkaline and Enzyme Methods. *Food Chemistry* 107: 929–38.

Shu, Y. Y., M. Y. Ko, and Y. S. Chang. 2003. Microwave-Assisted Extraction of Ginsenosides from *Ginseng* Root. *Microchemical Journal* 74: 131–39.

Sihvonen, M., E. Jarvenpaa, V. Hietaniemi, and R. Huopalahti. 1999. Advances in Supercritical Carbon Dioxide Technologies. *Trends in Food Science and Technology* 10: 217–22.

Silva L. V., D. L. Nelson, M. F. B. Drummond, L. Dufossé, and M. B. A. Glória. 2005. Comparison of Hydrodistillation Methods for the Deodorization of Turmeric. *Food Research International* 38: 1087–96.

Smith, R. M. 2002. Extractions with Superheated Water. *Journal of Chromatography A* 975: 31–46.

Soto, C., R. Chamy, and M. E. Zuniga. 2007. Enzymatic Hydrolysis and Pressing Conditions Effect on Borage Oil Extraction by Cold Pressing. *Food Chemistry* 102: 834–40.

Sowbhagya, H. B., K. T. Purnima, S. P. Florence, A. G. A. Rao, and P. Srinivas. 2009. Evaluation of Enzyme-Assisted Extraction on Quality of Garlic Volatile Oil. *Food Chemistry* 113: 1234–38.

Sowbhagya, H. B., P. Srinivas, and N. Krishnamurthy. 2010. Effect of Enzymes on Extraction of Volatiles from Celery Seeds. *Food Chemistry* 120: 230–34.

Spar Eskilsson, S., and E. Bjorklund. 2000. Analytical-Scale Microwave-Assisted Extraction. *Journal of Chromatography A* 902: 227–50.

Spar Eskilsson, S., E. Bjorklund, L. Mathiasson, L. Karlsson, and A. Torstensson. 1999. Microwave-Assisted Extraction of *Felodipine* Tablets. *Journal of Chromatography A* 840: 59–70.

Szentmihalyi, K., P. Vinkler, B. Lakatos, V. Illes, and M. Then. 2002. Rose Hip (*Rosa canina L.*) Oil Obtained from Waste Hip Seeds by Different Extraction Methods. *Bioresource Technology* 82: 195–201.

Toepfl, S., A. Mathys, V. Heinz, and D. Knorr. 2006. Review: Potential of High Hydrostatic Pressure and Pulsed Electric Fields for Energy Efficiency and Environmentally Friendly Food Processing. *Food Research International* 22: 405–23.

Toma, M., M. Vinatoru, L. Paniwnyk, and T. J. Mason. 2001. Investigation of the Effects of Ultrasound on Vegetal Tissues during Solvent Extraction. *Ultrasonics Sonochemistry* 8: 137–42.

Tomaniova, M., J. Hajslova, J. Pavelka Jr., V. Kocourek, K. Holadova, and I. Klimova. 1998. Microwave-Assisted Solvent Extraction—A New Method for Isolation of Polynuclear Aromatic Hydrocarbons from Plants. *Journal of Chromatography A* 827: 21–29.

Turner, C., J. W. King, and L. Mathiasson. 2001. Supercritical Fluid Extraction and Chromatography for Fat-Soluble Vitamin Analysis. *Journal of Chromatography A* 936: 215–37.

Vauchel, P., R. Kaas, and A. Arhaliass. 2008. A New Process for Extracting Alginates from *Laminaria digitata*: Reactive Extrusion. *Food and Bioprocess Technology* 1: 297–300.

Vilegas, J. H. Y., E. de Marchi, and F. M. Lancas. 1997. Extraction of Low-Polarity Compounds (with Emphasis on Coumarin and Kaurenoic Acid) from *Milania glomerata ("Guaco")* Leaves. *Phytochemical Analysis* 8: 266–70.

Vinatoru, M. 2001. An Overview of the Ultrasonically Assisted Extraction of Bioactive Principles from Herbs. *Ultrasonics Sonochemistry* 8: 303–13.

Vinatoru, M., M. Toma, and T. J. Mason. 1999. Ultrasound-Assisted Extraction of Bioactive Principles from Plants and Their Constituents. *Advances in Sonochemistry* 5: 209–47.

Vinatoru, M., M. Toma, O. Radu, P. I. Filip, D. Lazurca, and T. J. Mason. 1997. The Use of Ultrasound for the Extraction of Bioactive Principles from Plant Materials. *Ultrasonics Sonochemistry* 4: 135–39.

Wang, L. J. 2008. *Energy Efficiency and Management in Food Processing Facilities*. Boca Raton, FL: CRC Press.

Wang, L. J., G. M. Ganjyal, D. D. Jones, C. L. Weller, and M. A. Hanna. 2004. Simulation of Fluid Flow, Heat Transfer and Melting of Biomaterials in a Single–Screw Extruder. *Journal of Food Science* 69: 212–23.

Wang, L. J., D. D. Jones, C. L. Weller, and M. A. Hanna. 2006. Modeling of Transport Phenomena and Melting Kinetics of Biomaterials in a Twin-Screw Extruder. *Advances in Polymer Technology* 25: 22–40.

Wang, L. J., and C. L. Weller. 2006. Recent Advances in Extraction of Natural Products from Plants. *Trends in Food Science and Technology* 17: 300–12.

Wang, L. J., C. L. Weller, V. L. Schlegel, T. P. Carr, and S. L. Cuppett. 2008. Supercritical Carbon Dioxide Extraction of Grain Sorghum DDGS Lipids. *Bioresource Technology* 99: 1373–82.

Wang, L. J., C. L. Weller, V. L. Schlegel, T. P. Carr, S. L. Cuppett, and K. T. Hwang. 2007. Comparison of Supercritical CO_2 and Hexane Extraction of Lipids from Sorghum Distillers Grains. *European Journal of Lipid Science and Technology* 109: 567–74.

Wildman, R. E. C. 2001. *Handbook of Nutraceuticals and Functional Foods*. Boca Raton, FL: CRC Press.

Williams, O. J., G. S. V. Raghavan, V. Orsat, and J. Dai. 2004. Microwave-Assisted Extraction of Capsaicinoids from Capsicum Fruit. *Journal of Food Biochemistry* 28: 113–22.

Wu, J., L. Lin, and F. Chau. 2001. Ultrasound-Assisted Extraction of Ginseng Saponins from *Ginseng* Roots and Cultured Ginseng Cells. *Ultrasonics Sonochemistry* 8: 347–52.

Zarnowski, R., and Y. Suzuki. 2004. Expedient Soxhlet Extraction of Resorcinolic Lipids from Wheat Grains. *Journal of Food Composition and Analysis* 17: 649–64.

Zhang, F., B. Chen, S. Xiao, and S. Yao. 2005. Optimization and Comparison of Different Extraction Techniques for Sanguinarine and Chelerythrine in Fruits of *Macleaya cordata* (Willd) R. Br. *Separation and Purification Technology* 42: 283–90.

Zhang, Q. H., G. V. Barbosa-Canovas, and B. G. Swanson. 1995. Engineering Aspects of Pulsed Electric Field Pasteurization. *Journal of Food Engineering* 25: 261–81.

Zu, Y., Y. Wang, Y. Fu, S. Li, R. Sun, W. Liu, and H. Luo. 2009. Enzyme-Assisted Extraction of Paclitaxel and Related Taxanes from Needles of *Taxus chinensis*. *Separation and Purification Technology* 68: 238–43.

CHAPTER 3

Liquid–Liquid Extraction and Adsorption Applied to the Processing of Nutraceuticals and Functional Foods

Antonio J. A. Meirelles, Eduardo A. C. Batista, Mariana C. Costa, and Marcelo Lanza

CONTENTS

3.1 INTRODUCTION

Food processing involves a series of mass and heat transfer steps performed with the aims of obtaining the crude product, eliminating undesirable components present in this product, and pasteurizing or sterilizing it, so that the final product becomes appropriate for storage and human consumption. Several of these steps are carried out under conditions of high temperature that can diminish the nutritional value of the end product, either by thermal degradation or by evaporation of nutraceutical compounds.

A typical example is the refining of palm oil, an edible oil that, in its crude form, is rich in carotenes, tocopherols, and tocotrienols. When submitted to the extreme temperature conditions of the so-called physical refining process, carotenes are thermally degraded, and significant amounts of tocopherols and tocotrienols are evaporated. Recent estimations reported in the literature indicated that the amount of carotenes degraded worldwide every day during palm oil refining corresponds to the suggested daily intake of vitamin A for the whole world population (Mayamol et al. 2007).

In fact, losses of nutraceutical components occur in the refining of most edible oils, even in the case of chemical treatment, a purification process usually carried out under milder temperature conditions compared with physical refining. For example, crude rice bran oil contains large amounts of γ-oryzanol (0.9%–2.9%), a valuable antioxidant, but the corresponding final product, refined by the chemical method, contains less than 0.2% (Orthoefer 1996).

The loss of nutraceuticals represents a significant drawback of the traditional purification procedures for oil refining, especially from a nutritional point of view. These minor components certainly enhance the nutritional value of the final product, but their presence in the refined product is sometimes not desired, especially in the case of widespread industrial use in food formulations. This is exactly the case for palm oil, given that its natural color, caused by the carotene content, should be avoided in those fractions used as food constituents. Although the increasing consumer interest in healthy foodstuffs may diminish the pressure for refined oils and fatty products with a light yellow color, almost colorless fatty fractions will still be favored by the industry in the formulation of food products, as is the case nowadays for the use of palm stearin as a natural trans-fatty acid–free substitute for hydrogenated oils. Thus, the best technological option would probably be to extract and/or recover the natural antioxidants from the crude product before refining it.

This chapter is focused on the use of two different purification techniques for either recovering nutraceuticals from natural products or producing nutraceutical-rich refined foodstuffs, that is, functional foods. Liquid–liquid extraction is a technique based on the use of a selective solvent for extracting specific components from a liquid feed stream. The feasibility of its use for deacidifying crude oils without causing high losses of antioxidants has already been shown in the literature and will be discussed in this chapter. Adsorption involves the selective capture of components from a fluid stream by a solid phase with a large contact surface. This technique is frequently suggested for recovering nutraceutical components from natural extracts.

In the next section we will review the recent literature on the use of these techniques for processing functional foods and recovering nutraceuticals. The third and fourth sections show the fundamentals of these purification techniques, including information on phase equilibrium, equipment and its operation, the main aspects of their design, and an evaluation of their performance.

3.2 PROCESSING OF FUNCTIONAL FOODS AND RECOVERY OF NUTRACEUTICAL COMPOUNDS BY LIQUID–LIQUID EXTRACTION OR ADSORPTION

3.2.1 Liquid–Liquid Extraction

Liquid–liquid extraction can be used for purifying liquid foods without a loss of nutraceutical compounds, as well as for extracting those components from foods in the fluid state. Such possible applications of liquid–liquid extraction are discussed in this chapter in the context of edible oil deacidification and the recovery of nutraceuticals from fatty systems.

The crude oil extracted from oilseeds is a mixture of triacylglycerols, partial acylglycerols, free fatty acids, phosphatides, pigments, sterols, and tocopherols (compounds that present vitamin E activity) (Cheryan 1998). The usual oil extraction and refining processes involve either solid–liquid extraction or pressing, solvent stripping and recovery, degumming, dewaxing, bleaching, deacidification, and deodorization. Deacidification can be done by chemical refining, associated with caustic treatment, or by physical refining, associated with steam stripping. Of the whole refining sequence, the removal of free fatty acids (deacidification) is the most difficult step of the oil purification process because it normally results in losses of neutral oil and, for this reason, has the largest impact on the economic performance of the process. In physical refining the deacidification step is performed by stripping with direct steam injection and is carried out under extreme conditions of temperature (463.15–543.15 K) and low pressures (400–1333.2 Pa) (Ceriani and Meirelles 2006). Under such conditions, the free fatty acids and flavor materials are volatile components and can be stripped out of the edible oil. Nevertheless, under these extreme conditions some other components may also be removed from the oil phase, either by evaporation, as in the case of short-chain fatty acid triacylglycerols, partial acylglycerols, and nutraceutical compounds such as tocols (tocopherols + tocotrienols), or by thermal degradation, which occurs with carotenes, for example. Thus, the loss of natural antioxidants (e.g., carotenes, γ-oryzanol, and tocols) is a significant drawback of this purification procedure, especially from a nutritional point of view. Although these minor components surely enhance the nutritional value of the final product, their presence in the refined product is sometimes not desired, as in the case of palm oil, given that its natural color (red), caused by the carotene content, should be avoided in the fractions used as food constituents.

A loss of nutraceutical components occurs in the refining of most edible oils, even in the case of chemical treatments (Leibovitz and Ruckenstein 1983; Orthoefer 1996; Antoniassi, Esteves, and Meirelles 1998). Previous investigations on oil refining based on solvent extraction have usually indicated this mass transfer operation as a predeacidification step that can reduce the acidity and make the intermediary product more appropriate for a final deacidification by conventional methods, with a reduced loss of neutral oil. They have also suggested it is a good way to recover and/or maintain nutraceutical compounds. Although oil deacidification by solvent extraction is not a new concept, the effective development of an industrial process based on this procedure requires that a series of prior studies be developed, mainly involving the recovery of nutraceuticals. Some of these recently applied studies are summarized in this review, such as the liquid–liquid equilibrium data for vegetable oil deacidification and nutraceutical recovery, continuous deacidification in laboratory equipment, and process simulation for continuous deacidification.

The development of liquid–liquid extraction for the deacidification process and recovery of nutraceuticals requires a systematic study of the corresponding phase equilibria, involving several oils of commercial and nutritional interest, as well as studies with the continuous process on a pilot scale. Liquid–liquid equilibrium data can be found in the literature (Batista et al. 1999a, 1999b; Gonçalves et al. 2002, 2004, 2007; Rodrigues et al. 2003, 2004, 2005a, 2005b, 2006a, 2006b, 2007, 2008; Cuevas, Rodrigues, and Meirelles 2009) for a series of vegetable oils, including palm, rice bran, corn, soybean, canola, and cottonseed; Brazil and Macadamia nuts; grape, avocado, and sesame seeds; garlic, peanut, sunflower, and babassu oils; and for some pure triacylglycerols, such as triolein and tricaprylin. The complete set of equilibrium data found in the literature includes nutraceutical compounds such as tocols (tocopherols + tocotrienols), carotenes, and γ-oryzanol, as well as fatty acids such as lauric, palmitic, oleic, and linoleic acids. These data were determined in a temperature range of 283.15–328.15 K, and short-chain alcohols were used as the solvents (methanol, ethanol, propanol, and isopropanol). For the majority of the experimental data, ethanol, with varying degrees of hydration, was used as the extraction solvent. Part of the experimental data was determined with model systems obtained using refined oils and commercial fatty acids, but equilibrium data for different crude or semiprocessed (bleached or degummed) oils were also determined and published in the scientific literature (Rodrigues et al. 2003, 2004, 2005b; Gonçalves, Batista, and Meirelles 2004). Detailed information on the experimental and analytical methodologies commonly used in determining the liquid–liquid extraction and equilibrium data for oil deacidification and the recovery of nutraceuticals can be found in several selected reports and obtained from the literature used for this review (Batista et al. 1999a, 1999b; Gonçalves et al. 2002, 2004, 2007; Rodrigues et al. 2003, 2004, 2005a, 2005b, 2006a, 2006b, 2007, 2008; Cuevas, Rodrigues, and Meirelles 2009).

3.2.1.1 *Liquid–Liquid Equilibrium Diagrams for Oil Deacidification and the Recovery of Nutraceuticals*

As the first step in the design of a liquid–liquid extraction process, the solvent selection is fundamental to the development of an efficient process. Solvent selection depends on a series of specific features, such as the appropriate distribution coefficients for the components that should be extracted, high selectivity, low mutual solubility with the feed stream diluent, easy recoverability, nonreactivity, nontoxicity, availability on the market, cost, and adequate physical properties (e.g., density, viscosity, and vapor pressure). Batista et al. (1999a) evaluated different alcoholic solvents (ethanol, methanol, *n*-propanol, and isopropanol) for the deacidification of canola oil, and the main results are shown in Figure 3.1.

Although various solvents exhibit some of the features described previously, in the case of edible oil deacidification and the recovery of nutraceuticals, short-chain alcohols are the most frequently recommended ones. Isopropanol and *n*-propanol show considerable mutual solubility with oils (Batista et al. 1999a), requiring either the use of low temperatures or the addition of a polar cosolvent modifier, for example, small amounts of water. Methanol shows an appropriate region of phase splitting (see Figure 3.1) (Batista et al. 1999a), but concerns related to its toxicity might hinder its application as a solvent for processing food products, despite its high volatility and ease of stripping from the raffinate phase, characteristics that make it possible to obtain a final product containing solvent residues far below the safe limits. According to Figure 3.1, ethanol seems to be the best alternative for a series of reasons: (1) the distribution coefficients of the fatty acids, as well as the solvent selectivity and

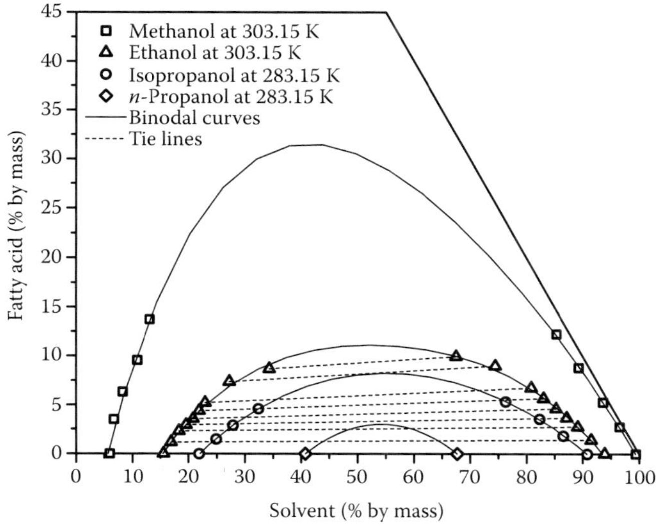

Figure 3.1 Liquid–liquid equilibrium diagrams for canola oil + oleic acid + alcoholic solvent. (From Batista, E., S. Monnerat, K. Kato, L. Stragevitch, and A. J. A. Meirelles, *J. Chem. Eng. Data*, 44, 1360, 1999. With permission.)

mutual solubility with neutral oil, show values appropriate for a liquid–liquid process, and these features can be further adjusted by adding small amounts of water to the solvent; (2) ethanol has adequate physical properties and can easily be stripped from the raffinate stream; (3) it is a biotechnological product available on the world market, mainly from Brazil and North America; and (4) ethanol residues, at least in low concentrations, can be considered as nontoxic. In fact, pure ethanol, which can be considered a food-grade solvent, is used in the formulation of some alcoholic beverages, as the extraction medium for natural products, and in many other applications in industries processing goods for direct human consumption. This product is known as extrafine or neutral alcohol (Decloux and Coustel 2005).

Gonçalves and Meirelles (2004) reported liquid–liquid experimental data for systems containing palm oil + palmitic–oleic acid + ethanol + water at 318.15 K, with the objective of evaluating the influence of different water contents in ethanol. Figure 3.2 shows the phase diagrams obtained for systems containing palmitic acid.

As shown in Figure 3.2, the size of the phase-splitting region increased with the water content of the solvent. In other words, the mutual solubility of the solvent and neutral oil decreased significantly with the addition of water to the ethanol. This addition only has a slight influence on the fatty acid distribution coefficient, provided the water content is not greater than 7%. For larger water contents, the distribution coefficient falls below 1 and decreases steadily.

The phase behavior obtained for the systems containing palmitic acid was also observed for the systems containing oleic acid: The regions of phase splitting are

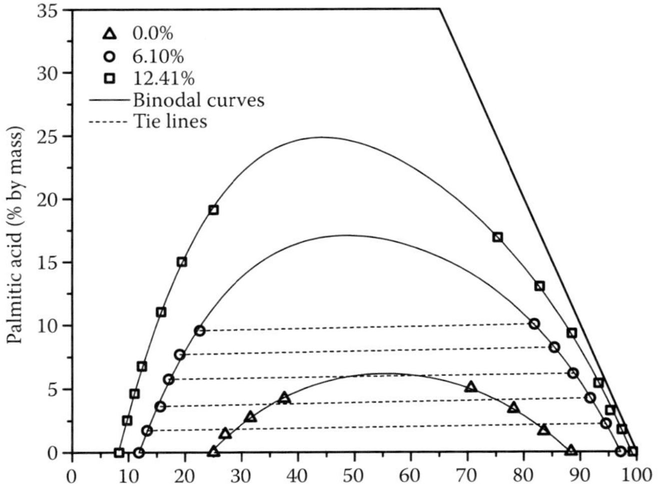

Figure 3.2 Liquid–liquid equilibrium diagrams at 318.15 K for palm oil + palmitic acid + ethanol with different water contents (0.0%, 6.10%, and 12.41% by mass). (From Gonçalves, C. B. and A. J. A. Meirelles, *Fluid Phase Equilib.*, 221, 139, 2004. With permission.)

similar, and the same occurs for the distribution coefficients whose results are shown in Figure 3.3 for the systems containing palmitic acid.

In terms of the net effect on its distribution between the alcoholic and oil phases, the unsaturation of the oleic acid compensates the negative effect of its larger carbon chain. Figure 3.4 shows the results for the distribution coefficients of the free fatty acids present in a bleached palm oil sample (free acidity equal to 3.9% by mass) in mixtures with an equal amount of hydrated ethanol (water content equal to 6.39% by mass) at 318.15 K.

As indicated in Figure 3.4, the double bond can more than compensate the size of the carbon chain. Based on the results obtained by Gonçalves and Meirelles (2004), one can conclude that hydrated ethanol containing 6%–8% by mass of water is the best solvent for deacidifying palm oil because it guarantees a distribution coefficient of around 1 and high selectivity. In fact, in the range of fatty acid contents relevant for the deacidification process (usually free acidity <4.0% by mass), such a solvent has selectivity values greater than 20 (Figure 3.5). This means that the extraction of fatty acids can occur without a significant loss of neutral oil, whose value in this case is restricted by its low solubility in hydrated ethanol.

Similar behavior was observed in the investigation of other oils (Batista et al. 1999a, 1999b; Gonçalves et al. 2002, 2004, 2007; Rodrigues et al. 2003, 2004, 2005a, 2005b, 2006a, 2006b, 2007, 2008; Cuevas, Rodrigues, and Meirelles 2009), but the highly unsaturated ones (linoleic oils) and those with shorter carbon chains (lauric oils) showed greater mutual solubility with hydrated ethanol than the palmitic and oleic oils. In these cases the best approach might be the use of ethanolic solutions with larger amounts of water.

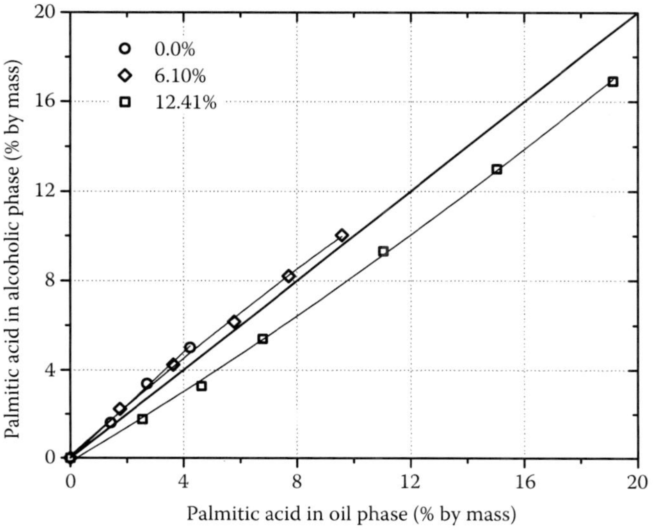

Figure 3.3 Distribution diagrams at 318.15 K for palmitic acid in palm oil + ethanol systems (0.0%, 6.10%, or 12.41% by mass of water in ethanol). (From Gonçalves, C. B. and A. J. A. Meirelles, *Fluid Phase Equilib.*, 221, 139, 2004. With permission.)

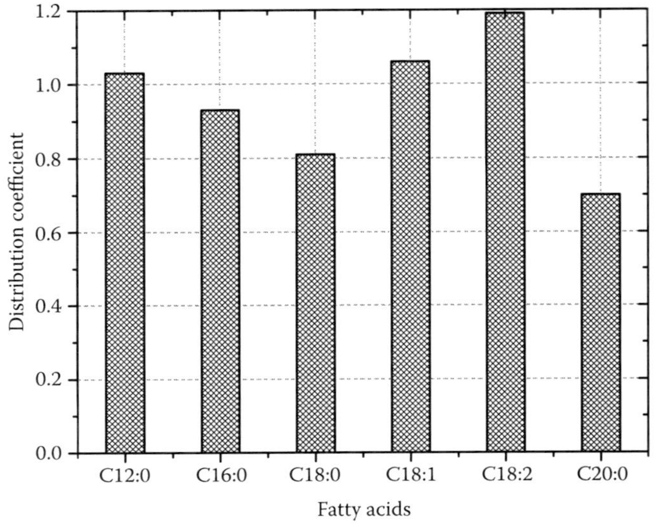

Figure 3.4 Distribution coefficients for the free fatty acids of palm oil in the system with aqueous ethanol containing 6.39% by mass of water. Fatty acids: C12:0 is lauric acid; C16:0 is palmitic acid; C18:0 is stearic acid; C18:1 is oleic acid; C18:2 is linoleic acid; and C20:0 is araquidic acid. (From Gonçalves, C. B. and A. J. A. Meirelles, *Fluid Phase Equilib.*, 221, 139, 2004. With permission.)

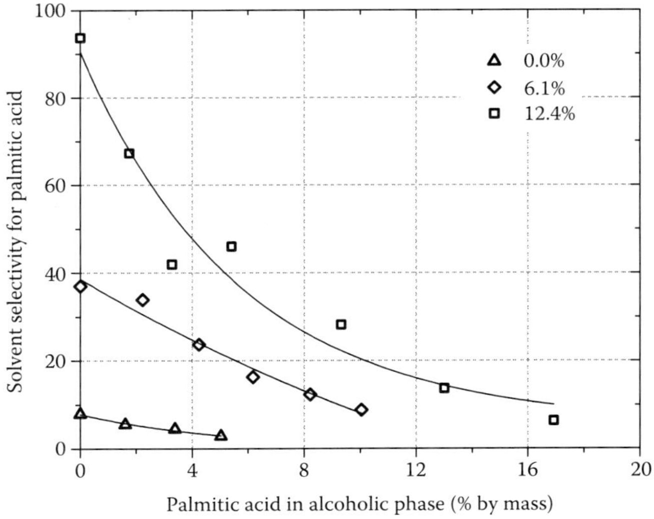

Figure 3.5 Solvent selectivity for palmitic acid as a function of the alcoholic phase acidity (0.0%, 6.10%, or 12.41% by mass of water in the ethanol). (From Gonçalves, C. B. and A. J. A. Meirelles, *Fluid Phase Equilib.*, 221, 139, 2004. With permission.)

Palm oil contains a considerable amount of tocopherols, tocotrienols, and carotenes. Although they do have nutritional value, the carotenoids are removed in the physical refining process to obtain a light yellow color oil, which has better acceptance for industrial purposes (Gonçalves, Pessôa Filho, and Meirelles 2007). In fact, physical refining is responsible for great losses of nutraceutical compounds during the processing of palm oil. The carotenoid content (~500–700 ppm in crude palm oil) is reduced by 50% during the bleached step of the physical refining process, the remainder being completely destroyed during the deacidification and/or deodorization steps as a result of the high temperatures (513.15–533.15 K) and low pressures (133.3–400 Pa) used. The tocopherols also are partially steam stripped during this stage of the refining process, their levels being reduced from 600–1000 ppm to 356–630 ppm (Gonçalves, Pessôa Filho, and Meirelles 2007). Thus, liquid–liquid extraction using appropriate solvents such as ethanol could be an alternative technique for refining palm oil. With this in mind, Gonçalves et al. (2007) studied the influence of deacidification by solvent extraction on the partition coefficients of carotenoids and tocopherols by measuring the equilibrium data for the system palm oil + fatty acids + ethanol + water + nutraceutical compounds at 318.15 K.

Figure 3.6 shows the distribution coefficients of the nutraceuticals between the alcoholic and palm oil phases, using ethanol with different water contents as the solvent (Gonçalves, Pessôa Filho, and Meirelles 2007).

As can be seen in Figure 3.6, the addition of water to the solvent decreases the distribution coefficients of both nutraceutical compounds. This means that if the water concentration is larger, the capacity of the solvent to extract the carotenoids and tocopherols is smaller.

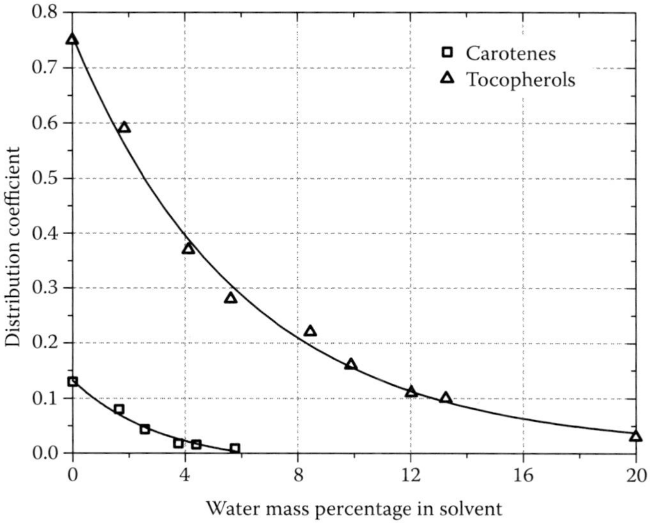

Figure 3.6 Distribution coefficients of the nutraceuticals in palm oil + hydrated ethanol systems at 318.15 K. (From Gonçalves, C. B., P. A. Pessôa Filho, and A. J. A. Meirelles, *J. Food Eng.*, 81, 21, 2007. With permission.)

The distribution coefficients of the carotenes decrease to values less than 0.01 for ethanol containing almost 6% by mass of water. For tocopherols, the distribution coefficient is much larger, approximately 0.25 for water contents in the range of 6%–8% by mass, but still well below 1. This indicates that in a deacidification step based on solvent extraction by hydrated ethanol, the carotenes will remain almost completely in the raffinate phase (oil-rich phase), although part of the tocopherols will be transferred to the extract stream (solvent-rich phase). It is important to emphasize that this effect is desirable because it demonstrates that most of such compounds remain in the oil refined by liquid–liquid extraction. As described previously, the tocopherols are extracted into the alcoholic phase to a larger extent than the carotenoids. Both the tocopherols and carotenoids are insoluble in water because they have a long apolar chain (which makes them liposoluble), but the OH group linked to the aromatic ring of the tocopherols enhances their solubility in ethanol. Gonçalves et al. (2007) concluded that the liquid–liquid extraction process carried out using an ethanolic solvent containing approximately 6% by mass of water, and with an oil/solvent mass ratio equal to 1:1, for example, allows for the maintenance of up to 99% by mass of the carotenoids and approximately 80% by mass of the tocopherols in the refined palm oil. In contrast, traditional physical refining usually provides a refined palm oil with approximately 0.03% by mass of tocopherols and exempt of carotenoids.

As in the case of palm oil, the physical refining of rice bran oil also partially removes important nutraceutical compounds such as γ-oryzanol and tocopherols/tocotrienols, hereafter referred to as *tocols*. Rice bran oil presents considerable potential as a nutraceutical food because of the health benefits that may be attributed to its high level of unsaponifiable matter, of which the most important nutraceutical component is γ-oryzanol, a complex mixture of ferulate esters with sterols and triterpene alcohols (Kim et al. 2001; Patel and Naik 2004; Rodrigues et al. 2006b), and tocols, a family of isomers that present vitamin E activity (Shin et al. 1997; Kim et al. 2001). Marshall and Wadworth (1994) showed a loss of up to 90% of the γ-oryzanol and tocotrienol contents of the crude oil throughout the processing steps. From these observations, it is evident that new techniques in the processing of crude oil must be developed to preserve the active components of rice bran oil. Therefore, Rodrigues et al. (2003, 2004) reported experimental equilibrium data for fatty systems containing rice bran oil, free fatty acids, ethanol, water, γ-oryzanol, and tocols at 298.15 K, providing important information on the phase equilibrium for designing separation processes involving fatty systems. The main objective of these data was to determine the distribution coefficients of the nutraceuticals (γ-oryzanol and tocols) in rice bran oil as a function of the water content of the ethanol. Figure 3.7 presents the main results obtained by these authors.

The behavior of the tocols (tocopherols + tocotrienols) is similar to that previously determined for the tocopherols of palm oil (see Figure 3.6), and in the case of solvents containing 6%–8% by mass of water, the corresponding distribution coefficients varied around the same value of 0.25. The distribution coefficient of γ-oryzanol is not much lower than the results obtained for the tocols; in fact, it shows a value of approximately 0.15 for the same range of water contents (6%–8% by mass). The authors observed that the partition coefficient of γ-oryzanol increased

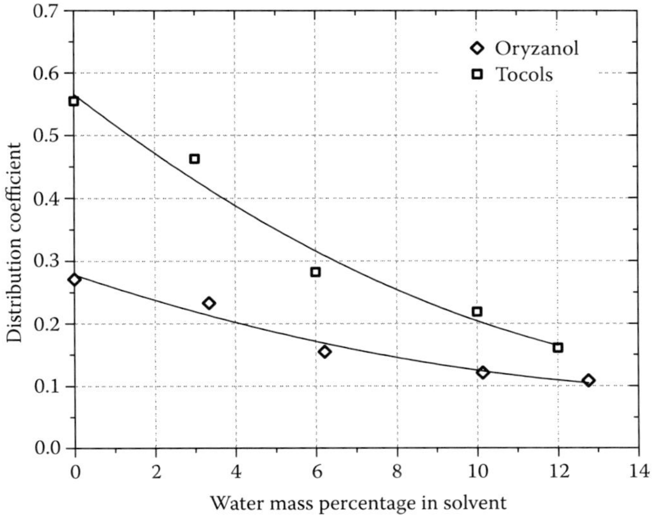

Figure 3.7 Distribution coefficients of the nutraceuticals in rice bran oil + hydrated ethanol systems at 298.15 K. (From Rodrigues, C. E. C., P. A. Pessôa Filho, and A. J. A. Meirelles, *Fluid Phase Equilib.*, 216, 271, 2004. With permission.)

when the free fatty acid content of the oil was higher. This can be attributed to the increase in mutual solubility between the oil and the solvent at higher free fatty acid concentrations. The oil transferred into the alcoholic phase carries part of the γ-oryzanol with it. On the other hand, the addition of water to the solvent decreases the γ-oryzanol partition coefficient and consequently reduces the loss of this nutraceutical compound during solvent extraction. As a consequence, in a deacidification process based on solvent extraction by hydrated ethanol, part of the nutraceuticals of rice bran oil will be maintained in the raffinate (oil-rich phase).

The authors determined the experimental and estimated selectivity for fatty acids and γ-oryzanol, and the results showed that the solvent selectivity, or the capacity of the solvent to extract free fatty acids and simultaneously preserve the γ-oryzanol in the oil, was more affected by the oil acidity than by the water content of the solvent. This means that both the water content of the solvent and the free fatty acid level in the oil influence the γ-oryzanol distribution coefficients in opposite directions, whereas increasing water content decreases extraction of oryzanol from the oil, so an increasing acidity value of the oil increases it. In this way, it is possible to achieve the same γ-oryzanol distribution coefficients for oils with different free fatty acid contents just by changing the water concentration in the ethanolic solvent. The authors concluded that the tocol partition coefficient increased slightly with increases in the crude oil/solvent mass ratio, but it is strongly influenced, in a negative way, by the addition of water to the solvent, a behavior similar to that already reported for the γ-oryzanol partition coefficient. Despite the same behavior for both nutraceutical compounds, it can be seen that the tocols are transferred to the alcoholic phase to a greater extent than γ-oryzanol. This can be attributed to structural differences between the molecules.

Karan (1998) investigated the recovery of γ-oryzanol from crude rice bran oil using liquid–liquid extraction. In contrast to edible oil deacidification by solvent extraction as described previously, the work of Karan (1998) intended to remove the γ-oryzanol from the crude oil and thus obtain a concentrate of the nutraceutical that could be used for other purposes, such as the enrichment of various food products. Organic solvents such as methanol, ethanol, and *N,N*-dimethylformamide (DMF) were tested as solvents. In the case of methanol, a mixture containing 10% of water on a volumetric basis was used, but even for a large solvent/oil ratio (5:1), only 3.4% of the γ-oryzanol was extracted from the crude oil. In relation to the amount of γ-oryzanol extracted, Karan (1998) obtained a much better result using pure ethanol in a sequence of five extraction steps. In each extraction step, a solvent/oil ratio of 4:1 was used, and the raffinate (oil) phase generated in each step was used as the oil source for the subsequent extraction step. In this way the author was able to extract 75% of the γ-oryzanol present in the original crude rice bran oil. Unfortunately, this scheme also caused the extraction of large amounts of rice bran oil, so that approximately 70% of the initial mass of oil was simultaneously extracted. For this reason, the γ-oryzanol concentration in the oil fraction at the end of the process was 0.93%, only slightly higher than its original concentration (0.871%) in the crude oil. It is evident that alcohol extraction for obtaining γ-oryzanol concentrates is a low efficiency process. Such results are compatible with the previous discussion on oil deacidification by extraction with alcoholic solvents. As that discussion indicated, alcoholic solvents do not usually extract large amounts of γ-oryzanol, and if these amounts were increased, larger amounts of other fatty components such as free fatty acids and acylglycerol would also be extracted.

Another approach tested by Karan (1998) involved the use of DMF as the solvent, and the rice bran oil source dissolved in hexane. The author observed an increase in the amounts of γ-oryzanol extracted into the DMF phase when the oil source was dissolved in relatively large amounts of hexane. Because some nonpolar compounds were also extracted from the oil by the DMF, a further purification step was applied by washing the DMF extract phase with hexane. The hexane was able to remove the nonpolar compounds from the DMF extract with only a minimal loss of the extracted γ-oryzanol contained in the DMF phase. Based on these results, the author investigated the following approach: a sequence of three to seven extraction steps using rice bran oil dissolved in hexane and DMF as the solvent. The combined DMF layers were then washed using hexane. According to Karan (1998), the best extraction scheme was that using five DMF extraction steps, allowing for 85.8% of the γ-oryzanol present in the original crude oil to be extracted and resulting in a DMF-combined layer with a γ-oryzanol concentration of 8.4% and a concentration factor of 9.7-fold.

3.2.1.2 Continuous Deacidification of Edible Oils by Liquid–Liquid Extraction

According to the results described previously, the oil deacidification process can be successfully performed via liquid–liquid extraction. To test whether this process would be a technically viable, continuous operation, the deacidification step was investigated by some authors (Antoniassi 1996; Pina and Meirelles 2000; Pina 2001; Reipert 2005; Sa 2007) on a laboratory scale, using different versions of the rotating disc contactor (RDC). RDC is a mechanically agitated liquid–liquid extractor composed of a cylindrical shell containing stator rings and a central rotating shaft carrying equally spaced perforated discs. The acidified oil stream is the heavy liquid and should be fed in at the top of the equipment, flowing downward as a dispersed phase of small oil droplets. The solvent stream represents the continuous phase, flowing upward and leaving the equipment as an acid-rich extract stream. The raffinate stream, containing neutral oil saturated with the solvent, leaves the equipment at the bottom. Continuous deacidification was tested for crude, pretreated, and model oil systems, including palm, rice bran, soybean, corn, and cottonseed oils (Antoniassi 1996; Pina and Meirelles 2000; Pina 2001; Reipert 2005; Sa 2007). The influence of the solvent/oil mass flow ratio, water content of the solvent, rotating disc speed, and oil mass flow were investigated. The final acidity, loss of neutral oil, solvent concentration in the raffinate stream, loss of nutraceuticals, percentage of fatty acids and partial acylglycerols transferred to the extract stream, and volumetric mass transfer coefficients were evaluated during the experiments, and detailed information can be found in the literature (Antoniassi 1996; Pina and Meirelles 2000; Pina 2001; Reipert 2005; Sa 2007).

The hydrodynamics and mass transfer performance in the continuous deacidification of edible oils were investigated by Pina and Meirelles (2000) using different versions of the RDC with an extraction zone length equal to 1.0 meter (Pina and Meirelles 2000). The best results for the dispersed phase holdup, and in consequence for the mass transfer area, were obtained using the RDC version equipped with perforated discs but without stator rings. This version was further tested in the deacidification of a model system composed of corn oil, with the free acidity obtained by the addition of commercial oleic acid. Hydrated ethanol was used as the solvent. The results shown in Figure 3.8 were obtained for an oil/solvent mass flow ratio equal to 1.0:2.0 and a rotating speed of 250 rpm.

As indicated in Figure 3.8, under such operational conditions an extraction zone of 1.0 meter was sufficient to deacidify edible oils (final acidity ≤0.3% by mass), provided the acidity in the feed stream was not greater than 4.0% by mass. In these experiments the loss of neutral oil varied in the range of 4.5%–4.9%, values much lower than those reported in the literature (Leibovitz and Ruckenstein 1983).

Further experiments on continuous deacidification were carried out with degummed and bleached corn oil, degummed soybean oil, crude and degummed cottonseed oils, bleached palm oil, and crude and degummed rice bran oils (Antoniassi 1996; Pina and Meirelles 2000; Pina 2001; Reipert 2005; Sa 2007), and similar results were found. The experiments were carried out using the same kind of extractor, and the results obtained confirmed the technical feasibility of the total deacidification of edible oils by solvent extraction.

With respect to the loss of nutraceuticals during the continuous deacidification by liquid–liquid extraction, some research studies have reported interesting results. In the case of bleached palm oil, the continuous deacidification experiments were

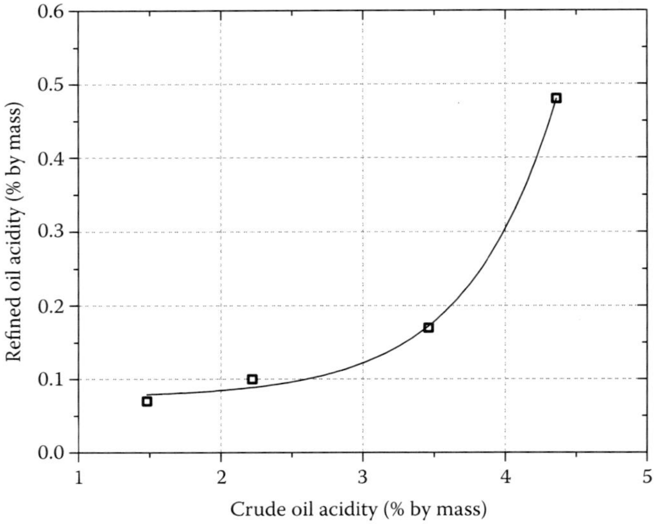

Figure 3.8 Refined oil acidity (composition on a solvent-free basis) as a function of the acidity of the feed stream to the RDC equipment. (From Pina, C. G., and A. J. A. Meirelles, *JAOCS*, 77, 553, 2000. With permission.)

carried out at 318.15 K using an ethanolic solvent with approximately 6% by mass of water (Gonçalves 2004). The oil free acidity was decreased to less than 0.2% by mass, and the observed carotene loss was less than 20%. In the case of degummed rice bran oil, the free acidity also decreased to a value less than 0.2% by mass, and the losses of tocols and γ-oryzanol were approximately 25% (Rodrigues 2004). These losses could be further reduced using an extractor with an appropriate extraction zone length. Because the initial acidity of both oils was high, they had to be reprocessed in the extractor more than twice, using a fresh solvent stream for each deacidification step, and part of the nutraceutical losses can be attributed to the use of a fresh solvent stream in each deacidification step.

Further investigations on continuous oil deacidification were carried out by Batista (2001) using process simulation based on the Newton–Raphson algorithm adapted for calculating liquid–liquid extractors. Corn oil was selected as the crude edible oil to be investigated. It was assumed that corn oil was composed of 9 different free fatty acids, 12 diacylglycerols, and 14 triacylglycerols, in a proportion equal to 4:4:92, respectively. The deacidification step was optimized based on the simulated results and a corresponding factorial design using the number of ideal stages of the extraction, the solvent/oil ratio, and the water percentage in the alcoholic solvent as the factors. Response surface methodology provided the following optimal result: an extractor equipped with 10 ideal stages was able to reduce the free acidity to a value less than 0.3% using aqueous ethanol as the solvent with 5.75% of water, and to guarantee a loss of neutral oil no higher than 2.4%, provided a solvent to oil ratio of 1.27 was used. Most of the neutral oil lost consisted of diacylglycerols, the amount extracted being close to 60% of the original amount in the crude oil (Batista et al. 2002).

3.2.1.3 Patents on Oil Processing by Liquid–Liquid Extraction

Various patents have been deposited around the world, mainly in Europe and the United States, presenting processes for refining edible oils without the loss of nutraceutical compounds or, alternatively, for concentrating these compounds in oil fractions, both processes being carried out by liquid–liquid extraction. The patents indicated various advantages of a purification process based on selective extraction. Swoboda (1985) reported a process for refining palm oil and palm oil fractions that could be carried out by extraction of the oil with an alcoholic solvent, or optionally with a hydrated alcoholic solvent (with 25% by mass of water), subsequently bleaching the raffinate stream of this solvent extraction or the oil derived therefrom. The solvent should either be a mixture of ethanol and water, or one of isopropanol and water, preferably with a composition close to the azeotropic one. Azeotropic mixtures are preferred because of the advantages of recycling the solvent. The experiments were carried out using a countercurrent configuration in a continuous process. According to the results reported by the authors, crude palm oil subjected to solvent extraction may produce a raffinate stream containing a carotenoid concentration similar to, or even larger than, the carotenoid concentration in the original oil. The palm oil obtained via alcoholic solvent extraction may be of considerable nutritional value by virtue of its high carotene content, absence of odor and flavor, and low free fatty acid level. Because the oil at this stage of refining generally has a pronounced red color, it may be used to provide a dietary source of vitamin A precursor (Daun 2005).

Hamm (1992) and Rodrigues et al. (2007) cited an application for a Japanese patent by Nippon Oils and Fats that presented the possibility of producing fish oil fractions enriched with eicosapentaenoic acid (EPA) by solvent extraction with aqueous acetone. According to this report, the EPA content was increased by 85% by extracting the original oil, containing 12.88% of this fatty compound, with 10 times its weight of a 9:1 acetone/water mixture.

In another invention suggested by Plonis and Trujillo-Quijano (1995), the deacidification of palm oil by liquid–liquid extraction produced an olein (the liquid fraction of palm oil, with a high content of unsaturated fatty acids) with a carotene content of 750–1000 mg/kg. The solvents used were short-chain alcohols, preferably ethanol or ketones, containing 1%–25% by volume of water and approximately 1% of citric acid. The authors also reported that the patented process could produce deacidified oil containing high levels of carotene and reduced amounts of diacylglycerols and free fatty acids. However, the deacidified oil presented enhanced flavor and aroma.

Cherukuri et al. (1999) deposited a patent to obtain rice bran oil enriched with high levels of tocols (tocopherols and tocotrienols) and γ-oryzanol, using a liquid–liquid extraction process employing lower aliphatic alcohols containing from one to six carbons, such as methanol, ethanol, or isopropanol. The process, developed based on experimental runs carried out in separation funnels on a laboratory scale, involves mixing rice bran oil and alcohol, separating the alcohol layer, and subsequently distilling this layer to recover the enriched rice bran oil.

3.2.2 Adsorption

Adsorption–desorption processes are generally used as separation techniques in the food, chemical, and pharmaceutical industries. The desired product, after a determined contact time, can be found free in the solution or, alternatively, adhered to the adsorbent. In the latter case, the desired product should be recovered by the passage of an appropriate solvent through the adsorbent, desorbing the desired product (elution).

Adsorption–desorption processes are used to separate, concentrate, or purify nutraceutical compounds present in aqueous or organic solutions, resulting from different steps in the production of juices and oils, for example. It is also possible to use this technique to recover nutraceuticals present in the wastewaters from some industries. One example is the production of olive oil, which generates a large amount of wastewater rich in polyphenols, sugars, lipids, and so on, and is responsible for environmental problems because of its high phytotoxicity. Agalias et al. (2007) suggested a system to recover the high polyphenol and lactone contents from olive oil production wastewaters using adsorbent resins after successive filtration steps to reduce the supernatants. This adsorption system made it possible to obtain an extract rich in polyphenols and lactones with wastewater in adequate conditions to be discarded.

Scordino et al. (2004) performed a test with a series of commercial resins to choose the best one to concentrate cyanidin 3-glucoside, a phenolic compound (anthocyanins), from an aqueous batch solution. When the authors used a methacrylic resin, the amount of cyanidin 3-glucoside adsorbed was approximately 6 mg/g resin, and when they used styrene-divinylbenzene copolymers as the adsorbent, the amount of cyanidin 3-glucoside adsorbed increased considerably to approximately 15 mg/g resin. Comparing the values adsorbed by each resin, it was clear that the cyanidin 3-glucoside had a greater affinity for the styrene-divinylbenzene copolymer, a strongly hydrophobic resin. In this study the authors also show that an increase in the pH value did not interfere significantly in the adsorption capacity of each resin. With adsorption onto styrene-divinylbenzene copolymers, a small fluctuation of approximately 2 mg/g resin was observed in the adsorption capacity with the increase in pH value. This fluctuation is not significant, indicating that the structural behavior of cyanidin 3-glucoside, which changes according to the pH value of the solution, did not interfere in the adsorption processes. Thus, in this case, the conformational form of the molecule, which depends on the pH value of the solution, was not a problem in the adsorption process.

Continuing the investigation, the authors compared the adsorption of cyanidin 3-glucoside and hesperidin, a compound also present in pigmented orange juice, using the same resins tested before to verify the selectivity of the resins. Of all the resins tested, just one, styrene-divinylbenzene copolymer, adsorbed the same amount of both components. This fact was an indication of the selective capacity of another resin that could be used to separate cyanidin 3-glucoside and hesperidin, cheaper than the first one, from the same solution. In this way, the authors tested the resin that presented the best adsorption capacity from aqueous solution (EXA-118) to remove cyanidin 3-glucoside and hesperidin from a pigmented orange juice. After elution with methanol, the amount of the anthocyanin cyanidin 3-glucoside recovered, was more than fourfold higher than the amount of hesperidin recovered, which was an excellent result considering that the amount of anthocyanins in the juice was approximately half that of hesperidin. The most concentrated extract was obtained using a solution of methanol/water (50%:50% v/v) as the eluting solution (Scordino et al. 2005).

Kammerer et al. (2007) used a polymethylmethacrylate resin to study the processing parameters of the apple juice adsorption process. The authors observed a decrease in the total amount of polyphenols adsorbed by each gram of the resin as the temperature used in the adsorption process was increased from 293.15 K to 353.15 K.

They also evaluated the amount of each polyphenolic compound recovered from the resin by elution using water–methanol or water–ethanol solutions with different concentrations. The best elution results were obtained with increased amounts of alcohol concentration in the solution, of approximately 70%–80% by volume of methanol and 60% for ethanol. Significant amounts of individual phenolic compounds, such as chlorogenic acid, 4-caffeoylquinic acid, phloridzin, and some quercetin derivatives, among others, were recovered, and the amount of each one was different according to the solvent used and its concentration in the solution. Based on these results the authors concluded that the resins exhibited higher affinity for specific phenolic compounds found in apple juice. In principle, this selective behavior by the resin makes it possible to separate and recover specific compounds with elevated purity. They also observed that the adsorption of phenolic compounds was improved by decreasing the solution pH value, whereas the desorption step was mainly dependent on the hydrophobicity of the different phenolic compounds.

Vinu, Hossian, and Srinivasu (2007) increased the selectivity of mesoporous carbon by way of the functionalization technique using ammonium persulfate. Functionalization is a technique used to enrich the surface of a material with a desired chemical group that can foment, for example, the adsorption of a group of molecules onto the surface of the material. The aforementioned authors performed the functionalization of the mesoporous carbon by an oxidation process using an ammonium persulfate solution and obtained a good adsorbent material for use with biomolecules. Another example of functionalization is the use of ionic liquids to modify a mesoporous siliceous substrate (Li et al. 2008). The functionalized resin was successfully used for extracting α-tocopherol from a model mixture of soybean oil deodorizer distillate, and it presented good reusability and selectivity.

The use of ion-exchange resins is another option to separate undesired components from desired ones by adsorption. Because of the presence of anionic or cationic sites along the solid surface, a compound with the opposite charge will be adsorbed by the resin. Such a mechanism was used to remove the acidity from juices (Chung et al. 2003; Lineback et al. 2003) and oleic acid from ethanol–water solutions (Cren and Meirelles 2005; Cren et al. 2009).

A series of synthetic adsorbents was tested to improve the recovery of carotene from crude palm oil (CPO), preserving the edible oil quality (Latip et al. 2000). The recovery process consisted of diluting the CPO with three parts of isopropanol, with subsequent adsorption onto synthetic adsorbents by mixing at a controlled temperature. The CPO was then eluted using isopropanol. The nonadsorbed carotene was removed from the adsorbent using hexane at a constant temperature. Both solvents were evaporated off under vacuum, so that one carotene-rich fraction and another CPO-rich fraction were obtained. In this way, the authors chose the best synthetic resins and the best process condition to recover carotene from CPO, preserving the oil quality.

In a subsequent study, Latip et al. (2001) investigated the influence of temperature, adsorption time, and elution time in the recovery of oil using isopropanol as the solvent. The temperature was varied from 313.15 K to 353.15 K, and they observed that at higher temperatures the concentration of carotene in the eluted oil increased, whereas the amount of carotene recovered by the subsequent resin elution with hexane decreased. In fact, the best recovery of carotene from the oil was obtained at 313.15 K. They also concluded that a contact time of at least 0.5 hours during the adsorption process was required to recover appropriate amounts of carotene. In the case of elution with isopropanol, they observed that increasing the contact time decreased the amount of carotene retained by the resin and, consequently, the quantity of carotene recovered by elution of the resin with hexane. Based on these results it can be concluded that by controlling the temperature of the adsorption process and the elution contact time with isopropanol, one can define the amount of carotene to be maintained in the palm oil and the amount to be recovered as an isolated nutraceutical dissolved in hexane.

As expected, adsorption is influenced by the kind of adsorbent used, and small changes in the adsorbent structure may have a significant effect on the adsorption process. Ahmad et al. (2009) observed that the adsorption of β-carotene from CPO using silica gel and florisil (Sigma-Aldrich, St. Louis, MO)—two silica-based absorbents—was greater using florisil, which includes magnesium oxide and a small portion of sodium sulfate in its structure in addition to a difference in its pore size. Independent of these differences, the author observed that the adsorption of β-carotene was dependent on the temperature, the contact time, and the initial concentration. Increasing these variables increased the adsorption of β-carotene. The adsorption process was also influenced by the adsorbent/solution mass ratio, the agitation speed in the case of batch adsorption, the temperature, and the adsorbent particle size, as observed by Ma and Li (2004) in the adsorption of β-carotene from soybean oil using clay as the adsorbent.

The adsorption of polyphenols obtained by extraction from crude *Inga edulis* leaves (an Amazonian tree) has also been investigated (Silva et al. 2007). The authors used macroporous resins and concluded that adsorption was influenced by the type of adsorbent and by the proportion of water in the ethanolic solutions, but not by the pH value of the solution. Isotherms for the two classes of polyphenols—phenolic and flavonoid compounds—were determined in a previously mentioned article by Scordino et al. (2004), and it was also observed that the adsorption of cyanidin

3-glucoside, a flavonoid, was not influenced by the solution acidity. Nevertheless, acidity is usually an important factor in the adsorption of biological molecules that can exhibit some structural change according to the pH value of the solution.

Advances in the development of mesoporous materials can result in the improvement of the adsorption and desorption process because these materials exhibit large specific surface areas, large specific pore volumes, and well-ordered pore structures (Ciesla and Schüth 1999; Taguchi and Schüth 2005). Although the mechanism for obtaining such materials is not completely understood, it is possible to produce mesoporous materials with controlled pore sizes and distribution (Zhao et al. 1998). The application of mesoporous materials to adsorption–desorption processes has been widely investigated, and they can diminish some of the difficulties caused by the pore size, especially when the adsorbate molecule is a large one, as in the case for proteins.

The adsorption of vitamin B2 (riboflavin) and two proteins (lysozyme and trypsin) by two different mesoporous molecular sieves was studied by Kisler et al. (2001). The amount of riboflavin adsorbed by the molecular sieves was smaller than that adsorbed by mesoporous activated carbon, but, as expected, it was greater than the amount of protein adsorbed as a result of the size differences of the molecules, riboflavin being more than 30 times smaller than lysozyme and 60 times smaller than trypsin. This means that these molecules were mainly separated by size exclusion.

Another investigation confirmed that the adsorption of vitamin E from a solution of *n*-heptane and *n*-butanol was dependent on the volume and the pore diameter of the adsorbent used (mesoporous material), as well as on the solvent polarity (Hartmann, Vinu, and Chandrasekar 2005).

Chen and Payne (2001) investigated the possibility of recovering and separating α- and δ-tocopherols from hexane by adsorption onto an acrylic ester resin. In a first stage of the study, a solution with α- and/or δ-tocopherols was added to the resin, and after reaching equilibrium, the hexane solution was analyzed. It was found that the acrylic ester resin adsorbed 3 to 4 times more δ-tocopherol than α-tocopherol. The authors used ethyl propionate to simulate the binding site of the acrylic ester adsorbent. The results obtained indicated the formation of an intermolecular hydrogen bond between the phenolic hydroxyl of the tocopherols and the ethyl propionate, in a way similar to that occurring between the tocopherols and the adsorbent. However, in the case of α-tocopherol, this hydrogen bonding is weak as a result of steric constraints by its methyl and hydroxyl groups. According to the authors, it is possible to separate similar molecules such as α- and δ-tocopherols, although it is a difficult task. However, it can become easier with an understanding of the mechanisms governing the adsorption processes. Bono, Ming, and Sundang (2007) also studied the adsorption of vitamin E (α-tocopherol) from ethanolic solutions onto activated carbon and concluded that this process appears to obey the monolayer theory.

The monolayer theory is one of the hypotheses assumed in deriving the Langmuir isotherm. Many investigations on adsorption begin by determining the adsorption isotherms to better understand the mechanism of adsorption and desorption (Chan, Baharin, and Man 2000; Sabah 2007; Ahmad et al. 2009). In Section 3.4 of this chapter ("Fundamentals of Adsorption Applied to the Recovery of Nutraceuticals"), the modeling of adsorption isotherms will be discussed, as well as some basic aspects of the batchwise operation of the adsorption process and the corresponding breakthrough curves.

3.3 FUNDAMENTALS OF LIQUID–LIQUID EXTRACTION APPLIED TO THE PROCESSING OF FUNCTIONAL FOODS

Liquid–liquid extraction or solvent extraction is a unit operation that brings into contact two insoluble liquids: a feed stream and a solvent. When the components of the original mixture distribute themselves in different ways in the two liquid phases, a degree of separation is obtained.

If a vegetable oil with certain free fatty acid content is shaken with a polar solvent such as a short-chain alcohol, part of the free fatty acid and relatively little of the vegetable oil will migrate to the alcoholic phase. When shaking stops, the two phases will decant as a result of the difference in their densities. The free fatty acid contents in the two phases will be different from each other and from the original content, and thus a degree of free fatty acid extraction will be obtained. This is an example of stagewise contact, which can be carried out in a batch or continuous way. If one wishes to reduce the free fatty acid content, for example, to produce edible oil, the residual oil phase can be put in contact with more solvent.

The original mixture containing the solute to be extracted is called the *feed*, and the other chemically different liquid is the *solvent*. The solvent-rich phase that leaves the equipment is the *extract stream*, and the residual liquid from which the solute was extracted is the *raffinate stream*.

3.3.1 Liquid–Liquid Equilibrium Diagrams for Fatty Systems

Liquid–liquid extraction involves at least three components that appear to different extents in the two phases. A liquid–liquid equilibrium diagram in which only one pair of components is partially soluble is presented in Figure 3.9. The diagrams, presented with triangular coordinates, are used at constant temperature and pressure. Liquid C (solute, in the present case, the fatty acid) is completely soluble in liquids A (diluent, in the present case, vegetable oil) and B (solvent), but A and B only dissolve into each other to a limited extent and are represented in the diagram by the saturated liquid binary solutions at L (rich in diluent – A) and at K (rich in solvent – B). On the rectangular coordinates, the abscissa and ordinate present the compositions of the solvent (component B) and solute (component C), respectively. Any binary mixture between L and K will separate into two immiscible liquids with the compositions shown at L and K. Point L represents the solubility of the solvent in the diluent and point K, the solubility of the diluent in the solvent at the temperature of the diagram.

The $LRPEK$ curve is the binodal curve and represents the change in solubility of the diluent-rich phase and the solvent-rich phase. Any ternary mixture on the outside of the curve will be a one-phase solution (homogeneous region), and any ternary

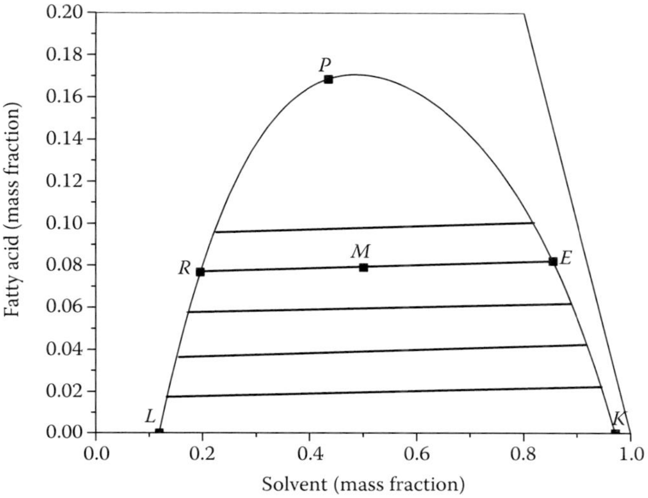

Figure 3.9 Liquid–liquid equilibrium diagram. *L* – *K*, base line; *R* – *E*, tie line; *M*, overall composition; *P*, plait point.

mixture underneath the curve (heterogeneous region), such as mixture *M*, will form two immiscible phases with the equilibrium compositions indicated at *R* (diluent-rich phase) and *E* (solvent-rich phase). The line *RE* is a tie line and must necessarily pass through point *M*, which represents the overall composition.

The point *P*, known as the *plait point*, is the last tie line, where the binodal curve converges and the compositions of the diluent-rich phase and the solvent-rich phase are equal. The *LRP* curve represents the diluent-rich phase, and the *KEP* curve represents the solvent-rich phase.

The distribution coefficient (k_i) of component *i* is defined as the ratio of its composition in the solvent-rich phase (*E*) to its composition in the diluent-rich phase (*R*):

$$k_i = \frac{y_{i,E}}{x_{i,R}} \tag{3.1}$$

According to Figure 3.9, the composition of *C* (solute) in solvent-rich phase ($y_{C,E}$) is larger than in the diluent-rich phase ($x_{C,R}$), and hence the solute distribution coefficient is larger than unity.

The capacity of solvent (*B*) to separate solute (*C*) from diluent (*A*) is measured by the ratio of the distribution coefficient of the solute (*C*) to the distribution coefficient of the diluent (*A*). This separation factor is known as the *selectivity* and represents the effectiveness of a solvent in extracting the solute from the diluent. The selectivity must exceed unity, and the greater the value, the better or easier the separation.

$$\beta_{CA} = \frac{k_C}{k_A} \tag{3.2}$$

If the selectivity is equal to unity, this means that the diluent and the solute are distributed in the same way in the phases, and thus their separation from each other is not possible.

3.3.2 Liquid–Liquid Extraction Equipment

The rate of mass transfer between two liquid phases is directly proportional to the overall mass transfer coefficient, the interfacial area, and the composition difference driving force. The rate may be increased by dispersing one of the liquids into smaller droplets immersed in the other one, with a consequent increase in the interfacial area. This favors eddy diffusion and improves the mass transference between the phases.

Liquid–liquid extraction equipment provides direct contact between two immiscible liquids that are not in equilibrium and involves dispersing one liquid in the form of small droplets (the dispersed phase) into the other liquid (continuous phase) in an attempt to bring the liquids to equilibrium. After this contact the resulting liquids are mechanically separated because of the difference in their densities. This equipment can be classified according to the type of phase contact, either in stagewise or in continuous (differential) contact, or by the type of its internal construction, in this case into mixer–settlers, columns without agitation, mechanically agitated columns, and centrifugal extractors (Frank et al. 2008).

3.3.2.1 Equipment for Stagewise Contact

This type of equipment is organized in different stages, each stage representing a step of mass transfer contact between the two phases and, subsequently, of its separation by density difference. Each stage should work as close as possible to an equilibrium stage, which means that the effluent streams should leave this stage as close as possible to the corresponding equilibrium concentrations.

Mixer–settler is the most typical and oldest extraction equipment, in which each stage presents two well-defined and delimited regions: the first region, the mixer, involves dispersing one liquid into the other, and the second, the settler, involves the mechanical separation. The equipment may be operated in a batch way or in a continuous one. If batch, the same vessel will be used for both mixing and settling, but if continuous, the mixing and settling are usually done in different vessels. The mixing vessel uses some form of rotating impeller placed at its center, which provides an effective dispersion of the phases. The basic unit of the mixer–settler may be connected to form a cascade for cross flow or, more commonly, for countercurrent flow. For economic reasons the use of several mixer–settler units is limited up to five theoretical stages (Blass and Goettert 1994), but it is mostly preferred in cases where no more than two stages are necessary.

The perforated-plate (sieve-plate) column is similar to a tray distillation column. For dispersed phases consisting of light liquids, the plates contain downspouts at their free extremity, which allow for the downward flow of the heavy liquid (continuous phase). Below each plate and outside the downspout, the droplets of the light phase coalesce and accumulate in a liquid layer. This layer of liquid flows through the holes of the plate and is dispersed in a large number of droplets within the continuous phase located above this plate. For dispersed phases consisting of heavy liquids, the flow configuration is the other way up.

3.3.2.2 Equipment for Continuous Contact

In such equipment, the liquids flow in continuous multistage countercurrent contact as a result of the difference in density of the liquid streams, without complete separation. The force of gravity acts to provide the flows, and the equipment is usually a vertical column with the light liquid entering at the bottom and the heavy one entering at the top. Complete separation of the phases only occurs at one extremity of the equipment, at the top if the dispersed phase is the light liquid, or at the bottom if the heavy liquid is the dispersed one.

The simplest equipment for differential contact is the spray column, which basically consists of an empty shell with provision for introducing and removing the liquids. If the light liquid is the dispersed one, the heavy liquid enters at the top through the distributor and fills the column, then flows downward as a continuous phase and leaves at the bottom. The light liquid enters at the bottom of the column through a distributor, which disperses it into small droplets. These droplets flow upward through the continuous phase, coalesce, and form an interface at the top of the column where the light liquid leaves the equipment. Although this column is easily constructed, it is not recommended for use with more than one or two theoretical stages because of its low mass transfer efficiency as a result of the absence of internal parts that would improve phase dispersion.

In packed columns the shell of the column is filled with random or structural packing arrangements. In the first case, the packing consists of small regular elements with size no larger than one-eighth of the column diameter. On the other hand, structured packing is formed from vertical corrugated thin sheets of ceramic, metal, or plastic materials, with the angle of the corrugations reversed in adjacent sheets to form an open honeycomb structure with inclined channels and a large surface area.

Extractors can also be mechanically agitated to disperse one liquid into the other and ensure rapid mass transfer. There are a great variety of mechanically agitated columns for continuous contact. The first example is the RDC column, which consists of a column with a rotating central shaft containing equally spaced flat discs. Each disc is positioned at the center of a chamber delimited by horizontal stator rings fixed to the column shell. Modifications of the original RDC column can be found in the literature, such as the ones that use perforated discs or columns without stator rings.

The Khüni column has a rotating central shaft with impellers that are fixed at the center of a compartment delimited by two adjacent perforated plates. Pulsed columns are a variation of agitated columns, where perforated plates move up and down or the liquids are pulsed in a stationary column by an outside mechanism. This last type of agitation is compatible with other extractors, such as packed or perforated-plate columns.

3.3.2.3 Centrifugal Extractors

The force of gravity may be replaced by centrifugal force in cases where the difference between the phase densities is small, or for mixtures with tendencies to form emulsions. The most important centrifugal extractor is the Podbielniak extractor, which consists of a cylindrical drum containing perforated concentric shells that rotate rapidly. Continuous centrifuges can also be used connected to a settler to accelerate separation of the phases.

More information about equipment for liquid–liquid extraction can be found in Treybal (1980), Godfrey and Slater (1994), and Robbins and Cusak (1997).

3.3.3 Mass Transfer Equations and the Types of Extraction

The mass balances for an extractor of the stagewise type are presented below. Each stage is a theoretical stage so that the leaving extract and raffinate streams are in equilibrium. The lever-arm rule is discussed first because this rule is required for understanding the mathematical calculations associated with each type of extraction.

Lever-arm rule: If a mixture containing A and C with F kilograms is shaken with a solvent B with S kilograms, a new ternary mixture is generated with M kilograms. When agitation stops, the system will separate into two phases, one phase rich in component A with R kilograms and other phase rich in component B with E kilograms.

The mixer and settler units are represented in Figures 3.10 and 3.11, respectively, and the lever-arm rule for mixing and settling processes is represented in Figure 3.12.

In case of the mixing process the global mass balance and the mass balances for components B and C are:

$$F + S = M \tag{3.3}$$

Component B:

$$Fx_{B,F} + Sy_{B,S} = Mx_{B,M} \tag{3.4}$$

Component C:

$$Fx_{C,F} + Sy_{C,S} = Mx_{C,M} \tag{3.5}$$

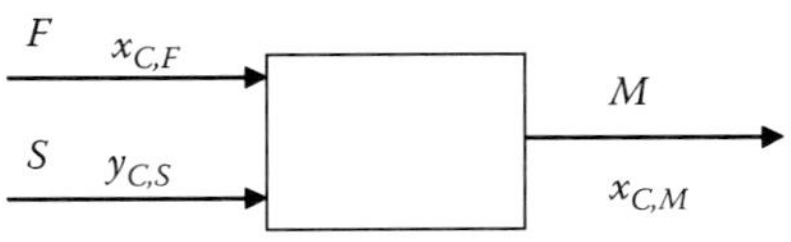

Figure 3.10 Mixing process.

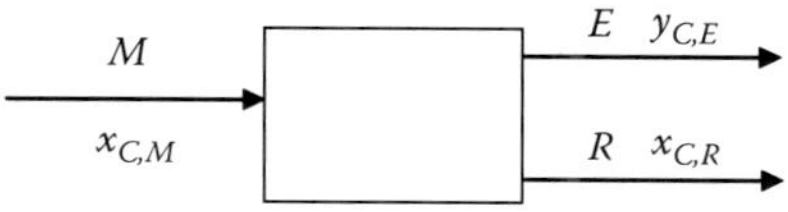

Figure 3.11 Settling process.

If one substitutes Equation 3.3 into Equation 3.4 and rearranges, the following result is obtained:

$$\frac{F}{S} = \frac{y_{B,S} - x_{B,M}}{x_{B,M} - x_{B,F}} = \frac{\overline{NS}}{\overline{ON}} \tag{3.6}$$

When Equation 3.3 is substituted into Equation 3.5, the result is:

$$\frac{F}{S} = \frac{x_{C,M} - y_{C,S}}{x_{C,F} - x_{C,M}} = \frac{\overline{MN}}{\overline{FP}} \tag{3.7}$$

Equations 3.6 and 3.7 can be combined and rearranged to obtain Equation 3.8:

$$\frac{x_{C,F} - x_{C,M}}{x_{B,M} - x_{B,F}} = \frac{x_{C,M} - y_{C,S}}{y_{B,S} - x_{B,M}} \tag{3.8}$$

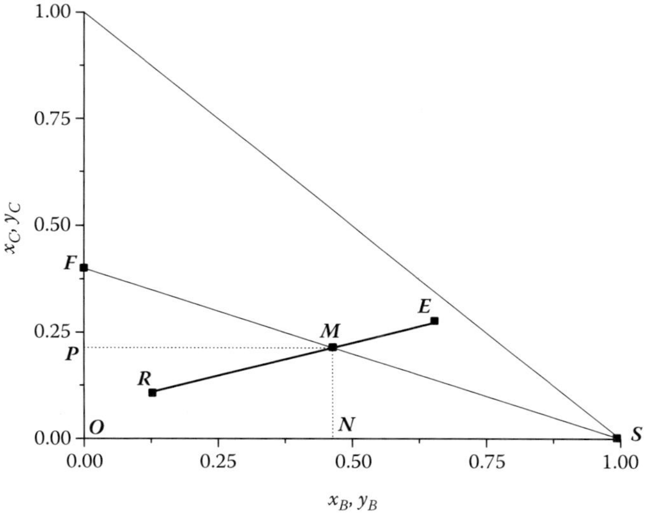

Figure 3.12 Lever-arm rule with rectangular coordinates.

This indicates that points F, M, and S must be lined up (see Figure 3.12). By using the similarity of triangles one can obtain Equation 3.9:

$$\frac{F}{S} = \frac{\overline{MN}}{\overline{FP}} = \frac{\overline{MS}}{\overline{FM}} \tag{3.9}$$

By analogy one can derive Equation 3.10:

$$\frac{R}{E} = \frac{\overline{ME}}{\overline{RM}} \tag{3.10}$$

For more details on the lever-arm rule, see Treybal (1980), Geankoplis (1993), and Batista et al. (2009).

As described earlier, there are a variety of equipment configurations, from a single-stage contact, such as a mixer–settlers operating in a batch or continuous way, to countercurrent multistage operations, such as columns of multistage type.

In the design of mass transfer stagewise operations, the first step is the calculation of theoretical stages solving mass balance equations and equilibrium conditions. In the sequence graphical methods are presented for calculating liquid–liquid extraction in single-stage and multistage countercurrent equipment.

3.3.3.1 Single-Stage Equilibrium Extraction

Consider the following example: 100 kg/h of a vegetable oil (A) containing 5 mass% of free fatty acids (C) and 100 kg/h of aqueous ethanol (B) enter a single equilibrium stage as that shown in Figure 3.13. The required equilibrium data were taken from Gonçalves and Meirelles (2004), who measured liquid–liquid equilibrium systems containing palm oil, free fatty acids, and aqueous ethanol at 318.15 K. Feed and solvent streams are mixed, and the exit streams R_1 and E_1 leave the stage in equilibrium.

The following result is obtained for the global mass balance:

$$F + S = E_1 + R_1 = M = 200 \text{ kg/h}$$

The lever-arm rule allows indicating the composition of the overall mixture M in Figure 3.14.

$$\frac{\overline{FM}}{\overline{FS}} = \frac{S}{M} = \frac{100}{200} = 0.5$$

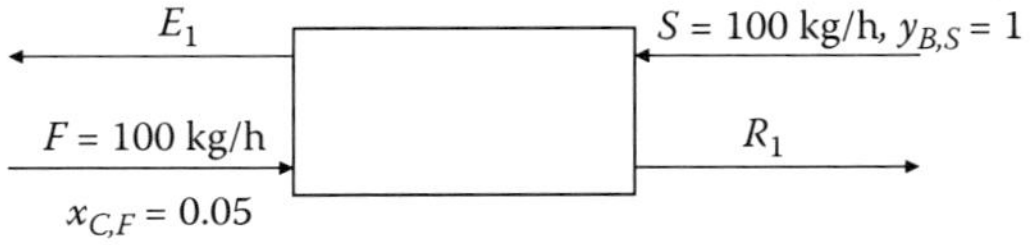

Figure 3.13 Single-stage extraction.

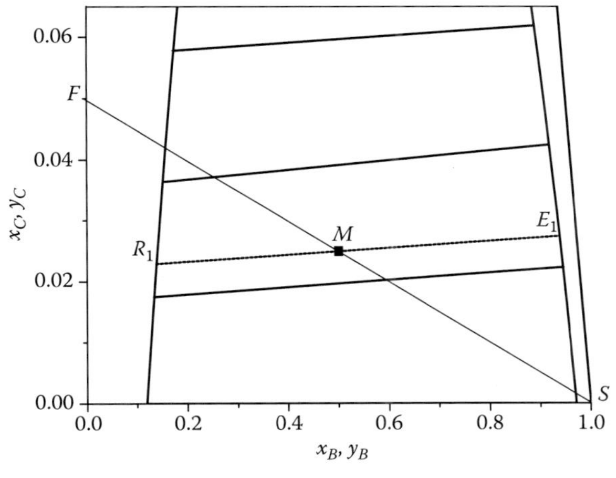

Figure 3.14 Phase diagram for a single-stage extraction.

This composition can also be calculated by the mass balances for component C and B:

$$x_{C,M} = \frac{x_{C,F}F + y_{C,S_1}S}{M}$$

$$x_{B,M} = \frac{x_{B,F}F + y_{B,S_1}S}{M}$$

The extract and raffinate mass flows can be calculated according to the lever-arm rule:

$$\frac{\overline{R_1M}}{\overline{E_1M}} = \frac{E_1}{R_1} = 0.83 \Rightarrow E_1 = 0.83 \cdot R_1$$

$$E_1 = 90.71 \text{ kg/h}$$
$$R_1 = 109.29 \text{ kg/h}$$

The composition of extract (E_1) and raffinate (R_1) streams can be obtained directly from the liquid–liquid diagram (see Figure 3.14):

E_1

$$y_{C,E_1} = 0.027$$
$$y_{B,E_1} = 0.938$$

$$y_{A,E_1} = 1-(y_{B,E_1} + y_{C,E_1})$$
$$y_{A,E_1} = 0.035$$

R_1

$$x_{C,R_1} = 0.023$$
$$x_{B,R_1} = 0.138$$

$$x_{A,R_1} = 1-(x_{B,R_1} + x_{C,R_1})$$
$$x_{A,R_1} = 0.839$$

3.3.3.2 Continuous Multistage Countercurrent Extractor

Consider the following case: 100 kg/h of a vegetable oil (A) with 5 mass% of free fatty acids (C) enters in the first stage of a countercurrent extractor, and 300 kg/h of aqueous ethanol (B) is fed into the opposite side of the extractor. Extract and raffinate streams flow in a countercurrent arrangement, such as that shown in Figure 3.15. All the raffinate and extract streams that leave the same stage are in equilibrium. Also suppose that the mass fraction of fatty acids in the final raffinate stream must be ≤0.005 (0.5 mass%). The procedure indicated below allows determining the number of theoretical stages required for deacidifying the edible oil according to these specified conditions.

The global mass balance for the whole extractor is given by:

$$F + S = M = R_N + S_1$$

The mass balances for each stage are:

Stage 1: $E_1 + R_1 = F + E_2 \Rightarrow E_1 - F = E_2 - R_1$

Stage 2: $E_2 + R_2 = R_1 + E_3 \Rightarrow E_2 - R_1 = E_3 - R_2$

Stage 3: $E_3 + R_3 = R_2 + E_4 \Rightarrow E_3 - R_2 = E_4 - R_3$

And for the N stage (last stage):

Stage N: $E_N + R_N = R_{N-1} + S \Rightarrow E_N - R_{N-1} = S - R_N$

The equations above indicated that the difference between the extract and raffinate streams that flows between the same two adjacent stages remains constant. This difference is indicated below by the symbol Δ:

$$E_1 - F = E_2 - R_1 = E_3 - R_2 = \ldots = E_N - R_{N-1} = S - R_N = \Delta$$

Using these sets of equations the following results can be obtained in the case of the deacidification by countercurrent extraction. The global mass balance for the extractor is

$$F + S = M = R_N + S = 400 \text{ kg/h}$$

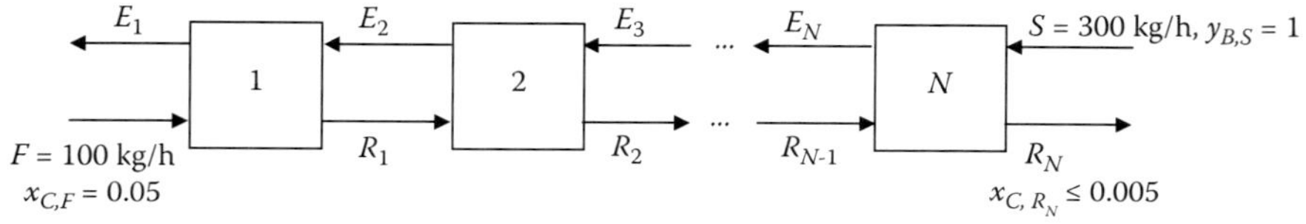

Figure 3.15 Flow sheet of countercurrent extraction.

By matching the points F and S and applying the lever-arm rule one can mark in the diagram the composition for the global mixture M:

$$\frac{\overline{FM}}{\overline{FS}} = \frac{S}{M} = \frac{300}{400} = \frac{3}{4}$$

This can also be obtained by the mass balances of components B and C:

$$x_{C,M} = \frac{x_{C,F}F + y_{C,S}S}{M} \qquad x_{C,M} = 0.0125$$

$$x_{B,M} = \frac{x_{B,F}F + y_{B,S}S}{M} \qquad x_{B,M} = 0.750$$

By matching point R_N to M one can determine the composition of E_1 in the binodal curve. Points R_N and E_1 are lined up by mass balance. To obtain point Δ, trace the lines $\overline{FE_1}$ and $\overline{R_N S}$. The interception of these two lines corresponds to point Δ.

According to the mass balance for the first stage, points F, E_1, and Δ are lined up. The same must occur in case of the last stage for points R_N, S, and Δ:

$$E_1 - F = S - R_N = \Delta$$

According to the mass balance for the second stage, point R_1 should be connected to Δ to obtain point E_2 in the binodal curve:

$$E_2 - R_1 = \Delta$$

This procedure should be used until $x_{C,R_N} \leq 0.005$. In this example, only two stages are required to reach this composition of component C in the raffinate stream.

The mass flows of raffinate and extract can be calculated by using the lever-arm rule, as indicated in Figure 3.16:

$$\frac{\overline{R_N M}}{\overline{E_1 M}} = \frac{E_1}{R_N} = 3.3 \Rightarrow E_1 = 3.3 R_N$$

$$E_1 + R_N = 400 \text{ kg/h}$$
$$R_N = 93.02 \text{ kg/h}$$
$$E_1 = 306.98 \text{ kg/h}$$

3.3.4 Retention of Nutraceuticals in the Deacidification of Vegetable Oils by Liquid–Liquid Extraction

When both operating and equilibrium curves are straight lines, the number of theoretical stages can be calculated without using a graphical method as that shown

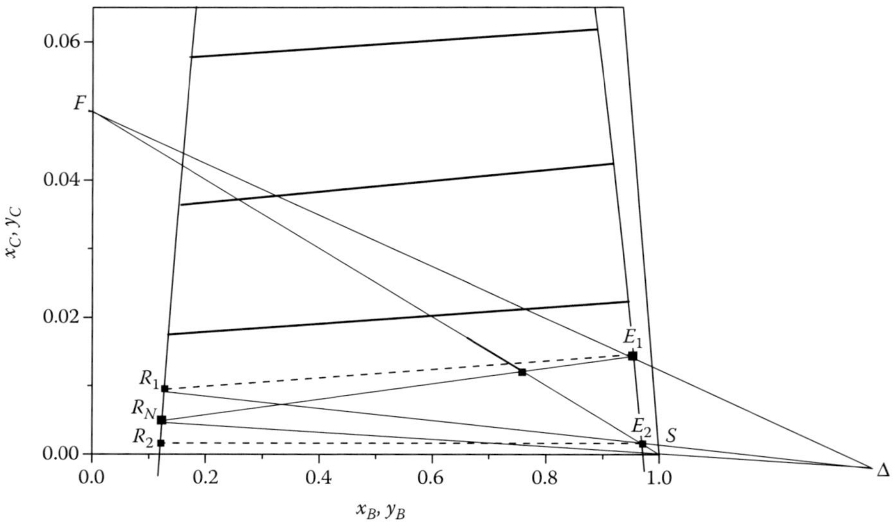

Figure 3.16 Phase diagram for countercurrent extraction.

in Figure 3.16. If the quantity of solute is small and only this component distributes between the two liquid phases, the extract (E) and raffinate (R) streams remain approximately constant along the whole equipment, and their values are equal to the solvent (S) and feed streams (F), respectively. In this case the operating curve will be a straight line. The equilibrium curve can be obtained from liquid–liquid equilibrium data, and for a diluted system, this line is also a straight one. Even for a relatively large quantity of solute, the equilibrium curve may still be a straight line if the solute concentrations in both liquid phases are calculated in a solute-free basis. In this concentration basis (solute-free), the operating line also remains straight, provided that diluent and solvent are immiscible liquids and only the solute is transferred from one phase to the other. In this case the number of theoretical stages can be calculated by Equation 3.11 (Treybal 1980).

$$N = \frac{\log\left[\dfrac{x_{C,F} - y_{C,S}/m}{x_{C,R_N} - y_{C,S}/m}(1-A) + A\right]}{\log(1/A)} \tag{3.11}$$

where N = number of theoretical stages
$x_{C,F}$ = solute (C) concentration in the feed stream F
x_{C,R_N} = solute concentration in the raffinate stream R_N
$y_{C,S}$ = solute concentration in the solvent stream
m = slope of the straight equilibrium line

A is the extraction factor, calculated by Equation 3.12, using the values of the feed (F) and solvent (S) streams.

$$A = \frac{F}{m \cdot S} \tag{3.12}$$

As an example, Equation 3.11 was used for estimating the number of theoretical stages required for palm oil deacidification by liquid–liquid extraction. Consider a CPO with 2 mass% of free fatty acids. In this case the solute concentration in the feed stream is $x_{FFA,F} = 0.02$, and the concentration in the raffinate stream should be decreased to $x_{FFA,R_N} = 0.003$ (0.3 mass%), so that the final product can be considered a refined oil. Equilibrium data for this system were measured by Gonçalves and Meirelles (2004) using ethanol with 6.10 mass% of water as the solvent at a temperature of 318.15 K. The corresponding equilibrium line is given by Equation 3.13, whose slope is $m = 1.067$ ($R^2 = 0.995$). Taking into account that aqueous ethanol contains no fatty acids ($y_{FFA,S} = 0.0$), the number of theoretical stages was calculated for different solvent stream/feed stream ratios (*S/F*) and is represented in Figure 3.17.

$$y_{FFA,E} = 1.067.x_{FFA,R} \tag{3.13}$$

Figure 3.17 shows that the number of stages decreases as the *S/F* ratio increases. In other words, for a higher number of stages less solvent is required to reach the desirable oil deacidification.

On the other hand, the results shown in Figure 3.17 should be considered as estimated values. As indicated by the equilibrium data (Gonçalves and Meirelles

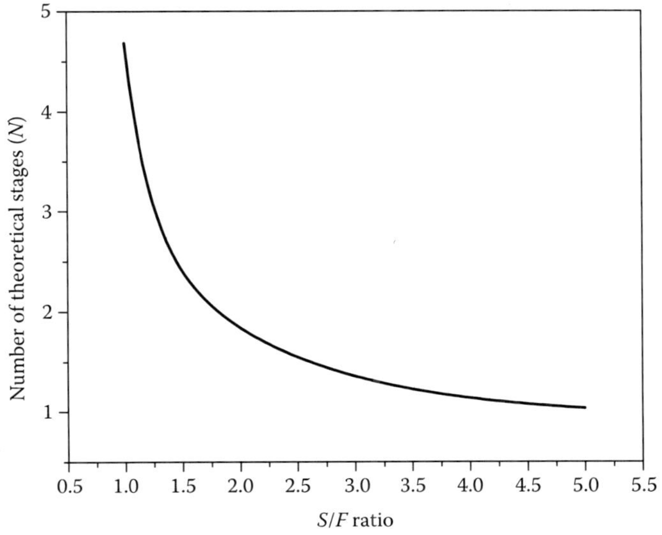

Figure 3.17 Number of theoretical stages as a function of *S/F* ratio in palm oil deacidification by liquid–liquid extraction.

2004), the equilibrium curve is a straight line, and the presence of water in the alcoholic solvent reduces the amount of neutral oil dissolved in the extract phase. Nevertheless, the quantity of ethanol transferred to the raffinate stream is not negligible, so the requirement that the operating curve be a straight line is not completely fulfilled.

For comparison we calculated the number of theoretical stages using both approaches for the following conditions: $x_{FFA,F} = 0.02$, $x_{FFA,R_N} = 0.002$, and *S*/*F* ratio = 1.25. According to Equation 3.11, a number of theoretical stages equal to 3.1 are required for reaching the desirable palm oil deacidification. Using the graphical method, a number equal to 3 was obtained. The deviation between the two methods was only 3%.

For the same conditions shown in Figure 3.17, we calculated the retention of nutraceuticals during palm oil deacidification by liquid–liquid extraction. The corresponding retention results should also be considered as estimated values. In fact, the aforementioned approach is based on the distribution of a single component between the two phases, and the inclusion of carotenes and tocopherols represents two further components, besides free fatty acids, whose transference is being estimated. However, these minor compounds are present in such a small quantity that the estimated retention values are probably reliable.

The partition of nutraceuticals (*i*) in the aforementioned liquid–liquid system was measured by Gonçalves et al. (2007). The corresponding equilibrium curves are also straight lines given by Equation 3.14. For carotenes, *m* is equal to 0.0173, and for tocopherols its value is 0.290.

$$y_{i,E} = mx_{i,R} \tag{3.14}$$

The approach based on two immiscible liquids also generates Equation 3.15, which allows the calculation of the solute composition in the raffinate. In this way the composition of nutraceuticals in the deacidified oil (x_{i,R_N}) was calculated and the corresponding retention estimated.

$$\frac{x_{i,F} - x_{i,R_N}}{x_{i,F} - y_{i,S} / m} = \frac{(1/A)^{N+1} - 1/A}{(1/A)^{N+1} - 1} \tag{3.15}$$

Figure 3.18 shows the retention of carotenes and tocopherols for palm oil containing initially 500 mg/kg carotenes and 1000 mg/kg tocopherols. Figure 3.18 shows that both retentions of nutraceuticals decrease as the *S*/*F* ratio increases. The impact in the retention of nutraceuticals is more significant in the case of tocopherols than for carotenes. This occurs because the distribution coefficient for tocopherols is much higher than for carotenes. In the case of using an *S*/*F* ratio equal to 1.25, the refined oil still retains 489 mg/kg carotenes (retention of 97.8%) and 648 mg/kg tocopherols (retention of 64.8%). These results indicate that liquid–liquid extraction is a promising technique for oil deacidification with the benefit of improving the quality of the final product.

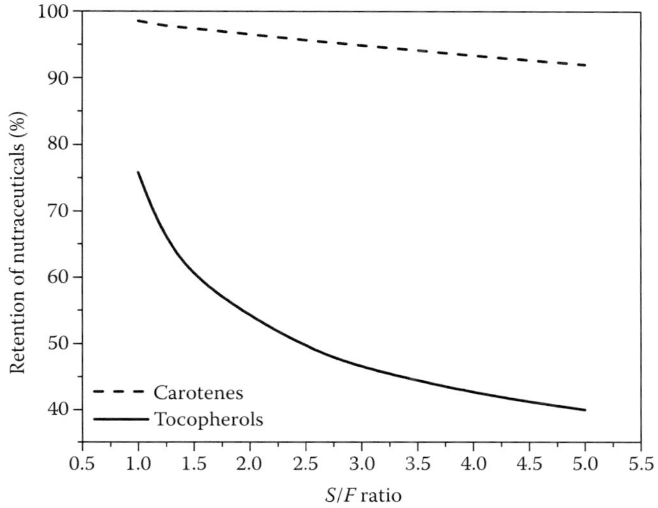

Figure 3.18 Retention of nutraceuticals in oil deacidification by liquid–liquid extraction.

3.4 FUNDAMENTALS OF ADSORPTION APPLIED TO THE RECOVERY OF NUTRACEUTICALS

Adsorption is a separation process in which a solute, originally dissolved in a fluid stream, is transferred to a solid phase and attached to the exposed surface of this solid by physical and/or chemical interactions. The ion-exchange processes, in which a reversible reaction occurs between the active ionic sites distributed along the solid phase of an ion-exchange resin and some of the ions dissolved in the fluid stream, are also included in this terminology.

The solid phase retaining the solute is called the *adsorbent*. There are several porous solids with a large interfacial area per unit of volume, including the surface area related to the pore walls inside the solid particles, which work as adsorbents, for example, the zeolites and molecular sieves used for dehydrating gaseous and vapor mixtures. Other examples are activated carbon, activated alumina, and silica gel. There are also synthetic resins, either of an exclusively adsorptive kind without specific active sites or ion-exchange ones containing ionic sites that can capture ions of opposite charge dissolved in the fluid stream.

The solute transferred from the fluid to the solid phase is called the *adsorbate*. In the processing of food and natural products, the fluid streams containing the adsorbate are usually liquid extracts, although in other areas adsorptive processes are also used for purifying or for recovering components from gaseous or vapor streams. The present item is focused on the use of adsorption for processing liquid extracts of interest in the production of nutraceuticals and nutritional foods.

The type of interaction that binds the adsorbate to the adsorbent is used for classifying different kinds of adsorption. In the case of physical adsorption (physisorption), the solute can be maintained at any point on the surface of the adsorbent by the relatively weak Van der Waals forces, and the adsorption can occur in multiple layers distributed over the whole solid surface. Chemical sorption, also called *chemisorption*, occurs when a chemical bond is formed between the molecules of solute and active sites present on the solid surface. In this case the solute molecules are attached to specific points on the surface of the adsorbent, not at any place on its surface. The heat of adsorption involved is large, similar to the energy of chemical bonds, and usually only a single layer of solute molecules is adsorbed.

Whereas some authors consider ion exchange as chemical adsorption (Ibarz and Barbosa-Canovas 2003), other researchers opt for classifying it as electrostatic adsorption (Inglesakis and Poupoulos 2006). In fact, coulomb attractive forces are the interactions responsible for attachment of the ionic species, transferred from the liquid stream to the charged functional groups that are part of the solid structure. Similar to chemical adsorption, the ionic species adsorbed during ion-exchange processes are connected only to the active sites of the solid phase. Furthermore, ion exchange involves a reversible reaction in which an ionic species linked to the active sites of the solid phase is displaced by ionic species dissolved in the liquid stream according to a stoichiometric relationship defined by the charges of both species. Ion-exchange resins are classified according to the type of ions they can capture from the liquid phase. For example, strong anionic resins are efficient in the adsorption of anions originating from the dissociation of weak acids, such as carbonate, bicarbonate, and acetate. On the other hand, weak anionic resins have greater adsorption capacity but are more appropriate for adsorbing anions derived from strong acids, such as sulfate and chloride. Cationic resins have anionic sites distributed over their surfaces that are able to capture cations from the liquid phase, such as calcium, magnesium, and sodium. The adsorption of anions displaces hydroxyl (OH^-) species originally linked to the anionic resin active sites, so that the pH of the liquid phase increases. In the case of cationic resins, H^+-ions are displaced by the cations present in the fluid stream, so that the pH decreases.

The design of new adsorptive processes and the evaluation of old ones requires a range of knowledge related to phase equilibrium, the kinetics of diffusion and convective mass transfer, and possible equipment configurations. Some of these aspects will be discussed in Sections 3.4.1 to 3.4.3.

3.4.1 Phase Equilibrium in Adsorptive Processes

The contact of an adsorbent with an adsorbate dissolved in a fluid medium generates the transference of the solute to the solid surface and its attachment onto this surface. If enough contact time is allowed—a situation more likely to occur in batch systems—stable solute concentrations in the liquid and solid phases are attained, corresponding to the adsorption equilibrium for the specific system under consideration, composed of adsorbent, adsorbate, and solvent medium and the corresponding fixed temperature selected for the experiments.

The function that relates the solute concentration in the solid phase to its remaining value in the fluid media at the selected equilibrium temperature is called the *adsorption isotherm*. The Langmuir and Freundlich isotherms are two of the most used equations found in the literature for representing adsorption equilibrium data. The Langmuir isotherm is given by Equation 3.16:

$$q_e = \frac{q_{max} \cdot K_L \cdot C_e}{1 + K_L \cdot C_e} \tag{3.16}$$

where q_e = equilibrium concentration of the adsorbate in the solid phase
q_{max} = maximum uptake capacity of the solid phase, that is, the maximum adsorbate concentration that can be retained by the adsorbent
C_e = equilibrium concentration of adsorbate in the liquid phase
K_L = Langmuir adsorption equilibrium constant

Usually q_e and q_{max} are given in milligrams of adsorbate per gram of solid phase, and C_e is expressed in milligrams of adsorbate per liter of solution. In this case K_L is expressed in $L \times mg^{-1}$.

The Langmuir model is based on the assumption that the adsorbent contains fixed individual sites able to adsorb a single molecule per site, so that q_{max} corresponds to the total monolayer capacity of the adsorbent. This model can be derived from kinetic considerations supposing that the net adsorption of the solute depends on the difference of its adsorption and desorption rates; when both rates are equal and the net adsorption attains a value equal to zero, a dynamic equilibrium between solid and liquid phases is obtained.

Another model for adsorption equilibrium is the Freundlich isotherm, given by Equation 3.17. This model does not assume the formation of a monolayer and can be used for adsorbents with heterogeneous surface-containing sites with different adsorption potentials.

$$q_e = K_F \cdot (C_e)^n \tag{3.17}$$

where q_e and C_e = same meaning already given previously
K_F and n = Freundlich constants

Using units such as $mg \times g^{-1}$ and $mg \times L^{-1}$ for q_e and C_e, respectively, n is dimensionless and K_F is expressed in $mg \times (g{\cdot}(mg \times L^{-1})^{1/n})^{-1}$.

Figure 3.19 shows the Langmuir isotherms for the adsorption of β-carotene from hexane solutions onto silica gel, according to data reported by Ahmad et al. (2009). Figure 3.20 shows the Freundlich isotherms for the adsorption of phenolic and flavonoid compounds from hydroalcoholic plant extracts onto the macroporous resin XAD-7, according to data reported by Silva et al. (2007).

Although ion-exchange equilibrium can also be represented in terms of the equilibrium constant associated with the exchange reaction, the corresponding equilibrium data are also frequently represented using one of the previously noted

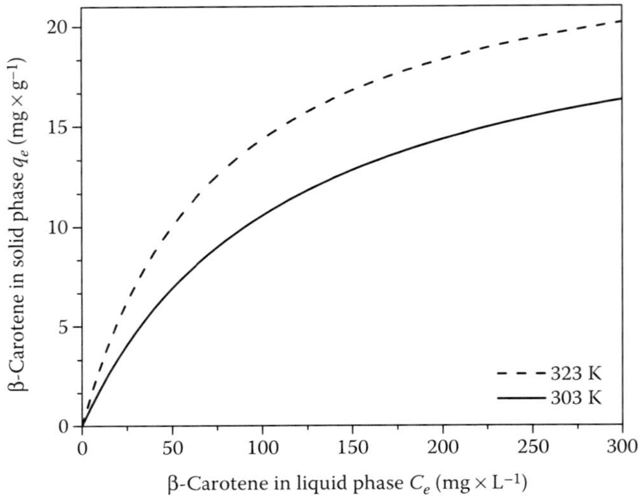

Figure 3.19 Langmuir isotherms for β-carotene adsorption onto silica gel. (From Ahmad, A. L., C. Y. Chan, S. R. A. Shukor, and M. D. Mashitah, *Chem. Eng. J.*, 148, 378, 2009. With permission.)

isotherms, especially the Langmuir equation. Cren and Meirelles (2005) used the Langmuir isotherm to describe the adsorption of oleic acid from alcoholic solutions onto the strong anionic resin Amberlyst A26-OH (Dow, Midland, Michigan). The exchange reaction is represented by Equation 3.18:

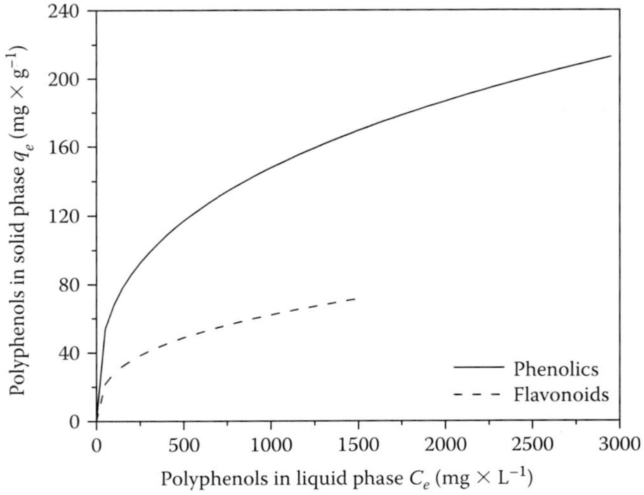

Figure 3.20 Freundlich isotherms for polyphenol adsorption onto resin XAD-7. (From Silva, E. M., D. R. Pompeu, Y. Larondelle, and H. Rogez, *Sep. Purif. Tech.*, 53, 274, 2007. With permission.)

$$E^{\alpha+}OH^-_{\alpha} + \beta R - COO^- + \beta H^+ \leftrightarrow E^{\alpha+}OH^-_{\alpha-\delta}R - COO^-_{\delta} + (\beta - \delta)\,R - COO^- + \beta H^+ + \delta OH^-. \quad (3.18)$$

where $E^{\alpha+}OH^-_{\alpha}$ = anionic resin in its OH-form, containing α active sites
$R - COO^-$ = fatty acid in its dissociated form
$E^{\alpha+}OH^-_{\alpha\text{-}\delta}R - COO^-_{\delta}$ = resin with δ sites occupied by the acid anion

The corresponding Langmuir isotherm for oleic acid adsorption by ion exchange is given in Figure 3.21. In this case, as well as in the case of the β-carotene adsorption shown in Figure 3.19, the solid-phase concentration is expressed in milligrams of adsorbate per gram of dry solid phase.

As can be seen in Figure 3.21, the ion-exchange isotherm is steeply inclined, indicating that adsorption of the oleate anions by the resin is favorable. This is normal behavior for ion-exchange adsorption, as a consequence of the high specificity of the resin active sites in capturing species of the opposite charge dissolved in the liquid medium. In fact, isotherms with an upward convex shape, such as those shown in Figures 3.19 to 3.21, are favorable to adsorption. In contrast, isotherms with an upward concave shape are unfavorable because in this case the corresponding adsorbents only work well with large solute concentrations in the liquid phase.

Because the equilibrium behavior in adsorption processes of interest in the processing of nutraceuticals and functional foods has been discussed, information concerning equipment used on an industrial scale and their operation modes will be discussed in the next section.

3.4.2 Equipment and Operation Modes

Adsorptive processes are usually carried out with the solid phase organized in a fixed bed that is continuously percolated by the liquid stream, as in the scheme

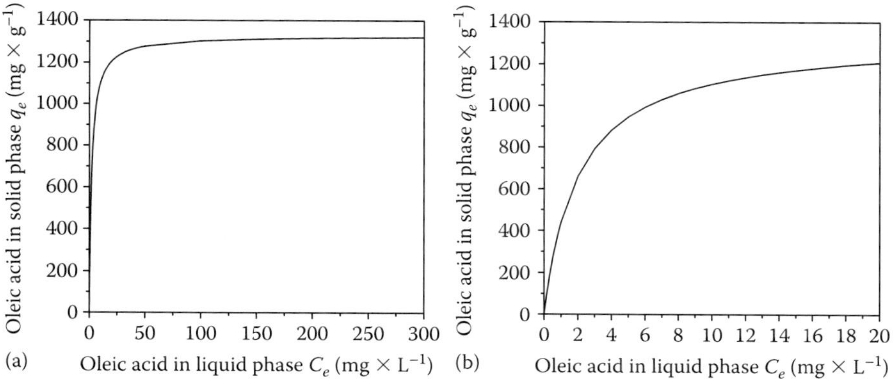

Figure 3.21 Langmuir isotherms for oleic acid adsorption onto the anionic resin Amberlyte A26-OH (a) and the corresponding zoom for lower concentrations (b). (From Cren, E. C., and A. J. A. Meirelles, *J. Chem. Eng. Data*, 50, 1529, 2005. With permission.)

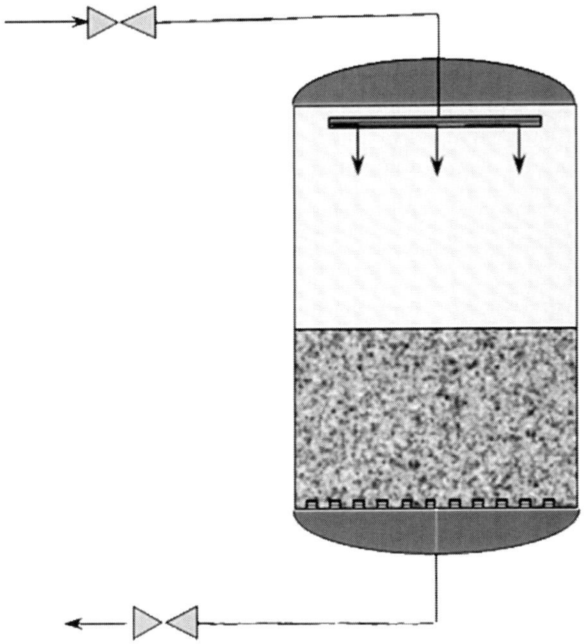

Figure 3.22 Scheme of typical equipment for fixed-bed adsorption.

shown in Figure 3.22. The equipment is a vertical cylindrical shell charged with the desired amount of solid phase, resting on a support with a series of small holes at the bottom of the cylinder, allowing for the exit of the fluid stream without carrying the small beads of the adsorbent with it. The most common operation mode is to feed the liquid stream in at the top of the equipment, so that the liquid phase flows downward, covering the entire solid bed. In fact, the solid bed should be totally immersed in the liquid phase, and air and gas bubbles must be avoided inside the bed to guarantee better contact between both phases.

The fixed-bed operation mode is also a batch operation in which the frontline of the solid-phase concentration migrates from the top of the adsorbent bed downward, as shown in Figure 3.23. During this migration the bed can be divided into three

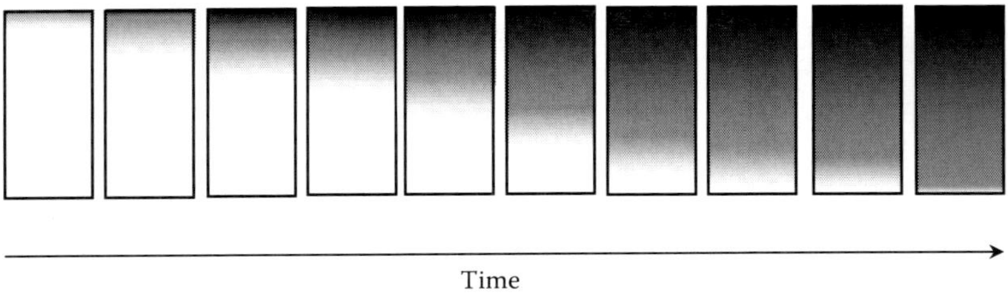

Figure 3.23 Migration of the solid-phase concentration as adsorption evolves.

concentration zones: the bottom one almost free of solute adsorption because this part is being percolated by a fluid stream that already contains no adsorbate; the top one practically saturated with the adsorbate and percolated by the fluid stream without net mass transfer; and the intermediate zone, which corresponds to the real adsorption or exchange zone, where the solute mass transfer is in fact occurring. At every instant the solid-phase concentration in this intermediate zone is changing from a value corresponding to saturation at its top borderline to an almost null concentration at its bottom borderline. This zone migrates downward with time as an adsorption wave and finally reaches the bottom of the bed, where the solute concentration begins to increase in the fluid stream that exits the equipment (Treybal 1980).

The line representing the exit concentration as a function of time or volume of the processed liquid stream is known as the *breakthrough curve*. Figure 3.24 shows two typical breakthrough curves with the exit concentration expressed as a dimensionless concentration obtained by dividing the instantaneous exit concentration, C_t, by the input concentration of the liquid phase, C_o. The solid curve represents an idealized step function, in which the exit concentration is equal to zero until the solute adsorption reaches and saturates the bottom of the solid bed, but then increases instantaneously to a value equal to the input concentration, or to 1.0, on the dimensionless concentration scale. The ideal step curve corresponds to a situation in which the intermediate zone mentioned previously reduces to a borderline between the top saturation region and the bottom zone with no adsorption. When this borderline reaches the bottom of the solid bed, the saturation of the adsorbent is complete and the exit concentration jumps to the input value. Such a situation requires an infinitely rapid mass transfer rate.

Real breakthrough curves show a behavior similar to that indicated by the dashed line in Figure 3.24. In this case the "leakage" of solute into the exit stream begins to occur well before the solid bed is saturated. The exact form of the breakthrough

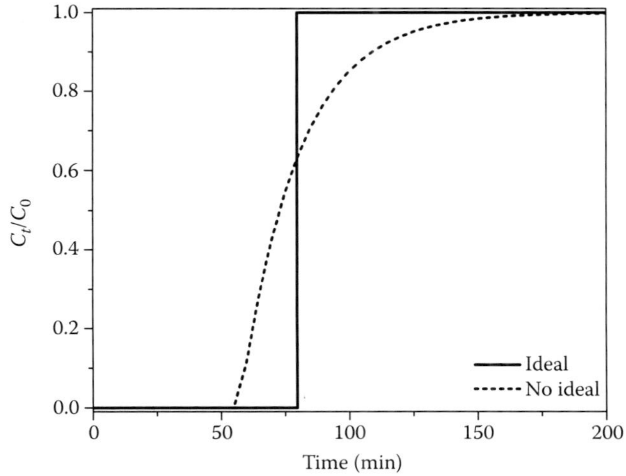

Figure 3.24 Typical breakthrough curves.

curve depends on a series of factors related to the mass transfer mechanism, the type of equilibrium between adsorbent and adsorbate, and the liquid-phase flow conditions through the solid bed because such aspects define the size of the intermediate (or mass transfer) zone mentioned previously.

Figure 3.25 shows an actual breakthrough curve divided into three subregions: the first one evolving from the start of the process up to the breakthrough point indicated by the time t_{break}, the second one limited by t_{break} and t_{end}, and the third one obtained after the solid bed achieved equilibrium with the liquid stream, indicated by the exhaustion time t_{end}. The exhaustion time t_{end} is defined by the first instant at which $C_t/C_o \cong 1.0$ and the continuation of the adsorption process after this instant has no practical purpose, but even in the second region the continuation of the adsorption process already compromises the recovery of the solute or the quality of the exit stream. For this reason, adsorption processes are only carried out until the breakthrough point is reached. The breakthrough point t_{break} is defined as the instant at which the outlet stream reaches a specific adimensional concentration, usually a low value within the range $0.01 \le C_{t_{break}}/C_o \le 0.10$, but the exact specification has some arbitrariness. In the present case we will assume that t_{break} corresponds to an adimensional concentration $C_{t_{break}}/C_o = 0.10$. Based on these definitions, the following equations for calculating the amount of solute can be formulated:

$$S_{break} = C_o \cdot Q \cdot \left(t_{break} - \int_0^{t_{break}} \frac{C_t}{C_o} \cdot \mathrm{d}t \right) \tag{3.19}$$

$$S_{end} = C_o \cdot Q \cdot \left(t_{end} - \int_0^{t_{end}} \frac{C_t}{C_o} \cdot \mathrm{d}t \right) \tag{3.20}$$

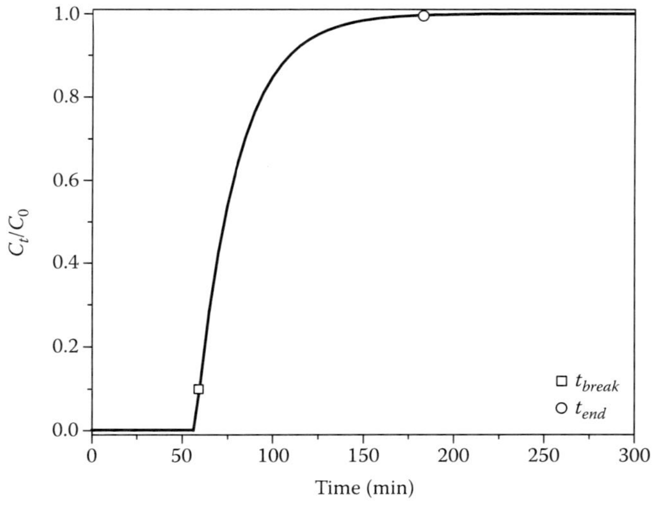

Figure 3.25 The breakthrough curve and the exhaustion and breakthrough points.

where S_{break} and S_{end} = amounts of solute adsorbed before the breakthrough and exhaustion points, respectively, were achieved, both expressed in milligrams of adsorbate
Q = volumetric flow rate ($L \times min^{-1}$)

Considering that fixed-bed adsorption is carried out using a constant volumetric flow rate of the liquid phase, the breakthrough curves can be represented as evolving in time or as a function of the accumulated volume of processed liquid stream, without any significant change in Equations 3.19 and 3.20. In fact, the volume of processed liquid stream, V_l, is given by the term $V_l = Q \cdot t$, so that Equation 3.21 can replace Equation 3.19.

$$S_{break} = C_o \cdot \left(V_{l,break} - \int_0^{V_{l,break}} \frac{C_t}{C_o} \cdot dV_l \right) \tag{3.21}$$

As is clear from the previous definitions, in real breakthrough curves part of the resin capacity cannot effectively be used because all the adsorption occurring after the breakthrough point should be avoided in industrial practice. On the other hand, S_{break} and S_{end} only coincide in the case of an idealized step function, whose corresponding breakthrough time t_{ideal} also coincides with $t_{break} = t_{end}$.

The exhaustion time t_{end} indicates that the total amount of solute adsorbed by the whole solid bed corresponds to the equilibrium concentration q_o with a feed stream concentration C_o. Considering that the step function is defined by the following conditions: $\forall t < t_{ideal} \rightarrow C_t/C_o = 0.0$ and $\forall t \geq t_{ideal} \rightarrow C_t/C_o = 1.0$, the integration of Equations 3.19 and 3.20 gives Equation 3.22. Furthermore, dividing the amount of solute adsorbed before the breakthrough point for an ideal curve, S_{ideal}, by the mass of adsorbent m_s should give the uptake capacity of the solid phase q_o under the prevailing equilibrium conditions, as indicated by Equation 3.23. Care should be taken in expressing the mass of adsorbent adequately because, as mentioned previously, the uptake capacity of adsorbents is sometimes given as the mass of adsorbate per mass of dry solid phase, and in this case m_s in Equation 3.23 must be replaced by the mass of dry adsorbent $m_{s,dry}$.

$$S_{ideal} = C_o \cdot Q \cdot t_{ideal} \tag{3.22}$$

$$\frac{S_{ideal}}{m_s} = \frac{C_o \cdot Q \cdot t_{ideal}}{m_s} = q_o \tag{3.23}$$

Note that Equation 3.23 allows for the estimation of the amount of adsorbent m_s for an ideal breakthrough curve, provided a value for the corresponding breakthrough time is selected, and the solid-phase uptake capacity under the prevailing equilibrium conditions is known.

The amount of adsorbent m_s and the volume of the solid phase are related by the adsorbent bulk density ρ_b, defined as the mass of solid divided by the total bed volume V_b. In this way, the bed length corresponding to an ideal breakthrough curve L_{ideal} can be determined from Equation 3.24:

$$L_{ideal} = \frac{V_b}{A} = \frac{m_s}{\rho_b \cdot A} \quad (3.24)$$

where A = cross sectional area of the equipment containing the solid bed

As already observed, if solid-phase concentrations are expressed on a dry weight basis, the mass of adsorbent in Equation 3.24 must take into account the humidity of the solid X_s, given in mass of solvent per mass of dry solid, according to Equation 3.25:

$$m_s = m_{s,dry} \cdot (1 + X_s) \quad (3.25)$$

A practical approach for scaling up adsorption processes makes use of the previous equations in the following way. Based on an actual breakthrough curve obtained using a pilot-scale column, the length of unused bed L_{nu} can be evaluated (Cooney 1999). This bed height corresponds to that part of the adsorption process that would occur between t_{break} and t_{end}, so that the actual solid bed should have this additional height to adsorb, up to the breakthrough time, the same total amount of solute S_{ideal} adsorbed in the case of an ideal breakthrough curve. Considering the aforementioned aspects, the actual bed length L_{actual} for a full-scale process, conducted under conditions similar to those used in the pilot test, can be estimated according to Equation 3.26:

$$L_{actual} = L_{ideal} + L_{nu} = L_{ideal} + (1 - \frac{t_{break}}{t_{ideal}}) \cdot L_{ideal} = L_{ideal} \cdot (2 - \frac{t_{break}}{t_{ideal}}) \quad (3.26)$$

Once the breakthrough point is reached, the adsorbent should be regenerated so that a new adsorption cycle can be initiated. For this reason fixed-bed operations frequently involve two parallel operating solid beds, one in the adsorption period and the other in the desorptive recovery of the adsorbent, so that the whole process can be conducted continuously.

The use of a larger number of fixed beds in the adsorption period, operating in series, so that the liquid stream percolates through the sequence of solid beds, allows for better usage of the adsorption capacity. Consider a set of n fixed beds, one in desorptive recovery and the other (n-1) beds in the adsorption step. The liquid phase should first be fed into the fixed bed closest to its exhaustion time and flow from this through the sequence of (n-1) beds in the direction of the fixed bed containing the most recently recovered adsorbent, so that the whole set of fixed beds simulates countercurrent contact with the best distribution of mass transfer driving force. Because the n adsorbent beds are in fact fixed beds, with no movement or flow of the solid phase, the inlet and outlet positions, through which the liquid streams are fed into and withdrawn from each bed, are switched at regular time intervals to simulate countercurrent contact, always guaranteeing that the liquid stream flows through the sequence of beds from the most saturated to the least saturated one. When a bed reaches complete exhaustion, it is excluded from the sequence and submitted to desorptive recovery, whereas the most recently recovered bed is reintroduced into the sequence as the last one through which the liquid stream percolates. At every switching of the liquid stream inlets and outlets, a determined bed "moves" one step closer to the point at which the liquid stream is fed into the apparatus, that is, the first step in the sequence of fixed beds. At the next switching, the bed that reaches the first step will be separated and directed to adsorbent recovery. In this way it is possible to use almost the whole adsorbent capacity and also avoid any risk of solute "leakage."

The use of similar schemes and even of more sophisticated ones, the latter case being applied to the fractionation and purification of mixtures containing two or more solutes of importance, is a topic of growing interest in the pharmaceutical and biotechnological areas. Such equipment, known as simulated moving beds, include two inlet streams, the mixture to be fractionated and the eluent for adsorbent recovery, and two outlet streams, extract and raffinate, which are rich in different solutes contained in the feed stream.

3.4.3 Modeling Breakthrough Curves

As indicated previously, most adsorption and ion-exchange processes are carried out in fixed-bed operations. Besides the practical approach for scaling up adsorption processes mentioned previously, more rigorous alternatives are based on the formulation and solving of differential mass balance equations for the adsorbate. The integration of such equations allows an appropriate representation of breakthrough curves experimentally determined under different conditions, as well as the prediction of fixed-bed operations for design purposes. In the present item we will present some basic formulations for the mass balance equation and discuss one of its mathematical solutions and the corresponding results in terms of the calculated breakthrough curves.

The phenomenological model was developed using the following assumptions: (1) there is a single adsorbable component in the liquid phase, the liquid stream is fed into the fixed bed under constant conditions of concentration and temperature, and the solute concentration in the feed stream is sufficiently small for the liquid-phase velocity through the solid bed to be unaffected by the adsorption process; (2) the process conditions are isothermal and isobaric; (3) the physical properties of the fluid and solid phases are constant; (4) the column porosity (extraparticle void fraction) is constant, and the particle porosity (intraparticle void fraction) is negligible; and (5) plug flow conditions prevail, so that axial and radial dispersions can be neglected.

Under such assumptions the mass balance equation for the solute, applied to an element of volume in the adsorption column, takes into account the depletion of the solute concentration in the liquid phase as a result of adsorption (first term in Equation 3.27), solute accumulation in the solid phase (second term), and the change in solute concentration resulting from liquid-phase flow (third term), assuming the following form:

$$\frac{\partial C}{\partial t} + \rho_b \frac{1}{\varepsilon}\frac{\partial q}{\partial t} = -v\frac{\partial C}{\partial z} \tag{3.27}$$

where C (mg × L^{-1}) and q (mg/g adsorbent) = solute concentrations in the liquid and solid phases, respectively

ε = column void fraction (volume voids per total bed volume)
ρ_b = bed bulk density (mass of adsorbent per total bed volume, g × L^{-1})
v = interstitial velocity of the liquid phase (dm × min^{-1})
t = time (min)
z = axial distance coordinate (dm)

Integration of Equation 3.27 allows one to calculate how the solute concentration in the liquid and adsorbent phases evolves with time and varies with the axial position in the bed. Evolution of the fluid-phase concentration at the fixed-bed exit, that is, the breakthrough curve, can also be obtained. To integrate Equation 3.27 one must define the initial and boundary conditions, and, in addition, relate the rate of solute uptake by the adsorbent $\partial q / \partial t$ to either the liquid- (C) or solid-phase (q) concentrations.

Considering that the adsorbent and liquid phases within the fixed bed are initially free of solute and the feed stream has a constant concentration C_o at the bed entrance, the following initial and boundary conditions can be defined: $\forall z$ and $t = 0 \Rightarrow C(z,0) = 0$ and $q(z,0) = 0$. and $\forall t$ and $z = 0 \Rightarrow C(0,t) = C_o$. On the other hand, the relation of $\partial q / \partial t$ to C or q defines the type of solution obtained and the kind of physical situation described.

The easiest assumption is that local equilibrium prevails at every axial position along the fixed bed, so that the concentrations of the solid and liquid phases are related by the equilibrium curve (isotherm) along the entire fixed bed. This assumption requires that the mass transfer rate between liquid and solid phases be necessarily fast. Nevertheless, in most practical situations, the mass transfer rate is controlled either by diffusion in the solid phase or by film resistance in the liquid phase, and, in the case of macroporous solids, pore diffusion can also be considered in an explicit way. Detailed discussions on the mass transfer mechanisms during adsorption and some of the corresponding ways of solving Equation 3.27 can be found in Perry and Green (1999) and Cooney (1999).

Our attention has been especially focused on an important case for the adsorption of relatively large molecules, in which case the mass transfer control by diffusion in the solid phase is more probable. In this case an approximate analytical solution is possible if one assumes a linear driving force (LDF) approach for describing the concentration profile within the solid phase and an adsorption isotherm with a constant separation factor F_s less than 1.

The LDF approach can be seen in Equation 3.28 and assumes that the rate of solute uptake by the adsorbent is proportional to the difference between the local adsorbent equilibrium concentration and the instantaneous concentration of the adsorbent at the corresponding axial position. The proportionality constant is the rate coefficient k_n, expressed as min^{-1}.

$$\frac{\partial q}{\partial t} = k_n \cdot (q_e - q) \tag{3.28}$$

The separation factor F_S is calculated using Equation 3.29:

$$F_S = \frac{X_e / (1 - Y_e)}{Y_e / (1 - X_e)} \tag{3.29}$$

where X and Y = dimensionless concentrations of the liquid and solid phases

X and Y are obtained by dividing the corresponding original concentrations by those of the reference, in the present case selected as the liquid-phase input concentration C_o and the corresponding equilibrium concentration in the solid phase q_o, as indicated by Equations 3.30 and 3.31:

$$X = \frac{C}{C_o} \tag{3.30}$$

$$Y = \frac{q}{q_o} \tag{3.31}$$

Note that in Equation 3.29 the separation factor is defined in a way similar to that of relative volatility in distillation and selectivity in liquid–liquid extraction. In fact, it corresponds to the reciprocal of these variables. The F_s values allow for the following isotherm classification: $F_s < 1$ corresponds to favorable isotherms, $F_s > 1$ to unfavorable ones, $F_s = 1$ is obtained for linear isotherms, and $F_s = 0$ for irreversible ones.

Thus, the dimensionless version of the Langmuir isotherm can be written as:

$$Y_e = \frac{X_e}{F_s + (1 - F_s) \cdot X_e} \tag{3.32}$$

In this case the separation factor is given by:

$$F_s = \frac{1}{1 + K_L \cdot C_o} \tag{3.33}$$

Note that the convex Langmuir isotherm, adimensionalized using a specific and constant feed stream concentration C_o, gives a constant separation factor that is less than 1. In this case the approximate solution for Equation 3.27, already expressed in terms of the dimensionless concentration of the liquid-phase exit stream (breakthrough curve), is given as:

$$\frac{1}{1 - F_s} \cdot \text{In} \left[\frac{(1 - X)}{X^{F_s}} \right] + 1 = N \cdot (1 - \tau) \tag{3.34}$$

where

$$N = \frac{k_n \cdot \wedge \cdot V_b}{Q} \tag{3.35}$$

$$k_n = \frac{15 \cdot \Psi \cdot D_s}{r_p^2} \tag{3.36}$$

$$\wedge = \frac{\rho_b \cdot q_o}{C_o} \tag{3.37}$$

$$\Psi = \frac{0.894}{1 - 0.106 \cdot F_s^{0.5}} \tag{3.38}$$

$$\tau = \frac{t - \frac{V_b \cdot \varepsilon}{Q}}{\frac{V_b \cdot \wedge}{Q}} \tag{3.39}$$

where N = solid-phase number of transfer units and corresponds to an adimensionalized form of the rate coefficient k_n

Λ = partition ratio and corresponds to the volumetric uptake capacity of the adsorbent divided by the input liquid concentration (i.e., the dimensionless uptake capacity of the solid phase)

Ψ = correction factor for improving the predictive capacity of this approximated solution

D_s = diffusion coefficient for the adsorbate in the solid phase ($dm^2 \times min^{-1}$)

r_p = radius of the adsorbent particles (dm)

τ = a kind of dimensionless time

The other variables were already defined previously.

The previous model was used for predicting breakthrough curves in the case of the adsorption of oleic acid by ion exchange with the strong anionic resin Amberlyst A26-OH (Dow). The following conditions were considered: a fixed-bed volume of V_b = 0.250 L, porosity ε = 0.39, and containing resin particles with an average radius of $r_p = 3.0 \cdot 10^{-3}$ dm. The bed is percolated by a solution containing oleic acid dissolved in azeotropic ethanol, with an initial concentration C_o = 8000 mg × L^{-1} (≅1 mass% of acidity), temperature of 298.15 K, and volumetric flow rate $Q = 2.5 \times 10^{-2}$ L × min^{-1}. The Langmuir isotherm is shown in Figure 3.21 and has the following parameters: K_L = 0.4925 L × mg^{-1} and q_{max} = 349.5 mg × (g of wet resin)$^{-1}$, corresponding to 1329 mg of acid × (g of dry resin)$^{-1}$.

Most values were taken from Cren et al. (2009), except for the liquid-phase concentration that was decreased to guarantee an almost constant interstitial velocity,

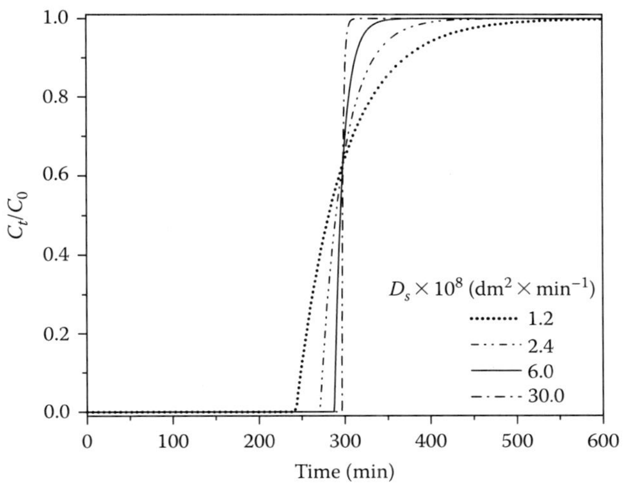

Figure 3.26 Breakthrough curves for different solid-phase diffusivities.

despite the solute transference to the solid phase. The separation factor can be calculated via Equation 3.33 and has the following value: $F_s = 2.5 \cdot 10^{-4}$.

Cren et al. (2009) also modeled experimental breakthrough curves for this system using the LDF approach. They solved the partial differential equation system, with no approximation, using the method of lines and adjusting a single rate coefficient k_n for the entire set of 11 curves, obtained for different oleic acid concentrations and volumetric flow rates. The k_n value obtained was $2.54 \cdot 10^{-2}$ min^{-1}. Considering that the average size of the resin was $r_p = 3.0 \times 10^{-3}$ dm, the solid-phase diffusivity was estimated by Equation 3.36 as $D_s = 1.5 \times 10^{-8}$ dm$^2 \times$ min^{-1}.

Figure 3.26 shows the breakthrough curves predicted for solid-phase diffusivities varying within the range $(1.2–30) \cdot 10^{-8}$ dm$^2 \times$ min^{-1}. As is clear from Figure 3.26, larger diffusion coefficients decrease the length of unused bed, improving utilization of the resin uptake capacity. For the largest value tested, the breakthrough curve showed behavior close to an idealized step function, whose ideal breakthrough time would be $t_{ideal} = 295$ minutes. In this case the behavior obtained was similar to the local equilibrium approach mentioned previously.

Figure 3.27 shows the breakthrough curves predicted for different separation factors F_s varying within the range from $2.5 \cdot 10^{-4}$ to $4.0 \cdot 10^{-1}$ for a diffusion coefficient $D_s = 6.0 \times 10^{-8}$ dm$^2 \times$ min^{-1}. In the case of adsorption by ion exchange, the Langmuir constant K_L usually has a high value, so that the separation factor tends to be small, such as the lowest value in the range indicated previously. On the other hand, in simple adsorptive cases, K_L is not so low and the separation factor is not so favorable. For example, in the case of the adsorption of β-carotene by silica gel, F_S would have a value of approximately 0.18 according to the data reported by Ahmad et al. (2009) (see Figure 3.19), and an initial concentration $C_o = 500$ mg $\times$ L^{-1}, a value similar to the carotene content of CPO. It should be observed that for separation

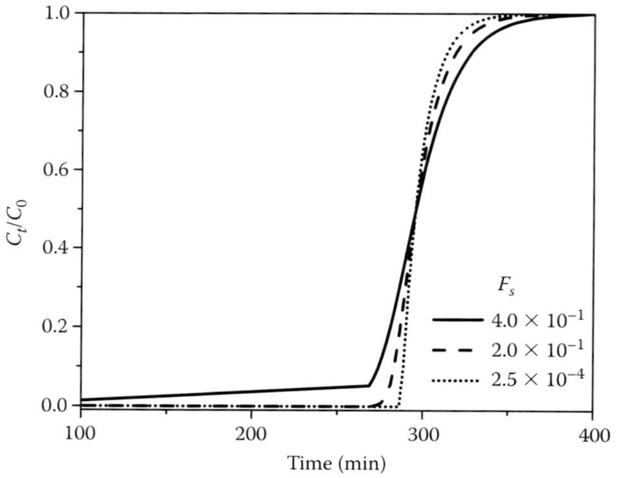

Figure 3.27 Breakthrough curves for different separation factors.

factors defined according to Equation 3.29, lower values mean isotherms more favorable for adsorption, and in the limit case of $F_s \rightarrow 0$, the isotherm tends to be an irreversible one. For this reason larger separation factors increase the length of unused bed and advance the breakthrough time significantly, as shown in Figure 3.27.

According to the prevailing mass transfer mechanism, breakthrough curves can be predicted in the way indicated previously or by other strategies of integrating Equation 3.27 as suggested in the literature (Cooney 1999; Perry and Green 1999). By this phenomenological approach or by using the concept of the length of unused bed, one has the main tools for designing and/or evaluating the performance of adsorption processes carried out in a batch way.

3.5 CONCLUDING REMARKS

In the present chapter, two different mass transfer unit operations, liquid–liquid extraction and adsorption, were presented as alternative processes for purifying liquid foods and/or recovering valuable components from liquid solutions with minimal losses of nutraceutical compounds. In fact, both separation processes are usually carried out under mild conditions of temperature, so that in this first step they are able to preserve components of nutritional value. Nevertheless, liquid–liquid extraction necessarily requires the use of an appropriate solvent, and adsorption processes are often carried out with the original solution dissolved in a selected solvent.

To recover the final product as a pure one, such solvent must be stripped of the obtained liquid stream. Fortunately, the most appropriate solvents are usually light components, such as anhydrous or hydrated short-chain alcohols in the case of liquid–liquid extraction, or water, alcohols, or hydrocarbons in the case of adsorption processes. This means that they can be easily stripped at relatively low temperatures, especially when the liquid streams are evaporated or distilled under vacuum conditions. Anyway, in the development of such processes special care should be taken during the solvent selection to guarantee that the subsequent stripping conditions are mild enough for preserving the quality of nutraceuticals and functional foods.

ACKNOWLEDGMENTS

The authors wish to acknowledge FAPESP (Fundação de Amparo à Pesquisa do Estado de São Paulo, 2008/56258-8 and 2008/09502-0), CNPq (Conselho Nacional de Desenvolvimento Científico e Tecnológico, 306250/2007 and 480992/2009-6), and CAPES/PNPD (Coordenação de Aperfeiçoamento de Pessoal de Nível Superior–Programa Nacional de Pós-Doutorado) for their financial support.

REFERENCES

Agalias, A., P. Magiatis, A. L. Skaltsounis, E. Mikros, A. Tsarbopoulos, E. Gikas, I. Spanos, and T. Manios. 2007. A New Process for the Management of Olive Oil Mill Waste Water and Recovery of Natural Antioxidants. *Journal of Agricultural and Food Chemistry* 55: 2671–76.

Ahmad, A. L., C. Y. Chan, S. R. A. Shukor, and M. D. Mashitah. 2009. Adsorption Kinetics and Thermodynamics of Beta-Carotene on Silica-Based Adsorbent. *Chemical Engineering Journal* 148: 378–84.

Antoniassi, R. 1996. *Deacidification of Corn Oil with Ethanol in a Rotating Disc Column (RDC)*. PhD diss., State University of Campinas, Brazil.

Antoniassi, R., W. Esteves, and A. J. A. Meirelles. 1998. Pretreatment of Corn Oil for Physical Refining. *Journal of the American Oil Chemists' Society* 75: 1411–15.

Batista, E., R. Antoniassi, M. R. Wolf-Maciel, and A. J. A. Meirelles. 2002. Liquid-Liquid Extraction for Deacidification of Vegetable Oils. Paper presented at the International Solvent Extraction Conference (ISEC 2002), Cape Town, South Africa, March 2002, 638–43.

Batista, E., S. Monnerat, K. Kato, L. Stragevitch, and A. J. A. Meirelles. 1999a. Liquid-Liquid Equilibrium for Systems of Canola Oil, Oleic Acid, and Short-Chain Alcohols. *Journal of Chemical & Engineering Data* 44: 1360–64.

Batista, E., S. Monnerat, L. Stragevitch, C. G. Pina, C. B. Gonçalves, and A. J. A. Meirelles. 1999b. Prediction of Liquid-Liquid Equilibrium for Systems of Vegetable Oils, Fatty Acids, and Ethanol. *Journal of Chemical & Engineering Data* 44: 1365–69.

Blass, E., and W. Goettert. 1994. Selection of Extractors and Solvents. *Liquid-Liquid Extraction Equipment*. Edited by J. C. Godfrey and M. J. Slater. Chichester, UK: John Wiley & Sons, 737–67.

Bono, A., C. C. Ming, and M. Sundang. 2007. Liquid Phase Adsorption of α-Tocopherol by Activated Carbon. *Journal of Applied Sciences* 7: 2080–83.

Ceriani, R., and A. J. A. Meirelles. 2006. Simulation of Continuous Physical Refiners for Edible Oil Deacidification. *Journal of Food Engineering* 76: 261–71.

Chan, K. W., B. S. Baharin, and Y. B. C. Man. 2000. Adsorption Isotherm Studies of Palm, Carotene Extraction by Synthetic Polymer Adsorbent. *Journal of Food Lipids* 7: 127–41.

Chen, T. H., and G. F. Payne. 2001. Separation of Alpha- and Delta-Tocopherols Due to an Attenuation of Hydrogen Bonding. *Industrial & Engineering Chemistry Research* 40: 3413–17.

Cherukuri, R. S. V., R. Cheruvanky, I. Lynch, and D. L. McPeak. 1999. Process for Obtaining Micronutrient Enriched Rice Bran Oil. US Patent: 5,985,344, filed August 31,1998, and issued November 16, 1999.

Cheryan, M. 1998. *Ultrafiltration and Microfiltration Handbook*, 2nd ed. Boca Raton, FL: CRC Press.

Chung, Y., O. Chu, D. Lineback, M. Pepper, and M. Perez Alvarez. 2003. Single Strength Juice Deadification by Ion-Exchange Column. Patent: WO 03/028485 A1, filed September 20, 2002, and issued April 10, 2003.

Ciesla, U., and F. Schüth. 1999. Ordered Mesoporous Materials. *Microporous and Mesoporous Materials* 27: 131–49.

Cooney, D. O. 1999. *Adsorption Design for Wastewater Treatment*. Boca Raton, FL: CRC Press.

Cren, E. C., L. Cardozo, E. A. Silva, and A. J. A. Meirelles. 2009. Breakthrough Curves for Oleic Acid Removal from Ethanolic Solutions Using a Strong Anion Exchange Resin. *Separation and Purification Technology* 69: 1–6.

Cren, E. C., and A. J. A. Meirelles. 2005. Adsorption Isotherms for Oleic Acid Removal from Ethanol Plus Water Solutions Using the Strong Anion-Exchange Resin Amberlyst A26 OH. *Journal of Chemical & Engineering Data* 50: 1529–34.

Cuevas, M. S., C. E. C. Rodrigues, and A. J. A. Meirelles. 2009. Effect of Solvent Hydration and Temperature in the Deacidification Process of Sunflower Oil Using Ethanol. *Journal of Food Engineering* 95:291–97.

Daun, H. 2005. Produce Color and Appearance. *Produce Degradation: Pathways and Prevention.* Edited by O. Lamikanra, S. H. Iman, and D. Ukuku. Boca Raton, FL: Taylor & Francis, 191–222.

Decloux, M., and J. Coustel. 2005. Simulation of a Neutral Spirit Production Plant Using Beer Distillation. *International Sugar Journal* 107: 628–43.

Frank, T. C., L. Dahuron, B. S. Holden, W. D. Prince, A. S. Seibert, and L. C. Wilson. 2008. Liquid-Liquid Extraction and Other Liquid-Liquid Operations and Equipment. *Perry´s Chemical Engineers´ Handbook.* Edited by R. H. Perry and D. W. Green. New York: McGraw-Hill, 1–105.

Geankoplis, C. J. (1993). *Transport Processes and Unit Operations.* Upper Saddle River, NJ: Prentice Hall PTR.

Godfrey, J. C., and M. J. Slater. 1994. *Liquid-Liquid Extraction Equipment.* Chichester, UK: John Wiley & Sons.

Gonçalves, C. B. 2004. *Phase Equilibrium of Systems Composed by Vegetable Oils, Fatty Acids and Hydrated Ethanol*. PhD diss., State University of Campinas, Brazil.

Gonçalves, C. B., E. Batista, and A. J. A. Meirelles. 2002. Liquid-Liquid Equilibrium Data for the System Corn Oil + Oleic Acid + Ethanol + Water at 298.15 K. *Journal of Chemical & Engineering Data* 47: 416–20.

Gonçalves, C. B., and A. J. A. Meirelles. 2004. Liquid–Liquid Equilibrium Data for the System Palm Oil + Fatty Acids + Ethanol + Water at 318.2 K. *Fluid Phase Equilibria* 221: 139–50.

Gonçalves, C. B., P. A. Pessôa Filho, and A. J. A. Meirelles. 2007. Partition of Nutraceutical Compounds in Deacidification of Palm Oil by Solvent Extraction. *Journal of Food Engineering* 81: 21–26.

Hamm, W. 1992. Liquid-Liquid Extraction in Food Processing. *Science and Practice of Liquid-Liquid Extraction*. Editedd By J. D. Thornton. Oxford, UK: Clarendon Press, vol. 2, 309–52.
Hartmann, M., A. Vinu, and G. Chandrasekar. 2005. Adsorption of Vitamin E on Mesoporous Carbon Molecular Sieves. *Chemistry of Materials* 17: 829–33.
Ibarz, A., and G. V. Barbosa-Canovas. 2003. *Unit Operations in Food Engineering*. Boca Raton, FL: CRC Press.
Inglesakis, V. J., and S. Poupoulos. 2006. *Adsorption, Ion Exchange, and Catalysis: Design of Operations and Environmental Applications*. Amsterdam: Elsevier.
Karan, A. 1998. Separation of Oryzanol from Crude Rice Bran Oil. https://tspace.library.utoronto.ca/bitstream/1807/15517/1/MQ40908.pdf.
Kim, J., J. S. Godber, J. King, and W. Prinyawiwatkul. 2001. Inhibition of Cholesterol Autoxidation by the Nonsaponifiable Fraction in Rice Bran in an Aqueous Model System. *Journal of the American Oil Chemists' Society* 78: 685–89.
Kisler, J. M., A. Dahler, G. W. Stevens, and A. J. O'Connor. 2001. Separation of Biological Molecules Using Mesoporous Molecular Sieves. *Microporous and Mesoporous Materials* 44: 769–74.
Latip, R. A., B. S. Baharin, Y. B. C. Man, and R. A. Rahman. 2000. Evaluation of Different Types of Synthetic Adsorbents for Carotene Extraction from Crude Palm Oil. *Journal of the American Oil Chemists' Society* 77: 1277–81.
Latip, R. A., B. S. Baharin, Y. B. C. Man, and R. A. Rahman. 2001. Effect of Adsorption and Solvent Extraction Process on the Percentage of Carotene Extracted from Crude Palm Oil. *Journal of the American Oil Chemists' Society* 78: 83–87.
Leibovitz, Z., and C. Ruckenstein. 1983. Our Experience in Processing Maize (Corn) Germ Oil. *Journal of the American Oil Chemists' Society* 60: 347A–51A.
Li, M., P. J. Pham, C. U. Pittman, and T. Y. Li. 2008. Selective Solid-Phase Extraction of α-Tocopherol by Functionalized Ionic Liquid-Modified Mesoporous SBA-15 Adsorbent. *Analytical Sciences* 24: 1245–50.
Lineback, D., O. Chu, Y. Chung, M. Pepper, and M. Perez Alvarez. 2003. Juice Deadification. Patent: WO 03/028483 A2, filed September 20, 2002, and issued April 10, 2003.
Ma, M. H., and C. I. Lin. 2004. Adsorption Kinetics of Beta-Carotene from Soy Oil Using Regenerated Clay. *Separation and Purification Technology* 39: 201–09.
Marshall, W. E., and J. I. Wadworth. 1994. *Rice Science and Technology*. New York: Marcel Dekker Inc.
Mayamol, P. N., C. Balachandran, T. Samuel, A. Sundaresan, and C. Arumughan. 2007. Process technology for the production of micronutrient rich red palm olein. *Journal of the American Oil Chemists' Society* 84: 587–96.
Orthoefer, F. T. 1996. Rice Bran Oil: Healthy Lipid Source. *Food Technology* 50: 62–64.
Patel, M., and S. N. Naik. 2004. Gamma-Oryzanol from Rice Bran Oil—A Review. *Journal of Scientific & Industrial Research* 63: 569–78.
Perry, R. H., and D. W. Green. 1999. *Perry's Chemical Engineers' Handbook*. New York: McGraw-Hill.
Pina, C. G. 2001. *Performance of a Rotating Disc Column in the Deacidification of Corn Oil*. PhD diss., State University of Campinas, Brazil.
Pina, C. G., and A. J. A. Meirelles. 2000. Deacidification of Corn Oil by Solvent Extraction in a Perforated Rotating Disc Column. *Journal of the American Oil Chemists' Society* 77: 553–59.
Plonis, G. F., and J. Trujillo-Quijano. 1993. Process for Obtaining an Oil with Adjustable Carotene Content from Palm Oil. Patent: EP0529107A1, filed June 19, 1991, and issued March 03, 1993.

Reipert, E. C. D'A. 2005. *Deacidification of Babassu and Cottonseed Oils by Liquid-Liquid Extraction*. Master's diss., State University of Campinas, Brazil.

Robbins, L. A., and R. W. Cusack. 1997. *Liquid-Liquid Extraction Operations and Equipment. Perry's Chemical Engineers' Handbook.* Edited R. H. Perry and D. W. Green. New York: McGraw-Hill, 1–47.

Rodrigues, C. E. C. 2004. *Deacidification of Rice Bran Oil by Liquid-Liquid Extraction* (in Portuguese). PhD diss., State University of Campinas, Brazil.

Rodrigues, C. E. C., R. Antoniassi, and A. J. A. Meirelles. 2003. Equilibrium Data for the System Rice Bran Oil + Fatty Acids + Ethanol + Water at 298.2 K. *Journal of Chemical & Engineering Data* 48: 367–73.

Rodrigues, C. E. C., A. Filipini, and A. J. A. Meirelles. 2006a. Phase Equilibrium for Systems Composed by High Unsaturated Vegetable Oils + Linoleic Acid + Ethanol + Water at 298.2 K. *Journal of Chemical & Engineering Data* 51: 15–21.

Rodrigues, C. E. C., and A. J. A. Meirelles. 2008. Extraction of Free Fatty Acids from Peanut Oil and Avocado Seed Oil: Liquid-Liquid Equilibrium Data at 298.2 K. *Journal of Chemical & Engineering Data* 53: 1698–704.

Rodrigues, C. E. C., M. M. Onoyama, and A. J. A. Meirelles. 2006b. Optimization of the Rice Bran Oil Deacidification Process by Liquid-Liquid Extraction. *Journal of Food Engineering* 73: 370–78.

Rodrigues, C. E. C., E. C. D. Peixoto, and A. J. A. Meirelles. 2007. Phase Equilibrium for Systems Composed by Refined Soybean Oil + Commercial Linoleic Acid + Ethanol + Water, at 323.2K. *Fluid Phase Equilibria* 261: 122–28.

Rodrigues, C. E. C., P. A. Pessôa Filho, and A. J. A. Meirelles. 2004. Phase Equilibrium for the System Rice Bran Oil + Fatty Acids + Ethanol + Water + γ-Oryzanol + Tocols. *Fluid Phase Equilibria* 216: 271–83.

Rodrigues, C. E. C., E. C. D. Reipert, A. F. Souza, P. A. Pessôa Filho, and A. J. A. Meirelles. 2005a. Equilibrium Data for Systems Composed by Cottonseed Oil + Commercial Linoleic Acid + Ethanol + Water + Tocopherols at 298.2 K. *Fluid Phase Equilibria* 238: 193–203.

Rodrigues, C. E. C., F. A. Silva, A. Marsaioli Jr., and A. J. A. Meirelles. 2005b. Deacidification of Brazil Nut and Macadamia Nut Oils by Solvent Extraction: Liquid-Liquid Equilibrium Data at 298.2 K. *Journal of Chemical & Engineering Data* 50: 517–23.

Sa, L. A. 2007. *Deacidification of Soybean Oil by Liquid-Liquid Extraction*. Master's diss., State University of Campinas, Brazil.

Sabah, E. 2007. Decolorization of Vegetable Oils: Chlorophyll-A Adsorption by Acid-Activated Sepiolite. *Journal of Colloid and Interface Science* 310: 1–7.

Scordino, M., A. Di Mauro, A. Passerini, and E. Maccarone. 2004. Adsorption of Flavonoids on Resins: Cyanidin 3-Glucoside. *Journal of Agricultural and Food Chemistry* 52: 1965–72.

Scordino, M., A. Di Mauro, A. Passerini, and E. Maccarone. 2005. Selective Recovery of Anthocyanins and Hydroxycinnamates from a Byproduct of Citrus Processing. *Journal of Agricultural and Food Chemistry* 53: 651–58.

Shin, T. S., J. S. Godber, D. E. Martin, and J. H. Wells. 1997. Hydrolytic Stability and Changes in E Vitamers and Oryzanol of Extruded Rice Bran during Storage. *Journal of Food Science* 62: 704–28.

Silva, E. M., D. R. Pompeu, Y. Larondelle, and H. Rogez. 2007. Optimisation of the Adsorption of Polyphenols from Inga edulis Leaves on Macroporous Resins Using an Experimental Design Methodology. *Separation and Purification Technology* 53: 274–80.

Swoboda, P. A. T. 1985. Refining of Palm Oils. Patent: GB2144143, filed June 28, 1984, and issued February 27, 1985.

Taguchi, A., and F. Schüth. 2005. Ordered Mesoporous Materials in Catalysis. *Microporous and Mesoporous Materials* 77: 1–45.

Treybal, R. 1981. *Mass Transfer Operations*. Singapore: McGraw-Hill.

Vinu, A., K. Z. Hossian, P. Srinivasu, M. Miyahara, S. Anandan, N. Gokulakrishnan, T. Mori, K. Ariga, and V. V. Balasubramanian. 2007. Carboxy-mesoporous Carbon and Its Excellent Adsorption Capability for Proteins. *Journal of Materials Chemistry* 17: 1819–25.

Zhao, D., Y. Peidong, H. Qisheng, F. C. Bradley, and D. S. Galen. 1998. Topological Construction of Mesoporous Materials. *Current Opinion in Solid State & Materials Science* 3: 111–21.

CHAPTER 4

Application of Enzyme-Assisted Oil Extraction Technology in the Processing of Nutraceuticals and Functional Food

María Elvira Zúñiga, Carmen Soto, Jacqueline Concha, and Eduardo Pérez

CONTENTS

4.1 INTRODUCTION

The market of nutraceuticals and functional foods has grown steadily in recent decades. Substances such as oils and fatty acids, peptides, phytoestrogens, phytosterols, carotenoids, dietary fiber, phenolic compounds, oligosaccharides, and probiotics, among many others, are associated with the prevention and treatment of diseases, and have led to the development of a new niche of consumers who demand products with additional health benefits.

Fat and oils are food components in high demand used in products such as margarine, creams, mayonnaise, shortening, and some functional foods (e.g., oils rich in omega-3 fatty acids). The use of enzyme-assisted processes, which improve release of free oil under mild operational conditions as an additional step in conventional techniques such as cold pressing, aqueous extraction, or organic solvent extraction, is fundamental to obtain oil of better quality and greater nutritional value. By-products of the disruption of cell walls by the enzymatic treatment also have a high valorization potential to obtain bioactive compounds, such as oligosaccharides, lignans, ferulic acids, phenolic compounds, peptides, and oligopeptides.

This chapter provides a background on the effects of enzymatic treatments in the extraction of edible and nonedible oil in terms of yield and quality. The first part is dedicated to health benefits associated with consumption of some oils, and the second part covers the effect of operational conditions during the enzymatic treatment on oil yield and quality. The third part deals with bioactive compounds recovered from defatted meal and their potential applications in human health.

4.1.1 Nutraceuticals

Isolation, purification, identification, and characterization of biologically active components with health benefits are of utmost interest in the food and nutraceutical industries. Benefits of functional foods and nutraceuticals include prevention and treatment of cardiovascular diseases, cancer, diabetes, and inflammation; retardation of the aging process; and enhancement of immune response.

Phytochemicals such as phenolics, carotenoids, and dietary fiber have been recognized because of their antioxidant, anticarcinogenic, and antimutagenic activities, among other health-promoting properties. Growing interest in the substitution of synthetic antioxidants with natural ones has led to tremendous development in the screening of natural antioxidants in residual sources from agricultural industries (Larrauri et al. 1996; Lu and Foo 2000; Moure et al. 2001; Chau and Huang 2003; Wolfe and Liu 2003; Ozkan et al. 2004; Alasalvar et al. 2009; Butsat et al. 2009).

Phenolic compounds are widely distributed in the plant kingdom as secondary metabolites, which are derivatives of pentose phosphate, shikimate, and phenylpropanoid pathways. Structurally, phenolic compounds comprise an aromatic ring, bearing one or more hydroxyl substituents, ranging from simple phenolic molecules to highly polymerized compounds. Despite this structural diversity, this group of compounds is often referred to as *polyphenols* (Balasundram et al. 2006).

4.1.2 Nutraceutical Oils

Health benefits associated with oil consumption are not limited to monounsaturated fatty acid or polyunsaturated fatty acid (PUFA) as often discussed. Current research indicates that several compounds such as polyphenols and carotenes have emerged as important players in the protection against several diseases.

4.1.3 Fish Oil

Marine fish oil constitutes the most abundant source of n-3 PUFAs such as eicosapentaenoic acid (EPA) and docosahexaenoic acid (DHA). Health benefits associated with fish oil consumption are mostly mediated by EPA and DHA according to fish supplementation studies and DHA/EPA clinical studies (Harris et al. 2003; Calder 2004; Lovegrove et al. 2004; Gebauer et al. 2006; Harris and Bulchandani 2006; Metcalf et al. 2007; Calzolari et al. 2009; Lee et al. 2009; Meyer et al. 2009). Reduction of cardiovascular risk by n-3 PUFAs is derived from protection against pathological processes leading to cardiovascular disease, myocardial infection, and stroke; however, the mechanism of action of these compounds is still unclear (Calder 2004). Taking into consideration the labile nature of unsaturated fatty acids, marine fish oil must be extracted under mild conditions.

4.1.4 Olive Oil

At first, oleic acid from olive oil was considered key to cardioprotection properties associated with the Mediterranean diet; however, when it was demonstrated that oleic acid intake is similar between Mediterranean and Western diets, it was clear that health benefits could not be attributed to oleic oil consumption alone (Dougherty et al. 1987; Carluccio et al. 2003; Choi 2003; Psaltopoulou et al. 2004; Visioli et al. 2006; Brill 2009). First-pressed extra virgin olive oil contains high amounts of polyphenols that have been shown to inhibit lipid oxidation, protecting against formation of lipid peroxides and atherosclerotic plaques, and consequently decreasing cardiovascular risk (Visioli et al. 1995; Visioli and Galli 1998, 2002). Polyphenols in extra virgin olive oil such as hydroxytyrosol, oleuropein, and others have been associated with anticancer properties and the ability to decrease cholesterol levels and platelet aggregation (Visioli and Galli 1998; Tripoli et al. 2005; García-Villalba et al. 2010). Phenolic compounds from olive oil also have been found in human low-density lipoprotein, supporting their function against lipoperoxidation in vivo (de la Torre-Carbot et al. 2007). The fact that olive oil extraction is mostly done in mild conditions suggests that most of its bioactive components are conserved functional at the time of consumption, leading to its known health benefits.

4.1.5 Grape Seed Oil

Grape seed oil is particularly rich in linoleic acid (72%), oleic acid (16%), and vitamin E. The most abundant saturated acids are palmitic acid (7%) and stearic acid

(4%) (Kamel et al. 1985). Because of its high level of unsaturated fatty acids, it is popularly used in raw food preparations, pharmaceuticals, cosmetics, and medicine.

4.1.6 Palm Oil

Palm oil is the second most consumed vegetable oil in the world, and its use in moderate amounts has been associated with a lower risk of arterial thrombosis and atherosclerosis, lower biosynthesis of endogenous cholesterol, and lower platelet aggregation and blood pressure (Rand et al. 1988; de Bosch et al. 1996; Ebong et al. 1999; Kritchevsky et al. 2002; Wilson et al. 2005). Palm oil contains saturated and unsaturated fatty acids in similar amounts, each with distinctive and opposite effects in terms of cardiovascular risk; therefore, benefits associated with palm oil consumption are largely dependent on methods of extraction, preservation, and preparation (Edem 2002). In particular, red palm oil has been reported to be rich in carotenoids, tocopherols, tocotrienols, and lycopenes, and to have protective properties against ischemia and reperfusion injury typically associated with cardiovascular diseases (Esterhuyse et al. 2005; Kruger et al. 2007; van Rooyen et al. 2008).

4.1.7 Canola and Rapeseed Oil

Canola oil contains 55% oleic acid, 25% linoleic acid, 10% α-linolenate, and only 4% saturated fatty acids (Dupont et al. 1989). Consumption of canola oil has been associated with a significant hypolipidemic effect and prevention and protection from stroke, given its high antioxidant activity (Gorinstein et al. 2003; Ohara et al. 2008; Noroozi et al. 2009; Nguemeni et al. 2010).

4.1.8 Sesame Oil

Sesame oil has been shown to help in recovery from renal and hepatic injury (Hsu et al. 2007; Chandrasekaran et al. 2009; Gupta et al. 2009; Kuhad et al. 2009; Hsu et al. 2010). In particular, health benefits associated with consumption of this oil are thought to be mediated by sesamol (3,4-methylenedioxyphenol), a potent antioxidant that is hypothesized to result in lower levels of lipid peroxidation and decreased mitochondrial oxidative stress (Hsu and Liu 2004; Chandrasekaran et al. 2009).

4.1.9 Rice Bran Oil

Health benefits of consumption of rice bran oil are associated with its fatty acid composition (80% oleic and linoleic fatty acids) and the presence of γ-oryzanol and ferulic acid. Rice bran oil produces a hypolipidemic effect as a combination of reduction in cholesterol levels and high antioxidant activity (Rukmini and Raghuram 1991; Cicero and Gaddi 2001; Wilson et al. 2007). Additionally, rice bran oil has been shown to suppress hyperinsulinemic response, as well as improve lipid profiles in rats with induced type 2 diabetes mellitus (Chou et al. 2009). Polyphenols, phytosterols, tocopherols, and tocotrienols have also been hypothesized to provide health benefits associated with consumption of this oil (Cicero and Gaddi 2001).

4.1.10 Oils with Potential Health Benefits

Some claim that almond oil has anti-inflammatory, immune-boosting, and antihepatotoxicity effects (Ahmad 2010). Interestingly, it has been determined that improvement of plasma lipid profiles associated with almond consumption is mediated by components of the oil fraction of these nuts; however, no exact mechanism of action has been proposed to date (Hyson et al. 2002).

Evening primrose, borage, and black currant oils are rich in γ-linolenic acid and have been recognized as beneficial to health. Intake of evening primrose oil aids in treatment of heart disease, arthritis, skin problems, and premenstrual syndrome, among other benefits (Barre 2001). Borage and black currant oils also have been reported to stimulate the immune function in elderly people and decrease dermatitis and effects of degenerative diseases (Barre 2001; Kast 2001; Ward and Singh 2005).

Rose hip oil is rich in all-trans retinoic acid, a compound known for its anticancer properties (Avvisati and Tallman 2003; Orlandi et al. 2003). However, this compound is mostly known for its benefits in skin health, and therefore is typically used in topical applications (Kang et al. 2009; Shamban 2009; Tsai and Hsu 2009).

4.2 VEGETABLE OIL EXTRACTION PROCESSES AND ITS EFFECT ON PRODUCT QUALITY AND BIOACTIVE COMPOUNDS

Oil extraction from seeds and fruits is mostly done by mechanical pressing and organic solvent extraction; however, other processes such as water steam distillation and supercritical fluid extraction are also available.

Quality and stability of the extracted oils depend on multiple factors, such as composition, production, and storage conditions. Usually oil quality is measured in terms of free fatty acid content (acidity value), peroxide index, fatty acid composition, and antioxidant content. For defatted meal, quality is determined according to proximal composition and digestibility.

Oil extraction by pressing is mainly used before solvent extraction on seeds with oil content higher than 20%, taking into account that the mechanical process leaves a residual cake with 3%–4% of oil (Figure 4.1). In this regard, organic solvent extraction is often recommended for seeds with oil content less than 10% (Norris 1982; Ullmann 1985).

Organic solvent extraction has several disadvantages, such as risk of environmental contamination with volatile organic compounds, odor emission, hexane emanation, risk of fire or explosion in the compressor room, lower quality of the oil and residual meal as a result of the high temperature applied during the process, and high energy costs (Norris 1982; Hoffmann 1989; Topallar 2000).

Despite solvent extractions producing higher yields, good quality oil obtained by pressing justifies its use. Pressing extraction involves mechanical recovery of oil,

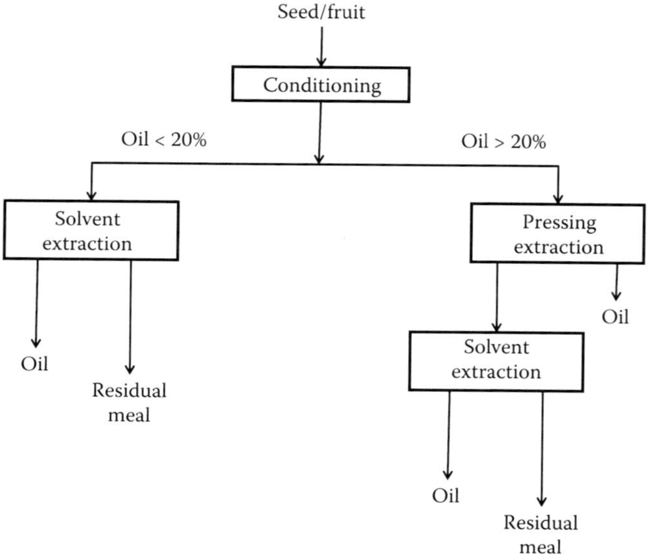

Figure 4.1 Vegetable oil extraction process.

which can be performed in batches (discontinuously) or in a continuous process. The main reasons for the popularity of mechanical oil extraction equipment are its simple construction and operation, its adaptability to different oilseed types, and a short time for oil production (Singh and Bargale 2000).

Because vegetable oil is located inside the cells and surrounded by other macromolecules, such as proteins and carbohydrates (e.g., starch, cellulose, hemicellulose, pectin), pretreatment is required to increase fluidity of the oil from inside the cell. This conditioning stage may include downsizing, rupture, milling, or heat/hydrothermal treatment by baking or steam application (Smith et al. 1993).

Commercial enzymes constitute an interesting alternative in oil extraction that allows improvement of yield and quality, obtaining oils with functional properties and a defatted meal that can be used as raw material for extraction of bioactive substances. Degradation of the cell wall polysaccharides accomplished with cellulases, hemicellulases, pectinases, and other enzymes results in decomposition of the structural polysaccharide matrix of the vegetable cell wall (Pinelo and Meyer 2008).

4.3 ENZYME INCORPORATION IN OIL EXTRACTION PROCESSES AND ITS EFFECT ON BIOACTIVE COMPOUNDS

Use of enzymes in the extraction of vegetable oils dates back to studies on olive oil in the 1950s (Martinez 1957). It was not until the early 1970s that several authors studied the effect of enzymes in the extraction of various vegetable oils (Montedoro

1973; Fantozzi 1977) and pointed out the potential application of cellulolytic enzymes, used in the food industry, in oil extraction processes (Weetall 1977).

Enzymes are highly efficient and specific biological catalysts that increase the speed of a natural reaction. They do not produce large amounts of undesired products and act generally under smooth environmental conditions, which is why complex systems of operation (that work under great security restrictions) are not required (Illanes 1994).

Enzymatic-aided processes to recover oil can be classified as enzymatic-assisted solvent extraction processes and enzymatic-assisted pressing processes. The first can be further categorized into enzyme-aided aqueous extraction processes and enzyme-aided organic solvent extraction processes. Additionally, enzyme incorporation can be done in a stage before the extraction or during the extractive stage (Dominguez et al. 1995; Rosenthal et al. 2001).

Aqueous oil extraction assisted by enzymes (Figure 4.2) is the most studied process because it has the advantage of not using toxic or dangerous compounds and produces high quality oils. Besides, its operation is uncomplicated, the discontinuous operation is easy, and the investment and operation costs are low. Using enzyme application in oil extraction by pressing (Figure 4.3), authors show that this process reduces the time of the mechanical treatment and provides higher yields of extraction and better quality of the products (Siniscalco 1988; Smith et al. 1993;

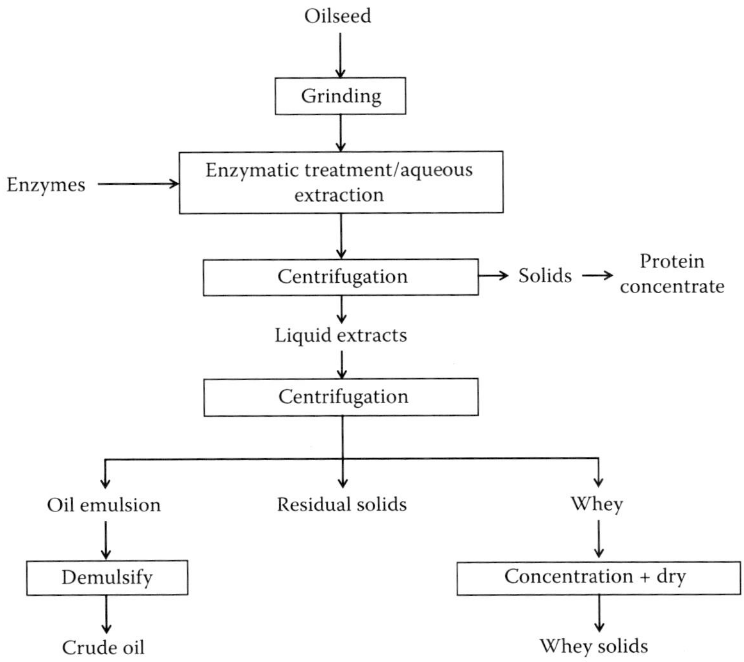

Figure 4.2 Enzyme-aided aqueous oil extraction process.

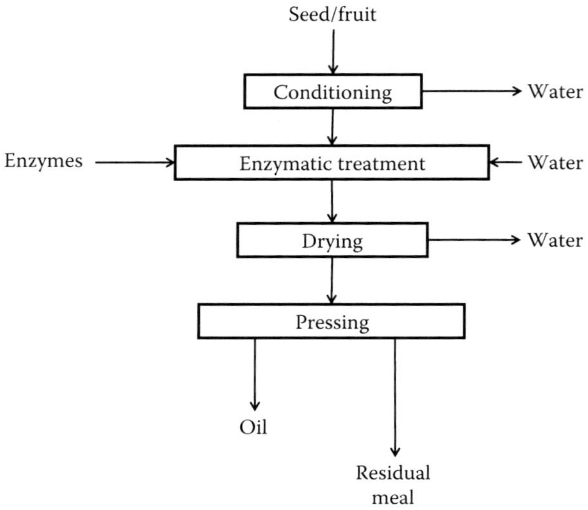

Figure 4.3 Oil extraction by enzyme-assisted cold pressing process.

Dominguez et al. 1994; Rosenthal et al. 2001; Vierhuis et al. 2001; Zúñiga 2001; Concha et al. 2004; Collao et al. 2007; Soto et al. 2007).

Soybean oil recovery using enzymes before organic solvent extraction increased oil availability when the treatment was performed at 45°C and a crude mixture of cellulose, hemicellulase, chitinase, xylanase, pectinase, and protease from *Aspergillus fumigatus* was used (Kashyap et al. 2007). The addition of proteases, especially ones that degrade oleosins (proteinic membranes) that enclose lipid inclusions, has proven to increase yield of conventional processes (Rosenthal et al. 2001). Fullbrook et al. (1983) evaluated soybean and rapeseed oil extraction by aqueous enzymatic hydrolysis, and it was shown that enzyme application increased oil extraction from 3.8% to 13% in rapeseed and from 2.6% to 9% in soybeans. Jung, Maurer, and Johnson (2009) determined that incorporation of 0.5% protease increased yield in oil extraction from 62% to 95%. Similarly, Smith et al. (1993) reported that application of a crude enzyme extract obtained from *A. fumigatus* increased yield by mechanical soybean oil extraction to 63.5%. The authors did not report data regarding oil quality, but they mentioned that enzymatic treatment maintains or improves oil quality.

Dandjouma et al. (2008) evaluated oil extraction from *Ricinodendron heudelotii* seeds using cellulose and protease before extraction with hexane. Best results were obtained with 0.3 g protease/100 g seed, increasing yield from 53% to 68% of extraction yield.

Application of cellulose, pectinase, or protease (0.5%) before steam distillation or hydrodistillation increased oil extraction yield from 0.28% to 0.57% in volatile garlic oil (Sowbhagya et al. 2009). Although authors did not report the evaluation of oil physicochemical parameters, they indicated that quality did not change with the application of enzymes. Sowbhagya et al. (2010) showed that viscozyme (pectinase plus β-glucanase) in a 0.5% enzyme/substrate (E/S) ratio produced higher increases in volatile oil extraction from celery seeds when the treatment was carried out during 60 minutes before extraction by distillation. The oil quality obtained was similar to those obtained by distillation alone, but oil recovery using cellulases showed a higher content of limonene, which has been considered an anticarcinogenic molecule (Uedo et al. 1999).

Aqueous enzymatic extraction of coconut oil has been reported by several authors. McGlone, López-Munguia, and Vern (1986) determined that it is possible to extract up to 80% of the oil using polygalacturonase, α-amylase, and proteases, obtaining a product with physicochemical characteristics within the required parameters demanded by the Mexican Official Standards for oil. Using the same seed, Olsen (1988) increased yield from 42% to 98% after incorporation of enzymes. Results obtained by Che Man et al. (1996) were not so successful but provided a higher yield than those obtained in the conventional process, with 41% efficiency after addition of four enzymes during extraction compared with 19% efficiency obtained in the control.

Conventional aqueous extraction of *Butyrospermum parkii* seed oil yields approximately 35%, producing a low quality product, molasses, and latex formation, which can clog equipment. Interestingly, incorporation of cellulases, hemicellulases, α-amylases, and proteases decreased these problems and increased yield of oil extraction up to 70% at the same time (Tano-Debrah and Ohta 1994). In *Jatropha curcas* oil extraction, a sequential use of ultrasonic conditioning followed by an aqueous treatment increased yield to 67%. This value increased to 74% when an alkaline protease was added and required less time (6 hours vs. 18 hours) (Shah et al. 2005).

Sunflower oil extraction by enzymatic-assisted aqueous processes has been evaluated (Latif and Anwar 2009). Use of viscozyme increased yield up to 87.7%, reflecting a 110% increase over control values. Enzyme addition improved substantially the oil quality, with a reduction of free fatty acid content (0.94 to 0.69) and peroxide index (1.78 to 1.36) compared with organic solvent extraction. In addition, the oil extracted by the enzymatic process had a higher content of total tocopherols (842 mg/kg_{oil}) than those extracted with organic solvent (799 mg/kg_{oil}) and those obtained in the control process without enzymes (778 mg/kg_{oil}).

García et al. (2001) reported that the formulations Olivex, Novoferm 12, and Glucanex (Novozymes, Bagsvaerd, Denmark) applied during the malaxation of olives before oil extraction slightly increased phenolic compounds, such as vanillic acid, vanillin, pinoresinol, and luteolin, in the oil.

Aqueous extraction of peanut oil with alcalase increased extraction yield to 73%, but there were no data regarding oil quality. In addition, a protein hydrolysate was obtained, which is composed of nearly 80% peptides with a molecular mass of less than 2000 Da, and most were oligopeptides (Jiang et al. 2010).

The extraction of rice bran oil has been evaluated using enzyme-assisted aqueous technology, increasing yield from 13% to 80% when a mixture of amylases, cellulases, and proteases is used (Sharma et al. 2001). Although the authors do not report the quality of the products obtained, the mild process operational conditions probably allowed the maintenance of a high content of polyunsaturated fatty acids

(40% linoleic acid) and other beneficial components such as vitamin E and tocotrienol, both characteristic of this oil.

Traditionally olive oil is extracted from fresh olives using a mechanical process without the use of excessive heat or any form of additives or solvents (Ranalli et al. 1999). One of the main steps of pressing extraction in olive oil is malaxation, which contributes to the activation of several enzymatic processes. Ranalli et al. (2001) reported that the addition of commercial enzyme preparations during malaxation reduced complexation of hydrophilic phenols with polysaccharides, increasing the concentration of free phenols in the olive paste and their consequent release into the oil and wastewaters during processing. A similar result was obtained by Aliakbarian et al. (2008), who studied the application of a mixture of three enzymes in the oil extraction process of Italian olive fruits, producing the best total phenolic amount (844.30 μg C_{AE}/g_{oil}), *o*-diphenols (120.80 μg C_{AE}/g_{oil}), antioxidant activity (24.93 μg 2,2-diphenyl-1-picrylhydrazyl [DPPH]/μL extract), and yield (15.72 g_{oil}/100 kg_{paste}). Another study carried out with aqueous extraction of virgin olive oil using industrial enzymes showed that enzyme concentration had a highly significant effect ($p < 0.01$) in the yield, color, turbidity, and total polyphenol content, but there were no significant effects on acidity, peroxide value, and iodine value ($p < 0.05$). Phenolic content in the oil increased from 13.9% to 72.6% (Najafian et al. 2009). Thus, enzyme incorporation in oil extraction has been shown to increase oil yield and sometimes quality, especially in terms of polyphenol concentration in the oil extracted (Najafian et al. 2009; Passos et al. 2009).

When commercial enzymes were used in an aqueous process on different species of canola seed, the enzymatic process produced an increase on the free fatty acid value (0.4%–0.6%) and peroxide index (0.3–0.4 nM/kg), but these changes were less severe than the ones caused, for example, by oil storage (Sosulski and Sosulski 1993). As summarized in Table 4.1, when enzymes are incorporated in an aqueous process, the oil yield is increased, but only in a few cases is the yield near 100%, as obtained using organic solvent extraction.

Regarding product quality, enzymatic-aided aqueous extraction produces oil with better organoleptic characteristics and higher oxidative stability. Domínguez et al. (1994) reported that in this type of enzymatic process, meals with high protein and low fiber content can be obtained from soybean seeds and also that antinutritional compounds could be observed in low quantities, although this fact is attributable to the aqueous process instead of the enzyme incorporation.

Commercial enzyme incorporation in extraction by cold pressing has been mainly focused in fruits and seeds, such as canola (rapeseed), soybean, rose hip, borage, Chilean hazelnut, evening primrose, and grape seed, among others (Sineiro et al. 1998; Moure et al. 2001; Zúñiga 2001a, 2001b, 2003; Guerra 2003; Santamaria et al. 2003; Concha et al. 2004, 2006; Soto et al. 2004, 2007; Tobar et al. 2005; Collao et al. 2007; Latif et al. 2007). Enzymatic treatment before pressing increases oil yield from 7% to 60% (as observed in Table 4.2), depending on the type of seed, its original oil content, and previous conditioning of the raw material, enzyme activity, and pretreatment conditions. The study of these variables allows establishment of the ability of enzymes to improve the extraction process. As will be discussed in

Table 4.1 Enzymatic Aqueous Oil Extraction

Seed or Fruit	Enzymatic Treatment	Oil Extraction Yield (%) Control	Oil Extraction Yield (%) With Enzyme	Reference
Coconut	Protease + polygalacturonase + amylase (1% E/S ratio, pH 4.0, 4 h, 40°C)	12	80	McGlone, López-Munguia, and Vernon (1986)
Coconut	Gamanase (1% E/S ratio, pH 4.5, 4 h, 50°C)	42	98	Olsen (1988)
Coconut	Protease + polygalacturonase + amylase + cellulase (0.1% E/S, pH 6.5, 0.5 h, 25°C)	19.3	41.67	Che Man et al. (1996)
Cocoa	Protease + hemicellulase (40°C)	41.6	65.5	Tano-Debrah and Ohta (1995)
Sunflower	Protease (2% E/S ratio, pH 5.0, 3 h, 50°C)	41.6	57.1	Badr and Sitohy (1992)
Cocoa	Protease + cellulose + hemicellulase (1% E/S, 6 h, 37°C)	58.7	72.7	Tano-Debrah and Ohta (1995)
Peanut	Protease (2.5% E/S ratio, 40°C, 6 h)	40	86	Sharma, Khare, and Gupta (2002)
Rice	Protease + cellulase + amylase (65°C, 20 h)	14	78	Sharma, Khare, and Gupta (2001)
Almond	Protease + cellulase + hemicellulase (1% E/S ratio, 4 h, 30°C)	40	75	Tano-Debrah, Yoshimura, and Ohta (1996)

E/S, enzyme/substrate

Section 4.4, enzymatic treatment can improve the oil quality in terms of recovery of labile compounds; however, simply increasing the amount of recovered oil with a low oxidative level is already a significant advantage to be considered.

Evening primrose oil is known for its high content of polyunsaturated γ-linolenic fatty acid. It has been reported in the use of five commercial enzymes in evening primrose oil extraction, the best results were obtained with Ultrazym 100G (Novozymes), with an oil yield of 68% at 45°C, 40% moisture, using 4% of E/S ratio over 15 hours. This result represented a 12% increase compared with control samples (Collao et al. 2007). The authors did not evaluate oil quality but determined a residual increase in meal quality.

Chilean hazelnut oil is mostly used in cosmetic formulations. It is easily absorbed by the skin because of its high oleic and palmitic fatty acid content, conferring a good potential as a vehicle of liposoluble compounds. Santamaría et al. (2003) evaluated conventional cold pressing and pressing with enzymatic pretreatment in Chilean hazelnuts using a mixture of Ultrazym (pectinase) and Celluclast (cellulase; Novozymes). Enzymatic treatment increased oil yield from 53% to 75%. Moreover, 98.6% of oil was recovered when a double warm pressing was applied on treated seeds. Additionally, no differences were observed by Zuñiga et al. (2001)

Table 4.2 Effect of Enzymatic Treatment before Cold Pressing Process on Fruit and Seed Oil Extraction Yield

	Enzymatic Treatment	Control (%)	With Enzymes (%)	Reference
Borage	0.25% E/S ratio, Celluclast–Olivex (1:1), 45°C, 20% moisture, 9 h	77.7	85	Soto et al. (2007)
Rose hip	1% E/S ratio, Cellubrix–Finizym (1:1), 45°C, 30% moisture, 9 h	47	63.9	Concha et al. (2004)
Rose hip	1.5% E/S ratio, Cellubrix–Olivex (1:1), 45°C, 30% moisture, 6 h	47	55	Concha et al. (2004)
Canola	1% E/S, Ultrazym 100G, 45°C, 30% moisture, 24 h	81.8	87.5	Zúñiga (2001)
Canola	0.1% E/S, SP-249, 50°C, 30% moisture, 6 h	72	91.8	Sosulski and Sosulski (1993)
Chilean hazelnut	1% E/S ratio, Ultrazym 100G–Cellulase 40.000 (1:1), 45°C, 35%–40% moisture, 9 h	53.4	72.6	Zúñiga (2001)
Chilean hazelnut	1% E/S, Olivex–Celluclast (1:1), 40°C, 45% moisture, 6 h	66	74	Zúñiga et al. (2003)
Grape seed	2% E/S, Ultrazym 100G, 50°C, 60% moisture, 4 h	60	72	Guerra and Zúñiga (2003)
Cottonseed	2% E/S, 40°C, 45% moisture, 6 h	8.5	12.9	Latif, Anwar, and Ashraf (2007)
Soybean	11.84% E/S, crude extract *Aspergillus fumigatus*, 45°C, 23% moisture, 13–24 h	39.6	63.5	Smith et al. (1993)

E/S, enzyme/substrate

regarding physical and chemical properties (e.g., refraction index, saponification index, iodine index, and acidity as oleic acid) of Chilean hazelnut oil extracted by pressing, applying the same aforementioned mixture (Ultrazym/Celluclast 1%), confirming that enzymatic treatment produces good quality products.

As was observed previously, grape seed oil is considered a nutraceutical oil. Guerra and Zuñiga (2003) and Tobar et al. (2005) studied grape seed oil extraction using Ultrazym 100G and Ultrazym/Celluclast (3:1) mixtures, respectively. Oil extraction yields of 70% and 45% were observed by both authors under operational conditions of 45°C–50°C and 50%–60% moisture. Passos et al. (2009) determined that a mix of proteases, cellulases, pectinases, and xylanases applied for 24 hours at 40°C before extraction of grape seed oil by supercritical fluid increased extraction efficiency up to 43%. In addition, Tobar et al. (2005) determined that pressing conditions were affected by the enzymatic treatment, as well as the antioxidant content in the residual meal, which increased more than 60% compared with the control process.

Studies here mentioned are deficient in oil quality analysis and the effect of enzymatic treatment on them; however, considering that mild operational conditions were used in these processes, it is suggested that the oil will keep the good characteristics of the conventional process. For example, grape seed oil obtained by cold pressing has 2.9 mg polyphenol/kg oil, with 1.3 mg/kg corresponding to catechin and epicatechin and 0.3 mg/kg to resveratrol (Maier et al. 2009).

Some authors consider that oil quality is determined by the composition of lipid profile, especially in those with high polyunsaturated fatty acid content; however, others assess the presence of other compounds that are beneficial to human health. In this context, both the increase in oil recovery with an adequate lipid profile and higher content of a specific compound of interest are discussed in Section 4.4.

Rose hip oil extraction was studied by Zúñiga et al. (2001) and Concha et al. (2004, 2006). To preserve rose hip oil's cosmetic and therapeutic properties, low temperature oil extraction processes are applied, keeping active compounds undamaged and maintaining the ratio between saturated and unsaturated fatty acids. The noted authors evaluated the hydrolysis and pressing conditions of the enzymatic pretreatment on rose hip oil extraction and product quality. When they used mixtures of commercial enzymes (cellulases and pectinases), a synergic effect was observed, with oil yield increases between 20% and 40%. Also, they observed that oil extraction yield increased up to 74% when pressing conditions were modified. In addition to these good results in oil recovery, cold pressing allowed conservation of tretinoin, a precursor of vitamin A (Pareja 1990). This compound increased to 0.357 mg/L compared with 0.051 mg/L of oil recovered using organic solvents.

Use of enzymes in borage oil extraction by cold pressing resulted in an 85% oil yield, an increase of 9% compared with the control sample. Furthermore, oil recovery reached 92% when the effect of enzyme treatment on pressing conditions was studied (Soto et al. 2007). The authors also determined that oil quality (e.g., saponification index, acidity, iodine number, and peroxide value) was not affected and that the fatty acid profile was similar to those obtained in a conventional extraction process, maintaining more than 70% of polyunsaturated fatty acid and 22% of γ-linolenic acid. Tocopherol content obtained by solvent extraction was 1328 mg/kg_{oil} (Soto et al. 2008). Slightly higher values were obtained with cold pressing processes without and with enzyme incorporation (1480 and 1494 mg/kg_{oil}, respectively) (Soto et al. 2008). Also, when borage oil was extracted by cold pressing, tocopherol content was obtained in the range of 711–1267 mg/L (Velasco and Goffman 1999; Khan and Shahidi 2002). These results suggest that besides the extraction process, geographical origin and maturity of the seeds extracted can affect the quality of the final product.

Latif, Anwar, and Ashraf (2007) evaluated the effect of enzyme incorporation on cottonseed oil extraction. Cotton is an important commercial crop and a good source of quality oil. They used a mixture of enzymes such as lipases, cellulases, proteases, xylanases, and amylases, among others, recovering between 59.7% and 79.32% of oil, which had low free fatty acid content and a better oxidative stability compared with oil extracted with solvent. Tocopherol content, determined as γ-tocopherol, reached up to 490 mg/kg when Phytezyme was applied in oil extraction. This value is comparable with those obtained by the control process (460 mg/kg); however, it is higher than those reported using solvent oil extraction (390 mg/kg).

Results in borage and cottonseed oil show the effect of the type of process in product quality. Although there were not significant differences in tocopherol of oils obtained by pressing, values obtained were higher in those extracted by solvent, demonstrating that the temperature of the process and/or the presence of organic solvents produce the denaturation of the biomolecule, as well as the need to develop mild extraction processes. Even though enzymatic treatment did not increase tocopherol content, it increased oil production efficiency and quality.

As observed, few researchers determine product quality and mainly focus on development of highly efficient processes. This implies that the use of oils and by-products as a nutraceutical or in functional food formulation can only be inferred according to its components, in most cases just reported in literature. The kind of extraction process affects considerably the quality of oil extracted, especially the content of compounds with antioxidant properties (tocopherols). High temperature during organic solvent extraction decreases these compounds in the oil and residual meal, as is possible to observe in Section 4.4. Also, application of this process is limited by oil characteristics because healthy oils are those with high unsaturated fatty acid content, but they are highly susceptible to oxidation, which is accelerated by temperature. On the other hand, the use of hydrolytic enzymes facilitates the recovery of oil from inside the cell, either by the increased accessibility of solvent (aqueous extraction) or by the greater number of channels formed, which increase oil flux (pressing extraction). The low temperature processing added to the release of molecules caused by cell fiber's hydrolysis results in a greater recovery of active compounds in the oil.

In addition to the high oil quality obtained using a pressing process, the good characteristics of the by-products obtained using enzymatic-aided processes make these techniques more valuable. Some authors indicated that defatted meal of oilseeds has phenolic compounds (Shahidi et al. 1997; Wettasinghe and Shahidi 1999b; Moure et al. 2001; Wettasinghe et al. 2001; Matthaus 2002a, 2002b; Cruz et al. 2004; Shahidi 2009). This fact suggested that defatted meals are an interesting source of bioactive compounds. On the other hand, other authors reported that cellulase and pectinase incorporation could increase phenolic compounds extracted from grape pomace (Pinelo and Meyer 2008), blackcurrant pomace (Landbo and Meyer 2001), evening primrose defatted meal (Collao et al. 2007), borage defatted seeds (Soto et al. 2008), and depleted grape seed (Tobar et al. 2005), becoming an alternative for oil extraction by-product valorization.

4.4 EFFECT OF OIL EXTRACTION PROCESSES IN BY-PRODUCT QUALITY

Two products are obtained from oil extraction processes: edible or nonedible oils and meal (cake), which is typically used in animal feeding and has been recognized in recent years as an important source of phytochemicals.

The main components of defatted meal are usually crude protein and fiber (Table 4.3). In addition, defatted meal or oil cake could be considered a lignocellulosic material because it is a composite polymeric material containing mostly cellulose, lignin, and hemicellulose.

Table 4.3 Proximate Composition of Defatted Meal

Oil Cake	Crude Protein (%)	Crude Fat (%)	Crude Fiber (%)	Ash (%)	References
Sunflower	57.4	0.8	21.0	6.4	Gandhi, Jha, and Gupta (2008)
Coconut	22.75 ± 0.22	2.89 ± 0.22	12.11 ± 0.24	54.84 ± 0.32	Moorthy and Viswanathan (2009)
Canola	36.13 ± 0.38	2.77 ± 0.08	11.54 ± 0.87	6.26 ± 0.11	Khattab and Arntfield (2009)
Soybean	48.29 ± 0.44	1.59 ± 0.04	3.50 ± 0.14	6.53 ± 0.12	Khattab and Arntfield (2009)
Flaxseed	38.96 ± 0.14	2.17 ± 0.07	5.27 ± 0.06	7.01 ± 0.07	Khattab and Arntfield (2009)
Sesame	23–46	1.4–27	5–16	7.5–17.5	Omar (2002)

There are edible oil cakes, with high nutritional value, especially in protein content with ranges of 15%–50%. Composition varies depending on their variety, growing conditions, and extraction methods. Because of their rich protein content, they are used as animal feed, especially for ruminants and fish. On the other hand, nonedible oil cakes, such as castor cake, karanja cake, and neem cake, are used as organic nitrogenous fertilizers because of their nitrogen, phosphorus, and potassium content. Some of these oil cakes improve nitrogen uptake of plants because they retard the nitrification of soil. They also protect plants from soil nematodes, insects, and parasites, thereby offering great resistance to infection (Ramachandran et al. 2007).

Residual meal can be considered an important source of bioactive molecules because of its high content of pectin, a galacturonic acid polymer attached to units of glucose, xylose, arabinose, and galactose; cellulose; hemicelluloses; and lignin. Also, it has been reported that phenol compounds can be linked to these polysaccharides, increasing potential health benefits of this by-product.

Recently there has been an increase in studies related to defatted meal because it has been shown that nondigestible carbohydrates (dietary fiber), proteins, and polyphenols with antioxidant activity and biologically active substances that have positive health effects can be obtained from this source (Garrote et al. 2004).

Other important bioactive molecules that can be obtained from depleted meals are antioxidant compounds, as reported by Schmidt and Pokorny (2005). Extraction of phenolic antioxidants from defatted seed residues avoids the expensive process of drying the material (Peschel et al. 2006). Highly active extracts and single compounds have already been obtained from various oilseed meals (Matthaus 2002b; Shahidi 2004, 2009), but the majority of research has focused on sesame seed cake (Ohtsuki et al. 2003; Suja et al. 2004).

Maier et al. (2009) studied the content of individual phenolic compounds in press residues from grape seed oil production and the effects of different solvents on the yields of phenolic compounds. The authors demonstrated that press residues from grape seed oil production are polyphenol-rich by-products with high antioxidant activity. However, oil extraction applied in this study needs to be optimized because the temperature increased above 60°C during pressing, and losses of target compounds could not be avoided. Thus, a cold pressing process might increase the phenolic content and the antioxidant capacity of the seed oil press residues. Polyphenols can easily be extracted from these by-products in high amounts, enabling their application as ingredients of functional or enriched foods.

In the case of analysis of borage residual meal quality, when oil extraction was done by cold pressing using 0.25 g_{enzyme}/100 g_{seed} cellulase–pectinase mixture, a 15% fiber reduction was observed (Soto et al. 2007); also, a 60.8% increase in phenolic compounds and a better antioxidant activity were observed in alcohol extracts obtained from residual meal after an enzymatic-aided oil extraction process (Soto et al. 2008), demonstrating the effectiveness of enzyme treatment to release bioactive molecules. A similar behavior was observed with an evening primrose oil extraction process (Collao et al. 2007), showing an important cellulose decrease as a result of enzyme action and an increase in antioxidant capability (determined as DPPH scavenging) and phenolic compounds to approximately 250–276 mg catechin/g meal.

Incorporation of an enzyme treatment in oil extraction changes recovery of antioxidant compounds from depleted meal, probably as a result of the breakdown of the antioxidant compound–vegetable matrix or directly as a result of hydrolysis of vegetable fiber compounds with antioxidant capacity. In both cases, enzyme treatment makes these compounds more accessible to solvent extraction.

As mentioned previously, Concha et al. (2006) studied rose hip oil extraction and evaluated the physicochemical properties of the products, with and without enzymatic pretreatment. The authors reveal that in addition to good oil quality (high tretinoin content), rose hip defatted meal obtained was free of organic solvent, allowing its use as a dietary fiber supplement. Franco et al. (2007) reported that ethanolic extracts of rose hip meals have a high antioxidant activity, determined as DPPH inhibition percentage. In particular, it has been determined that the antioxidant capacity of rose hip extract is higher than those observed with commercial antioxidants such as butylated hydroxytoluene and butylated hydroxyanisole, making this by-product an interesting alternative to replace other synthetic molecules.

In general, the antioxidants are isolated by solvent extraction, and both extraction yield and antioxidant activity of the extracts are strongly dependent on the solvent because of the variant antioxidant potential of compounds with different polarity. Nonpolar solvents (e.g., hexane, petroleum ether) can be used for the recovery of tocopherols and certain phenolic terpenes. Ethyl ether and ethyl acetate are efficient for the recovery of flavonoid aglycones, low molecular weight phenols, and phenolic acids. Solvents of higher polarity (ethanol or ethanol–water mixtures) additionally can extract flavonoid glycosides and higher molecular weight phenols, resulting in higher yields of total extracted polyphenols. However, many undesirable substances are frequently coextracted in the latter case, and purification is necessary to isolate the antioxidant fraction.

In addition to the solvent used, other extraction parameters that affect yield and antioxidant activity of the extract are temperature, time of extraction, and pH in the case of aqueous ethanol solutions (Moure et al. 2001; Oreopoulou 2003). According to Wettasinghe and Shahidi (1999a), extraction of antioxidants from an oilseed meal with aqueous ethanol shows the highest yield at an ethanol concentration in the range of 50%–60%, whereas antioxidant activity of the extract increases with temperature and time, up to 70°C and 60 minutes, respectively, and decreases afterward. In general, temperatures less than 60°C–70°C are suggested for extraction and drying of the raw material because they induce less damage to the antioxidant compounds.

Although the aforementioned works do not report the effect of extracted compounds on human health, some researchers have reported biological activities of phenolic compounds, such as antioxidant properties and radioprotective effects (Castillo et al. 2000), prevention of cataracts (Yamakoshi et al. 2002), antihyperglycemic effects (Pinent et al. 2004), postprandial lipidemia enhancement (Del Bas et al. 2005), modulation of the expression of antioxidant enzyme systems (Puiggros et al. 2005), improvement of insulin sensitivity and prevention of hypertriglyceridemia (Al-Awwadi et al. 2005), inhibition of aromatase and suppression of its expression (Kijima et al. 2006), inhibition of protein kinase activity of the epidermal growth factor receptor, protective effects against oxidative damage in mouse brain cells (Guo et al. 2007), and anti-inflammatory effects (Terra et al. 2007).

Proteins are the main source of nitrogen and essential amino acids in the human diet. Commercial dietary proteins are obtained from animal and vegetable sources and then are used as functional ingredients (Periago et al. 2009; Jiang et al. 2010). Protein content of defatted meals from dehulled oilseeds depends on the seed used and ranges between 35% and 60% (dry basis). As a general trend, meals contain antinutritional compounds, such as oligosaccharides, trypsin inhibitors, phytic acid, and tannins, and have low protein solubility, limiting its use in food applications. On the other hand, some authors have reported an improvement in functional and nutritional protein when enzymatic-aided aqueous oil extraction is carried out (Dominguez et al. 1994).

Application of enzymatic treatment on oilseed cell walls, besides enhancing oil extractability from seeds, also improves the protein digestibility and quality of the residual meal. Studies by Moure et al. (2002) in pressed cakes of *Guevina avellana* show that the protein is more easily extractable from meals when enzyme treatment is applied and maximal recovery of protein (percentage solubilized) is 85.8%.

Soluble carbohydrates from hemicellulose are the main components (25%–55%) of defatted meal in olive oil extraction processes after extraction of phenolic compounds (Rodriguez et al. 2008).

One of the most important characteristics of residual meal obtained from oil extraction processes is its fiber content, especially hemicelluloses, because an enzymatic saccharification of hemicellulose would result in the production of various sugars, for example, xylobiose (derived from xylan), which is found in the market as a low-calorie sweetener for dietetics (Saha 2003).

In this point, the hemicellulose obtention is crucial; however, the lignocellulosic material associated is hardly hydrolyzed. On the other hand, oligosaccharides can be produced during the hydrolysis of higher molecules as xylan (Saha 2003). For this reason, the application of enzymes with exo-xylanase, β-xylosidase, endo-β-xylanases, xylan acetylesterase, and glucuronidase activities is an advantage (Jeffries 1994; Saha 2003; Mosier et al. 2005) and an alternative to be incorporated into treatment before oil extraction.

Considering the previous background, oil extraction processes with incorporation of enzymes can be used to obtain oils with good cosmetic and nutritional qualities, as well as a residual cake with several compounds, which have potential benefits for human health.

In the future, nutraceutical and functional food ingredients (oil and meal molecules) will be considered as precursors of 21st century foods. Great changes in food preparation are occurring: the food industry is becoming an assembly industry, where the trend is the separation of raw materials into simple compounds but with high interest.

REFERENCES

Ahmad, Z. 2010. The Uses and Properties of Almond Oil. *Complementary Therapies in Clinical Practice* 16: 10–12.

Alasalvar, C., M. Karamac, A. Kosinska, A. Rybarczyk, F. Shahidi, and R. Amarowicz. 2009. Antioxidant Activity of Hazelnut Skin Phenolics. *Journal of Agricultural and Food Chemistry* 57: 4645–50.

Al-Awwadi, N. A., C. Araiz, A. Bornet, S. Delbosc, J. P. Cristol, N. Linck, J. Azay, P. L. Teissedre, and G. Cros. 2005. Extracts Enriched in Different Polyphenolic Families Normalize Increased Cardiac NADPH Oxidase Expression While Having Differential Effects on Insulin Resistance, Hypertension, and Cardiac Hypertrophy in High-Fructose-Fed Rats. *Journal of Agricultural and Food Chemistry* 53: 151–57.

Aliakbarian, B., D. De Faveri, A. Converti, and P. Perego. 2008. Optimisation of Olive Oil Extraction by Means of Enzyme Processing Aids Using Response Surface Methodology. *Biochemical Engineering Journal* 42: 34–40.

Avvisati, G., and M. S. Tallman. 2003. All-Trans Retinoic Acid in Acute Promyelocytic Leukaemia. *Best Practice & Research Clinical Haematology* 16: 419–32.

Badr, F. H., and M. Z. Sitohy. 1992. Optimizing Conditions for Enzymatic Extraction of Sunflower Oil. *Grasas y Aceites* 43: 281–83.

Balasundram, N., K. Sundram, and S. Samman. 2006. Phenolic Compounds in Plants and Agri-Industrial By-Products: Antioxidant Activity, Occurrence, and Potential Uses. *Food Chemistry* 99: 191–203.

Barre, D. E. 2001. Potential of Evening Primrose, Borage, Black Currant, and Fungal Oils in Human Health. *Annals of Nutrition and Metabolism* 45: 47–57.

Brill, J. B. 2009. The Mediterranean Diet and Your Health. *American Journal of Lifestyle Medicine* 3: 44–56.

Butsat, S., N. Weerapreeyakul, and S. Siriamornpun. 2009. Changes in Phenolic Acids and Antioxidant Activity in Thai Rice Husk at Five Growth Stages during Grain Development. *Journal of Agricultural and Food Chemistry* 57: 4566–71.

Calder, P. C. 2004. N-3 Fatty Acids and Cardiovascular Disease: Evidence Explained and Mechanisms Explored. *Clinical Science* 107: 1–11.

Calzolari, I., S. Fumagalli, N. Marchionni, and M. Di Bari. 2009. Polyunsaturated Fatty Acids and Cardiovascular Disease. *Current Pharmaceutical Design* 15: 4094–102.

Carluccio, M. A., L. Siculella, M. A. Ancora, M. Massaro, E. Scoditti, C. Storelli, F. Visioli, A. Distante, and R. De Caterina. 2003. Olive Oil and Red Wine Antioxidant Polyphenols Inhibit Endothelial Activation: Antiatherogenic Properties of Mediterranean Diet Phytochemicals. *Arteriosclerosis, Thrombosis, and Vascular Biology* 23: 622–29.

Castillo, J., O. Benavente-Garcia, J. Lorente, M. J. Alcaraz, A. Redondo, A. Ortuno, and J. A. Del Rio. 2000. Antioxidant Activity and Radioprotective Effects against Chromosomal Damage Induced in Vivo by X-Rays of Flavan-3-Ols (Procyanidins) from Grape Seeds (Vitis Vinifera): Comparative Study versus Other Phenolic and Organic Compounds. *Journal of Agricultural and Food Chemistry* 48: 1738–45.

Chandrasekaran, V. R. M., D. Z. Hsu, and M. Y. Liu. 2009. The Protective Effect of Sesamol against Mitochondrial Oxidative Stress and Hepatic Injury in Acetaminophen-Overdosed Rats. *Shock* 32: 89–93.

Chau, C. F., and Y. L. Huang. 2003. Comparison of the Chemical Composition and Physicochemical Properties of Different Fibers Prepared from the Peel of Citrus Sinensis L. Cv. Liucheng. *Journal of Agricultural and Food Chemistry* 51: 2615–18.

Che Man, Y., Suhardiyono, A. Asbi, M. Azudin, and L. Wei. 1996. Aqueous Enzymatic Extraction of Coconut Oil. *Journal of the American Oil Chemists' Society* 73: 683–86.

Choi, S. B. 2003. Benefits of Mediterranean Diet Affirmed, Again. *CMAJ* 169: 316.

Chou, T. W., C. Y. Ma, H. H. Cheng, Y. Y. Chen, and M. H. Lai. 2009. A Rice Bran Oil Diet Improves Lipid Abnormalities and Suppress Hyperinsulinemic Responses in Rats with Streptozotocin/Nicotinamide-Induced Type 2 Diabetes. *Journal of Clinical Biochemistry and Nutrition* 45: 29–36.

Cicero, A. F. G., and A. Gaddi. 2001. Rice Bran Oil and Gamma-Oryzanol in the Treatment of Hyperlipoproteinaemias and Other Conditions. *Phytotherapy Research* 15: 277–89.

Collao, C. A., E. Curotto, and M. E. Zuniga. 2007. Enzymatic Treatment on Oil Extraction and Antioxidant Recuperation from Oenothera Biennis by Cold Pressing. *Grasas y Aceites* 58: 10–14.

Concha, J., C. Soto, R. Chamy, and M. E. Zuniga. 2006. Effect of Rosehip Extraction Process on Oil and Defatted Meal Physicochemical Properties. *Journal of the American Oil Chemists' Society* 83: 771–75.

Concha, J., C. Soto, R. Chamy, and M. Zúñiga. 2004. Enzymatic Pretreatment on Rose-Hip Oil Extraction: Hydrolysis and Pressing Conditions. *Journal of the American Oil Chemists' Society* 81: 549–52.

Cruz, J. M., H. Dominguez, and J. C. Parajo. 2004. Assessment of the Production of Antioxidants from Winemaking Waste Solids. *Journal of Agricultural and Food Chemistry* 52: 5612–20.

Dandjouma, A. K. A., C. Tchiegang, C. Kapseu, M. Linder, and M. Parmentier. 2008. Enzyme-Assisted Hexane Extraction of Ricinodendron Heudelotii (Bail.) Pierre Ex Pax Seeds Oil. *International Journal of Food Science and Technology* 43: 1169–75.

de Bosch, N. B., V. Bosch, and R. Apitz. 1996. Dietary Fatty Acids in Athero-Thrombogenesis: Influence of Palm Oil Ingestion. *Pathophysiology of Haemostasis and Thrombosis* 26: 46–54.

de la Torre-Carbot, K., J. L. Chávez-Servín, O. Jaúregui, A. I. Castellote, R. M. Lamuela-Raventós, M. Fitó, M.-I. Covas, D. Muñoz-Aguayo, and M. C. López-Sabater. 2007. Presence of Virgin Olive Oil Phenolic Metabolites in Human Low Density Lipoprotein Fraction: Determination by High-Performance Liquid Chromatography-Electrospray Ionization Tandem Mass Spectrometry. *Analytica Chimica Acta* 583: 402–10.

Del Bas, J. M., J. Fernandez-Larrea, M. Blay, A. Ardevol, M. J. Salvado, L. Arola, and C. Blade. 2005. Grape Seed Procyanidins Improve Atherosclerotic Risk Index and Induce Liver Cyp7a1 and Shp Expression in Healthy Rats. *Faseb Journal* 19: 479–81.

Dominguez, H., M. J. Nunez, and J. M. Lema. 1994. Enzymatic Pretreatment to Enhance Oil Extraction from Fruits and Oilseeds—a Review. *Food Chemistry* 49: 271–86.

Dominguez, H., M. J. Nunez, and J. M. Lema. 1995. Aqueous Processing of Sunflower Kernels with Enzymatic Technology. *Food Chemistry* 53: 427–34.

Dougherty, R. M., C. Galli, A. Ferro-Luzzi, and J. M. Iacono. 1987. Lipid and Phospholipid Fatty Acid Composition of Plasma, Red Blood Cells, and Platelets and How They Are Affected by Dietary Lipids: A Study of Normal Subjects from Italy, Finland, and the USA. *American Journal of Clinical Nutrition* 45: 443–55.

Dupont, J., P. J. White, K. M. Johnston, H. A. Heggtveit, B. E. McDonald, S. M. Grundy, and A. Bonanome. 1989. Food Safety and Health Effects of Canola Oil. *Journal of the American College of Nutrition* 8: 360–75.

Ebong, P. E., D. U. Owu, and E. U. Isong. 1999. Influence of Palm Oil (Elaesis Guineensis) on Health. *Plant Foods for Human Nutrition (Formerly Qualitas Plantarum)* 53: 209–22.

Edem, D. O. 2002. Palm Oil: Biochemical, Physiological, Nutritional, Hematological and Toxicological Aspects: A Review. *Plant Foods for Human Nutrition (Formerly Qualitas Plantarum)* 57: 319–41.

Esterhuyse, A. J., E. F. du Toit, A. J. S. Benadè, and J. van Rooyen. 2005. Dietary Red Palm Oil Improves Reperfusion Cardiac Function in the Isolated Perfused Rat Heart of Animals Fed a High Cholesterol Diet. *Prostaglandins, Leukotrienes and Essential Fatty Acids* 72: 153–61.

Fantozzi, P., G. Petruccioli, and G. Montedoro. 1977. Enzymatic Treatment of Olive Pastes after Single Pressing Extraction: Effect of Cultivar, Harvesting Time, and Storage. *La Rivista Italiana delle Sostanze Grasse* 54: 381–88.

Franco, D., M. Pinelo, J. Sineiro, and M. J. Nunez. 2007. Processing of Rosa Rubiginosa: Extraction of Oil and Antioxidant Substances. *Bioresource Technology* 98: 3506–12.

Fullbrook, P. 1983. The Use of Enzymes in the Processing of Oilseeds. *Journal of the American Oil Chemists' Society* 60: 476–78.

Gandhi, A. P., K. Jha, and V. Gupta. 2008. Studies on the Production of Defatted Sunflower Meal with Low Polyphenol and Phytate Contents and Its Nutritional Profile. *ASEAN Food Journal* 15: 97–100.

García, A., M. Brenes, M. J. Moyano, J. Alba, P. García, and A. Garrido. 2001. Improvement of Phenolic Compound Content in Virgin Olive Oils by Using Enzymes during Malaxation. *Journal of Food Engineering* 48: 189–94.

García-Villalba, R., A. Carrasco-Pancorbo, C. Oliveras-Ferraros, A. Vázquez-Martín, J. A. Menéndez, A. Segura-Carretero, and A. Fernández-Gutiérrez. 2010. Characterization and Quantification of Phenolic Compounds of Extra-Virgin Olive Oils with Anticancer Properties by a Rapid and Resolutive Lc-Esi-Tof Ms Method. *Journal of Pharmaceutical and Biomedical Analysis* 51: 416–29.

Garrote, G., J. M. Cruz, A. Moure, H. Dominguez, and J. C. Parajo. 2004. Antioxidant Activity of Byproducts from the Hydrolytic Processing of Selected Lignocellulosic Materials. *Trends in Food Science & Technology* 15: 191–200.

Gebauer, S. K., T. L. Psota, W. S. Harris, and P. M. Kris-Etherton. 2006. N-3 Fatty Acid Dietary Recommendations and Food Sources to Achieve Essentiality and Cardiovascular Benefits. *American Journal of Clinical Nutrition* 83: S1526–35.

Gorinstein, S., H. Leontowicz, M. Leontowicz, A. Lojek, M. Cíz, R. Krzeminski, Z. Zachwieja, Z. Jastrzebski, E. Delgado-Licon, O. Martin-Belloso, and S. Trakhtenberg. 2003. Seed Oils Improve Lipid Metabolism and Increase Antioxidant Potential in Rats Fed Diets Containing Cholesterol. *Nutrition Research* 23: 317–30.

Guerra E. G., and M. E. Zúñiga. 2003. Tratamiento Enzimático en la Extracción de Aceite de Pepa de Uva, Vitis Vinifera, por Prensado en Frío. *Grasas y Aceites* 54: 53–57.

Guo, L., L. H. Wang, B. S. Sun, J. Y. Yang, Y. Q. Zhao, Y. X. Dong, M. I. Spranger, and C. F. Wu. 2007. Direct in Vivo Evidence of Protective Effects of Grape Seed Procyanidin Fractions and Other Antioxidants against Ethanol-Induced Oxidative DNA Damage in Mouse Brain Cells. *Journal of Agricultural and Food Chemistry* 55: 5881–91.

Gupta, A., S. Sharma, I. Kaur, and K. Chopra. 2009. Renoprotective Effects of Sesamol in Ferric Nitrilotriacetate-Induced Oxidative Renal Injury in Rats. *Basic & Clinical Pharmacology & Toxicology* 104: 316–21.

Harris, W. S., and D. Bulchandani. 2006. Why Do Omega-3 Fatty Acids Lower Serum Triglycerides? *Current Opinion in Lipidology* 17: 387–93.

Harris, W. S., Y. Park, and W. L. Isley. 2003. Cardiovascular Disease and Long-Chain Omega-3 Fatty Acids. *Current Opinion in Lipidology* 14: 9–14.

Hoffmann, G. 1989. Water and Heat-Promoted Fat Separation from Animal and Plant "Fatty Tissues." The Chemistry and Technology of Edible Oils and Fats and Their High Fat Products. *Food Science and Technology.* San Diego: Academic Press Inc.

Hsu, D. Z., K. T. Chen, T. H. Lin, Y. H. Li, and M. Y. Liu. 2007. Sesame Oil Attenuates Cisplatin-Induced Hepatic and Renal Injuries by Inhibiting Nitric Oxide-Associated Lipid Peroxidation in Mice. *Shock* 27: 199–204.

Hsu, D. Z., C. T. Liu, Y. H. Li, P. Y. Chu, and M. Y. Liu. 2010. Protective Effect of Daily Sesame Oil Supplement on Gentamicin-Induced Renal Injury in Rats. *Shock* 33: 88–92.

Hsu, D. Z., and M. Y. Liu. 2004. Effects of Sesame Oil on Oxidative Stress after the Onset of Sepsis in Rats. *Shock* 22: 582–85.

Hyson, D. A., B. O. Schneeman, and P. A. Davis. 2002. Almonds and Almond Oil Have Similar Effects on Plasma Lipids and Ldl Oxidation in Healthy Men and Women. *Journal of Nutrition* 132: 703–07.

Illanes, A. 1994. *Biotecnología de Enzimas. Serie de Monografías Científicas del Programa Regional de Desarrollo Científico y Tecnológico de la Organización de los Estados Americanos.* Washington, DC: Ediciones Universitarias de Valparaíso, 256.

Jeffries, T. W. 1994. Biochemistry of Microbial Degradation. *Biodegradation of Lignin and Hemicelluloses.* Edited by C. Ratledge. Amsterdam: Kluwer Academic Publishers.

Jiang, L., D. Hua, Z. Wang, and S. Xu. 2010. Aqueous Enzymatic Extraction of Peanut Oil and Protein Hydrolysates. *Food and Bioproducts Processing* 88: 233–238.

Jung, S., D. Maurer, and L. A. Johnson. 2009. Factors Affecting Emulsion Stability and Quality of Oil Recovered from Enzyme-Assisted Aqueous Extraction of Soybeans. *Bioresource Technology* 100: 5340–47.

Kamel, B., H. Dawson, and Y. Kakuda. 1985. Characteristics and Composition of Melon and Grape Seed Oils and Cakes. *Journal of the American Oil Chemists' Society* 62: 881–83.

Kang, H. Y., L. Valerio, P. Bahadoran, and J. P. Ortonne. 2009. The Role of Topical Retinoids in the Treatment of Pigmentary Disorders an Evidence-Based Review. *American Journal of Clinical Dermatology* 10: 251–60.

Kashyap, M. C., Y. C. Agrawal, P. K. Ghosh, D. S. Jayas, B. C. Sarkar, and B. P. N. Singh. 2007. Oil Extraction Rates of Enzymatically Hydrolyzed Soybeans. *Journal of Food Engineering* 81: 611–17.

Kast, R. E. 2001. Borage Oil Reduction of Rheumatoid Arthritis Activity May Be Mediated by Increased Camp That Suppresses Tumor Necrosis Factor-Alpha. *International Immunopharmacology* 1: 2197–99.

Khan, M. A., and F. Shahidi. 2002. Photooxidative Stability of Stripped and Non-Stripped Borage and Evening Primrose Oils and Their Emulsions in Water. *Food Chemistry* 79: 47–53.

Khattab, R. Y., and S. D. Arntfield. 2009. Functional Properties of Raw and Processed Canola Meal. *LWT - Food Science and Technology* 42: 1119–24.

Kijima, I., S. Phung, G. Hur, S. L. Kwok, and S. U. Chen. 2006. Grape Seed Extract Is an Aromatase Inhibitor and a Suppressor of Aromatase Expression. *Cancer Research* 66: 5960–67.

Kritchevsky, D., S. A. Tepper, S. K. Czarnecki, and K. Sundram. 2002. Red Palm Oil in Experimental Atherosclerosis. *Asia Pacific Journal of Clinical Nutrition* 11: S433–37.

Kruger, M. J., A.-M. Engelbrecht, J. Esterhuyse, E. F. du Toit, and J. van Rooyen. 2007. Dietary Red Palm Oil Reduces Ischaemia-Reperfusion Injury in Rats Fed a Hypercholesterolaemic Diet. *British Journal of Nutrition* 97: 653–60.

Kuhad, A., A. K. Sachdeva, and K. Chopra. 2009. Attenuation of Renoinflammatory Cascade in Experimental Model of Diabetic Nephropathy by Sesamol. *Journal of Agricultural and Food Chemistry* 57: 6123–28.

Landbo, A. K., and A. S. Meyer. 2001. Enzyme-Assisted Extraction of Antioxidative Phenols from Black Current Juice Press Residues (Ribes Nigrum). *Journal of Agricultural and Food Chemistry* 49: 3169–77.

Larrauri, J. A., P. Ruperez, B. Borroto, and F. SauraCalixto. 1996. Mango Peels as a New Tropical Fibre: Preparation and Characterization. *LWT - Food Science and Technology* 29: 729–33.

Latif, S., and F. Anwar. 2009. Effect of Aqueous Enzymatic Processes on Sunflower Oil Quality. *Journal of the American Oil Chemists' Society* 86: 393–400.

Latif, S., F. Anwar, and M. Ashraf. 2007. Characterization of Enzyme-Assisted Cold-Pressed Cottonseed Oil. *Journal of Food Lipids* 14: 424–36.

Lee, J. H., J. H. O'Keefe, C. J. Lavie, and W. S. Harris. 2009. Omega-3 Fatty Acids: Cardiovascular Benefits, Sources and Sustainability. *Nature Reviews Cardiology* 6: 753–58.

Lovegrove, J. A., S. S. Lovegrove, S. V. M. Lesauvage, L. M. Brady, N. Saini, A. M. Minihane, and C. M. Williams. 2004. Moderate Fish-Oil Supplementation Reverses Low-Platelet, Long-Chain N-3 Polyunsaturated Fatty Acid Status and Reduces Plasma Triacylglycerol Concentrations in British Indo-Asians. *American Journal of Clinical Nutrition* 79: 974–82.

Lu, Y. R., and L. Y. Foo. 2000. Antioxidant and Radical Scavenging Activities of Polyphenols from Apple Pomace. *Food Chemistry* 68: 81–85.

Maier, T., A. Schieber, D. R. Kammerer, and R. Carle. 2009. Residues of Grape (Vitis vinifera L.) Seed Oil Production as a Valuable Source of Phenolic Antioxidants. *Food Chemistry* 112: 551–59.

Martinez, J. M., C. Gómez, and C. Janer del Vale. 1957. Estudios Fisico-Químcios sobre las Pastas de Aceitunas Molidas. *Grasas y Aceites* 8: 151–61.

Matthaus, B. 2002a. Antioxidant Activity of Extracts Isolated from Residues of Oilseeds, Such as Rapeseed or Sunflower. *Agro Food Industry Hi-Tech* 13: 22–25.

Matthaus, B. 2002b. Antioxidant Activity of Extracts Obtained from Residues of Different Oilseeds. *Journal of Agricultural and Food Chemistry* 50: 3444–52.

McGlone, O. C., A. López-Munguia, and J. Vernon. 1986. Coconut Oil Extraction by a New Enzymatic Process. *Journal of Food Science* 51: 695–97.

Metcalf, R. G., M. J. James, R. A. Gibson, J. R. M. Edwards, J. Stubberfield, R. Stuklis, K. Roberts-Thomson, G. D. Young, and L. G. Cleland. 2007. Effects of Fish-Oil Supplementation on Myocardial Fatty Acids in Humans. *American Journal of Clinical Nutrition* 85: 1222–28.

Meyer, B. J., A. E. Lane, and N. J. Mann. 2009. Comparison of Seal Oil to Tuna Oil on Plasma Lipid Levels and Blood Pressure in Hypertriglyceridaemic Subjects. *Lipids* 44: 827–35.

Montedoro, G., and Y. G. Petruccioli. 1973. Aggiornamenti Sui Trattamenti con Additivi Enzimaticci Nell'estrazione Dell'olio di Oliva. *La Rivista Italiana delle Sostanze Grasse* 50: 331–44.

Moorthy, M., and K. Viswanathan. 2009. Nutritive value of extracted coconut (Cocos nucifera) meal. *Research Journal of Agriculture and Biological Sciences* 5: 515–17.

Mosier, N., C. Wyman, B. Dale, R. Elander, Y. Y. Lee, M. Holtzapple, and M. Ladisch. 2005. Features of Promising Technologies for Pretreatment of Lignocellulosic Biomass. *Bioresource Technology* 96: 673–86.

Moure, A., J. M. Cruz, D. Franco, J. M. Dominguez, J. Sineiro, H. Dominguez, M. J. Nunez, and J. C. Parajo. 2001. Natural Antioxidants from Residual Sources. *Food Chemistry* 72: 145–71.

Moure, A., H. Dominguez, M. E. Zuniga, C. Soto, and R. Chamy. 2002. Characterisation of Protein Concentrates from Pressed Cakes of Guevina avellana (Chilean Hazelnut). *Food Chemistry* 78: 179–86.

Moure, A., D. Franco, R. I. Santamaria, C. Soto, J. Sineiro, R. Dominguez, M. E. Zuniga, M. J. Nunez, R. Chamy, A. Lopez-Munguia, and J. M. Lema. 2001. Enzyme-Aided Alternative Processes for the Extraction of Oil from Rosa Rubiginosa. *Journal of the American Oil Chemists' Society* 78: 437–39.

Najafian, L., A. Ghodsvali, M. H. H. Khodaparast, and L. L. Diosady. 2009. Aqueous Extraction of Virgin Olive Oil Using Industrial Enzymes. *Food Research International* 42: 171–75.

Nguemeni, C., B. Delplanque, C. Rovère, N. Simon-Rousseau, C. Gandin, G. Agnani, J. L. Nahon, C. Heurteaux, and N. Blondeau. 2010. Dietary Supplementation of Alpha-Linolenic Acid in an Enriched Rapeseed Oil Diet Protects from Stroke. *Pharmacological Research* 61: 226–233.

Noroozi, M., R. Zavoshy, and H. J. Hashemi. 2009. The Effects of Low-Calorie Diet with Canola Oil on Blood Lipids in Hyperlipidemic Patients. *Journal of Food and Nutrition Research* 48: 178–82.

Norris, F. 1982. Solvent Extraction. *Bayley's Industrial Oil and Fats Products*. Edited by M. Formo, R. Allen, R. Krishnamurthy, G. McDemmort, F. Norris, and N. Sonntag. Hoboken, NJ: John Wiley & Sons Inc., 215–17.

Ohara, N., Y. Naito, T. Nagata, S. Tachibana, M. Okimoto, and H. Okuyama. 2008. Dietary Intake of Rapeseed Oil as the Sole Fat Nutrient in Wistar Rats—Lack of Increase in Plasma Lipids and Renal Lesions. *Journal of Toxicological Sciences* 33: 641–45.

Ohtsuki, T., J. Akiyama, T. Shimoyama, S. Yazaki, S. Ui, Y. Hirose, and A. Mimura. 2003. Increased Production of Antioxidative Sesaminol Glucosides from Sesame Oil Cake through Fermentation by Bacillus Circulans Strain Yus-2. *Bioscience Biotechnology and Biochemistry* 67: 2304–06.

Olsen, H. S. 1988. Aqueous Enzymatic Extraction of Oil from Seeds. *Asian Food Conference Proceedings*. Bangkok, Thailand: Novo Industry A/S.

Omar, J. M. A. 2002. Effects of Feeding Different Levels of Sesame Oil Cake on Performance and Digestibility of Awassi Lambs. *Small Ruminant Research* 46: 187–90.

Oreopoulou, V. 2003. Extraction of Natural Antioxidants. *Extraction Optimization in Food Engineering*. Edited by C. Tzia and G. Liadakis. New York: Marcel Dekker, 329–46.

Orlandi, M., B. Mantovani, K. Ammar, E. Avitabile, P. Dal Monte, and G. Bartolini. 2003. Retinoids and Cancer: Antitumoral Effects of Atra, 9-Cis Ra and the New Retinoid Iif on the Hl-60 Leukemic Cell Line. *Medical Principles and Practice* 12: 164–69.

Ozkan, G., O. Sagdic, N. G. Baydar, and Z. Kurumahmutoglu. 2004. Antibacterial Activities and Total Phenolic Contents of Grape Pomace Extracts. *Journal of the Science of Food and Agriculture* 84: 1807–11.

Pareja, B., and H. Kehl. 1990. Contribución a la Identificación de los Principios Activos en el Aceite de Rosa Aff Rubifinosa L. *Anales de la Real Academia de Farmacia* 56: 283–94.

Passos, C. P., R. M. Silva, F. A. Da Silva, M. A. Coimbra, and C. M. Silva. 2009. Enhancement of the Supercritical Fluid Extraction of Grape Seed Oil by Using Enzymatically Pre-Treated Seed. *Journal of Supercritical Fluids* 48: 225–29.

Periago, M. J., J. Garcia-Alonso, K. Jacob, A. B. Olivares, M. J. Bernal, M. D. Iniesta, C. Martinez, and G. Ros. 2009. Bioactive Compounds, Folates and Antioxidant Properties of Tomatoes (Lycopersicum esculentum) during Vine Ripening. *International Journal of Food Sciences and Nutrition* 60: 694–708.

Peschel, W., F. Sanchez-Rabaneda, W. Diekmann, A. Plescher, I. Gartzia, D. Jimenez, R. Lamuela-Raventos, S. Buxaderas, and C. Codina. 2006. An Industrial Approach in the Search of Natural Antioxidants from Vegetable and Fruit Wastes. *Food Chemistry* 97: 137–50.

Pinelo, M., and A. S. Meyer. 2008. Enzyme-Assisted Extraction of Antioxidants: Release of Phenols from Vegetal Matrixes. *Electronic Journal of Environmental, Agricultural, and Food Chemistry* 8: 3217–20.

Pinent, M., M. Blay, M. C. Blade, M. J. Salvado, L. Arola, and A. Ardevol. 2004. Grape Seed-Derived Procyanidins Have an Antihyperglycemic Effect in Streptozotocin-Induced Diabetic Rats and Insulinomimetic Activity in Insulin-Sensitive Cell Lines. *Endocrinology* 145: 4985–90.

Psaltopoulou, T., A. Naska, P. Orfanos, D. Trichopoulos, T. Mountokalakis, and A. Trichopoulou. 2004. Olive Oil, the Mediterranean Diet, and Arterial Blood Pressure: The Greek European Prospective Investigation into Cancer and Nutrition (Epic) Study. *American Journal of Clinical Nutrition* 80: 1012–18.

Puiggros, F., N. Llopiz, A. Ardevol, C. Blade, L. Arola, and M. J. Salvado. 2005. Grape Seed Procyanidins Prevent Oxidative Injury by Modulating the Expression of Antioxidant Enzyme Systems. *Journal of Agricultural and Food Chemistry* 53: 6080–86.

Ramachandran, S., S. K. Singh, C. Larroche, C. R. Soccol, and A. Pandey. 2007. Oil Cakes and Their Biotechnological Applications—a Review. *Bioresource Technology* 98: 2000–09.

Ranalli, A., S. Contento, C. Schiavone, and N. Simone. 2001. Malaxing Temperature Affects Volatile and Phenol Composition as Well as Other Analytical Features of Virgin Olive Oil. *European Journal of Lipid Science and Technology* 103: 228–38.

Ranalli, A., M. L. Ferrante, G. De Mattia, and N. Costantini. 1999. Analytical Evaluation of Virgin Olive Oil of First and Second Extraction. *Journal of Agricultural and Food Chemistry* 47: 417–24.

Rand, M., A. Hennissen, and G. Hornstra. 1988. Effects of Dietary Palm Oil on Arterial Thrombosis, Platelet Responses and Platelet Membrane Fluidity in Rats. *Lipids* 23: 1019–23.

Rodriguez, G., A. Lama, R. Rodriguez, A. Jimenez, R. Guillen, and J. Fernandez-Bolanos. 2008. Olive Stone an Attractive Source of Bioactive and Valuable Compounds. *Bioresource Technology* 99: 5261–69.

Rosenthal, A., D. L. Pyle, K. Niranjan, S. Gilmour, and L. Trinca. 2001. Combined Effect of Operational Variables and Enzyme Activity on Aqueous Enzymatic Extraction of Oil and Protein from Soybean. *Enzyme and Microbial Technology* 28: 499–509.

Rukmini, C., and T. C. Raghuram. 1991. Nutritional and Biochemical Aspects of the Hypolipidemic Action of Rice Bran Oil—a Review. *Journal of the American College of Nutrition* 10: 593–601.

Saha, B. C. 2003. Hemicellulose Bioconversion. *Journal of Industrial Microbiology & Biotechnology* 30: 279–91.

Santamaria, R. I., C. Soto, M. E. Zuniga, R. Chamy, and A. Lopez-Munguia. 2003. Enzymatic Extraction of Oil from Guevina Avellana, the Chilean Hazelnut. *Journal of the American Oil Chemists' Society* 80: 33–36.

Schmidt, S., and J. Pokorny. 2005. Potential Application of Oilseeds as Sources of Antioxidants for Food Lipids—a Review. *Czech Journal of Food Sciences* 23: 93–102.

Shah, S., A. Sharma, and M. N. Gupta. 2005. Extraction of Oil from Jatropha Curcas L. Seed Kernels by Combination of Ultrasonication and Aqueous Enzymatic Oil Extraction. *Bioresource Technology* 96: 121–23.

Shahidi, F. 2004. Functional Foods: Their Role in Health Promotion and Disease Prevention. *Journal of Food Science* 69: R146–49.

Shahidi, F. 2009. Nutraceuticals and Functional Foods: Whole versus Processed Foods. *Trends in Food Science & Technology* 20: 376–87.

Shahidi, F., R. Amarowicz, H. A. AbouGharbia, and A. A. Y. Shehata. 1997. Endogenous Antioxidants and Stability of Sesame Oil as Affected by Processing and Storage. *Journal of the American Oil Chemists' Society* 74: 143–48.

Shamban, A. T. 2009. Current and New Treatments of Photodamaged Skin. *Facial Plastic Surgery* 25: 337–46.

Sharma, A., S. K. Khare, and M. N. Gupta. 2001. Enzyme-Assisted Aqueous Extraction of Rice Bran Oil. *Journal of the American Oil Chemists' Society* 78: 949–51.

Sharma, A., S. K. Khare, and M. N. Gupta. 2002. Enzyme-Assisted Aqueous Extraction of Peanut Oil. *Journal of the American Oil Chemists Society* 79: 215–18.

Sineiro, J., H. Dominguez, M. J. Nunez, and J. M. Lema. 1998. Optimization of the Enzymatic Treatment during Aqueous Oil Extraction from Sunflower Seeds. *Food Chemistry* 61: 467–74.

Singh, J., and P. C. Bargale. 2000. Development of a Small Capacity Double Stage Compression Screw Press for Oil Expression. *Journal of Food Engineering* 43: 75–82.

Siniscalco, V., and G. F. Montedoro. 1988. Estrazione Maccanica Dell'olio di Oliva Mediante L'impliego di Coadiuvanti Tecnologic. *La Rivista de lle Sostanze Grasse* 65: 675–78.

Smith, D. D., Y. C. Agrawal, B. C. Sarkar, and B. P. N. Singh. 1993. Enzymatic-Hydrolysis Pretreatment for Mechanical Expelling of Soybeans. *Journal of the American Oil Chemists' Society* 70: 885–90.

Sosulski, K., and F. Sosulski. 1993. Enzyme-Aided vs. Two-Stage Processing of Canola: Technology, Product Quality and Cost Evaluation. *Journal of the American Oil Chemists' Society* 70: 825–29.

Soto, C., J. Concha, and M. E. Zuniga. 2008. Antioxidant Content of Oil and Defatted Meal Obtained from Borage Seeds by an Enzymatic-Aided Cold Pressing Process. *Process Biochemistry* 43: 696–99.

Soto, C. G., R. Chamy, and M. E. Zuniga. 2004. Effect of Enzymatic Application on Borage (Borago officinalis) Oil Extraction by Cold Pressing. *Journal of Chemical Engineering of Japan* 37: 326–31.

Soto, C., R. Chamy, and M. E. Zuniga. 2007. Enzymatic Hydrolysis and Pressing Conditions Effect on Borage Oil Extraction by Cold Pressing. *Food Chemistry* 102: 834–40.

Sowbhagya, H. B., K. T. Purnima, S. P. Florence, A. G. Appu Rao, and P. Srinivas. 2009. Evaluation of Enzyme-Assisted Extraction on Quality of Garlic Volatile Oil. *Food Chemistry* 113: 1234–38.

Sowbhagya, H. B., P. Srinivas, and N. Krishnamurthy. 2010. Effect of Enzymes on Extraction of Volatiles from Celery Seeds. *Food Chemistry* 120: 230–34.

Suja, K. P., A. Jayalekshmy, and C. Arumughan. 2004. Free Radical Scavenging Behavior of Antioxidant Compounds of Sesame (Sesamum indicum L.) in DPPH Center Dot System. *Journal of Agricultural and Food Chemistry* 52: 912–15.

Tano-Debrah, K., and Y. Ohta. 1994. Enzyme-Assisted Aqueous Extraction of Fat from Kernels of the Shea Tree, Butyrospermum Parkii. *Journal of the American Oil Chemists' Society* 71: 979–83.

Tano-Debrah, K., and Y. Ohta. 1995. Application of Enzyme-Assisted Aqueous Fat Extraction to Cocoa Fat. *Journal of the American Oil Chemists' Society* 72: 1409–11.

Tano-Debrah, K., Y. Yoshimura, and Y. Ohta. 1996. Enzyme-Assisted Extraction of Shea Fat: Evidence from Light Microscopy on the Degradative Effects of Enzyme Treatment on Cells of Shea Kernel Meal. *Journal of the American Oil Chemists' Society* 73: 449–53.

Terra, X., J. Valls, X. Vitrac, J. M. Merrillon, L. Arola, A. Ardevol, C. Blade, J. Fernandez-Larrea, G. Pujadas, J. Salvado, and M. Blay. 2007. Grape-Seed Procyanidins Act as Antiinflammatory Agents in Endotoxin-Stimulated Raw 264.7 Macrophages by Inhibiting Nfkb Signaling Pathway. *Journal of Agricultural and Food Chemistry* 55: 4357–65.

Tobar, P., A. Moure, C. Soto, R. Chamy, and M. E. Zuniga. 2005. Winery Solid Residue Revalorization into Oil and Antioxidant with Nutraceutical Properties by an Enzyme Assisted Process. *Water Science and Technology* 51: 47–52.

Topallar, H., and U. Gecgel. 2000. Kinetics and Thermodynamics of Oil Extraction from Sunflower Seeds in the Presence of Aqueous Acidic Hexane Solutions. *Turkish Journal of Chemistry* 24: 247–53.

Tripoli, E., M. Giammanco, G. Tabacchi, D. Di Majo, S. Giammanco, and M. La Guardia. 2005. The Phenolic Compounds of Olive Oil: Structure, Biological Activity and Beneficial Effects on Human Health. *Nutrition Research Reviews* 18: 98–112.

Tsai, K. Y., and H. C. Hsu. 2009. Successful Treatment of Topical Tretinoin Cream in a Patient with Unilateral Nevoid Hyperkeratosis of Nipple and Areola. *Dermatologica Sinica* 27: 128–31.

Uedo, N., M. Tatsuta, H. Iishi, M. Baba, N. Sakai, H. Yano, and T. Otani. 1999. Inhibition by D-Limonene of Gastric Carcinogenesis Induced by N-Methyl-N'-Nitro-N-Nitrosoguanidine in Wistar Rats. *Cancer Letters* 137: 131–36.

Ullmann., F. 1985. *Ullmann's Encyclopedia of Industrial Chemistry*, 5th ed. Weinheim, Germany: Verlagsgesellchaft.

van Rooyen, J., A. J. Esterhuyse, A. M. Engelbrecht, and E. F. du Toit. 2008. Health Benefits of a Natural Carotenoid Rich Oil: A Proposed Mechanism of Protection against Ischaemia/Reperfusion Injury. *Asia Pacific Journal of Clinical Nutrition* 17: 316–19.

Velasco, L., and F. D. Goffman. 1999. Chemotaxonomic Significance of Fatty Acids and Tocopherols in Boraginaceae. *Phytochemistry* 52: 423–26.

Vierhuis, E., M. Servili, M. Baldioli, H. A. Schols, A. G. J. Voragen, and G. Montedoro. 2001. Effect of Enzyme Treatment during Mechanical Extraction of Olive Oil on Phenolic Compounds and Polysaccharides. *Journal of Agricultural and Food Chemistry* 49: 1218–23.

Visioli, F., G. Bellomo, G. Montedoro, and C. Galli. 1995. Low Density Lipoprotein Oxidation Is Inhibited in Vitro by Olive Oil Constituents. *Atherosclerosis* 117: 25–32.

Visioli, F., and C. Galli. 1998. Olive Oil Phenols and Their Potential Effects on Human Health. *Journal of Agricultural and Food Chemistry* 46: 4292–96.

Visioli, F., and C. Galli. 2002. Biological Properties of Olive Oil Phytochemicals. *Critical Reviews in Food Science and Nutrition* 42: 209–21.

Visioli, F., S. Grande, P. Bogani, and C. Galli. 2006. Antioxidant Properties of Olive Oil Phenolics. *Olive Oil and Health*. Edited by J. L. Quiles, M. C. Ramírez-Tortosa, and P. Yaqoob. Wallingford, UK: CABI, 110–18.

Ward, O. P., and A. Singh. 2005. Omega-3/6 Fatty Acids: Alternative Sources of Production. *Process Biochemistry* 40: 3627–52.

Weetall, H. 1977. Industrial Application of Immobilized Enzymes: Present State of the Art. *Biotechnological Application of Proteins and Enzymes*. Edited by Z. Bohak and S. Nathand. New York: Academic Press.

Wettasinghe, M., and F. Shahidi. 1999a. Antioxidant and Free Radical-Scavenging Properties of Ethanolic Extracts of Defatted Borage (Borago officinalis L.) Seeds. *Food Chemistry* 67: 399–414.

Wettasinghe, M., and F. Shahidi. 1999b. Evening Primrose Meal: A Source of Natural Antioxidants and Scavenger of Hydrogen Peroxide and Oxygen-Derived Free Radicals. *Journal of Agricultural and Food Chemistry* 47: 1801–12.

Wettasinghe, M., F. Shahidi, R. Amarowicz, and M. M. Abou-Zaid. 2001. Phenolic Acids in Defatted Seeds of Borage (Borago officinalis L.). *Food Chemistry* 75: 49–56.

Wilson, T. A., R. J. Nicolosi, T. Kotyla, K. Sundram, and D. Kritchevsky. 2005. Different Palm Oil Preparations Reduce Plasma Cholesterol Concentrations and Aortic Cholesterol Accumulation Compared to Coconut Oil in Hypercholesterolemic Hamsters. *Journal of Nutritional Biochemistry* 16: 633–40.

Wilson, T. A., R. J. Nicolosi, B. Woolfrey, and D. Kritchevsky. 2007. Rice Bran Oil and Oryzanol Reduce Plasma Lipid and Lipoprotein Cholesterol Concentrations and Aortic Cholesterol Ester Accumulation to a Greater Extent than Ferulic Acid in Hypercholesterolemic Hamsters. *Journal of Nutritional Biochemistry* 18: 105–12.

Wolfe, K. L., and R. H. Liu. 2003. Apple Peels as a Value-Added Food Ingredient. *Journal of Agricultural and Food Chemistry* 51: 1676–83.

Yamakoshi, J., M. Saito, S. Kataoka, and S. Tokutake. 2002. Procyanidin-Rich Extract from Grape Seeds Prevents Cataract Formation in Hereditary Cataractous (Icr/F) Rats. *Journal of Agricultural and Food Chemistry* 50: 4983–88.

Zúñiga, M. E., R. Chamy, and J. M. Lema. 2001a. Canola and Chilean Hazelnut Products Obtained by Enzyme-Assisted Cold-Pressed Oil Extraction. *Proceedings of the World Conference on Oilseeds Processing and Utilization*. Edited by R. F. Wilson. Urbana, IL: AOCS Press, 203–10.

Zuñiga, M. E., J. Concha, C. Soto, and R. Chamy. 2001b. Enzyme Formulation Effect of the Rosehip Oil Cold-Pressed Extraction Process. *Proceedings of the World Conference and Exhibition on Oilseed Processing and Utilization*. Edited by R. F. Wilson. Urbana, IL: AOCS Press, 210–13.

Zuñiga, M. E., C. Soto, A. Mora, R. Chamy, and J. M. Lema. 2003. Enzymatic Pre-Treatment of Guevina Avellana Mol Oil Extraction by Pressing. *Process Biochemistry* 39: 51–57.

CHAPTER 5

Nutraceutical Processing Using Mixing Technology

Theory and Equipment

Sri Hyndhavi Ramadugu, Vijay Kumar Puli, and Yashwant Pathak

CONTENTS

5.1 DEFINITION

Mixing is one of the common unit operations used in the preparation of different types of formulations in the nutraceutical industry, such as liquids, tablets, capsules, and compound powders. Mixing is most widely used when two or more substances are combined. Every product needs perfect mixing, in which each particle of one material lies as close as possible to a particle of another material. Mixing of nutraceuticals depends on the type of the product and the objective of mixing. It is a unit operation in which a uniform mixture is obtained from two or more components by dispersing one into the others. The larger component of the mixture is called the *continuous phase*, and the smaller component is called the *dispersed phase*.

5.1.1 Objectives of Mixing

Mixing of nutraceuticals such as solids, liquids, and gases is one of the most common processes in the food and nutraceutical industry. Mixing increases the homogeneity of the product by reducing the nonuniformity of ingredients in the composition of the product. Mixing plays a major role in the preparation of many products because it ensures the delivery of the product with reproducible properties (Cullen 2009).

Several mechanisms are involved in the process of mixing. The process of mixing ranges from simple diffusion of the particles to shear deformation flow and then redistribution of the particles. During the shear flow of materials, collision of particles takes place, which results in rearrangements of particles. Finally, the process of mixing takes place. Mixing is an important technique, but at the same time overmixing of the particles results in undesirable properties that lead to an unacceptable product (Lindley 1991). There are many factors that influence the quality of mixing, such as mixing time, speed of mixing rotation, type of mixer, and dry or wet mixing processes (Kung et al. 2009).

5.1.2 Importance of Mixing

Mixing is a critical process because the quality of the final product and its attributes are derived by the quality of the mix. Improper mixing results in a nonhomogeneous product with respect to the desired attributes like chemical composition, color, texture, and also particle size. In many industrial applications, the process of mixing requires careful selection, design, and scale-up to get effective and efficient mixing of the product. The terms *mixing* and *blending* are often used interchangeably, but they are slightly different. Blending is a process of combining materials; it is a relatively gentle process compared with mixing. Blending is mostly useful for solid–solid mixing or mixing of bulk solids with a small quantity of liquid. Mixing and blending are the most demanding unit operations in the nutraceutical and pharmaceutical industries (Tekchandaney 2009).

5.2 TYPES OF MIXTURES

Depending on the behavior of the particle, mixtures are derived into three types, which are outlined below.

5.2.1 Positive Mixtures

Positive mixtures are defined as mixtures that mix spontaneously and irreversibly by diffusion and obtain a perfect mix. This type of mixture does not require any energy. An example of positive mixtures includes miscible liquids such as corn and soybean oil.

5.2.2 Negative Mixtures

Negative mixtures are defined as the components that will tend to separate out, and this requires a continuous input of energy to keep the components adequately dispersed. These negative mixtures require a higher degree of mixing efficiency than positive mixtures. Examples include corn oil and water.

5.2.3 Neutral Mixtures

These mixtures remain static in nature. These mixtures have no tendency to mix spontaneously or separate spontaneously. These mixtures are capable of demixing, but energy input is required. Examples include mixed powders and pastes (Cullen 2009).

5.3 PROCESS OF MIXING

Mixing is the combination of different materials. Four mixing operations are possible based on the mixing of different phase materials in the food and nutraceutical industry.

5.3.1 Solid–Solid Mixing

Mixing of solids is an important phenomenon in many industries, especially in the nutraceutical and food industry. In solid–solid mixing, a product with too low an active ingredient will be ineffective, and the product with too high an active ingredient may be lethal. To provide good solid–solid mixing, the tendency to segregate is a phenomenon to be avoided or overcome. Segregation occurs when a system contains particles of different sizes or densities (Holdich 2002).

5.3.2 Solid–Liquid Mixing

Among various industrial applications of stirred vessels, the agitation of solid–liquid mixing is most common. In this type of mixing, mechanical agitation is used to suspend the particles in a liquid to promote mass transfer or a chemical reaction. The liquids used in these applications usually have low viscosity, and the particles will settle down when agitation stops. One of the most important aspects of solid–liquid mixing is the distribution of solid particles inside the volume because in most cases it may affect the apparatus performance and the process efficiency.

5.3.3 Gas–Liquid–Solid Mixing

In some processes of mixing, such as catalytic hydrogenation of vegetable oils, slurry reactors, three-phase fluidized beds, froth flotation, and fermentation process, the success and efficiency of mixing are directly influenced by the extent of mixing between the three phases of gas, liquid, and solid (Chhabra and Richardson 2008).

5.3.4 Liquid–Liquid Mixing

Liquid–liquid extraction is a process where two immiscible liquids are stirred together and one liquid becomes dispersed as droplets in the second liquid, which forms a continuous phase. Liquid–liquid extraction, a process using successive mixing and settling stages, is an important example of this type of mixing. Agitation plays an important role in the liquid–liquid systems. It controls the breakup of drops, which is called *dispersion*, the combining of drops, known as *coalescence*, as well as the suspension of drops within the system (Calabrese, Leng, and Armenante 2006).

5.4 FACTORS AFFECTING THE PROCESS OF MIXING

Product mixing depends on various factors, such as particle shape, particle size, particle attraction, density of material, and proportion of materials.

5.4.1 Particle Shape

Shape of the material is important for the perfect mix. The ideal particle shape is spherical for uniform mixing. Spherical particles always have a better flowability and therefore are more easily mixed, but at the same time these particles also tend to segregate more easily than nonspherical particles.

5.4.2 Particle Size

Particle size is also considered one of the factors affecting the process for a perfect mix. Variation in particle size leads to the separation because small particles move downward through the space present between the larger particles.

5.4.3 Particle Attraction

During mixing, many particles exert attractive forces as a result of electrostatic charges present in them. This leads to separation of particles.

5.4.4 Density of the Material

Powders of different density are difficult to mix because dense materials always move downward and tend to settle down at the bottom.

5.4.5 Proportions of Materials

The perfect product mix is best achieved if the two powders are mixed in equal proportions by weight or volume. This type of mixing is always done in the ascending order of their weights.

5.5 MIXING OF POWDERS

Mixing of powders such as solids has a specific property because particles need to move faster to be reactive to each other. Mixing of powders takes place through several mechanisms, such as convective mixing, shear mixing, and diffusion mixing.

5.5.1 Convective Mixing

Convective mixing takes place with the bulk movement of the groups of particles from one part of the powder bed to another. This type of mixing occurs by an inversion of the powder bed using blades and paddles. In convective mixing, a circulating flow of powders is usually caused by the rotational motion of a mixer vessel, an agitating impeller. This type of mixing is more widely used for mixing bulk quantities of the product (Weighing & Batching Systems 2006).

5.5.2 Shear Mixing

Shear mixing occurs when a layer of one material flows over another material. During the shear mixing of products, when shear forces occur, it reduces the scale of segregation by thinning the dissimilar layers of solid material. High-shear mixers rotate with a high mixing tool speed such that both the mixing and grinding effect are achieved (Daunmann et al. 2009).

5.5.3 Diffusion Mixing

Diffusion mixing takes place when there is a random motion of particles taking place within a powder bed that causes them to change their position relative to one another. Diffusion mixing has the potential to produce a random mix, resulting in low speed of mixing (Kung et al. 2009).

5.6 MIXING OF LIQUIDS

Mixing of liquids is an important phenomenon because it involves two liquids and follows three mechanisms: turbulent mixing, bulk transport, and molecular diffusion.

Bulk transport of liquids is the same as that of convective mixing of solids and involves the movement of large volume of materials from one position in the mixer to another. Bulk transport of liquids produces a high degree of mixing and is considered a quick process.

Turbulent mixing of liquids takes place when the movement of particles happens in a turbulent manner. During the process of mixing, the constant changes in the speed and direction of particles mean that the induced turbulence is a highly effective mechanism for mixing of particles.

Molecular diffusion results in the mixing of individual molecules of the product. This type of mixing mostly occurs with miscible fluids whenever the concentration gradient exists, which results in the formation of a perfect mix (Feder 2008).

5.7 MIXING OF SEMISOLIDS

The mechanism involved in mixing semisolids depends on the character of the material, which can show considerable variation. Solids and liquids undergo perfect uniform mixing, whereas semisolids rarely undergo uniform mixing. Many semisolids, including neutral mixtures, have no tendency to segregate, although sedimentation may occur. The most commonly used semisolid mixers are listed below:

1. Planetary mixes: This type of mixer has a mixing arm rotating about its own axis and also about a common axis usually at the center of the mixing wheel. The blades provide a kneading action, and the narrow passage between the blades and the wall of the mixer can provide shear forces.

2. Triple roller mill: The differential speed and narrow clearance between the rollers develop high shear over small volumes of material. The roller mills are generally used to grind and complete the homogeneity of ointments.
3. Sigma blade mixer: These mixers contain two blades that operate in a mixing vessel with a double trough shape and the blades moving at different speeds toward each other. These types of mixers are used for products like ointments, probiotics, and prebiotics.

5.8 TECHNIQUES, METHODS, AND EQUIPMENT USED FOR THE MIXING OF NUTRACEUTICALS

Mixing is the combination of different materials. In mixing, the spatial distribution of separate components is reduced to obtain a certain degree of homogeneity. Various mixing techniques are distinguished in the nutraceutical industry. Solid–solid mixing is mainly used for the preparation of food products such as blends of tea, coffee, and malt, whereas solid–liquid mixing is carried out during the production of many spices and herbs and is mainly used when the ingredients are mixed in more or less a liquid state and then solidify on cooling. Liquid–liquid mixing is used during the production of emulsions and mixtures of miscible liquids. Liquid–gas mixing is used during spray drying, and the liquid phase is mixed in a stream of gas (Daunmann et al. 2009; Mixing, Blending, Stirring n.d.).

Several types of mixing equipment are used for the separation of different products in large quantities.

5.8.1 Agitators: Mixers for Liquid–Liquid and Liquid–Gas Mixing

These mixers carry out a variety of process functions such as blending miscible liquids, contacting or dispersing immiscible liquids, dispersing a gas in a liquid, heat transfer in agitated liquid, and suspension of solids in liquids. These are carried out in agitated vessels with the help of rotating impellers. Depending on the angle that the impeller blade makes with the plane of impeller rotation, these impellers are classified into two types: axial and radial. These have many applications in the nutraceutical and pharmaceutical industry, including antibiotics, vitamins and minerals, mixing of grain weeds, cereals, herbal supplements, nutritional supplements, spices and herbs, and gelatins.

- Axial flow impellers: The impeller blade makes an angle of less than 90° with the plane of impeller rotation. As a result, the locus of flow occurs along the axis of the impeller.
- Radial flow impellers: The impeller blade in radial flow impellers is parallel to the axis of the impeller. As a result, the radial flow impeller discharges flow along the impeller radius in distinct patterns.

5.8.2 Rotary Batch Mixers

Rotary batch mixers are used for the mixing of large quantities and produce uniform particle distribution, uniform liquid coatings, and gentle product handling with low energy. These types of mixtures are mainly used for powder substances having the specific features of four-way mixing action, such as tumble, cut, turn, and fold. This equipment provides equal efficient mixing at any capacity and is ideal for encapsulation, drying, heating, and cooling processes. It handles dry powder and granular products. These mixers provide rapid and uniform distribution of particles without any breakdown. These mixers are used in the nutraceutical industry for production of antibiotics, vitamins and minerals, mixing of grain weeds, cereals, herbal supplements, nutritional supplements, spices and herbs, and gelatins. These rotary batch mixers are classified into standard rotary batch mixers, rotary glass batch mixers, rotary batch dryers or coolers, and mini rotary batch mixers.

5.8.3 Ribbon Blenders, Paddle Blenders, and Plow Blenders

These types of mixers are most widely used in the nutraceutical and food industries. These mixers have high mixing spin, efficiency, and a wide range of features that are useful for more demanding applications. The capacity of these blenders ranges from 1–100 ft.^{3}, with two blender styles for heavy to extra heavy duty mixing. In this type of mixer, every particle is subjected to agitation during blending, loading of materials, blending, and discharge. Mixing capacity in this equipment is higher and calculated approximately as 70%–80% of total vessel volume. Some special features of this equipment include the ability to blend 1% concentrate to perfect analysis, in batches more than 6 tons, with a capacity of 1–1000 ft.^{3} It also has broad design versatilities to meet customer requirements. These blenders have many applications in production of herbal supplements, minerals, nutritional supplements, pharmaceuticals, starches, soybean oil, spices, and gelatins.

5.8.4 Cylindrical Plow Benders

These blenders are widely used for fibrous, interlocking, dense, abrasive, and moist or oily materials at higher speeds with higher intensity and with more uniform results. The horizontal-orientated mixing shaft rotates approximately 4 times faster than the shafts of trough rates, minimizing impact and degradation. The particles blend faster than trough-style horizontal mixers, achieve 100% uniform blends, and handle a greater variety of materials. These blenders are not suitable for mixing pastes. These blenders are configured with single or multiple charging ports located on top of the vessel. Several applications include spices, herbal products, vitamins, minerals, nutraceuticals, and pharmaceuticals (Mixers, Blenders and Size Reduction Equipment for Bulk Solid Materials n.d.).

5.8.5 Rotary Continuous Blenders

Continuous blenders mix the particles with highest mixing homogeneity, providing 100% uniform dust-free blending at rates ranging 25–5000 ft.^{3}/h. These are mostly used for mixing bulk quantities in the food and pharmaceutical industries.

Continuous blenders have maximum blending action with minimum requirements for product residence time. A series of internal, stationary circular flights is arranged with pitches diametrically opposed to provide optimum blending action. These machines have staggered right- or left-hand blending flights, which creates a unique splitting and combining action that produces a uniform blend of ingredients of varying particle sizes and densities. These mixers ensure unlimited versatility and maintenance-free operation. Most suitable for liquid additions and coatings, rotary continuous mixers are rugged, compact, easy to maintain, and simple to clean. These mixers are capable of completely discharging any remaining material and offer an optional reversing discharge feature. The capacity of these mixers ranges from 25–5000 ft.^{3}/h and consumes low power.

These mixers have both dry and liquid mixing applications, and they are mostly used for antibiotic and vitamin remixes, blending grains and cereals, herbal supplements, minerals, nutritional supplements, pharmaceuticals, soybean meal, and starches (Mixers, Blenders and Size Reduction Equipment for Bulk Solid Materials n.d.).

5.8.6 High-Intensity Continuous Blenders

These blender mixers are specifically designed for intensive mixing. These are particularly useful when the scale of scrutiny, which is the scale at which mixing is required, is small compared with batch size, as well as when there are more powders to be mixed and some of the ingredients are small in quantity. These are also used to mix small quantities of liquid materials. High-intensity continuous blenders provide the added shear needed for high-speed blending and homogenizing and declumping of dry ingredients, slurries, and pastes. The blender is equipped with spray ports for large additions of liquids that can also serve to mix dry materials. The machine has a high capacity that operates on small working volumes, with minimum residence times of 30 to 50 seconds. Some specific features of these high-intensity continuous blenders are a single agitator style with high-speed continuous processor and a capacity up to 2800 ft.^{3}/h. It has a high-impact, pug mill–type, pin-and-paddle agitator for intensive blending, and spray ports are available for large additions of liquids. Typical applications of these blenders include antibiotic and vitamin remixes, blending grains and cereals, herbal supplements, minerals, nutritional supplements, pharmaceuticals, soybean meal, and starches.

5.9 THEORIES INVOLVED IN MIXING TECHNOLOGY

5.9.1 Theory of Solid Mixing

Compared with liquids and semisolids, it is difficult to achieve a uniform or perfect mix with solid materials. The degree of mixing of solid particles depends on the relative particle size, shape, and density of each component, the moisture component, surface and flow characteristics of each component, and also the tendency of the materials to aggregate. Materials having similar size, shape, and density tend to form a uniform or a perfect mix more easily than dissimilar materials. During mixing of particles, the difference in these properties leads to separation of particles. The uniformity of the final product depends on the equilibrium achieved between the mechanisms of mixing and unmixing.

If two component mixtures are sampled at the start of mixing, most samples will consist entirely of one of the components. As the mixing process continues, the composition of each sample becomes more uniform and approaches the average composition of the mixture. One method of determining the changes in composition is to calculate the standard deviation of each fraction in successive samples:

$$\sigma_m = \sqrt{\left[\frac{1}{(n-1)}\sum[(c-c)]^2\right]}$$

where σ_m = standard deviation
n = number of samples
c = concentration of the component in each sample
$\bar{c}$ = mean concentration of the samples

5.9.2 Theory of Liquid Mixing

Mixing of liquids includes mixing of liquids with low viscosity and with high viscosity. The component velocities induced in low viscosity liquids are longitudinal velocity, rotational velocity, and radial velocity. To mix low viscosity liquids, most frequently turbulence must be induced throughout the bulk of the liquids to mix slow-moving parts within the faster-moving parts. In high viscosity liquids, mixing occurs by kneading (the material against the vessel wall or into another material), folding (unmixed food into the mixed part), and shearing to stretch the material (Hui 2006).

The rate of mixing is characterized by a mixing index. The mixing rate constant equation depends on the characteristics of both the mixers and the liquids. The effect of the mixer's characteristics on k is given by:

$$\mathrm{k}\alpha = \frac{D^3 N}{D_t^2 Z}$$

where D (m) = diameter of the agitator
N (rev$^{S^{-1}}$) = agitator speed
D_t (m) = vessel diameter
Z (m) = height of the liquid

5.10 FUTURE ASPECTS IN MIXING OF NUTRACEUTICALS

Mixing technology has wide applications in the nutraceutical industry, where it is used to combine many ingredients to achieve different functional properties or characteristics. It is most often used primarily to develop desirable product characteristics rather than simply ensure homogeneity. It is multicomponent, containing ingredients of different physical properties and quantities. The nutraceutical industry must continuously seek improvements in process design to increase efficiency and facilitate the development of novel products. Despite ubiquity of the mixing process and large quantities of materials mixed every day, mixing processes are not fully understood. Advances in computational techniques have facilitated a more fundamental understanding of mixing process for both complex fluids and mixer designs. Perfect mixing is rarely possible; consequently, mixing of nutraceuticals will be a source of variability in the manufacturing processes. There is an increasing trend in the nutraceutical industry to adopt a process analytical technology (PAT) framework for innovative process manufacturing and quality assurance. PAT is initiated by the Food and Drug Administration and is specially designed to ease the uptake of new analytical technologies (Freeman 2008). PAT is a system for designing, analyzing, and controlling manufacturing process to achieve a good quality of the material. PAT is now being used in the food and nutraceutical industry, where it is seen as a technology that helps many industries to improve their conformity with good mixing technology. The emphasis in PAT is on the manufacturing process to increase the basic premise of the current mixing quality system because quality cannot be imparted into products—it should be built in or should be by design.

REFERENCES

Calabrese, R. V., D. E. Leng, and P. M. Armenante. 2005. Liquid-Liquid Mixing in Agitators. *Encyclopedia of Chemical Processing*. Boca Raton, FL: Taylor and Francis.

Campbell, G. M. 1995. New Mixing Technology for the Food Industry. *Food Technology International Europe*. Edited by A. Turner. London: Sterling International, 282–96.

Chhabra, R. B., and J. F. Richardson. 2008. *Non-Newtonian Flow and Applied Rheology: Engineering Applications*. Oxford, UK: Butterworth-Heinemann, 1176–82.

Cullen, P. J. 2009. *Food Mixing: Principles and Applications*. Hoboken, NJ: Wiley Blackwell, 148–242.

Daunmann, B., X. Sun, M. Anlauf, S. Gerl, and H. Nirschl. 2009. Mixing Agglomeration in a High Shear Mixer with a Stirred Mixing Vessel. *Chemical Engineering and Technology* 33: 42–56.

Eder, D. 2008. Trends in Mixing and Blending. *Food Processing*. http://www.foodprocessing.com/articles/2008/064.html.

Fellows, P. J. 2000. *Food Processing Technology Principles and Practice*. Cambridge, UK: Woodhead Publishing, 132–34.

Freeman, T. 2008. Embracing QbD and PAT. http://www.pharmaceuticalonline.com/article.mvc/Embracing-QbD-And-PAT-0002?user=20&source=nl:21678.

Holdich, R. 2002. *Fundamentals of Particle Technology*. Nottingham, UK: Midland Information Technology & Publishing. http://midlandit.co.uk/particletechnology/mixing.pdf.

Hui, Y. H. 2006. *Handbook of Food Science Technology and Engineering*. Boca Raton, FL: Taylor and Francis, 102–09.

Kung, C., T. Liao, K. Tseng, K. Chen, and M. Chaung. 2009. The Influences of Powder Mixing Process on the Quality of W-CU Composites. *Transactions of the Canadian Society for Mechanical Engineering* 33: 361–76.

Lindley, J. A. 1991. Mixing Processes for Agricultural and Food Materials. Part 2: Highly Viscous Liquids and Cohesive Materials. *Journal of Agricultural Engineering Research* 48: 9–15.

Mixers, Blenders and Size Reduction Equipment for Bulk Solid Materials. n.d. http://www.munsonmachinery.com.

Mixing, Blending, Stirring. n.d. http://www.hyfoma.com/en/content/processing-technology/size-reduction-mixing-forming/mixing-blending-stirring.

Process Analytical Technology. n.d. http://processanalyticaltechnology.com.

Process Engineering Solutions. n.d. http://www.process.gr/index.asp.

Tekchandaney, J. 2009. Introduction to Mixing Technology. *Mechanical Engineering, Mixing Technology*. http://www.brighthub.com/engineering/mechanical/articles/41798.aspx.

Weighing & Batching Systems. 2006. *Powder Bulk Solids, the Source for Dry Processing and Bulk Handling Technology*. http://www.powderbulksolids.com/editorial/detail-article.php?id=2137.

CHAPTER 6

Separation Technologies in Nutraceutical Processing

Vijay Kumar Puli, Sri Hyndhavi Ramadugu, and Yashwant Pathak

CONTENTS

6.1 INTRODUCTION

Since 1990, there has been a lot of interest in functional foods and nutraceuticals, particularly because of their impact on human health and the prevention of certain diseases. Apart from growing good quality natural products, there are challenges in extracting, refining, analyzing active ingredients, and evaluating safety before an effective, reliable product is made. Special attention needs to be focused on separation and purification mainly because 40%–80% of the cost of production is assigned to separation and purification unit operations in making functional food and nutraceuticals. There are also several carotenoids and polyphenol products that have been identified from various plants and a single plant could contain highly complex profiles of these compounds. These are labile to heat, air, and light, and they may exist at low concentrations in the plants. This makes the separation and detection of nutraceuticals a challenging task. In this chapter, we will focus on the techniques used for the separation of nutraceuticals from the natural products, as well as their applications and influencing factors.

6.2 PRESSURE-DRIVEN PROCESSES

6.2.1 Membrane Basics and Principles

In the past two decades, a number of separation processes have emerged in which membranes play a major role in the isolation and enrichment of functional compounds by concentrating, fractionating, and purifying solutions or colloidal dispersions. Membrane separations have led to the development of innovative products with unique functional, nutritional, and health attributes.

Membrane processing permits concentration and separation without the use of heat. Particles are separated based on molecular size, shape, and or charge via pressure and specially designed semipermeable membranes. The most commonly used membrane processes in the food industry are pressure driven. Depending on the size of the particles to be separated, membrane filtration can be classified as microfiltration (MF), ultrafiltration (UF), nanofiltration (NF), and reverse osmosis (RO) (Table 6.1).

Membrane separations operate at relatively low temperatures and pressures, resulting in minimal denaturation and degradation of temperature-sensitive ingredients. Compared with thermal processes (e.g., evaporation), membrane filtration is an energy-efficient method of clarifying, concentrating, and purifying food-processing streams. Membrane filtration improves product quality by retaining flavors and vitamins and minimizing protein denaturation (Pilot Plant Membrane Filtration Opportunities for the Food Processing Industry n.d.).

6.2.2 Selecting a Membrane System

Membrane is the most important part of the separation process. The challenge for the process engineer is to develop a process giving the highest yield of the desired product with minimum investment and operational costs. Simple guidelines for implementing membrane filtration into new operational processes are outlined.

- Know the characteristics of the particles and liquids (e.g., dissolved solid content, molecular weight of dissolved species, nature and loading of any suspended material) and the process conditions (e.g., temperature, pressure, and capacity requirements).
- Select a membrane pore size (MF, NF, UF, or RO).
- Ensure chemical compatibility between feed liquid, membrane, and potential membrane fouling.
- Identify design criteria: capacity of the plant, operating costs, nature and desired composition of permeate and retentate.
- Pilot-scale tests using membrane material, pore size, configuration, and intended use on the production unit should be performed. Testing should establish the

Table 6.1 Classification of Membrane Technology by Driving Force

Driving Force			
Pressure	**Concentration**	**Temperature**	**Electrical Potential**
Gas and vapor separation	Membrane extraction dialysis	Membrane distillation	Electrodialysis
Nanofiltration			Electro-osmosis
Ultrafiltration			
Microfiltration			
Pervaporation			

Source: Separation Technologies. http://www.wellnesswest.ca/dmdoc/tw_seperationtechnologies.pdf, 2005.

operating parameters (e.g., batch or continuous operation, permeation rate, pressure drop, and retention levels as concentration increases), process reproducibility, quality of end product, degree of membrane fouling, effectiveness of cleaning regimen, and estimating membrane life under repeated cycles of usage and cleaning.

Cellulosic membranes are most commonly used for bioprocessing applications because of their low fouling characteristics and low protein adsorption. Synthetic polymers such as polysulfone and polyvinylidene fluorides are widely used because of their greater chemical and mechanical stability (Zydney 2000).

6.2.3 Ultrafiltration

UF is used in a variety of industries mainly because of its low energy requirements, nonthermal nature, and simplicity. It covers the region between MF and RO to remove particles in the size range of 0.001–0.02 μm (Separation Technologies 2005). Solvents and salts of low molecular weight will pass through, whereas large molecules are retained. UF requires applied pressure of 1–7 bar to overcome viscous resistance of liquid permeation through the membrane.

UF is ideal for the removal of antinutritional factors such as oligosaccharides, phytic acid, and some trypsin inhibitors from vegetable protein processing to produce purified protein isolates or concentrates with superior functional properties.

Many commercial UF membranes are designed for cross-flow filtration. In cross-flow filtration, the liquid to be filtered flows at high speed along the membrane. The pressure on the membrane is the driving force for the passage of the liquid through the membrane. The high flow prevents the depositing of solids (fouling) and therefore the blockage of the membrane. The concentrated solid matter is called *retentate*, and the filtrate is called *permeate* (Figure 6.1).

Cross-flow membrane filtration technology is gaining acceptance as an important manufacturing step in many industries, from food and dairy to pharmaceutical and biotechnology sectors. Its ability to produce specific separations at low temperatures with no phase change makes the technology a cost-effective solution to conventional methods. In cross-flow filtration the feed flows parallel to the membrane surface, and with an appropriate driving force the permeating species passes through the membrane. This permeate is then collected as a "second" phase. The feed, gradually reduced in concentration of the permeating species, exits the unit as the retentate.

UF technology was tested to be most promising for bioactive peptide separation. The common application of UF in the field of bioactive peptides is the peptide enrichment from protein hydrolysates. Adding UF membranes in the production process of bioactive peptides by the use of enzymatic membrane reactors allows the continuous production of specific peptide sequences with functional and nutritional properties.

Several research reports have described the continuous extraction of bioactive peptides, especially from milk proteins. Apart from their uses in the continuous production of bioactive peptides, UF membranes have also been used to

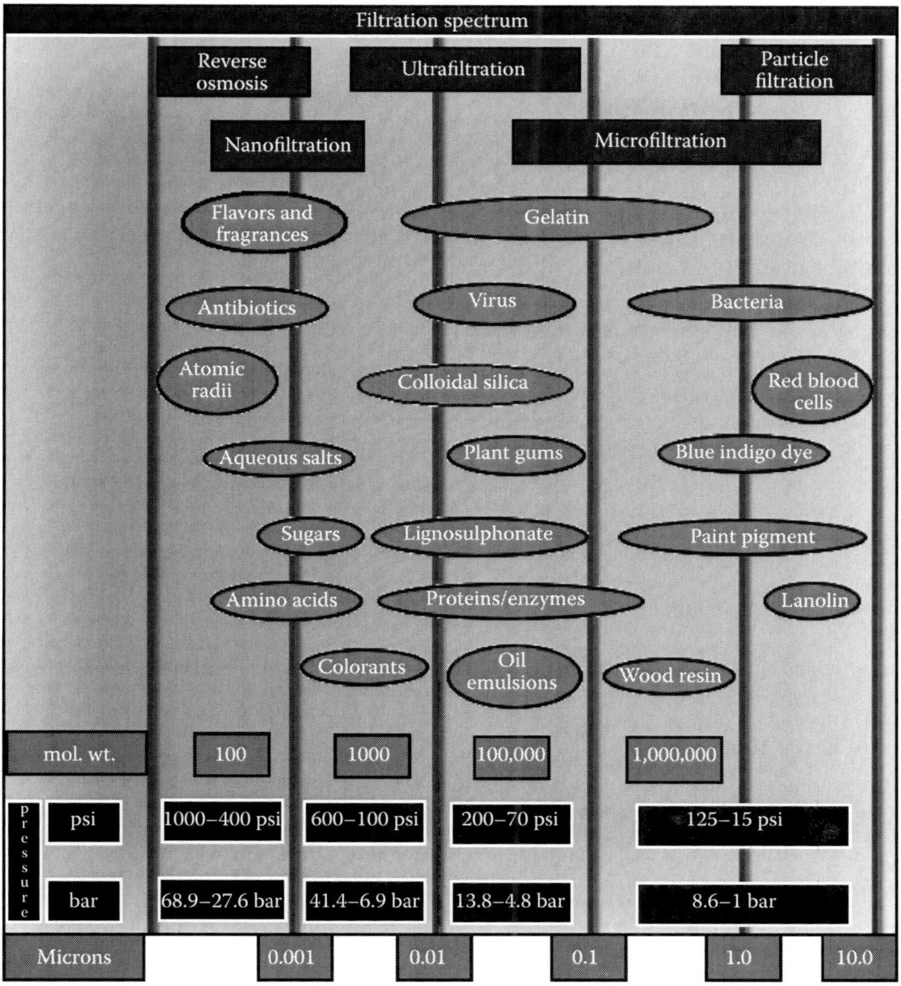

Figure 6.1 Filtration spectrum chart of cross-flow filtration.

fractionate proteins, peptides, and amino acids contained in hydrolysates. More recent studies reported the enrichment of antioxidant peptides by UF from an alfalfa leaf protein hydrolysate with a 3-kDa molecular weight cutoff (MWCO) polysulfone membrane and wheat gluten hydrolysates with UF membranes (10, 5, and 3 kDa). Other studies (Pihlanto-Leppälä, Akkanen, and Korhonen 2008) reported the feasibility of fractionation by UF to produce fractions exhibiting angiotensin-converting enzyme–inhibitory and antioxidant activities from potato protein hydrolysates, and UF also was used in the purification of peptides that had antimicrobial activity against several pathogens, such as *Candida albicans*, which is the most common cause of oral, esophageal, vaginal, and urinary candidiasis (Kim et al. 2005, 2006).

6.2.3.1 *Diafiltration*

Diafiltration is a specialized type of UF process where the retentate is diluted with water and reultrafiltered to improve the recovery of the material in the permeate or to improve the purity of the retained stream. Typical uses are recovery of antibiotics from the fermentation broth and enhancing the proteins. Recent experiments have demonstrated the production of protein products from partially defatted soybean meal and undefatted lupin meal using UF followed by diafiltration.

In one recent experiment, peroxidase and Bowman-Birk inhibitor, valuable products that have potential medicinal uses, were produced from soybean hulls using UF, followed by discontinuous diafiltration, a membrane-based method that removes low molecular weight solutes such as salts while continuously replacing the solvent lost with the permeate (Hojilla-Evangelista, Sessa, and Mohamed 2004).

6.2.3.2 *Ultrafiltration Applications*

Protein processing:

- One of the earliest applications of UF is the recovery and concentration of protein from cheese whey. There is considerable commercial interest in the preparation of whey proteins for nutraceutical applications. Recent developments in membrane filtration have provided exciting new opportunities for large-scale protein fractionation.
- With UF, various types of amino acids can be processed more uniformly despite the strong characteristic differences of each kind of broth.
- UF in the field of bioactive peptides enriches peptides from protein hydrolysates. Furthermore, adding UF membranes in the production process of bioactive peptides using enzymatic membrane reactors allows production of specific peptide sequences with functional and nutritional properties.
- UF has become the primary means for protein concentration and buffer exchange in the production of therapeutic proteins and industrial enzymes (Table 6.2).

6.2.4 Microfiltration

MF is a low-pressure (10–100 psig) process for separating larger-size solutes from aqueous solutions using a semipermeable membrane. This process is carried out by having a process solution flow along a membrane surface under pressure. MF removes suspended solids and colloids, bacteria, and viruses in the range of 0.05–10 μm and retains color and taste of compounds, alcohol, and sugars.

The MF filter is usually made from a thin polymer film with a uniform pore size and a high pore density of approximately 80%. MF products are designed for crossflow separation, where a feed stream is introduced into the membrane element under pressure and passed over the membrane surface in a controlled flow path. A portion

Table 6.2 Applications of Ultrafiltration in Processing Bioactive Peptides

Driving Force	Process	Protein Hydrolysate Source	Biological Activity
Pressure-driven membrane processes	Ultrafiltration	Alfalfa white protein	ACE inhibitor
		Alfalfa leaf protein	Antioxidant
		Wheat gluten	ACE inhibitor
		Soybean protein	ACE inhibitor
		Soybean β-conglycinin	ACE inhibitor
		Soybean glycinin	ACE inhibitor
		Sea cucumber gelatin	ACE inhibitor
		Potato	Antioxidant
			ACE inhibitor
			Antimicrobial

ACE, angiotensin-converting enzyme.

of the feed passes through the membrane and is called *permeate*. The rejected materials are flushed away in a stream called the *concentrate*. Cross-flow membrane filtration uses high cross-flow rates to enhance permeate passage and reduce membrane fouling. MF is used for fermentation, broth and biomass clarification, and recovery.

6.2.4.1 Microfiltration Applications

- From functional and nutritional points of view, MF can produce high-purity whey protein isolates (WPIs). Microfiltered WPI tends to be low in fat content, which designates the "nutritional" or "nutraceutical" value (Akpinar-Bayisit, Ozcan, and Yilmaz-Ersan 2009).
- MF is used in the retention of fat and large molecular weight proteins from small whey proteins, lactose, and minerals of low molecular weight.

6.2.5 Reverse Osmosis

RO is a moderate to high pressure-driven process (80–1200 psig) for separating larger solutes from aqueous solutions by means of a semipermeable membrane (Separation Technologies 2005). Mechanisms are based on size, shape, ionic charge, and membrane interactions. The membrane pore size can vary from 0.1–1.0 nm.

The process is carried out by flowing a process solution along a membrane surface under pressure. Retained solutes such as particulate matter and dissolved salts leave with the flowing process stream and do not accumulate on the membrane surface; this is achieved through cross-flow separation, where a feed stream is introduced into the membrane element under pressure and passed over the membrane surface in a controlled flow path. A portion of the feed passes through the membrane and is called *permeate*. The rejected materials are flushed away in a stream called the *concentrate*. Cross-flow membrane filtration uses a high cross-flow rate to enhance permeate passage and reduce membrane fouling.

RO was initially used for the desalination of water. Today RO is widely used to concentrate many streams in the dairy and food-processing industries.

6.2.5.1 *Reverse Osmosis Applications*

- Used in the recovery of whey proteins from whey concentrates
- Used in the recovery of oilseed protein
- Used in the production of pure water and concentration of maple sap

6.2.6 Nanofiltration

NF is selected when RO and UF are not efficient choices for separation. NF has an intermediate cutoff range of 200–1000 Da, between RO and UF (Separation Technologies 2005; Wellness West 2005).

NF operates at low to moderately high pressure (50–450 psig) with higher flow rates of water. In NF, mechanisms underlying separation selectivity are more complex. The selectivity is regulated by both size exclusion and electrostatic repulsion mechanisms, which are dependent on membrane type and feed composition.

NF membrane allows diffusion of certain ionic species, predominantly monovalent ions. Larger ionic species, including divalent and multivalent and more complex molecules, are highly retained (GEA Filtration: Nanofiltration n.d.).

The process is carried out by having a process solution flow along a membrane surface under pressure. Cross-flow membrane filtration uses a high cross-flow rate to enhance permeate passage and reduce membrane fouling. Retained solutes (such as dissolved salts) leave with the flowing process stream and do not accumulate on the membrane surface.

It is difficult to predict the performance of NF membranes, especially if more than three solutes are present in the solution because membrane rejection is influenced by the size, structure, and charge of the components in solution. As a result, scaling-up and pilot studies are highly recommended for NF applications.

6.2.6.1 *Nanofiltration Applications*

- NF is used in the partial desalination of whey-reducing recovery costs of lactose and whey protein solids.
- NF is also used in the desalination of concentrated antibiotics, increasing the value of the pharmaceutical product.
- NF is used for color reduction or manipulation of food products and concentration of food, dairy, and beverage products.
- NF has also been used to fractionate complex protein hydrolysates that contain bioactive peptides. NF membranes with a range of MWCO between 1000 and 5000 Da are used to separate peptides present in hydrolysates of whey proteins (Pouliot et al. 1999, 2000).
- More recently a research study used the combination of UF and NF to fractionate peptides contained in sweet whey (Butylina, Luque, and Nyström 2006) (Table 6.3).

Table 6.3 Applications of Nanofiltration in Processing Bioactive Peptides

Driving Force	Process	Protein Hydrolysate Source	Biological Activity
Pressure-driven membrane processes	Nanofiltration	Bovine whey proteins	ACE inhibitor, opioid, antimicrobial
		Bovine whey β-lactoglobulin	ACE inhibitor, opioid, antimicrobial

ACE, angiotensin-converting enzyme.

6.2.7 Membrane Process Limitations

One of the major limits of conventional pressure-driven processes in offering large-scale production of bioactive compounds is its fouling and poor selectivity.

6.2.7.1 Membrane Fouling

Conventional pressure-driven processes are limited in their applications because of fouling characteristics. Because the solutions to be treated and the operating conditions vary from application to application, a conclusion cannot be drawn on which materials are the best to use to inhibit membrane fouling. The chemical composition and physical morphology of the membrane vary from manufacturer to manufacturer, depending on the process to be treated, operating conditions, and economic factors.

6.2.7.2 Membrane-Fouling Mechanisms

The main mechanisms of membrane fouling are shown to be bacterial growth, clogging of pores, adsorption of feed components, and chemical interaction between solutes and membrane material. One of the most serious forms of membrane fouling is bacterial adhesion. Once formed, biofilms are difficult to remove. Microbial fouling of RO membranes is one of the major factors in flux decline (Goosen et al. 2004).

6.2.7.3 Preventive Means

6.2.7.3.1 Membrane Surface Modification

One way to reduce membrane fouling is by modifying the membrane surface. Recent studies reported a foul-resistant RO membrane that reduces microbial adhesion. These studies suggested that presence of urea groups in the membrane reduces microbial adhesion, perhaps through charge repulsion (Jenkins and Tanner 1998).

Few studies (Kabsch-Korbutowicz, Majewska-Nowak, and Winnicki 1999) demonstrated that the hydrophilic ends of the membranes tested had the lowest proneness to fouling by organic colloids. Other research works (Tu et al. 2001) further confirmed that membranes with a higher negative charge and greater hydrophilicity were less prone to fouling as a result of fewer interactions between the chemical groups in the organic solute and the polar groups on the membrane surface.

6.2.7.3.2 Cleaning Membrane

Once a membrane surface has been fouled, it must be cleaned, which may lead to eventual loss of membrane properties. Therefore, rinsing water quality is of special importance in cleaning because the impurities present in water may affect the membrane performance and life of the membrane. Membrane manufacturers generally recommend high quality filtered and demineralized water. Alternatively, water treatment chemicals such as ethylenediaminetetraacetic acid and polyphosphates can be added to low quality water to increase the solubility of metal ions. Cleaning agents such as sodium dodecyl sulfate and sodium hydroxide can be used in combination to reach the optimum recovery of polysulfone membranes used in milk industries (Goosen et al. 2004).

6.3 PRESSURE-DRIVEN PROCESSES ASSISTED BY ELECTRIC FIELD

The ability to produce functional food ingredients from natural sources becomes increasingly attractive to the food industry. To improve the efficiency of conventional pressure-driven membrane filtration, the application of an external electrical field, which acts as an additional driving force to the pressure gradient, has been carried out (Bazinet and Firdaous 2009). Electrically enhanced membrane filtration (EMF) seems to be a breakthrough technology for the isolation of charged nutraceutical ingredients from natural sources. EMF combines the separation mechanisms of membrane filtration and electrophoresis, and it is believed that EMF is more selective than membrane filtration and less costly than chromatography (Bargeman et al. 2002).

Besides many applications of EMF in functional foods and nutraceuticals, the production of bioactive peptides also finds interesting uses for EMF. Processes combining an electric field as the driving force to porous membranes have been developed for the separation of bioactive peptides.

These pressure-driven processes assisted by electric field have shown to be promising in the separation and purification of bioactive peptides. Most recently, electrodialysis (ED) with UF membranes has been developed to fractionate peptides using conventional ED cells (Bazinet and Firdaous 2009).

6.3.1 Electrophoretic Membrane Processes

6.3.1.1 Electrophoresis

Electrophoresis is an electrically driven operation with no pressure applied and is a basic technique used in the separation of particles of colloidal size. Electrophoresis is generally carried out in supporting media like gels and is known for its high resolution (Bazinet and Firdaous 2009). However, when electrophoresis was used for production purpose, poor resolution was obtained compared with analytical scale, and finally it was found that the productivity could not be increased over a particular limit, typically about a few milligrams per hour.

To overcome these limitations, different technologies such as ED and fixed boundary electrophoresis were used. These electromembrane operations have two common characteristics. First, the selectivity comes from different solute flow rates through the membrane. Second, the productivity can be increased by increasing the number of separation chambers without significantly damaging the selectivity.

Fixed-boundary electrophoresis is another separation technology that has gained importance in the functional food industry, although this technique has never been tested for bioactive peptide separation. However, its operation conditions are limiting factors for large-scale bioactive peptide production (Bazinet and Firdaous 2009).

6.3.2 Electrodialysis

6.3.2.1 Principle

ED is a membrane separation process in which ionic (electrically charged) species are transported from one solution to another by crossing one or more selective permeable membranes under the influence of an electric potential. When the ED is running, the direct current field affects the flow of charged species in water solutions in such a way that cations moving toward cathode pass through the cation exchange membranes and cannot go through the anion exchange membranes, whereas the anions drawn to the anode pass through the anion exchange membranes but stop at cation exchange membranes (Electrodialysis n.d.).

6.3.2.2 Electrodialysis with Filtration Membranes

Recently the new technology ED with UF appears to be a more selective method for the separation of bioactive peptides from raw hydrolysates. ED with filtration membranes (EDFM) couples size exclusion properties of UF membranes with the charge selectivity of ED. EDFM acts by attracting the molecules, according to their charge, through filtration membranes with the proper MWCO range according to the fraction of interest.

The feed solution containing the peptides to be separated flows through a compartment, and the pH of the feed solution induces a charge-dependent selective migration of the peptides under the influence of the electric field. At the same time,

Table 6.4 Applications of Electrodialysis in Processing Bioactive Peptides

Driving Force	Process	Protein Hydrolysate Source	Biological Activity
Electrophoretic membrane processes	Electrodialysis with filtration membranes	Bovine whey β-lactoglobulin	ACE inhibitor, opioid, antimicrobial
		Alfalfa white protein	ACE inhibitor

ACE, angiotensin-converting enzyme.

the filtration membrane performs the separation of the peptides by molecular weight exclusion or passage permission.

EDFM has several advantages. It allows the selective separation at a large scale, and in one process we can separate one or more charged molecules of interest. However, the separation of one or more compounds in the same process is limited by the design of ED cells. Because EDFM stands as a reliable and cost-effective separation technique, it constitutes a major breakthrough in the purification and separation of bioactive peptides. The technology has already been tested for the separation of tobacco polyphenols, green tea catechins, proanthocyanidins, chitosan oligomers, and different bioactive peptides (Bazinet and Firdaous 2009) (Table 6.4).

6.4 THERMAL SEPARATIONS

6.4.1 Evaporation Techniques

6.4.1.1 Principle

Evaporation is a process used to remove a portion of water from a solution, suspension, or emulsion by boiling some of the liquid. This reduces the bulk and weight for subsequent processing, increases the concentration, helps preserve the product, and concentrates color and flavor. Evaporation starts with a liquid product and ends up with a more concentrated component. Evaporators generally operate at low temperatures, without vacuums, to rapidly and gently concentrate sensitive, delicate, and valuable products (MCD Technologies Inc. n.d.; PG&E n.d.).

Evaporators may operate singly or in series. Each one is referred to as an effect, and in multiple-effect systems, the product output from one effect is the feed for the following effect. Similarly, higher-temperature vapor driven out of the product in one effect is used to heat lower-temperature products in another. Efficiency is gained by using multiple-effect systems.

6.4.1.2 Advances in Evaporation Technology

Advances in evaporation and spray drying technology, along with the development of ED technology, led to the improvement in functionality of value-added products used in the food and nutraceutical industry. One such improvement was observed in functionality in whey-based powders.

There is considerable evidence that whey protein is the only dietary protein shown to boost glutathione, which is a building block for the body's antioxidant and immune defense systems (American Dairy Science Association 2009).

Using right evaporation and spray drying technologies, functional and nutritional properties of whey products such as whey protein concentrate and WPI can be preserved (American Dairy Science Association 2009). Recently, emphasis on the isolation of major proteins from whey, such as β-lactoglobulin, and nutritional supplements, such as α-lactalbumin, was observed.

Initial evaporation and drying equipment in the dairy industry were designed primarily for the purpose of preparing milk powder, but drying whey proved to be much more difficult because of the hygroscopic nature of the lactose component. Improvements in both evaporation and spray drying technology improved the quality of whey powder to the point that it can be used in food applications.

6.4.1.3 Criteria for Selection of Evaporator Plant Concept

During the design of evaporation plants, numerous requirements have to be considered. They determine the type of construction and arrangement to be chosen and the resulting process and economic data. The most important requirements are as follows (Courtesy GEA Evaporators Plant Concept; GEA Process Engineering Inc. n.d.):

- Capacity and operational data, including quantities, concentrations, temperatures, annual operating hours, change of product, and control of automation
- Product characteristics, including heat sensitivity; viscosity and flow properties; foaming tendency; fouling and precipitation; and boiling behavior
- Required operating media, such as steam, cooling water, electric power, cleaning agents, and spare parts
- Capital and other financial costs
- Personnel costs for operation and maintenance
- Standards and conditions for manufacture delivery and acceptance
- Choice of materials of construction and surface finishes
- Site conditions, such as available space, climate (for outdoor sites), connections for energy and product, and service platforms
- Legal regulations covering safety, accident prevention, sound emissions, environmental requirements, and others, depending on the specific project

6.4.2 Different Evaporator Systems

6.4.2.1 Rising Film Evaporators

These evaporators work on a "thermo-siphon" principle. Hot liquid is brought into the bottom of the unit and brought to boil as it ascends long, vertical tubes. The feed product enters the bottom of the heating tubes, and as it heats, steam begins to form. The ascending force of the vapor bubbles, which are produced during boiling, causes liquid and vapors to flow upward in parallel flow. Simultaneously, the increase in the vapor production causes the product to be pressed as a thin film on the walls of the tubes, and the liquid rises upward. For this reason, there is a higher turbulence in the liquids, which is advantageous for heat-sensitive, highly viscous, and moderately scaling liquid products.

These evaporators are not extensively used for processing bioactive protein components because a differential of 20°C is required to maintain circulation, which increases the rate of fouling. Increased fouling may in turn lead to increased protein denaturation and reduced solubility. The effects in rising film evaporators are limited by the large temperature differential required for the equipment to function properly.

6.4.2.2 Falling Film Evaporators

In falling film evaporators, the liquid flows downward forming a film along with the vapors in a parallel flow. The liquid to be evaporated is preheated and enters the heating tubes via distribution plates on the top bonnet of the evaporator. The gravity-induced downward flow is increasingly augmented by the cocurrent vapor flow. The separation of the concentrated product from its vapor is undergone in the lower part of the heat exchanger and separator.

Falling film evaporators can handle more viscous liquids and have a residence time of 20–30 seconds, compared with 3–4 minutes for rising film equipment. In addition, they also have low product hold-up during shutdown. Evaporators of such design have several advantages in terms of low temperature differentials (5°C), ability to concentrate to high solids, low liquid content, and reduced residence time.

The film mechanism and low residence time makes this evaporator ideally suited to heat-sensitive nonsalting and noncrystallizing solutions. However, falling film evaporators must be designed carefully for each operating condition; sufficient wetting (product film thickness) of the heating surface by liquid is extremely important for trouble-free operation of the plant. If the heating surfaces are not wetted sufficiently, dry patches and incrustations will occur; in the worst cases, the heating tubes will completely clog (GEA Process Engineering Inc. n.d.).

6.4.2.3 Scraped Surface Evaporators

Scraped surface evaporators are particularly suited for concentrating viscous, sticky, or heat-sensitive products (CONVAP Scraped-Surface Evaporator n.d.) and are also used for products with the tendency to scale or salt out during evaporation. These evaporators are generally operated under vacuum conditions at the final evaporation stage to achieve high concentration of solids.

To prevent burn-on and to promote mixing of highly viscous products, many companies use scraper blades in the calandria for proper mixing, thus preventing the thermal degradation of the material. These revolving blades also provide mechanical agitation, creating the conduction and convection conditions required for the heat transfer. The product is pumped into the lower end of the heat-exchange cylinder.

Heating media flow in the annular space between the heat-transfer wall and the insulated jacket (CONVAP Scraped-Surface Evaporator n.d.). The scraping blades, centrifugal action of the rotor, and vacuum all contribute to vaporization, which occurs in the custom-designed vapor dome.

The scraping blades continuously remove the thin product film from the precision-finished cylinder wall. The centrifugal action of the rotor, driven by a motor on the upper end of the unit, spins the heavier liquid droplets toward the cylinder wall. This action ensures a continuous rewetting of the heat transfer surface and prevents product burn-on (GEA Filtration: Membrane Filtration n.d.; Genomic Food Processing Equipment n.d.).

6.4.2.4 Plate Evaporators

Plate evaporators are compact systems with only 3–6 mm between the plates. This provides a high degree of turbulence and good heat transfer. During the plate evaporation process, liquid is pumped between thin plates with the heating medium on the mating surfaces. Instead of tube or shell heat exchangers, framed plates are used as a heating surface. The product is evaporated with the vapor generated forming a high-velocity core.

These systems use rising or falling film operation, which results in even and gentle evaporation of the product, suitable to handle medium-sized production runs of heat-sensitive products (American Dairy Science Association 2009; APV n.d.; GEA Process Engineering: Food & Dairy Industry n.d.).

Benefits of these evaporators include eliminating product deterioration even when highly heat-sensitive liquids are involved, allowing custom concentration to be produced and holding little product in the evaporator at any time.

6.4.2.5 Applications of Evaporators

- Used in the processing of milk proteins such as whey and lactose, which are rich in functional properties
- Used in the processing of coffee, tea, herbal extracts, and protein hydrolysates (S. S. Techno Services Pvt. Ltd. n.d.)
- Used in the processing of high-protein juices such as soya whey

6.4.3 Wiped Film Evaporators

These evaporators are designed for difficult concentration, evaporation, and reaction applications. Wiped film evaporators are operated under high vacuum and have a specifically designed condenser for the least solvent loss.

In these evaporators, feed generally enters from the top. The rotating blades create high centrifugal force, which keeps the feed against the heated cylinder wall, and a turbulent thin film that covers the entire heated surface all the time.

This thin film creates high heat transfer efficiency, minimizing the required area for evaporation, and is continuously renewed by the incoming feed as the more concentrated material moves toward the bottom discharge nozzle. Major advantages of this evaporator include short residence time, excellent heat transfer, handling of viscous fluids, and operation under high vacuum (Pfaudler Reactor Systems n.d.).

6.4.3.1 Applications

- Used for the concentration of herbal extracts such as green tea extract and lycopene from tomato
- Successfully used to separate volatile from less volatile components, such as oils, nutraceuticals, and fragrances (Advanced Synthesis n.d.)

6.4.4 Spray Dryers

Spray drying is the continuous process whereby fluid is converted into fine droplets and is exposed to hot drying media, so as to achieve the continuous conversion of solution, emulsion, or pumpable suspension into free-flowing powder. Conversion of the feed into fine droplets is achieved either by pumping the feed under high pressure through an orifice or by spraying through a high-speed rotating perforated disc.

Spray drying is an ideal choice to comply with the end-product quality standards such as particle size, moisture content, bulk density, and particle shape. Benefits of the spray dryers include the complete control of the product, which includes moisture content, particle size and structure, solubility, wettability, and retention of natural aromas and flavors (Anhydro n.d.).

6.4.4.1 Applications of Spray Dryers

- Spray drying is one of the best processes used for decades to encapsulate food ingredients such as flavors, lipids, and carotenoids (Gharsallaoui 2008).
- Continuous single-stage spray drying converts the final formulations into free-flowing powder.
- Used in the encapsulation of various formulations having special properties.
- Used in the microencapsulation of various compounds enabling taste masking (or) controlled release of functional coatings.

6.5 NEW PATENTED TECHNOLOGIES

Many technologies in drying and evaporation have been emerging to process a variety of products ranging from bioactive lactobacillus to nutraceuticals. Recently MCD Technologies (n.d.) developed a patented technology called *refractance window*, which is environmentally friendly using little water and preserving air quality. These evaporators operate at low temperatures without vacuum to rapidly and gently concentrate sensitive, delicate, and valuable products, retaining flavors, aromas, and colors and nutrients. Rapid evaporation occurs when infrared energy from circulating hot water (≤210°F/99°C) is transmitted at the speed of light directly into the product. Assisted also by conducted heat, product moisture rapidly evaporates and is carried away by mechanically boosted airflow. This greatly speeds this gentle concentration process, where product temperatures remain far lower (≤140°F/60°C) than circulating water (198°F/92°C).

6.5.1 Advantages

- Retains complex, subtle flavors and aromas, as well as colors and nutrients
- Low temperatures
- No vacuum
- Energy efficient
- Reduced labor costs, quick cleanup, and minimal maintenance

6.6 OPERATIONAL FEATURES OF EVAPORATORS THAT INFLUENCE THE PRODUCT QUALITY

The operational characteristics and design features of an evaporator have a direct influence on the functionality of the product. Several changes can be made to the operational practices and evaporator systems to improve the efficiency and reduce the energy consumption in an evaporator.

Mechanical vapor recompression (MVR) evaporators are generally considered as the evaporators that minimize product damage, followed by thermal vapor recompression evaporators. Because of preheating time differences and corresponding temperature differences during evaporation, MVR evaporators can produce a higher grade powder for a given solid concentration. Besides the quality, MVR evaporators are more energy efficient and reduce the opportunity for microbial growth. A few changes in the evaporation systems and operating practices can improve the efficiency of the system, some of which are listed below (American Dairy Science Association 2009).

6.6.1 Energy Efficiency Measures

- Preheating feed product reduces the heat required to achieve boiling in the evaporation vessel. This requires addition of heat exchanger appropriate to the characteristics of the feed.
- Vapor recompression takes advantage of the significant heating value of the vapor driven out of the product. Vapor can be reused in the same evaporator by increasing its temperature and pressure close to those of the steam injected into the heat exchanger. This can be done using a steam jet or a compressor.
- Installation of additional effects will generally improve the efficiency of an evaporator system (PG&E n.d.).

6.6.2 Operational Tips

- Maintain the optimum pressure profile as provided in the evaporator's design. Excess pressure inhibits evaporation by raising boiling point.

- Preconcentration of the feed will reduce the energy required to operate an evaporator. For some applications, preconcentration with separation membranes can save up to 90% of the energy consumption.
- Optimize the venting rate of noncondensable gases to reduce steam waste while maintaining appropriate evaporation vessel pressure. Noncondensable gases in the evaporation vessel increase pressure, which increases the boiling point and heat requirements.
- Condensed steam can be used to preheat feed product or used in the next effect of a multieffect system. It can also be fed back to the boiler to offset the use of cold makeup water (PG&E n.d.).

6.7 CURRENT AND FUTURE DEVELOPMENTS OF SEPARATION TECHNIQUES

Membrane technology continues to evolve rapidly, providing distinct separations and higher flux rates. Many advances in this technology have been observed, including cascading membrane technology and the use of gradient diafiltration technologies.

Also, new-generation membranes with ion exchange groups and charged membranes are being developed, which have created the opportunity for the production of specific components rich in medicinal and nutritional properties.

Production of bioactive peptides from protein hydrolysate sources has shown to be one of the emerging fields but was limited because of the lack of large-scale technologies. Bioactive peptides can be incorporated in the form of ingredients in functional and novel foods and dietary supplements to deliver specific health benefits.

Because of its high selectivity, ED with filtration membranes appears to be a novel technology. However, further scale-up developments are necessary to confirm its feasibility on a large scale.

There is growing commercial interest of milk-derived protein peptides in context to their potential health benefits. A few commercial developments have been made, and this trend is likely to continue, with further exploitation of functionalities of the peptides.

6.8 CONCLUSION

There is much growing interest in the development of separation techniques for the preparation of nutraceuticals and pharmaceuticals. Membrane processes offer tremendous opportunities in the nutraceutical industry, providing high throughput and fine-tuning to give high selectivity. These processes will be able to accomplish distinct separation of high-value products, resulting in the advancement of both product and quality. Development of this membrane technology has significantly advanced the selection of ingredients and expanded its application into the functional foods and nutraceutical industry.

Advances in membrane technology are primarily responsible for the separation of value-added products from many components, which are known for their medicinal and nutritional properties. Similarly, technological advancements in the development of new membranes and understanding the functionality of milk constituents have extended the applications of membrane separations in the dairy industry.

Developments in evaporation and spray drying technologies have produced a variety of functional ingredients, which initially suffered denaturation during evaporation. Technological advances in evaporation and spray drying techniques, along with the development of demineralization and ED technology, also led to the improvement in functionality of whey-based powders.

However, challenging developments in membrane technology and evaporation techniques should focus on efficient fractionation of functional components to get products of desired nutraceutical properties. New advances in membrane materials, evaporation techniques, modules, and processes should lead to the rapid development of separation techniques processing value-added components, making possible the cost-effective production of biological products.

REFERENCES

Advanced Synthesis. n.d. For Sales. http://www.advancedsynthesis.com/Surplus_Chemical_Equipment.htm.

Akpinar-Bayisit, A., T. Ozcan, and L. Yilmaz-Ersan. 2009. Membrane Processes in Production of Functional Whey Components. *Mljekarstvo* 59: 282–88.

American Dairy Science Association. 2009. Pioneer Paper: Value-Added Components Derived from Whey. http://www.adsa.org/speakerdocs/pp1_modler.pdf.

Anhydro. n.d. Spray Dryers. http://www.anhydro.com/content/us/products/dryers/spray_dryers.

APV. n.d. Plate Evaporators. http://www.apv.com/us/products/heatexchangers/plateevaporator/plate+evaporators.asp.

Bargeman, G., J. Houwing, I. Recio, G. H. Koops, and C. van der Horst. 2002. Electro-membrane Filtration for the Selective Isolation of Bioactive Peptides from an α(s2)-Casein Hydrolysate. *Biotechnology and Bioengineering* 80: 599–609.

Bazinet, L., and L. Firdaous. 2009. Membrane Processes and Devices for Separation of Bioactive Peptides. *Recent Patents on Biotechnology* 3: 61–72.

Butylina, S., S. Luque, and M. Nyström. 2006. Fractionation of Whey-Derived Peptides Using a Combination of Ultrafiltration and Nanofiltration. *Journal of Membrane Science* 280: 418–26.

CONVAP Scraped-Surface Evaporator. n.d. http://inoxpumps.com/pdf/convap.pdf.

Electrodialysis. n.d. http://www.mega.cz/electrodialysis.html.

GEA Filtration. n.d. Membrane Filtration. http://www.geafiltration.com/filtration_library/cGMP_membrane_filtration_sys.pdf.

GEA Filtration. n.d. Nanofiltration. http://www.geafiltration.com/technology/nanofiltration.asp.

GEA Process Engineering Inc. n.d. Evaporation Technologies. http://www.niroinc.com/evaporators_crystallizers.

Genomic Food Processing Equipment. n.d. Convap Scraped Surface Evaporator. http://www.genemco.com/catalog/ME2051convapevaporator.pdf.

Gharsallaoui, A. 2008. Microencapsulation of Food Ingredients by Spray-Drying. http://www.scitopics.com/microencapsulation_of_food_ingredients_by_spray_drying.html.

Goosen, M. F. A., S. S. Sablani, H. Al-Hinai, S. Al-Obeidani, R. Al-Belushi, and D. Jackson. 2004. Fouling of Reverse Osmosis and Ultrafiltration Membranes: A Critical Review. *Separation Science and Technology* 39: 2261–98.

Hojilla-Evangelista, M. P., D. J. Sessa, and A. Mohamed. 2004. Functional Properties of Soybean and Lupin Protein Concentrates Produced by Ultrafiltration-Diafiltration. *Journal of the American Oil Chemists' Society* 81: 1153–57.

Jenkins, M., and M. B. Tanner. 1998. Operational Experience with a New Fouling Resistant Reverse Osmosis Membrane. *Desalination* 119: 243–50.

Kabsch-Korbutowicz, M., K. Majewska-Nowak, and T. Winnicki. 1999. Analysis of Membrane Fouling in the Treatment of Water Solutions Containing Humic Acids and Mineral Salts. *Desalination* 126: 179–85.

Kim, J.-Y., S.-C. Park, M.-H. Kim, H.-T. Lim, Y. Park, and K.-S. Hahm. 2005. Antimicrobial Activity Studies on a Trypsin–Chymotrypsin Protease Inhibitor Obtained from Potato. *Biochemical and Biophysical Research Communications*330: 921–27.

Kim, M.-H., S.-C. Park, and J.-Y. Kim. 2006. Purification and Characterization of a Heat-Stable Serine Protease Inhibitor from the Tubers of New Potato Variety "Golden Valley." *Biochemical and Biophysical Research Communications* 346: 681–86.

MCD Technologies, Incorporated. n.d. http://www.nnni.ca/pdf/education/mcdbrochure.pdf.

PG&E. n.d. http://www.pge.com/includes/docs/pdfs/about/edusafety/training/pec/inforesource/food_processing_evaporator_systems.pdf.

Pfaudler Reactor Systems. n.d. Wiped Film Evaporators. http://www.pfaudler.com/wiped_film_evaporators.php.

Pihlanto-Leppälä, A., S. Akkanen, and H. J. Korhonen. 2008. ACE-Inhibitory and Antioxidant Properties of Potato *(Solanum tuberosum)*. *Food Chemistry* 109: 1004–12.

Pilot Plant Membrane Filtration Opportunities for the Food Processing Industry. n.d. http://fpc.unl.edu/pilotplant/membrane_filtration_opportunities.pdf.

Pouliot, Y., M. C. Wijers, S. F. Gauthier, and L. Nadeau. 1999. Fractionation of Whey Protein Hydrolysates Using Charged UF/NF Membranes. *Journal of Membrane Science* 158: 105–14.

Pouliot, Y., S. F. Gauthier, and J. L'Heureux. 2000. Effect of Peptide Distribution on the Fractionation of Whey Protein Hydrolysates by Nanofiltration Membranes. *Lait* 80: 113–22.

Separation Technologies. 2005. http://www.wellnesswest.ca/dmdoc/tw_seperationtechnologies.pdf.

S. S. Techno Services Pvt. Ltd. n.d. Evaporators for Food/Pharma/Natural Products. http://www.sstechno.com/evaporators.html.

Tu, S.-C., V. Ravindran, W. Den, and M. Pirbazar. 2001. Predictive Membrane Transport Model for Nanofiltration Processes in Water Treatment. *AIChE Journal* 47: 1346–62.

Wellness West. 2005. Membrane Processing: State of the Art Technology. http://www.wellnesswest.ca/dmdoc/tw_membraneprocessing.pdf.

Zydney, A. L. 2000. Membrane Bioseparations. *Encyclopedia of Separation Science*. London: Academic Press, 1728–55.

CHAPTER 7

Size Reduction, Particle Size, and Shape Characterization of Nutraceuticals

Sri Hyndhavi Ramadugu, Vijay Kumar Puli, and Yashwant Pathak

CONTENTS

7.1 INTRODUCTION

Nutraceutical, a combination of "nutrition" and "pharmaceutical," refers to extracts of foods claimed to have a medicinal effect on human health. Nutraceuticals are usually contained in a medicinal format such as a capsule, tablet, or powder and administered in a prescribed dose. More rigorously, nutraceutical implies that the extract or food is demonstrated to have a physiological benefit or provide protection against chronic disease. *Functional foods* are defined as being consumed as part of the usual diet but are demonstrated to have physiological benefits and reduce the risk of chronic disease beyond basic nutritional functions. Examples of some nutraceuticals are flavonoids, antioxidants, α-linolenic acid from flax seeds, β-carotene from marigold petals, and anthocyanins from berries (Karla 2003).

7.1.1 Particle Size

Solubility and bioavailability of pharmaceutical ingredients are affected considerably by their particle size and shape distribution. This also applies to nutraceuticals. Quantitative particle size analysis provides information that has a direct effect or correlation to issues such as bioavailability, solubility, and quality control. Before discussing methods for particle sizing, we have to understand how particle size distributions are defined. Particles are three-dimensional objects for which three parameters—length, breadth, and height—are required to provide a complete description. It is not possible to describe a particle using a single number that equates to the particle size. Therefore, most sizing techniques assume the material being measured is spherical because the sphere is the only shape that can be described by a single number (its diameter). This equivalent sphere approximation is useful and simplifies the way particle size distributions are represented. However, it does mean that different sizing techniques can lead to different results when measuring nonspherical particles (Jillavenkatesa, Dapkunas, and Lum 2001). Particle size analysis is the scientific and analytical technique used for the measurement and reporting of size distribution in any sample of particulate material. The particulate material could be in any form—solid, liquid, or gaseous. Particle size analysis plays a significant role in determining the performance of the final product. There are a number of analytical techniques and measurement methods available for the measurement of particle size (DiMemmo et al. 2005).

7.1.2 Solid–Liquid Extraction of Nutraceuticals

Solid–liquid extractions involve the separation of two phases, solid and liquid, from a suspension. It is applied in the nutraceuticals with the aim of:

- Protein extraction: recovering the valuable solids (the liquid is discarded)
- Oil extraction: recovering the liquids (solids are discarded)
- Oil and meal: recovering both liquid and solid
- Recovering neither

Solid–liquid extractions are used to recover food components such as sucrose in cane or beets, lipids from oilseeds, proteins in oilseed meals, photochemical from plants, and functional hydrocolloids from algae. The extraction may also be used to remove undesirable contaminants and toxins present in foods and feeds. There are four stages to solid–liquid separations, although not all stages may be present in all processes. These stages include:

- Pretreatment: to increase particle size and reduce viscosity
- Solids concentration: to reduce the volume of material to process
- Solids separation: to separate solids from liquid, to form porous cake, or to produce particle-free liquid
- Posttreatment: to remove soluble solids, remove moisture, reduce cake porosity, or prepare material for downstream processing

The cost of solid–liquid separations is directly related to volume of the material that must be processed.

7.2 ROLE OF PARTICLE CHARACTERIZATION

Knowledge of particle sizes and shapes and the size distribution of a powder system is a prerequisite for most production and processing operations in the nutraceutical and food industry. Hence particle characterization is important.

7.2.1 Particle Size

The study of the particle size technology is commonly used and is most important as most of the solids appear in powder or granular form. Some examples include foods (e.g., grain, flour, and sugar) and pharmaceuticals (e.g., drugs and excipients). The physical dimensions of these particles influence their physical properties and production processes, including

1. Particle size influences dissolution: small particles dissolve more rapidly than large ones, and particle size dissolution not only determines the behavior of nutraceuticals in vivo but also in various manufacturing processes.
2. The flow properties of powders are strongly dependent on particle size and shape, which affects the quality of the product (Chou, Chien, and Chau 2008).

7.2.2 Particle Shape

The shape of the particle is defined by the roundness or smoothness of the surface texture, which depends on the external morphology. Particle shape is important because it influences many critical properties such as flowability, compatibility, content uniformity, dissolution, drug release, bioavailability, and stability; all these factors ultimately affect the safety and efficacy of the dosage form. Particles with different shapes can have the same size but different properties. Particle shape can also influence particle size analysis dissolution and efficacy of the product (Zhenhua et al. 2000).

7.2.3 Zeta Potential

The zeta potential is a measure of the surface charge of dispersed particles in relation to the dispersing medium. In other words, zeta potential is the charge that a particle acquires in a particular medium. It is dependent on the pH, ionic strength, and concentration of a particular component. The magnitude of the measured zeta potential is an indication of the repulsive force that is present and can be used to predict the long-term stability of the product. If all the particles in a suspension have a large negative or positive zeta potential, then these particles tend to repel each other and there will not be any tendency to agglomerate (Swarbrick 2007).

7.3 SELECTION OF PARTICLE SIZE REDUCTION METHODS

Different mills can produce different end products from the same starting material. The subsequent usage of a powder usually controls the degree of size reduction. An important factor of size reduction is the cost, which can be economically undesirable when reducing particles to a finer degree if fine sizes are not needed for the processing. Once the particle size has been established, the selection of mills will also depend on the particle properties, such as hardness and toughness.

7.4 PARTICLE SIZE SEPARATION

7.4.1 Objectives of Size Separation

Size separation varies from solids, liquids, and semisolids. Solid separation is a process whereby particles are removed from gaseous and liquids. It has two main aims:

- To recover valuable products or by-products
- To prevent environmental pollution

Its main aim is to classify powders into different particle size ranges known as *particle size analysis*.

7.4.2 Size Separation Efficiency

Size separation efficiency is determined as a function of the effectiveness of a given process in separating particles into oversize and undersize fractions. If the separation process is 100% efficient, then all the oversize particles end up in the oversize product stream, and the undersize particles end up in the undersize product stream. Invariably, the industrial separation process produces an incomplete size separation, so that some undersize stream material is retained in the oversize stream and some oversize material may find its way into the undersize stream.

7.4.3 Size Separation Methods

Different methods are used for the separation of solids, liquids, and semisolids and can depend on the particle diameter (in micrometers). Size separation methods include size separation by sieving and size separation by fluid classification.

7.4.3.1 Size Separation by Sieving

This technique is used for measuring the particles whose size ranges from a particle diameter of 100–1000 μm. Size separation by sieving is different than size analysis. The use of sieving in size separation is done during processing of large volumes of powder. The sieves used for size separation are larger in area compared with that of size analysis. There are several techniques used to separate particles into appropriate size fractions efficiently. In the dry sieving process, these are based on mechanical disturbances in the powder bed. These specific methods are detailed in the following sections.

7.4.3.1.1 Agitation Methods

Size separation is achieved by electrically induced oscillation, mechanically induced vibration of the sieve meshes, or by gyration in which sieves are fitted to a flexible mounting. These are connected to an out-of-balance flywheel that imparts a rotary movement of small amplitude and high intensity to the sieve. It causes the particles to spin, thereby continuously changing their orientation and increasing their possibility to pass through a sieve aperture. The output from gyratory sieves is often considerably greater than that obtained using vibration or oscillation methods. These agitation methods can be converted to continuous processing by inclination of the sieve, leading to separate outlets for the undersize and oversize powder stream.

7.4.3.1.2 Brushing Methods

In this method a brush is used to reorient particles on the surface of a sieve and to prevent the apparatus blockage. A single brush can be rotated in a circular sieve, or for large-scale processing a horizontal cylindrical sieve is used with a spiral brush rotating about its longitudinal axis.

7.4.3.1.3 Centrifugal Methods

In this method particles are thrown outward to a vertical cylindrical sieve under the action of a high-speed rotor inside the cylinder. The current of air created by the rotor movement also assists in sieving, especially when fine powders are being processed.

7.4.3.2 Size Separation by Liquid Classification Sedimentation

This technique is used for the separation of fluid particles ranging from 0.1–1000 μm. It has two processes: centrifugal sedimentation ranging from 0.1–5-μm particle diameter and gravitational sedimentation ranging from 5–1000 μm.

This method uses a chamber containing a suspension of solid particles in a liquid, usually water. After the predetermined time, particles with less than a given diameter can be recovered using a pipette placed a fixed distance below the surface of liquid. Different size fractions can be pumped continuously instead of using pipette.

7.4.3.2.1 Elutriation Methods

Elutriation is a technique in which the fluid flows in an opposite direction to sedimentation movement, so that in gravitational elutriators particles move vertically downward, whereas the fluid travels vertically upward. If the upward velocity of the fluid is less than the settling velocity of the particle, sedimentation occurs and the particle moves downward against the flow of fluid. If the settling velocity of the particle is less than the upward fluid velocity, the particle moves upward with the flow of fluid. The elutriation particles are divided into different size fractions, depending on the velocity of the fluid.

7.5 SIZE REDUCTION METHODS

Particle size reduction is frequently used to improve an existing active ingredient of the component. Size reduction of the particle is more important because the small size particles always show better properties and are easy to handle. Some of the techniques used for size reduction of nutraceutical products are discussed in the following sections.

7.5.1 Large Bulk Size Reduction

This type of size reduction is mostly used in nutraceutical industries where large quantities are produced. Specific types are detailed in the following sections.

7.5.1.1 CX Heavy Duty Cutters

The CX Heavy Duty Cutter (Munson Machinery Company, Utica, NY) is a single-rotor high-torque cutter/hog with a positive, high profile raking action (unlike a shredder) for primary size reduction of large, bulky materials that typically would not fit into an SCC Screen Classifying Cutter (Munson Machinery Company) and/or require greater torque than other cutters provide. The CX Heavy Duty Cutter delivers high horsepower at low speeds to produce tremendous torque for seemingly effortless reduction of high density, high tensile bulk materials up to 36 in. in size, including plastic purging, extrusions, and related scrap products. It is available with electric or hydraulic drives to 125 hp to satisfy extremely demanding applications with ease and can be equipped with a gravity feed system or pneumatic feed stocks with a polycarbonate product feedback cover. The standard CX Cutter design uses a series of interconnected cylindrical discs, each incorporating two replaceable tool steel cutter inserts (optional four-insert disc design is available) in a staggered configuration. The inserts incorporate positive rake angle blade design, minimizing horsepower requirements and increasing shear efficiency. These cutters are mostly used in the food and nutraceutical industry with applications in the production of spices, herbs, and starches (Particle Characterization. n.d.).

7.5.1.2 Rotary De-Clumper

The product of more than a century of size reduction experience, the Rotary De-Clumper (Munson Machinery Company) is most widely used in the manufacture of different foods and nutraceuticals in large quantities. It is designed for dry powder deagglomeration or reducing scrap product or compacted, lumpy, and hard friable materials from silos, bags, or bulk sacks. The Rotary De-Clumper is a heavy-duty, dual-rotor mill that is inexpensive, simple to maintain, and trouble free. It is designed for cost-effective, rapid, in-line lump breaking, deagglomerating, or situations where high speed and high horsepower are not required. It is available in a variety of materials such as carbon steel, 304 and 316 stainless steel, abrasion-resistant steel (AR-235), and in a sanitary construction. Sizes range from a 3-hp 15 × 15–in. model to a 40-hp 36 × 48-in. model, with classifying screen sizes from 0.03125–2.5 in. Feed hoppers and support structures are available as options (Munson Machinery n.d.).

The Rotary De-Clumper is a modular dual-rotor design with rugged, three-point single-piece breaking heads constructed of abrasion steel (standard). The rotors are synchronized to operate approximately 100 rpm to avoid product heat increase. Bearings are isolated from the product processing area with a multiple labyrinth air-purge seal design. The rotor assembly can be quickly removed from the housing from one end for replacement.

The Rotary De-Clumper's smooth yet ultratough synchronized drive is one of the features that set it apart from its competition. No oil bath is needed. There are no intermeshing gears to wear out or bind. Only the shaft bearings require lubrication, and this is easily accomplished from external access points, making this mill a maintenance staff favorite. Larger units feature a redundant dual-drive configuration to eliminate any concern about upstream material backup in the event of motor or power failure—an important consideration when dealing with high-volume material flows. Application in the nutraceutical industry includes starches, spices, herbs, herbal supplements, and nutritional supplements.

7.5.1.3 Titan Shedders

Titan shedders have special features and can cut costs by reducing bulk waste and scrap from specific materials up to 50%–80%. These shredders handle the toughest size reduction problem materials with low revolutions per minute, high torque, and little or no temperature increase, and they reduce bulk volume up to 75%–80%. Shedders are designed to create an intensive shearing and cutting action with self-feeding hopper design. The Titan shredder is a double-rotor shredder design with rugged extended cutter teeth that chop and shred large solids while self-cleaning with each rotation. The cutter blades are mounted along two hex-shaped, heat-treated parallel shafts. Hex shafts are stronger and more fatigue resistant than traditional keyed bar shafts. This design allows heavy volumes of throughput with minimal power consumption. Cutter blades are constructed of a heat-treated and through-hardened 41/40 material and are resharpenable. They contain isolated bearings and gearing with a multiple labyrinth and lip seal design to prevent damage to critical bearing areas. These Titan shredders possess typical applications in nutraceuticals such as in the production of different spices, herbs, pharmaceutics, starches, vitamins, and food products (Munson Machinery n.d.).

7.5.1.4 Maxum Large Block Shredders

The special feature of Maxum large block shredders is to reduce large blocks and bails of material into particles of narrow size ranges with minimum fines. The Maxum 30 shredder (Munson Machinery Company) reduces large blocks and bails of material into particles of narrow size ranges with minimum fines. Each of two rotors is configured with a staggered array of specially treated, solid carbide tips able to shred a broad variety of materials, including the most difficult to shred products. The conical tips have an interference fit into receiver slots for rapid replacement.

Driven by two 20-hp (15-kW) motors, the counter-rotating shafts can grind blocks of material up to 2 × 2 × 6 ft. (610 × 610 × 3658 mm) in size and 1600 lb. (726 kg) in weight into uniform particles within narrow size ranges determined by the diameter of bed screen perforations, which range from 0.03125–1.5 in. (0.79–38 mm). The two 14-in. (355-mm) diameter shafts rotate at 25–35 rpm, producing up to 240 $ft.^3$ (6.8 m^3) of sized product per hour, depending on the material. The cutter handles large pieces of solid material, blocks of frozen products, compounds that cold flow, and highly abrasive products.

Material is fed through the top intake chute, and discharge is via gravity, pneumatic transition, or independently powered belt or screw conveyor. The unit is equipped as standard with two direct-coupled shaft-mounted gear reducers. These shredders include some specific features, such as each disc incorporates replaceable carbide inserts, and it also contains a classifying screen to control particle size. The Maxum 30 Large Block Shredder on an elevated stand accommodates a belt conveyor to remove discharged material.

7.5.2 Coarse Size Reduction

7.5.2.1 Screen Classifying Cutters

The screen classifying cutter (SCC) is highly effective in cutting hard, soft, and fibrous materials into controlled particle sizes with minimal fines at high rates. It features a patented helical rotor design with dozens of cutter blades attached to a helical array of staggered holders called *interconnected parallelograms* to continuously shear oversize materials against twin, stationary bed knives.

Unlike conventional granulators containing a small number of angled rotor blades that slice materials into strips in a scissor-like fashion, the SCC is configured with cutter blades along the entire shaft, with no gaps between blades, making total contact with the product. As a result, the material is cut into uniform pieces with minimum fines or imperfections in granulate and little to no heat generation. The blades are available in stainless steel, tool steel, and tungsten carbide, and with replaceable cutter tips and retaining socket-head screws for rapid replacement.

Primarily used for coarse grinding of materials into particles of uniform sizes ranging from granules down to 20–30 mesh, the cutter has widths ranging from 10–72 in. Bed screen perforations range from 0.03125–1.5 in. (0.79–38 mm) in diameter and up to 3-in. (76-mm) square, according to material characteristics and desired particle size. The 11-in. (279-mm) shaft rotates at 30–3600 rpm, producing up to 1000 $ft.^3$ (28 m^3) of sized product per hour, depending on application. Mostly used in industrial, food-grade construction, the cutter handles fiberglass insulation, potash, herbs and spices, flake materials, chalks, clays, filter cake, sugar, wood chips, foods, foundry materials, grains, minerals, pharmaceuticals, pigments, powdered metals, medical waste, and general chemical products. Material is fed through the top of an adjustable, double-baffled intake chute, or directly into the front of the chute through a hinged door. An independently powered, variable-speed pinch roll feeder is offered for horizontal-rolled product feed. Discharge is via gravity, pneumatic transition, or independently powered belt or screw conveyor. Typical applications include starches, sugar, spices, herbs, and tobacco leaves (Munson Machinery n.d.). These SCCs are further divided into two types: SSC magnum cutters and mini SSC cutters.

7.5.2.1.1 SSC Magnum Cutters

SSC Magnum Cutters (Munson Machinery Company) are specialized screen cutters for large-scale applications. The SCC Magnum Cutter is an ultraheavy-duty version of the SCC cutter, itself an exceptionally rugged machine.

Although the SCC Magnum Cutter handles the same types of materials as the SCC cutter, its extremely robust design enables it to accommodate infeed of significantly larger sizes and achieve much higher rates with ease.

7.5.2.1.2 Mini SSC Cutters

Mini SSC cutters are made of coarse materials and are industrial-grade heavy-duty rotary cutters mated to 15-hp drives, designed for low to moderate production rates. These are used to get the materials of similar size reduction characteristics but at lower capacities. Like SCC cutters, they are effective at cutting hard, soft, and fibrous materials into controlled particle sizes with minimal fines at high rates and feature a patented helical rotor design with multiple cutter blades attached to a helical array of staggered holders called interconnected parallelograms to continuously shear oversize materials against twin, stationary bed knives.

7.5.2.2 Knife Cutters

These rotary knife cutters are specially designed for grains, spices, roots, and related food products. One of the more popular applications is cutting corn for chicken feed. It delivers controlled end-product sizes with minimum fines at ultra-high rates, achieved by means of five rotating blades cutting against four stationary blades and propelling on-size particles through a perforated screen with 270° of exposure. An extensive selection of screen sizes allows tight control over end-product sizes.

7.5.2.3 Centrifugal Impact Mills and Pin Mills

Centrifugal impact mills are used for fine and coarse grinding of friable materials and wet or dry materials. This mill gives controllable particle size distribution and is easy to clean. The versatile centrifugal impact mill provides a simple, inexpensive means of grinding, sizing, deagglomerating, and homogenizing material, and can be used as an effective infestation destroyer. It yields particle sizes down to 325 meshes and is primarily used for friable materials. Available in 304 stainless steel, sanitary, abrasive-resistant, and standard carbon steel designs, the centrifugal impact mill operates without screens, hammers, knives, or rolls. Metered material is gravity fed through the centrally located inlet of the *stator disc*. Centrifugal forces accelerate the material and launch it into the impact zone. The action created by the stationary and rotating pins creates a *treacherous path* for material to pass through. Achieving the desired tight particle size distribution is obtained by controlling the rotor speed. Varying the rotor speed between a few hundred revolutions per minute and up to 5400 rpm provides the flexibility to use the machine as a coarse grinding or deagglomerating unit as well as a fine grinding mill. The machine is used for cracking wheat, grinding iron oxide powder, creating powdered sugar, or as a cellulose fiber conditioner; large production throughput rates can be attained in an inexpensive, compact machine. It also fits in neatly after most flaking operations. Applications include minerals, tobacco leaves, sugars, antibiotics, spices, herbs, and vitamins (McGlinchey 2005).

7.5.3 Fine Size Reduction

7.5.3.1 Attrition Mills

The attrition mill produces granular particles having a relatively narrow particle size spectrum ranging from 10–200 mesh. It is used primarily for reduction of fibrous materials, but is also suitable for friable products. It is available in three sizes with single- or dual- (counter-rotating) powered discs and an extensive selection of plate designs for optimum performance. In addition to general process applications, attrition mills are often used for conditioning of materials before packaging. Attrition mills produce a relatively narrow particle size spectrum, variable to 200 meshes. It has a high throughput/horsepower ratio. It is available with single- and double-powered runners. Interference press-fit runner and shaft assembly is supported by radial and thrust load with rated pillow block ball bearings. It is adjustable, and constant grinding plate spring pressure ensures product uniformity and has provision for quick-release grinding plate separation. Applications include food and pharmaceuticals, grains, cocoa, nuts, corn grinding, spices, and herbs.

7.5.3.2 Hammer Mills

The hammer mill reduces a broad variety of friable and fibrous materials into fine products in to uniform size ranges from 20–300 meshes. If raw material consists of large pieces exceeding 2–3 in. (50–75 mm), a precrusher may be used ahead of the hammer mill for initial reduction and uniform feeding of small pieces. This particular instrument is engineered for economical operation, extreme durability, and low horsepower per ton of output. Hammer mills also offer compactness, flexibility, and ease of operation. Designed for top efficiency, these hammer mills have an extra quarter screen located in the hinged section of the top case, increasing the screen area by approximately 50% over that of conventional designs, for a total of 270°. Some specific features of the hammer mill include a motor drive that is directly coupled to a rotor assembly. This is a large, obstruction-free product feed opening. It consists of a peripheral hammer moving at a velocity of up to 24,000 ft./min to ensure efficient and rapid particle size reduction. Applications include grain processing, spices, herbs, and corn oil.

7.6 THEORIES OR TECHNIQUES INVOLVED

Depending on the size, shape, and zeta potential of particles, different theories have evolved.

7.6.1 Particle Size

The study of particle dimensions is of critical importance in many areas of technology. Size separation varies for solids, liquids, and semisolids. Different techniques are used to study particle size analysis, and some are detailed in the following sections.

7.6.1.1 Laser Diffraction Theory

Laser diffraction has become a popular technique in many fields for measuring particle size distribution. The instruments widely used in this technique are most attractive for the capability to analyze over a broad size range in a variety of dispersion media. Laser diffraction-based particle size analysis relies on the fact that particles through a laser beam will scatter light at an angle that is directly related to their size. As particle size decreases, the observed scattering angle increases logarithmically. Scattering intensity is also dependent on particle size, diminishing with particle volume. Therefore, large particles scatter light at narrow angles with high intensity, whereas small particles scatter at wide range but with low intensity (Kippax n.d.). In laser diffraction theory, particle size distributions are calculated by comparing a sample scattering pattern with an optical model. There are two different models that clearly explain the size analysis: Fraunhofer approximation and Mie theory. Fraunhofer approximation assumes that the particles being measured are opaque and scatter light at narrow angles. As a result, it is applicable only to the large particles. Mie theory provides a more rigorous solution for the calculation of particle size distribution from light scattering data. It is applicable to both large and small particles. It allows for primary scattering from the surface of the particle, with the intensity predicted by the refractive index difference between the particle and the dispersion medium. It also predicts the secondary scattering of the particles having a size less than 50 μm in diameter (Zhenhua et al. 2000).

Laser diffraction is a nondestructive, nonintrusive method that can be used for either dry or wet samples. These techniques offer a wide dynamic measuring range, flexibility, generation of volume-based particle size distributions, high repeatability, and ease of verification (Cooper 1998).

7.6.1.2 Dynamic Light Scattering

Particle size analysis systems have been designed using a method of characterization called *dynamic light scattering*, which provides rapid detection of the precise size and size distribution of solid and agglomerated materials with nano- and micro-range particles. The dynamic light scattering technique measures the time-dependent fluctuations in the intensity of scattered light, which occur because the particles are undergoing random motion. Analysis of these intensity fluctuations enables the determination of the distribution of diffusion coefficients of the particles, which are converted into a size distribution. The upper size limit of the dynamic light scattering is sample density dependent because dynamic light scattering requires that particles are randomly diffusing, allowing the upper size limit as the point where the sedimentation of the particles dominates the diffusion process. The lower size limit of dynamic light scattering depends on the excess scattered light the sample generates compared with the suspending medium (Simon et al. 2004). Several factors will contribute to this lower size limit, including the sample concentration, the relative refractive index, laser power, and laser wavelength. The experiments using dynamic light scattering were performed at an angle of 90°. The samples should be much diluted to avoid multiple scattering phenomena (Available Particle Characterization Technologies n.d.).

7.6.1.3 Static Light Scattering

The static light scattering technique measures time-averaged intensity of scattered light as a function of sample concentration. The intensity of the scattered light that a macromolecule produces is proportional to the product of the weight-average molecular weight and the concentration of the macromolecule. For particles with smaller size showing isotropic scattering, static light scattering can be used to make accurate molecular weight measurements at a single angle, whereas for particles with larger size, static scattering may need to be done over a wide range of concentrations and a range of angles for accurate molecular weight determination (Saveyn et al. 2009).

7.6.2 Particle Shape

Particle shape characterization using image analysis is increasingly important to many industrial sectors, especially in the food and nutraceutical industries. Important information in terms of size, shape, and transparency of statistically significant number of particles may be obtained in a relatively short amount of time. Image analysis of particles can be divided into two broad types: static image systems, where the sample is stationary in the field of view, and dynamic imaging systems, where the sample is flowing through the field of view (Cartex and Yan 2005).

7.6.2.1 Dynamic Image Analysis

The dynamic image processing particle shape analysis system provides rapid and precise particle shape distributions for dry powders and bulk material in the size range from 30 μm to 30 mm. Dynamic image analysis uses an automated device containing a falling sample to a charge-coupled device camera, which takes digital images of the particles as they fall. When the image is captured, we can determine the size of the particle by assuming either a spherical- or cubical-shaped geometry. For this we use a product called the Optimiser 5400 for this measurement, and the particle measurement ranges from 100 μm to 5 mm (Particle Size n.d.).

7.6.2.2 Static Image Analysis

Static image analysis provides accurate particle size and shape distribution of particles ranging from 0.5–1000 μm. Nutraceutical industries widely use this technique for a variety of applications, including characterization of particles. In this technique the sample remains stationary.

7.6.3 Zeta Potential

The zeta potential of a particle is the overall charge that the particle acquires in a particular medium and can be measured using Zetasizer Nano series (Malvern Instruments Ltd., Malvern, UK). The magnitude of the measured zeta potential is an indication of the repulsive force that is present and can be used to predict the long-term stability of a product. Two methods are used to characterize particles possessing zeta potential: electroacoustic spectroscopy and acoustic spectroscopy (Malvern n.d.).

7.6.3.1 Electroacoustic Spectroscopy

Electroacoustic spectroscopy measures the interaction of electric and acoustic fields from which the zeta potential can be determined. As the name suggests, it is half acoustic and half electric. It probes the interaction between acoustic and electric fields. There are two ways to perform this technique. In the first method, the slurry is excited with pulses of ultrasound. The small fluid displacements produced by the sound wave cause small periodic displacements of the electrical double layer surrounding each charged particle. This distortion produces an electric field to the particles. In the second method, the reverse of the first method is performed, that is, excite the slurry with an electric field, which then interacts with the particles to create an acoustic response (Dukhin and Goetz 1996).

7.6.3.2 Acoustic Spectroscopy

Acoustic spectroscopy measures the attenuation and sound of ultrasound pulses as they pass through continental slurries. This technique is widely used in the nutraceutical industry, with a particle size distribution over a range from 10 nm to more than 10 μm. Acoustic attenuation is used for the entire particle size measurement range. Sound pulses are transmitted through the sample. The attenuation of these pulses is measured over a wide range of ultrasonic frequencies and at a range of detector spacings. The attenuation depends on the particle size, which can be calculated from the measured spectra (Goetz and Dukhin 2001).

REFERENCES

Available Particle Characterization Technologies. n.d. http://www.malvern.com/LabEng/support/technologies.htm.

Cartex, R. N., and Y. Yan. 2005. Measurement of Particle Shape Using Digital Imaging Techniques. *Journal of Physics: Conference Series* 15: 177–82.

Chou, S. Y., P. J. Chien, and C. F. Chau. 2008. Particle Size Reduction Effectively Enhances the Cholesterol Lowering Activities of Carrot Insoluble Fibre and Cellulose. *Journal of Agricultural and Food Chemistry* 56: 10994–98.

Cooper, J. 1998. Particle Size Analysis, Karer Diffraction Theory. *Materials World* 6: 5–7.

DiMemmo, L. N., B. Crooks, J. Hilden, C. Y. Lin, G. Young, and B. Sarsfield. 2005. Particle Shape Characterisation and Laser Light Scattering Measurements of Pharmaceutical Materials. *Microscopy and Microanalysis* 11: 1232–33.

Dukhin, A. S., and P. J. Goetz. 1996. Acoustic and Electroacoustic Spectroscopy. *American Chemical Society* 12: 4336–44.

Goetz, P. J., and A. S. Dukhin. 2001. Acoustic and Electroacoustic Spectroscopy for Characterizing Concentrated Dispersions and Emulsions. *Advances in Colloid and Interface Science* 392: 73–132.

Jillavenkatesa, A., S. J. Dapkunas, and L. H. Lum. 2001. Particle Size Characterisation. *Materials Science and Engineering Laboratory* 1: 960.

Karla, E. K. 2003. Nutraceutical—Definition and Introduction. *AAPS PharmSci* 3: 1–5.

Kippax, P. n.d. Measuring Particle Size Using Modern Laser Diffraction Techniques. http://www.chemeurope.com/en/whitepapers/61205/measuring-particle-size-using-modern-laser-diffraction-techniques.html.

Malvern. n.d. About Us. http://www.bioresearchonline.com/ecommcenter.mvc/malvern?vnetcookie=no.

McGlinchey, D. 2005. *Characterisation of Bulk Solids, Shape Characterisation of Irregular Particles*. Oxford, UK: Blackwell Publishing, 25–27.

Munson Machinery. n.d. http://www.munsonmachinery.com.

Particle Characterization. n.d. http://www.horiba.com/us/en/scientific/products/particle-characterization.

Particle Size. n.d. Micromeritics Analytical Services. http://www.particletesting.com/particle_size.aspx.

Saveyn, H., T. L. Thu, R. Govoreance, P. V. Meeven, and P. A. Vanrolleghem. 2009. Inline Comparison of Particle Sizing by Static Light Scattering, Time of Transition and Dynamic Image Analysis. *Particle and Particle Systems Characterisation* 23: 145–53.

Simon J., D. J. Croft, K. P. Samanthae, and H. Wilson. 2004. *Particle Size Analysis by Laser Diffraction.* London: Geological Society Special Publications 232: 63–73.

Swarbrick, J. 2007. Zeta Potential. *Encyclopedia of Pharmaceutical Technology*. Zug, Switzerland: Informa Health Care, 2386–87.

Zhenhua, M., H. G. Merkus, J. G. Desmet, C. Heffels, and B. Scarlett. 2000. New Development in Particle Characterisation by Laser Diffraction: Size and Shape. *Powder Technology* 111: 66–78.

CHAPTER 8

Application of Spray Drying Technology for Nutraceuticals and Functional Food Products

Donald Chiou and Timothy A. G. Langrish

CONTENTS

8.1 INTRODUCTION

Spray drying has been a widely used technique since its invention in 1878 by Samuel Perey. In the modern era, the scope for spray-dried products ranges from powdered milk, juices, and foods to pharmaceutical drugs and nutraceuticals (Hayashi 1989; Bhandari et al. 1993; Chan and Chew 2003). The demand for spray-dried foods in the past several decades has been steadily increasing because of the benefits from spray drying, such as longevity in product shelf life and the ease of convenience in transport and storage of the powdered product form (Hayashi 1989).

The operation of a spray drying system is a relatively simple process at first sight. Typically a feed liquid is sprayed into a chamber with hot gas flow (air or other gases), and the dryer shape is of cylindrical or conical form. The liquid feed material to be dried is normally pumped through an atomizer to form fine droplets. These droplets are passed into the drying gas flow before a particle separator (typically a cyclone) is used to remove the dried material. A basic schematic diagram of a typical spray dryer can be seen in Figure 8.1. Automation of the spraying system means operators mainly manipulate the drying variables remotely (Bajsic and Kranjcevic 2001). The operating conditions for spray drying include the inlet heat flow, fluid feed rate, and feed fluid atomization. However, research into the drying processes and fluid flow modeling has shown that the spray drying process is extremely complex (Langrish et al. 1993; Frydman et al. 1998; Guo et al. 2002; Harvie et al. 2002).

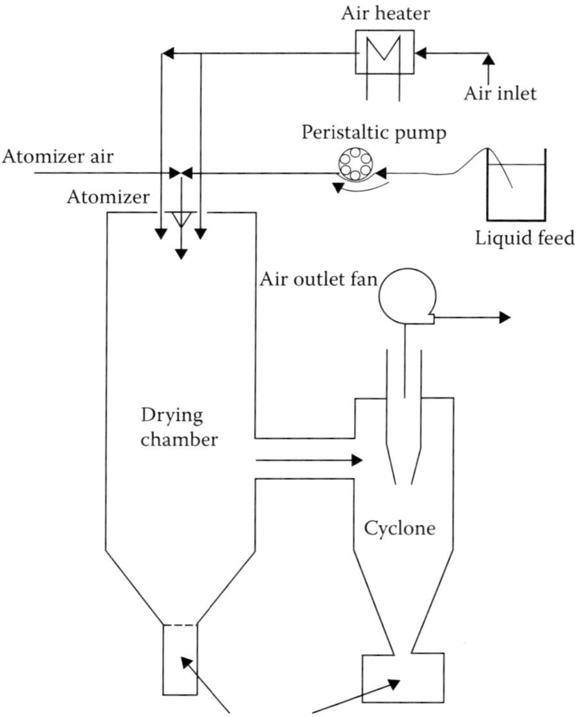

Figure 8.1 Basic schematic of a spray dryer.

As a method for weight reduction and increased shelf life, the use of spray drying is simple and generally considered to be a quick one-step process to make the powders or particles of the desired product. Water reduction plays a pivotal part in the attractiveness for spray drying processing because it reduces the overall weight of the product, allowing for savings in transportation and storage costs, and it reduces the water activity of the produced powders, enabling the product to have a potentially longer shelf life (Fontana 1998). Spray drying technology is also used in the area of encapsulation because the droplet drying process permits mixtures of compounds to be encapsulated inside a carrier material. This method of encapsulation has been used for the production of flavors and essential oils, pharmaceutical compounds, and other exposure-sensitive materials like antioxidant materials (Soottitantawat et al. 2005).

The application of this technology for the production of functional foods and nutraceuticals has emerged in the past decade and has shown itself to be a solid mainstream production method (Masters 2004). The use of this technology is not without certain considerations, however, because the operating conditions that are used greatly affect both the yield and the product quality through this simple process.

Although product quality is important in an industrial process for the end consumer of the material, if the process yield is not sufficiently high, then the production method is not viable when compared with the cost of production. Factors in spray drying that relate to the yield of the product are not only operational but also require understanding of the materials that will be dried through the process.

Operating conditions for yield and quality have been recently examined, with work conducted by several different researchers producing helpful results. One of the main issues in regard to selecting appropriate operating conditions is stickiness, where stickiness is an inherent property of the droplets that are drying, and also of the dried or semidried particles as they are carried to the collection system. Particles that are sticky have a greater tendency to adhere to the dryer and equipment walls and reduce yield; they also have the ability to agglomerate into larger particles. This becomes problematic when the larger particles that have adhered to each other and to the wall come off the dryer walls at a later time and contaminate the collected product. Particles that are exposed to the hot drying conditions for extended periods may have damage caused to the material through overdrying, browning, or even burning of the product.

The presence of these kinds of particles in the final product then becomes a problem for product quality. For example, spray-dried milk powder is typically an off-white or cream color; when milk droplets adhere to the dryer walls and are exposed to the high drying temperatures for a time, they undergo browning. If these particles then recirculate from the dryer walls into the collection system, the end product will contain brown specks of powder, and consumers are likely to reject this. Therefore, it is important to ensure that less wall deposition occurs, if possible, so contamination does not happen.

Understanding the fundamental processes that occur during the spray drying process enables the spray dryer design and operation to be tailored to suit the desired product properties and manufacture. Some of the concepts involved in spray drying technology will be examined in more detail and how they can be used to affect product quality and yield, as well as their applications for nutraceuticals and functional foods, will be discussed.

8.2 SPRAY DRYING FUNDAMENTALS

Within the spray drying process, there are some operational factors to be considered that affect the end product. The main five operational factors that will be considered and discussed are detailed in the following sections.

8.2.1 Main Gas (Aspirator) Atomizing Gas and Other Gases

This is the drying medium used to both extract moisture from the feed solution and then carry the particles through the dryer into the collection system. The gas can range from air, reactive gases (e.g., carbon dioxide), inert gases (e.g., nitrogen), and even solvents loaded (e.g., evaporated solvents using a sealed gas loop). This gas can be passed through both cocurrent and countercurrent flow paths to the atomization of the equipment, depending on the setup required of the system. The volume and speed of the gas movement through the system can also be changed and controlled as desired. From mass and energy balances, the gas flow rates will affect the humidity of the gas through the dryer and the moisture contents of the products.

8.2.2 Gas Temperature

This factor is generally known as the drying temperature because this is the primary driving force that encourages evaporation and drying of the liquid feed to the final dry product. This can range from below room temperature (as seen with freeze spray drying processes) to as high as desired, as long as the heating system can appropriately heat the main gas flow in a stable manner.

8.2.3 Atomization

Atomization as a term is slightly misleading in the sense that the liquid feed is not reduced to atomic scale particles, but rather is broken up into a fine spray or mist of droplets (thus, the term *spray drying*). The atomization can be achieved through a variety of methods such as a single-fluid nozzle, two-fluid nozzle, rotary or spinning disc, and ultrasonic atomizers. Atomization gases are used in the two-fluid nozzle and ultrasonic atomizers (typically a high-pressure gas stream across or on a tongue of material that vibrates at ultrasonic speeds), whereas the rotary or spinning disc atomization places a stream of feed liquid onto the surface of a rapidly spinning disc that disperses the feed into the spray through centrifugal forces.

8.2.4 Liquid Feed Rate

The liquid feed rate is the volume of the mixture being fed into the atomizer over a particular timeframe. This is typically controlled through a pumping system that ensures the desired rate of feed material is delivered to the appropriate drying conditions inside the dryer through the atomization device. This feed can be either a solution or slurry, depending on the desired process behavior required. For small-scale dryers such as laboratory scale, the feed rates are typically measured in $mL \cdot min^{-1}$, whereas industrial-sized dryers typically produce $L \cdot h^{-1}$ rates.

8.2.5 Liquid Feed Composition

The inherent physical and chemical properties of the feed are important to the process because the solid concentration in the feed affects the evaporative load in the dryer, and properties like the viscosity of the feed affect the particle size and sheer stress during pumping and atomization. These variables for the liquid feed can also be controlled appropriately as required.

The major issues in appropriately spray drying materials can be addressed by altering these five variables within the spray drying process. Understanding the effects that these variables have on the drying of the feed material is essential.

8.3 THE ISSUE OF STICKINESS

Stickiness is the major contributor to poor product quality and low yields. The production of powders through the spray dryer typically results in hot and sticky products, especially when the materials are high in sugars or carbohydrates that can break down to sugars. These sticky particles then adhere to the equipment walls, reducing yield and making cleaning difficult at the end of the process, and as previously mentioned, the recirculation of damaged materials is also a potential problem. This is especially problematic for the food industry, in which many products contain added sugar, natural sugars, or are starchy in composition.

Carriers are used in the spray drying process as a method for reducing the effects of stickiness in production. Carriers typically can encapsulate the sticky material to prevent it from adhering to the equipment or forming larger agglomerations that may not stick when they make contact with the equipment walls. Studies into spray drying of sugary materials like honey have found that up to 60% weight by weight of maltodextrin carrier is required to make the mixture sufficiently nonsticky when spray-dried into powder form (Bhandari et al. 1999). This is a high content of material as a carrier, and not all materials are feasible as carriers. In certain situations, the use of carriers is also undesired (e.g., where consumers can detect and taste the presence of excess carriers in the product); therefore, it is better to understand the problem to find alternatives in production methods than rely on carriers to reduce the stickiness problem.

The aspect that needs to be understood in this situation is the material properties that are being manipulated in the drying process, namely, the rubbery amorphous state of the feed material that is created during the drying process. Because spray drying is a rapid process, the feed material will be subjected to a high heating gradient as the moisture content of the produced droplet evaporates. The particle is also then potentially exposed for a period to a high temperature with potentially high humidity conditions in the main gas flow. These effects may play a part in producing an amorphous solid particle that has a sticky surface because traditionally it is considered that there is insufficient time during the drying process for a crystalline structure and surface to be formed with most materials.

Two key factors are necessary when dealing with the issue of stickiness in materials: moisture content and glass transition temperature.

8.3.1 Glass Transition Temperature and Moisture

All polymers that undergo melting and crystallization behavior have a glass transition temperature (T_g) (Wunderlich 2006). The glass transition temperature is typically recorded as the midpoint in a temperature range that describes when the material first begins to exhibit plastic behavior until it is in an amorphous state where the molecularly bonded structure of the material is in disarray (Liu et al. 2007). Above this temperature, the material is now prone to phase transformation behaviors, including melting and crystallization, depending on the energies applied to the material.

It has been suggested that at temperatures of 20° above the glass transition temperature, materials can begin to undergo crystallization behavior if the conditions allow the materials to appropriately crystallize (such as nucleation and being undisturbed from physical stimuli that would reduce the material back to the amorphous phase) (Bhandari et al. 1997). This means that if the glass transition temperature of the material is known, appropriate conditions of the spray drying process may be manipulated accordingly to produce properties in the powders as desired. By keeping the particle temperatures above or below the glass transition temperature range, it may be possible to control particles being produced in an amorphous or a crystalline form.

In addition to this, the rate of crystallization in relation to the glass transition temperature needs to be considered. For droplets that are formed immediately after atomization, if the droplets dry quickly and the particle temperatures rise well above the glass transition temperature, then the particles will become rubbery and sticky. These particles then may collide with the walls of the dryer. If the rate of crystallization is slow with these set of conditions, then the sticky particles may stick to the walls. A faster rate of crystallization with the same conditions would enable the particles to crystallize before the wall collisions and "bounce" off the wall. Therefore, the rate of crystallization in relation to the glass transition temperature is critical in determining the particle-to-wall interaction.

Matters are not that easy, however, and this is where the importance of moisture content, or the presence of water in the material, needs to be mentioned. The glass transition temperature of a material is an inherent property of its constituents. The

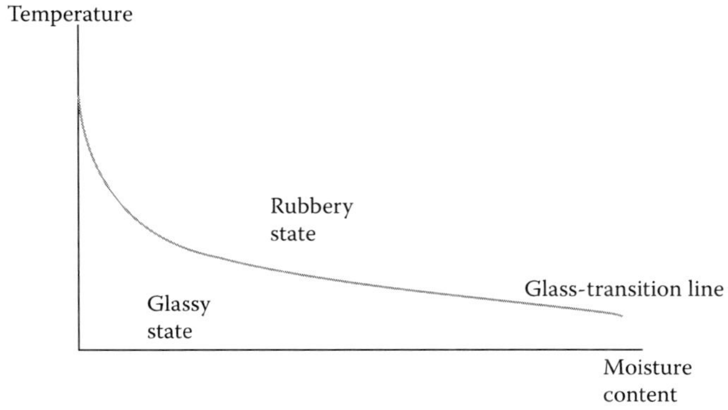

Figure 8.2 Schematic diagram of amorphous states in relation to glass transition temperatures.

Gordon–Taylor equation (Gordon and Taylor 1952) (Equation 8.1) is an example of how the glass transition temperature can be calculated for a homogeneous mixture of more than one material. The proportions of the different glass transition temperatures are averaged to find the overall glass transition temperature, taking into account the presence of water.

$$Tg = \frac{w_1 T_{g1} + k w_2 T_{g2}}{w_1 + k w_2} \tag{8.1}$$

Water is known as a plasticizer, and in this situation water has a glass transition temperature of approximately −137°C (pure water) (Yue and Angell 2004). If it is present in any mixture, depending on the proportion present in the mixture, the overall glass transition temperature can be reduced greatly. This can be seen in Figure 8.2 with a schematic representation of the relationship between the glass transition temperature and the moisture content. While in a static situation (where there is no change in water content), this can be calculated and taken into consideration; however, during the drying process, regardless of what kind of dryer (e.g., spray drying, drum drying, or tray drying), the loss of water (as moisture content) means that the glass transition temperature is continually changing in the material until all the moisture present is removed or the drying process ceases. This constant changing behavior in the glass transition temperature makes it slightly more difficult to tailor the operational conditions to produce accurate and reproducible powder properties unless the drying behavior during the drying process is examined in depth and known.

8.4 UNDERSTANDING DRYING BEHAVIOR

Spray drying behavior can be modeled to some extent using simple mass and heat balances (Truong et al. 2005; Hanus and Langrish 2007). The process is more complex, but the basic principles of the process can be applied quickly and easily. From the atomization nozzle, a mist of fine droplets is produced, and these are exposed to the drying gas. At this point, heat transfer between the gas and droplet occurs, and this in turn causes mass transfer. The droplet increases in temperature, and the moisture content turns from liquid into vapor as it evaporates.

The feed material may also contain other volatiles such as oils and solvents, and during this process of heat and mass transfer, they too may be lost. This process alters the properties of the drying speed and glass transition behavior of the droplet or particle as it travels through the dryer. The drying kinetics of this process is complex and has had the attention of many different approaches, including the use of fluid-flow models and CFD to determine the particle movement and gas–particle interactions (Kieviet et al. 1997; Frydman et al. 1999; Guo et al. 2004; Fletcher et al. 2006).

If the drying temperature of the system is at, or is higher than, the glass transition temperature of the material, the produced particles may turn rubbery and amorphous, having a sticky surface and become prone to adhering to the equipment. However, as previously mentioned, the particles may undergo crystallization behavior if the temperatures are sufficiently high and the rate of crystallization rate is sufficiently high. At this point, if the particle is undergoing crystallization, several factors are important in determining the extent of crystallization.

The residence time and bulk humidity of the drying system have to be considered once the particle is undergoing crystallization behavior after initial drying has occurred. Residence time is a factor influenced by both the aspirator air flow rate and the physical dimensions of the dryer. The greater the residence time, the longer the particles will remain within the dryer. The flow rate of the drying gas medium in the system will determine the speed that the particles are carried through, whereas obviously, the larger the size of the dryer, the longer the inherent residence time is. The relationship between these two, however, is not so straightforward. For example, using a simulation of spray drying behavior, when a laboratory scale spray dryer with a chamber size of 0.48 m length is expanded by a factor of 60 times to 28.8 m, the residence time at a set aspirator flow rate increases from 0.6 seconds to 36 seconds. To maintain the same the residence time of 0.6 seconds in a 28.8-m chamber, the aspirator flow rate would have to be increased by a factor of 70 (Chiou 2008). The importance of this aspect is that scale-up consideration of processes is not a simple matter, and the residence time has a significant effect on the spray drying process.

Particles that remain in the dryer longer are exposed to the conditions longer. This means that they may be imparted with a higher amount of thermal energy that could promote crystallization behavior, melt behavior, or other behavior such as browning and burning while the particle travels through a normal transport path without sticking to the equipment. This means that although crystallization may be promoted as the residence time is increased, damage to the particle is also possible, and an appropriate balance between the two is required.

The second consideration for particles traveling through a dryer besides the drying temperature of the aspirator gas and residence time is the bulk humidity of the gas itself. The bulk humidity of the drying gas has two main effects on the drying behavior of the droplets. If the bulk gas humidity is low, this promotes faster drying as a result of a greater driving force for mass transfer (equilibrium behavior), whereas a humid gas hinders the phase change of the droplet from liquid to gas because the

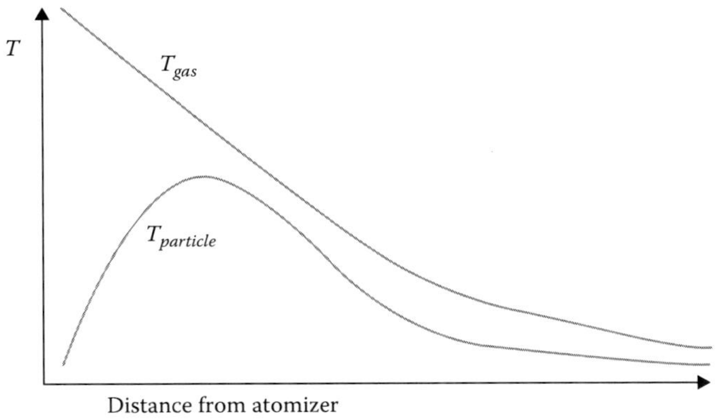

Figure 8.3 Temperature of the main drying gas and particle as function of distance from the atomizer.

gas flow cannot have a greater relative humidity than 100%. The second issue that the bulk humidity provides is the rewetting of the particles in transit.

With high humidity generated by the drying process, if the initial humidity is high, there is a risk of condensation of the moisture onto the particles as they experience cooling while in transit to the collection system. The gas temperature decreases because energy is required to evaporate moisture from the droplets and particles; this in turn causes the particle temperature to also decrease as they experience evaporative cooling (Figure 8.3). These two effects combined with the increased humidity of the gas from the drying may form the condensate. This may redissolve the surface of the particle and render it once more sticky with agglomeration effects that may appear as caking and further moisture-induced degradation. This consideration of condensation has been examined briefly using modeling and may be a problem in spray drying (Chiou 2008).

The residence time can be controlled by altering the main gas flow rates (because the particles rapidly relax to the gas velocity), whereas the bulk humidity of the dryer can be manipulated by using an appropriate initial bulk humidity or a sealed loop with a condensing system to remove or increase the humidity.

8.5 CRYSTALLIZATION BEHAVIOR

Crystallization has been mentioned several times without much explanation in regard to what is meant in this chapter and the actual behavior of the particles during the drying process. Crystallization traditionally is seen as a process, often slow, whereby growth of a material occurs on nuclei in solution as a result of attraction forces, applied energy, or changes in concentration and solubility (Piorkowska et al. 2006). The continuation of this growth of material forms structures in an ordered lattice, hence being a crystalline material. These materials are also typically classified as polymers, having repeated units of the same compounds that can be ordered into repeatable lattice structures (Sajkiewicz 2002). In the situation of amorphous to crystalline transformation, the starting material is the polymer but in an amorphous state where the molecular components are in disarray.

Amorphous to crystalline transformation traditionally is viewed as an energy-intensive process in which the material is given sufficient energy to internally rotate its structures to fall into the lower energy state of the crystalline lattice structure. This, however, usually entails melt behavior, where the material breaks its solid bonding form so the molecules are free to rearrange themselves.

In terms of spray drying crystallization, the process of crystallization behavior can be said to fall under both liquid formation and solid transformation. The initial formation of the solid in spray drying generally produces an amorphous solid as a result of the rapid evaporation and formation. There are exceptions to this, such as sodium chloride where the crystals crystallize almost immediately, but the reasons for this will be discussed further later. A gentler and slower drying process would have a greater likelihood of producing a crystal, but this is not always possible with spray drying because of residence times and appropriate end moisture content requirements. Work recently conducted on the crystallization behavior of these particles has found that it is possible to crystallize the products during the spray drying process through increasing the rates and understanding the material properties and operating conditions (Chiou et al. 2007).

Following the traditional idea of crystallization, as a result of the rapid drying of the droplet to form an amorphous solid, the particle then is exposed to a high temperature, high humidity environment. The presence of the high drying gas temperature provides potentially enough kinetic energy to cause molecular rearrangement, but the lack of enough time during the process means that crystallization cannot occur easily. However, it is still possible for partial crystallization behavior to begin, and this is what is more important to understand and control because having a control over the degree of crystallinity that the produced powders have is a useful practical application. This can be summarized in Figure 8.4 as a two-phase

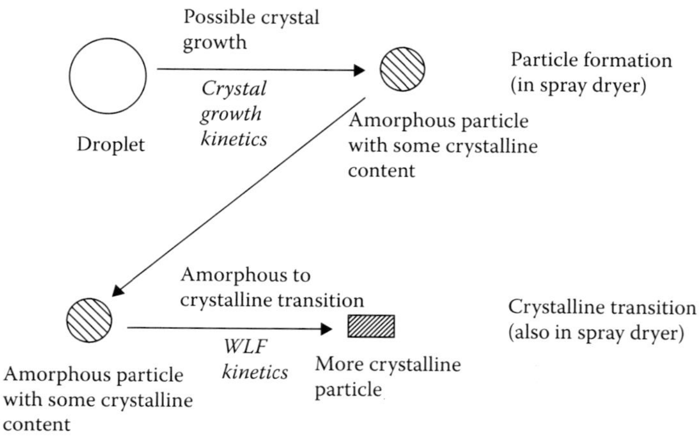

Figure 8.4 Crystallization and "recrystallization" phases in droplet drying (Chiou 2008).

process, which occurs during the one-step drying and particle production process within spray drying.

8.6 AMORPHOUS AND CRYSTALLINE MATERIALS

Amorphous and crystalline materials have differing properties that may be beneficial or a nuisance depending on the desired behavior of the product. For amorphous particles, there is a tendency toward them having easier dissolution and a faster or greater bioavailability and reactivity as a result of the molecular arrangement being in an unordered state (Choi et al. 2004; DiNunzio et al. 2008; Kim et al. 2008). This is because the reactive groups on the surface are more readily available and unhindered. Materials in the crystalline state are bound in a structure that has a closer formation, hindering reactivity and bioavailability because it restricts the space for compounds to react in and the bond formations are stronger as a result of the tighter packing orientation.

In regard to particle engineering, with the right knowledge and technical ability, it is possible to produce fully amorphous, fully crystalline, and partially crystallized materials. It is possible then to consider that two differently engineered particles may be useful and desired for different applications with only the crystallinity of the material being manipulated. The particles in consideration are what could be described as an "eggshell" formation and the reverse of that, a crystalline "core" particle as seen in Figure 8.5 (Lekhal et al. 2001).

In the scenario of the eggshell particle, the surface "shell" of the particle is in the crystalline state. The inner core of the particle is in the amorphous state. The purpose and use of this particle would be to provide a limited and slower reaction, protecting the majority of the contents for a later stage when a faster and rapid delivery is desired. This could be used in drug delivery where the crystalline shell protects and delays the delivery of a particular compound until it reaches particular areas

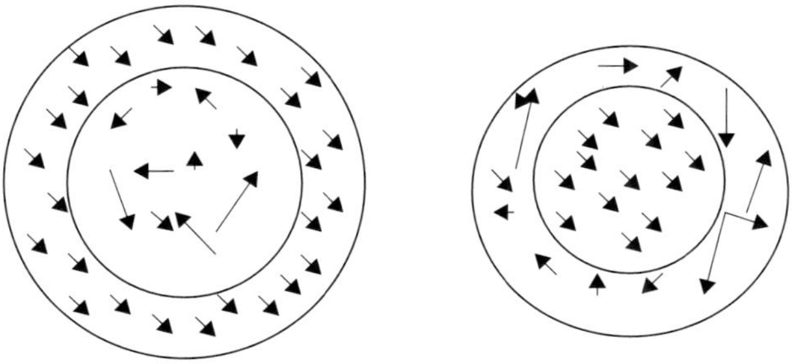

Figure 8.5 Diagrammatic representation of particle cross sections for potentially engineered particle crystallinity (eggshell and crystalline core particles).

where by then the crystalline shell has been dissolved, allowing a fast delivery of the contents. This is also useful for helping compounds that might normally have a short shelf life to keep their reactivity and bioavailability longer because the crystalline shell has a less reactive barrier to the internal contents.

For the reverse situation where the core of the particle is in the crystalline state with an amorphous exterior, it would allow a rapid initial distribution of the compound and then a much slower release, providing a longer distribution of the particle contents. This might be useful for delivery of the particle contents to multiple sites, or prolonging the effects of a material such as a digestive inhibitor, for example.

It may be possible to produce both of these types of particles using spray drying or other drying technology with the right application of materials, encapsulation carriers, and operation conditions. For example, if moisture can be absorbed onto the outer surface of the particle, allowing the outside to crystallize, an eggshell particle might be produced. Mechanical agitation of the particles, by contrast, is likely to produce an amorphous outer shell while keeping the interior crystalline (Huang et al. 2007).

8.7 CREATING ENGINEERED PARTICLES USING SPRAY DRYING

With these particular properties in mind, how can we use spray drying technology to engineer these particles and powders for nutraceutical, pharmaceutical, and functional food products? It is possible through the combined knowledge of the drying fundamentals being applied in spray drying, the material properties such as the glass transition temperature, and the effects of varying the operating conditions of the drying process. If we break down the whole procedure into stages of the process, we can examine the considerations involved in generating these specific particles. For ease, the eggshell type of particle will be called $P_{(ca)}$ for Particle (crystalline-amorphous), whereas the crystalline core particle will be $P_{(ac)}$ for Particle (amorphous-crystalline). A diagram displaying the differences in process to give specific outcomes is shown in Figure 8.6.

Feed slurry preparation is where the process begins. The desired material to be converted into a powder or particle needs to be prepared as either a solution or suspension. The initial starting concentrations and viscosity of the feed can play an important part in the process. Higher concentrations of material in the feed slurry are normally desired because they carry less solvent to be evaporated, leaving the drying gas drier, and also reduce the temperature decrease of the drying gas. This means the droplets formed also dry faster because there is less solvent to evaporate off the droplets. The viscosity of the feed is important because it can affect the pumping speed and consistency to the dryer. An even flow of feed into the system provides a stable drying environment where there are fewer fluctuations in the gas humidity and drying temperature, resulting in a more consistent final powder in terms of properties and moisture content. The viscosity also affects the atomization of the feed into droplets, with higher viscosity fluids typically creating larger droplets as a result of a stronger surface tension being present.

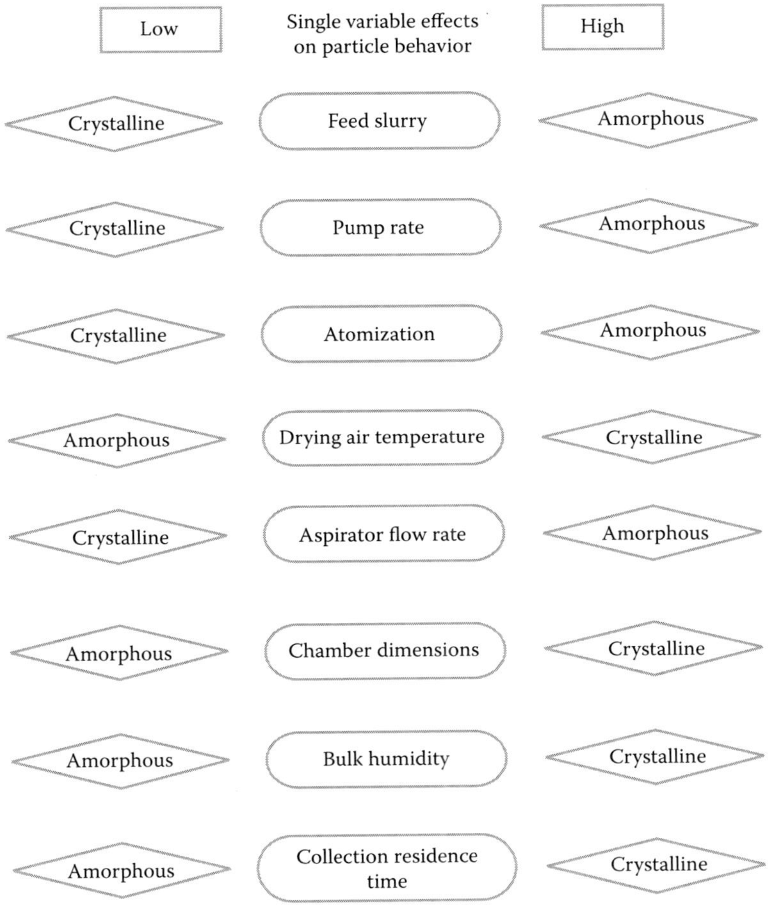

Figure 8.6 Single variable effects on particle behavior.

For the $P_{(ca)}$, having a high concentration feed is desirable because this both increases the size of the droplet and causes a more rapid "snap drying" of the droplets into a highly amorphous state. The $P_{(ac)}$ particles would benefit more from a lower concentration feed slurry because the droplets would dry slower, allowing greater crystallization time in a more gradual changing environment so the molecules have time to rearrange into the appropriate lower energy structure, and the particles would also form into smaller, more readily crystallized particles.

The feed slurry is then put through the feed pump into the atomizer. The rate of the slurry feeding into the atomizer may affect the droplet size, depending on the type of atomizer. For two-fluid nozzles, this is more so than for ultrasonic nozzles because the ultrasonic atomization relies on the vibration rate more than the feed rate. For rotary disc atomizers, the feed rate is partially responsible for droplet size, and rate of disc rotation also plays a part. With a higher rate of feed entering the atomizer with a constant atomization flow gas (for a two-fluid nozzle) or disc rotation rate (for a rotary or spinning disc nozzle), the droplets formed will be larger, and the same considerations as before for the feed slurry concentration and viscosity should be considered. The converse where a lower feed rate to the atomization system would produce smaller droplets is also true, and the same considerations given to the feed slurry concentration occur. Therefore, a high feed rate would promote $P_{(ca)}$ particles, whereas a low feed rate would be more likely to produce $P_{(ac)}$ particles.

Once an atomized spray of droplets is formed, they are in contact with three aspects of the dryer system: drying gas temperature, aspirator gas flow rate, and bulk humidity. These three aspects are interlinked at this point because of the drying process being so rapid. The drying kinetics of the process is also important at this stage. Droplets entering the moving gas flow will begin to experience an increased temperature gradient as a result of the drying temperature of the gas, whereas a high flow rate gas causes increased heat and mass transfer to the particles while reducing the bulk humidity, as shown in Figure 8.3.

As the droplets begin to increase in temperature, the moisture content is evaporated from the surface of the droplet to the main gas flow. The rate of this evaporation depends on several aspects, and depending on how fast or slow this process occurs, it will change how the particle is formed and in what form it will be. The transfer of heat through particle components is complicated, and much work in drying kinetics has been done. The rate of droplet drying affects the crystallization behavior because the rate of glass transition temperature change is affected by the moisture content as it changes through the dryer, and the change in moisture content is affected by the operating conditions.

For a slow evaporation process, typically produced by a lower drying temperature, slower gas flow rate, and higher bulk humidity, the particles will be more likely to begin crystallization during the drying process. This is limited, however, by the actual temperature of the particle because the higher the temperature, the greater the likely crystallization. The lower evaporation temperature also permits more water to be present to act as a lubricant (Chiou et al. 2008) for molecular rearrangement into the more crystalline structure, while also depressing the glass transition temperature so that the overall temperature required for crystallization behavior to begin is lower. Thus, whereas lower drying temperatures promote crystallization, insufficiently high temperatures will not permit crystallization to occur, or they may limit how much or fast the crystallization behavior does occur. The crystallization that occurs in a particle formed at the lower evaporation conditions may not be complete and may be inconsistent across the particle, which might be desirable for the formation of egg yolk or eggshell particles.

For the opposite case to slow evaporation, at higher drying temperatures, faster gas velocities, and lower bulk humidity, the droplets will have their moisture content evaporate faster and be more likely to generate an amorphous solid because there is less moisture for molecular rearrangement, an increased glass transition temperature with less water-induced plasticization, and a greater driving force to a dry gas that promotes evaporation. However, this does not mean that no crystallization is able to occur because at the higher drying temperature, although the evaporation process is considered to be higher, the rate of reaction is greater, and these two competing effects would possibly still allow some crystallization behavior to occur depending on the exact combination of variables. This then results once more in the scenario that a partially crystallized particle could be formed as part of this stage in the spray drying process.

Following this evaporative process is the remainder of the transit path that occurs while the droplet or particle is in transit from the atomizer to the collection system. The residence time of the particle is now significant. Drying of a fine mist in a hot gas is almost instantaneous, and although for small laboratory scale dryers (e.g., Büchi B290; Büchi Laboratory Equipment, Flawil, Switzerland) the residence time is typically 1 second or less, this is still sufficient for post-drying effects to potentially occur. During this transit, although the particle has been formed into its amorphous, partially crystalline or crystalline state as a solid, it can still undergo further transformation before reaching the collection system.

The transit of the particle is aided by the fact that the gas flow has sufficient momentum to move the small mass of the particle through the equipment while suspending the particle in the gas. This means that the particle once formed is still being exposed to the drying and evaporation conditions of the gas temperature, gas flow, and bulk humidity. This is a significant aspect, at least for the drying temperature and bulk humidity of the gas, because they have the ability to affect the particle in transit.

Residual moisture present in the particle (because the drying process rarely removes all the moisture present, including bound water in crystalline structures) still acts as a plasticizer and assists in the reduction of the glass transition temperature. In dry particles (<5% moisture content), this translates to nearly 7° depression of the glass transition temperature. From the Williams–Landel–Ferry kinetics (Chiou 2008), the increase of crystallization rate is super-Arrhenius, more than a 10 times factor for a 10°C increase in the difference between the particle temperature and glass transition temperature. This means that once the majority of the moisture has been evaporated and the dry particle remains, the particle temperature will increase rapidly to be at, or near, the drying gas temperature. This increased temperature could promote further crystallization in the particle (but less in scale as a result of lower moisture being available to act as a lubricant for molecular rearrangement). The rate at which the particle temperature increases is dependent on the size of the particle (larger particle classes being slower), the drying temperature, and the main gas flow rate. The second aspect of the transit portion of the drying process is also more related to the bulk humidity of the drying gas.

If higher bulk humidity is present in the drying gas, there is the possibility that the particles traveling in the gas may have condensate form on them. The saturation level of moisture in the gas changes according to the temperature of the gas. Heat loss through the system both in evaporation of the droplets to form the particles and through the walls of the equipment occurs continuously, and this decrease in temperature has the potential to cause condensation of the gas humidity either on the walls of the equipment, and/or on the particles themselves traveling within the gas flow (Chiou 2008). Should condensate form on the surface of the particles, there is the real possibility of surface dissolution and re-evaporation or drying that could form both amorphous and/or crystalline surfaces depending on the subsequent drying conditions.

To expand on this further, with a particle that has a redissolved surface, should the conditions promote crystallization (sufficiently high temperature but not rapid evaporation) then the outer surface of the particle may form into a crystalline state. The converse, where the conditions would promote amorphous material to form (rapid drying with high temperature) on the surface of the particle, is also possible. This means that both cases for the $P_{(ca)}$ and $P_{(ac)}$ conditions are available with the system.

In the final stage of the spray drying process, the particles are moved to the collection system. Regardless of the collection system in place, the particles will commonly reside there until they are either removed or passed to the next step of the processing system, such as fluidized bed crystallization or packaging, for example. This time spent waiting for the next step also plays a part in the engineering of the desired properties. When the particles are at rest in the collection system, it is normal that they are continually being exposed to the gas flow that delivered them into the system. What this means is, once more, they are further exposed to the previously mentioned temperatures and bulk humidity conditions. The difference at this stage of the process is that the drying temperature has normally decreased considerably, and the particles that are the most affected by this will be the outer layers of the deposited powders. Progressively, as more particles are being collected, the less the effect of the temperature and humidity will be on the inner layers until they are completely unaffected by the conditions.

With this lower temperature affecting them, condensate is still possible from the bulk humidity of the gas, but now at the lower temperatures, it may be that the temperature is insufficient to cause further crystallization, even with moisture in the particles causing plasticization of the material. This poses a problem for further processing and may be why current processes often include post-drying steps like fluidized bed crystallization and drying to remove and reduce further caking and agglomeration effects.

As can be seen, the multiple stages involved in the drying process all have the potential to implement change to the characteristics of the particles as they form. This is beneficial because it provides flexibility to the operator by altering specific variables when specific characteristics are desired. For example, if the material being produced is temperature sensitive (e.g., volatile flavors and essential oils), then it is critical that the drying temperatures do not exceed their boiling points to prevent the loss of these compounds. The drying temperature could be reduced, but by altering the atomization and feed rates, adequate drying of the finer and smaller-sized droplets is still possible. With changed residence time and bulk humidity conditions, it may then still be possible for these compounds to have desired amorphous or crystalline structures controlled and created. For a variety of given situations where there are limitations on the conditions required for processing, having the flexibility and understanding of these stages enables the operator to maximize the ability to produce the desired outcomes.

Taking this to the next consideration, with appropriate knowledge and understanding of the drying operations, there are particular situations where, regardless of how the drying process is operated, it does not always result in a stable compound with specific bioactivity (like antioxidants) because the drying process may decrease or damage the compounds in question. Particular volatiles may also be difficult to dry because their vaporization temperatures may be too low for sufficient drying even when all the conditions are appropriately implemented. To counteract this problem, carriers can be introduced into the feed material.

8.8 CARRIERS AND ENCAPSULATION

As previously mentioned in the introduction, carriers and encapsulates are often involved in spray drying technology; their purpose is to reduce the stickiness and loss or deactivation of sensitive materials in the produced particles and powders. There is an extensive range of these carriers and encapsulating agents, and it is also important to understand how they contribute to the process.

Carriers are typically used to reduce the stickiness in spray-dried products, such as the previously mentioned honey (Bhandari et al. 1999) but also in juices (Bhandari et al. 1993) and other products. There are two main reasons why carriers may assist in the reduction of stickiness. The first one is that carriers (e.g., maltodextrin) typically increase the glass transition temperature of the mixture. This means that the particles forming do not reach the rubbery amorphous and sticky state and dry without being sticky by being in the glassy amorphous form. The second method is through encapsulating the sticky material inside a nonsticky or less sticky matrix of material so when collisions occur against the dryer or in the collection system and postproduction, the particles remain in a free-flowing state.

The issue of glass transition temperature has been previously mentioned, but with regard to the use of carriers, the basis for increasing the glass transition temperature is the molecular weight of the materials. Larger molecular weight compounds and molecules have a greater ability to absorb energy before undergoing molecular change and interactions. A simple formula relating to the molecular weight of maltodextrins was determined where the dextrose equivalents of maltodextrin were correlated with the glass transition temperature (Bhandari and Howes 1999). It was seen that with increasing starch chain length and weight, the glass transition temperature also increased.

By binding a low glass transition temperature material into a higher glass transition temperature material, the overall mixture ends up with a weight fraction–based glass transition temperature, including any moisture content present as per the Gordon–Taylor equation. This means that the higher the content of high glass transition material, the more likely it will be in a nonrubbery glassy amorphous state.

This inherent ability of heavy molecular weight materials to have a higher glass transition temperature means that a large range of compounds have the potential to be carriers for the purpose of reducing stickiness in products. This could be a process-simplifying factor because combining ingredients that are normally mixed in a post-drying process in a predrying feed could potentially result in a more efficient product because the mixing phase still occurs but the higher molecular weight compound is less sticky, therefore leading to better yield and product quality. There is a downside in terms of using heavy molecular weight materials for carriers, however, and it is related to crystallization behavior.

It has been found that with increased molecular weights, the rate of crystallization is also slower because the heat and moisture against which the carrier material acts to prevent stickiness also inhibit crystallization behavior (Chiou and Langrish 2006, 2007a). The heavier molecular weights mean that the materials require more energy to reach the appropriate temperatures and molecular interactions to rearrange into crystalline form. This is a trade-off situation where being able to form nonsticky particles or crystalline particles needs to be considered before using such materials in the feed mixture.

The other aspect of carriers is the ability for them to encapsulate the material being produced. Encapsulation is a key method for protection of the active compounds in the mixture. By having a matrix of material surrounding the active compounds, this reduces the heat stress and oxidative stress from the drying process (Yoshii et al. 2008). It further acts by being a barrier to the environmental conditions after production that maintains the active ingredient until its desired release conditions are met.

Carriers can encapsulate through the formation, morphology, or the specific material that is used in their formulation. To illustrate this, when mixed into an emulsion and spray dried, essential oils form particles, with the oils distributed throughout the particle. The emulsified state forms pockets inside the carrier matrix, and although the surface-exposed pockets are lost, the internal pockets are encapsulated when almost instantaneous drying and solidification occurs (Soottitantawat et al. 2007). Migration of particular components to the surface of the droplets can also be used for producing an encapsulation matrix. Recent work done (Wang and Langrish 2009) has demonstrated the potential use of proteins as a carrier and encapsulation agent where the proteins present may migrate to the surface of the droplet before drying, forming a "skin." Morphology plays a part in protecting the contents of the particle. If the particle puffs up and then collapses during the formation process, it will crumple inward, reducing the surface area that is directly exposed to the environmental conditions that may degrade or damage the product. The disadvantage of this, however, is that the morphology can work against the desired bioavailability and reactivity of the particle by reducing the active sites for reaction after drying, although low reactivity during drying may be beneficial for reducing the extent of degradation reactions.

Different carrier materials also have potentially unique abilities to act as carriers and encapsulating agents. Work by some researchers (Chiou and Langrish 2007b) has shown that the use of fiber material from a variety of sources (fruit fibers, sugarcane fibers) has the potential to act as carriers and successfully encapsulate bioactive materials. The moisture absorption properties of these fibers have been shown to

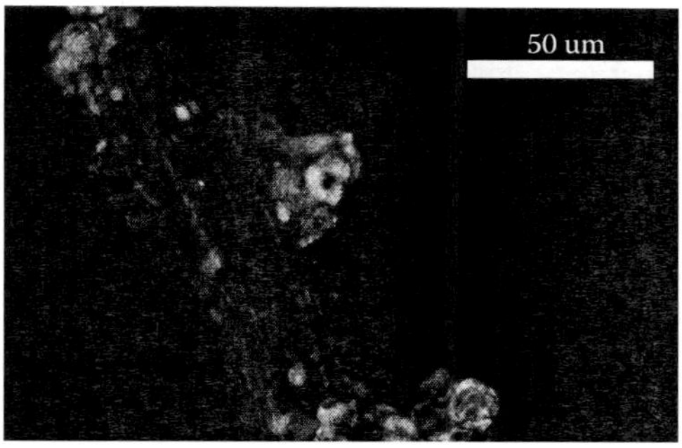

Figure 8.7 Confocal image of bioactive material present within fiber carrier.

absorb and adsorb the active compounds into and onto the fiber during the formulation and feed slurry stage, while the drying process then traps the compounds inside the fiber particles by removing their means of movement (the solvent). Through the use of confocal microscopy, it was seen that the bioactive materials were absorbed into the water channels of the fiber, where they remained post-drying. An example of this is seen in Figure 8.7, where the white signal spots are of hibiscus extract present within the water channels of citrus fibers and also on the surface of the fiber particles.

The use of natural fibers as encapsulating agents also has added benefit for the issue of stickiness and use as carriers because they are in a "crystalline" state. Heat and energy applied to these fibers do not convert them into an amorphous phase, and the particles remain in an inherently nonsticky state. Higher temperature and energy, however, will result in browning and burning of these fibers, so care must be taken when determining the appropriate drying temperatures in use. This non-amorphous property reduces the amount of stickiness that the particle has because the fiber particles reduce the amount of sticky material present in direct exposure to the equipment. The fibers are also thought to offer protection from the environment and may also be useful for slowing down the release of the encapsulated contents because digestion or prolonged rehydration is required to mobilize the compounds dried within the water channels inside. Work is being done to assess the extent of damage from heat and oxidative stress on the active compounds encapsulated in this manner. The carriers (fibers) also appear to stabilize the other bioactive material by inducing crystallization in this bioactive material, stabilizing the moisture sorption characteristics of this bioactive material.

Carriers can also act as bulking agents or excipients by being unreactive in nature and increasing the volume of the particles. This is common with pharmaceutical agents because the dose of drug compounds is on the milligram scale, and thus difficult to dose without bulking agents like lactose. With this kind of material (natural food fibers), it is possible that low-dose active compounds can be distributed through products that have added fiber content, which would further provide pro-health benefits in terms of soluble and insoluble dietary fiber. Examples of this kind of use would be producing a functional food or nutraceutical with a low-dose active ingredient in a loaded spray-dried fiber product being delivered into a cake, bread, or biscuit mix. Alternatively, for nonfiber materials, active nutritional ingredients could be bound into sugary compounds and sweeteners and then added to confectionary products as a method of counteracting the normally "unhealthy" side effects of excess consumption.

8.9 SPRAY AGGLOMERATION

Spray drying can be also used as an alternative method for agglomeration and granulation processing instead of conventional post-drying methods. Being a single-step process of droplet drying, it is possible to introduce dry materials into the atomized spray at the start of the drying process to form larger agglomerated particles. Depending on the direction of the dry feed being inserted (cocurrent, countercurrent, or some other in-between direction), the type of particle interaction will be different. Previous and current work done by students at the School of Chemical & Biomolecular Engineering at the University of Sydney has found that changing the distance of the inlet pipe for the dry powder feed into the liquid droplets affects the agglomeration rate and sizes (Langrish et al. 2008), with further work currently being investigated for crystallization seeding of droplets with dry feeds. Examples of spray agglomeration behavior can be seen in Figure 8.8.

The interaction that is being encountered in this process is that the wet droplet comes into contact with a dry particle. An example of the possible forms of agglomeration is seen in Figure 8.9. Surface adhesion of the smaller wet droplets onto the larger dry particle may occur and then dry to form agglomerates. Besides this, there is also the possibility that the dry particle adheres to the droplet, the difference being of particle or droplet size considerations, where the smaller of the two is the one being adhered onto the other. Further consideration of this is that the force of

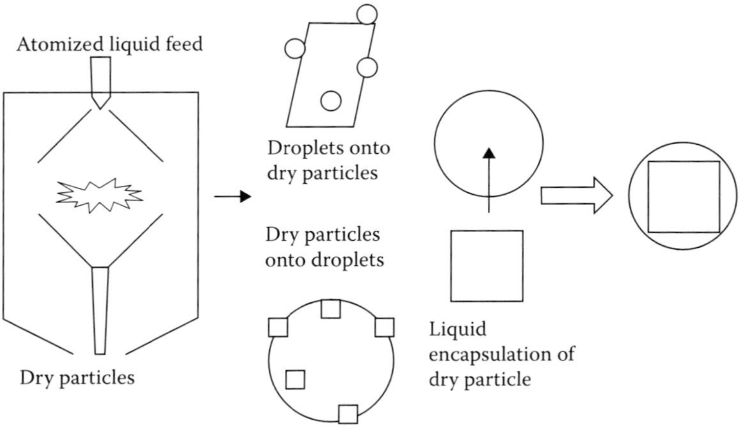

Figure 8.8 Droplet and particle interactions from spray agglomeration.

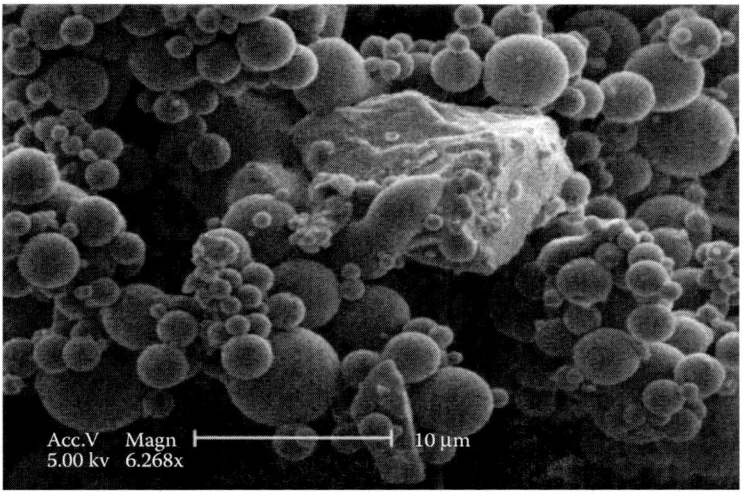

Figure 8.9 An example of spray agglomeration with small droplets adhered to the surface of a larger crystalline particle. In this situation, lactose solution was sprayed countercurrently to crystalline lactose powder.

collision and droplet or particle sizes may result in liquid encapsulation of dry particles before the drying of the droplet occurs, as shown in Figure 8.8. Compared with the $P_{(ac)}$ particle, the outcome is similar once the droplet dries, but with a lower degree of predictable and controlled storage behavior.

The useful aspect of spray agglomeration in the spray dryer is that often difficult to dry, and sticky, materials can be agglomerated in a semiencapsulated manner. This increases particle size and momentum for wall surface contact, reducing the likelihood that particle stickiness enables adhesion to the equipment; however, the higher momentum may give greater likelihood of adhesion because particles may deform more with higher momentum at impact, "splattering" more, and giving a larger contact area at impact, resulting in more adhesion. In addition to this stickiness effect, the drying process still occurs, meaning that amorphous powders can also be used to coat liquid droplets that have a rapid drying and crystallization behavior that would prevent the formation of $P_{(ac)}$-type particles.

8.10 ASSESSING POWDER CHARACTERISTICS

Any produced particle needs to be assessed for functional characteristics before it can be considered for its intended purpose. Characterization of particles is not new, and there are many different ways of assessing the particle properties. For investigating the properties of spray-dried particles, a few characteristics can be used to allow adjustments for the drying process when improvements are required.

The most important characteristic for these powders is moisture content. As previously discussed, the presence of moisture has many effects, ranging from allowing crystallization behavior to dissolution leading to amorphous surfaces, and also importantly for food products, moisture affects the water activity that restricts and limits harmful growth of toxic pathogens in foods. Standard thermogravimetric methods with a high accuracy and precision balance can be used to assess the moisture content to reasonable levels. Standard protocols indicate oven drying at 100°C–120°C for 24 hours before measuring the weight loss (Bhandari et al. 1997). This, however, is not necessarily accurate for determining only moisture loss because volatiles may be lost, and at such temperatures reactions can also lead to mass loss. There are alternative methods for determining moisture content, such as the Karl Fischer titration method, that use chemical reactions to detect, with high levels of accuracy, the moisture present (Keey 1992), and these need to be taken into consideration when measuring moisture contents.

Particle size measurements are useful in determining the atomization control of the process. Sieves and screen meshes can be used to manually sift through powder samples and determine their size distribution. This is only generally used if the particles size is relatively large (>50 μm). In many situations, use of a laser light diffraction technique is common, where the particles are either suspended in a solution or passed through the laser beam through a gas stream. If the particles are suspended, the suspension solution needs to be nonreactive with the material so no dissolution or agglomeration occurs, and the material remains in its native state (Iacocca and German 1997). The laser beam creates a diffraction pattern that is measured by the device, and the sizes of the particles are calculated according to the diffraction pattern. With mathematical calculations, it is also possible to determine an approximate shape factor, known as the *fractal dimension* (Bushell et al. 2002). The fractal dimension provides additional information regarding the shape of the particles, where one represents a singular point, two represents a sheetlike object, and three represents a spherical object, as indicated by Figure 8.10. This means that a fractal dimension of 2.5 would indicate a sheetlike object with a raised surface, tending toward a spherical appearance.

Measuring the amorphous or crystalline state of the powder is something that may be a good indicator for how well the operating conditions suit the desired properties. Crystalline particles have some distinct physical characteristics that can be used to distinguish them from amorphous particles. A standard technique for this characterization is differential scanning calorimetry (DSC) or its

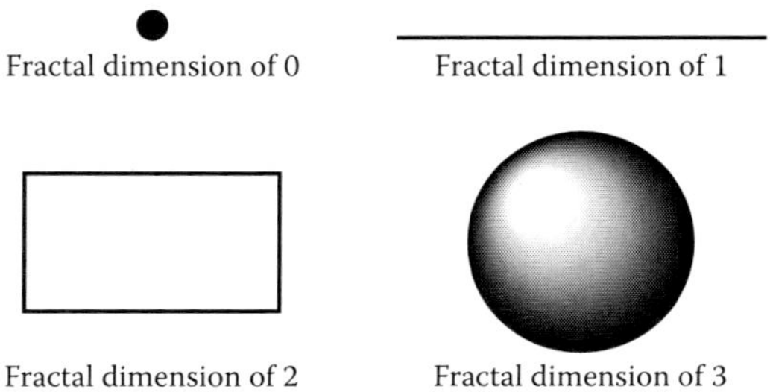

Figure 8.10 Shapes according to fractal dimension value.

modulated form (mDSC), whereby a heating gradient is applied (and in the case of mDSC, the gradient is modulated between a heating and cooling cycle) and the thermal behavior of the sample is measured. Exothermic and endothermic behaviors and events can determine the physical state of the material. With endothermic behavior, where energy is absorbed and desorbed indicates crystallization behavior because heat energy is absorbed into the material, crystallization occurs, and when the material settles into the lower energy state, the excess energy is released. The value of this energy event is an indicator of the amorphous content; for example, with α-lactose monohydrate, it has been reported that pure amorphous lactose has a peak energy of 112 $J \cdot g^{-1}$, whereas 50% crystalline material has a peak energy of 39 $J \cdot g^{-1}$, and 80% crystalline has only a 14 $J \cdot g^{-1}$ energy peak (Gombás et al. 2002).

An additional method for determining the amorphous and crystalline state of the particles is the porosity of the material. Porosity measurements can be done with gas sorption techniques that measure the surface area of materials and by comparing them with the volume, determine the porosity (Jagiello and Betz 2008). This indicator is useful because crystalline materials have a more compact structure and lower porosity. By comparisons of porosity, inferences can be made about the crystalline structure at the micropore level (Chiou et al. 2007).

Water-induced crystallization is another alternative method for comparisons between the different levels of crystallinity for different materials. This behavior was recently highlighted as a potential universal behavior for all spray-dried powders (Chiou and Langrish 2007a), where a variety of materials (maltodextrin, coffee, and tea) were all examined and compared with the known behavior of milk powders that underwent crystallization in controlled humidity conditions. This work and further works (Chiou et al. 2007, 2008; Imtiaz-Ul-Islam and Langrish 2008, 2009) have shown that particles can undergo crystallization if left in conditions with sufficient temperature and humidity with traceable changes through data-logging balances. The particles will absorb moisture readily, undergo conversion from amorphous to crystalline states as a result of molecular lubrication from the absorbed water, and then desorb the water once the material crystallizes. The peak height in the sorption behavior can be used to compare samples, with a greater peak height indicating a greater amorphous content, as seen in Figure 8.11.

X-ray diffraction (XRD) can be used to determine the degree of crystallinity for different materials, including powders. A radiation source emits a beam of x-rays that bounce off the sample, and these scattered rays are collected in a detector. The angles of the signal collected are distinctive, depending on the crystalline structure of the material (interplane spacing of array), and this can be an accurate method of determining the composition of the material. For measuring crystallinity, the peak heights of the signals can be used to determine the physical state. When the material is in an amorphous condition, compared with the crystalline sample under the same operating conditions, the peaks will have lower intensity and produce greater measurement noise. This is only one of the problems with using XRD because mechanical damage imparted to the powders is possible when packing the sample into the specimen holder, causing physical damage and reverting crystalline materials back

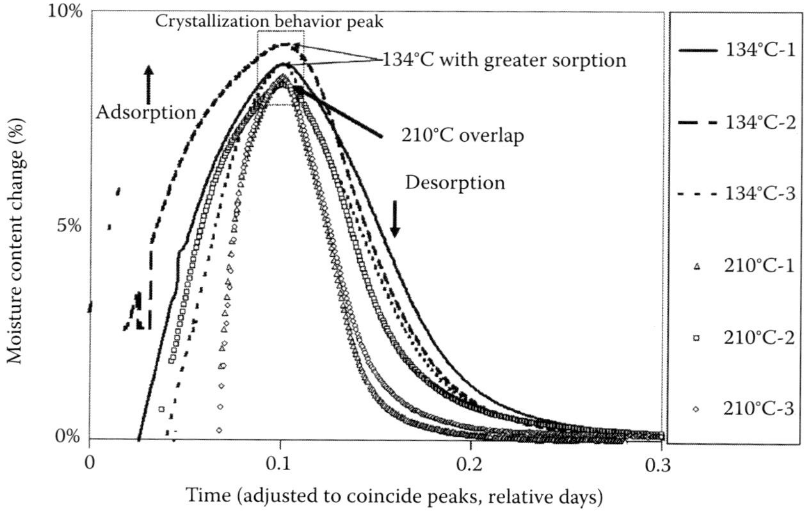

Figure 8.11 An example of powder sorption behavior for lactose produced at two different inlet air temperatures: 134°C and 210°C.

into an amorphous form. This can be seen in Figure 8.12, where the mechanical damage has reduced the signal clarity of the powders examined.

Finally, although not an exhausted list of characterization, the last method for determining characteristics of particles and powders is microscopy. Both optical and electron microscopy techniques are highly useful in characterization. Optical microscopy can be used to determine crystallinity through polarization of light; it can determine absorbance of materials through confocal microscopy and also determine particle sizes. Electron microscopy also offers similar benefits but allows a greater scope of magnification and examination of surface details. Recent work by Wang and Langrish (2007) has shown that it is possible to see crystals of sucrose

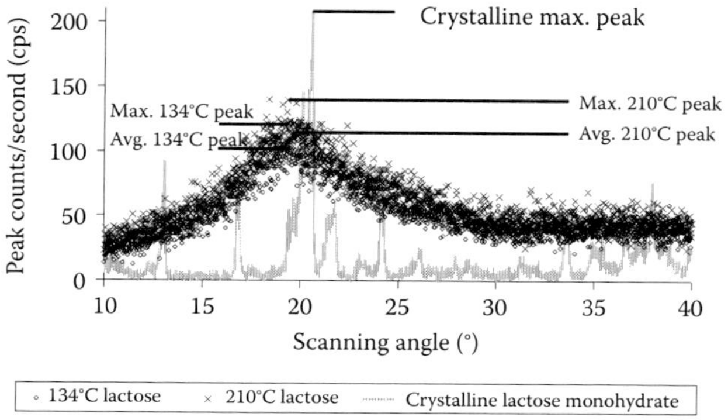

Figure 8.12 X-ray diffraction scan of spray-dried lactose powders with two different inlet air temperatures: 134°C and 210°C.

forming on amorphous sucrose produced by spray drying using scanning electron microscopy (SEM). Particle sizing and better morphology are also possible with SEM, whereas transmission electron microscopy can be used to examine cross-sectional samples of particles if desired. There are additional methods for assessing characteristics of particles such as Raman spectroscopy and a variety of alternative calorimetric methodologies (Lehto et al. 2006), but they will not be discussed here.

Once characterization is complete for the desired aspects, it is possible to plan the next step in the process. If by determining the particles are too wet, then by adjusting the drying temperature higher or by increasing the feed concentration, it is easily resolved. If the particle sizes are too large or small, then adjustments to the feed rate or atomization variables can be done to regulate the desired size range. If the crystallinity content is not what was expected or desired, then by adjusting a different variable on its own or multiple variables concurrently, it would be possible to alter the final characteristics. When these changes are implemented, characterization is once more required to confirm that the desired characteristics are seen, and if not, the variables will require change again until the right properties and characteristics are obtained.

8.11 LIMITATIONS OF SPRAY DRYING

Although spray drying technology has much to offer in terms of being a rapid one-step process for the production of particulates and powders, the limitations of using spray drying need to be taken into consideration. Spray drying technology can be an inefficient and energy-intensive process. Heat loss through the spray drying system is one of the major disadvantages because the entire process requires intensive insulation to prevent energy loss via the equipment. Heat loss through the equipment tends to become less of a problem as the scale of equipment increases because of the inverse relationship between heat loss and capacity. Because heat loss is proportional to the surface area of the loss and the capacity is proportional to volume of the process, as the size of operation increases, the effect of heat loss compared with the increase of capacity diminishes and becomes less of an issue. However, heat that is deliberately lost still remains a problem.

The drying medium gas is normally passed through the equipment and out, meaning that the heat is lost and new drying gas needs to be heated continuously. Even if the system is sealed into a loop so that the drying medium is not exhausted out of the system, the gas temperature is traditionally reduced to form condensates so that the drying gas is able to carry a greater evaporative load in its next pass. This then requires the reheating of the medium to the operational temperature. If a gas moisture removal system at high temperatures is introduced, it can greatly assist in the continual running of the process by reducing the energy required to heat the drying gas. This is also one of the reasons why high concentration feed slurries are used because the evaporative load required is lower for the lower amounts of moisture.

Although spray dryers can exist in a variety of sizes, for industrial production levels of high-volume output, they are typically large-sized structures. This is because of the need for a sufficiently long residence time for drying a large feed rate in relation to the drying temperature and particle size. They can range in the order of tens of meters in height and meters in diameter. The separation process (typically a cyclone and bag filters as previously mentioned) would then also be attached, with further product removal and processing required. Having sufficient space to build a dryer that meets the production yield requirements is an important consideration because a smaller-scale dryer may not necessarily meet the demand.

The size considerations and relation to the residence time have further implications. As hinted in our research, the relationship between the particle temperature and glass transition temperature has significant effects on the ability to crystallize or maintain amorphous behaviors. The residence time of a spray drying system must be sufficiently long (or short) to enable efficient drying of the particles while allowing time for the effect of the T-T_g to occur.

For high glass transition temperatures, it is important that enough residence time is available because it is not always possible to drive the drying temperatures high enough so the gap of T-T_g is sufficiently large for a fast transformation. It is thought that as long as there is enough time available when particles are in the right conditions, a reduced gap of particle temperatures and glass transition temperatures can still affect the particles enough to undergo crystallization. Drying at higher temperatures may cause the faster conversion, but this is not always feasible depending on the sensitivity of the material being produced.

It has been seen with some work on vitamin C that producing amorphous particles is extremely difficult, whereas our work has shown it is now possible to change the scale of process to enable the crystallization of lactose (which normally is considered to be amorphous when produced by spray drying). The challenge in producing deliberately amorphous products for materials that are traditionally crystalline is to be able to dry at a low enough temperature so that moisture or solvent is removed, while keeping the T-T_g as low as possible. This is difficult because water plasticizes the T_g further; for example, with vitamin C having a natural T_g of 233 K (−40°C) (Migliardo et al. 2001), this would mean to produce amorphous ascorbic acid, the drying temperature must be kept under −20°C to prevent crystallization behavior.

Although carriers can be added to amorphous materials to improve their glass transition behavior for crystallization purposes, carriers that increase the glass transition temperature can also be used for materials like ascorbic acid to increase the glass transition temperature to values that enable better drying, but maintain the ability to form amorphous products (if desired). An example of this has been seen where mixtures of lactose and vitamin C have been formed into amorphous powders.

The issue of scale-up is a current limitation. Although the understanding of the spray drying process is being furthered and deepened, the direct relation to increased scale is not fully known. Laboratory- and pilot-scale experimental runs are thought to apply to industrial-scale dryers also, but it is still unknown to what extent and how these factors can be "fine-tuned" for an industrial dryer scale. One of the most significant points for scale-up development is the increase in residence time for the particles, where crystalline transformation or amorphous reformation (from high bulk humidity and condensation effects) can occur. When taken into consideration, the ability to scale-up better has widespread implications, where a change in process that causes an increase in yield of 1% can translate to hundreds of tons of material and potentially millions of dollars worth of value in both product sales and waste reduction. This is even more important for pharmaceutical and nutraceutical products that contain potentially expensive active ingredients.

Productivity of spray drying systems can also be affected by the materials being produced. Unlike some processes where the system may have hundreds of operational hours before a shutdown and clean procedure is required, spray drying typically produces much shorter run cycles because the buildup of deposits within the equipment (as a result of stickiness) can cause problems in product quality and operational safety (Masters 1976). Having to turn off and cool down the equipment, clean, and then restart is a significant part of the cycle and may increase product costs because of production downtime. By working to reduce the issue of wall deposits and stickiness, production runs can be sustained for longer periods, leading to increased productivity levels.

8.12 CONCLUDING REMARKS

For the manufacture of nutraceuticals or value-added powders, spray drying is one of the potential methods of production. Although it has limitations in the types of materials that can be processed through the system (as liquid slurries), the potential to combine and create highly engineered particles is available. Through operational conditions and appropriate choice of carriers or encapsulation, specific properties and behaviors may be controlled to desired levels, such as time-delayed release mechanisms or sustained release of active ingredients within the formulated product.

Besides conventional droplet drying, it is also possible to generate agglomerated particles, making the spray drying chamber a multipurpose reactor where one-step reactions can occur, improving the flexibility and increasing the range of manufacturing options the operator of the system has. This means that with knowledge and experience of the correct conditions, a multitude of products can be produced using the same dryer setup.

Continued research is being applied to investigate the problems associated with scale-up, and along with work on degradation of antioxidant properties of materials produced, the overall direction and future for spray drying in nutraceuticals and functional food development and production is continuing to expand.

REFERENCES

Bajsic, I., and E. Kranjcevic. 2001. Automation of Industrial Spray Dryer. *Instrumentation Science and Technology* 29: 201–13.

Bhandari, B. R., B. D'arcy, and C. Kelly. 1999. Rheology and Crystallization Kinetics of Honey: Present Status. *International Journal of Food Properties* 2: 217–26.

Bhandari, B. R., N. Datta, and T. Howes. 1997. Problems Associated with Spray Drying of Sugar-Rich Foods. *Drying Technology* 15: 671–84.

Bhandari, B. R., and T. Howes. 1999. Implication of Glass Transition for the Drying and Stability of Dried Foods. *Journal of Food Technology* 40: 71–79.

Bhandari, B. R., A. Senoussi, E. D. Dumoulin, and A. Lebert. 1993. Spray Drying of Concentrated Fruit Juices. *Drying Technology* 11: 1081–92.

Bushell, G. C., Y. D. Yan, D. Woodfield, J. Raper, and R. Amal. 2002. On Techniques for the Measurement of Mass Fractal Dimension of Aggregates. *Advances in Colloid and Interface Science* 95: 1–50.

Chan, H.-K., and N. Y. K. Chew. 2003. Novel Alternative Methods for the Delivery of Drugs for the Treatment of Asthma. *Advanced Drug Delivery Reviews* 55: 793–805.

Chiou, D. 2008. The Development of New Carrier Technologies for Spray-Dried Fruit Extracts and Crystallisation of Amorphous Spray-Dried Powders. PhD thesis, School of Chemical & Biomolecular Engineering, University of Sydney, Sydney, Australia.

Chiou, D., and T. A. G. Langrish. 2006. Crystallization of Amorphous Spray-Dried Powders. *15th International Drying Symposium 2006 (IDS06)*. Edited by I. Farkas. Budapest, Hungary, August 2006, vol. A, 562–69.

Chiou, D., and T. A. G. Langrish. 2007a. Crystallisation of Amorphous Components of Spray-Dried Powders. *Drying Technology* 25: 1423–31.

Chiou, D., and T. A. G. Langrish. 2007b. Development and Characterisation of Novel Nutraceuticals with Spray Drying Technology. *Journal of Food Engineering* 82: 84–91.

Chiou, D., T. A. G. Langrish, and R. Braham. 2007. Partial Crystallisation of Materials in Spray Drying: Simulations and Experiments. *Drying Technology* 26: 27–38.

Chiou, D., T. A. G. Langrish, and R. Braham. 2008. The Effect of Temperature on the Crystallinity of Lactose Powders Produced by Spray Drying. *Journal of Food Engineering,* 86: 288–93.

Choi, W. S., H. I. Kim, S. S. Kwak, H. Y. Chung, H. Y. Chung, K. Yamamoto, T. Oguchi, Y. Tozuka, E. Yonemochi, and K. Terada. 2004. Amorphous Ultrafine Particle Preparation for Improvement of Bioavailability of Insoluble Drugs: Grinding Characteristics of Fine Grinding Mills. *International Journal of Mineral Processing* 74: S165–72.

Dinunzio, J. C., D. A. Miller, W. Yang, J. W. Mcginity, and R. O. Williams. 2008. Amorphous Compositions Using Concentration Enhancing Polymers for Improved Bioavailability of Itraconazole. *Molecular Pharmaceutics* 5: 968–80.

Fletcher, D. F., B. Guo, D. J. E. Harvie, T. A. G. Langrish, J. J. Nijdam, and J. Williams. 2006. What Is Important in the Simulation of Spray Dryer Performance and How Do Current CFD Models Perform? *Applied Mathematical Modelling* 30: 1281–92.

Fontana, A. J. 1998. Water Activity: Why It Is Important for Food Safety. First NSP International Conference on Food Safety, Albuquerque, NM, November 1998.

Frydman, A., J. Vasseur, F. Ducept, M. Sionneau, and J. Moureh. 1999. Simulation of Spray Drying in Superheated Steam Using Computational Fluid Dynamics. *Drying Technology, Proceedings of the 1998 11th International Drying Symposium IDS '98.* Halkidiki, Greece, August 1998, 1313–26.

Frydman, A., J. Vasseur, J. Moureh, M. Sionneau, and P. Tharrault. 1998. Comparison of Superheated Steam and Air Operated Spray Dryers Using Computational Fluid Dynamics. *Drying Technology* 16: 1305–38.

Gombás, Á., P. Szabó-Révész, M. Kata, G. Regdon, and I. Erős. 2002. Quantitative Determination of Crystallinity of α-Lactose Monohydrate by DSC. *Journal of Thermal Analysis and Calorimetry* 68: 503–10.

Gordon, M., and J. S. Taylor. 1952. Ideal Copolymer and the Second-Order Transition of Synthetic Rubbers. I. Non-crystalline Copolymers. *Journal of Applied Chemistry* 2: 493–500.

Guo, B., D. F. Fletcher, and T. A. G. Langrish. 2002. Flow Patterns in Sudden Expansions and Their Relevance to Understanding the Behaviour of Spray Dryers. *Developments in Chemical Engineering and Mineral Processing* 10: 305–22.

Guo, B., D. F. Fletcher, and T. A. G. Langrish. 2004. Simulation of the Agglomeration in a Spray Using Lagrangian Particle Tracking. *Applied Mathematical Modelling* 28: 273–90.

Hanus, M., and T. A. G. Langrish. 2007. Re-entrainment of Wall Deposits from a Laboratory-Scale Spray Dryer. *Asia-Pacific Journal of Chemical Engineering* 2: 90–107.

Harvie, D. J. E., D. F. Fletcher, and T. A. G. Langrish. 2002. A Computational Fluid Dynamics Study of a Tall-Form Spray Dryer. *Food and Bioproducts Processing: Transactions of the Institution of Chemical Engineers, Part C* 80: 163–75.

Hayashi, H. 1989. Drying Technologies of Foods—Their History and Future. *Drying Technology* 7: 315–69.

Huang, Z.-Q., J.-P. Lu, X.-H. Li, and Z.-F. Tong. 2007. Effect of Mechanical Activation on Physico-chemical Properties and Structure of Cassava Starch. *Carbohydrate Polymers* 68: 128–35.

Iacocca, R. G., and R. M. German. 1997. Comparison of Powder Particle Size Measuring Instruments. *International Journal of Powder Metallurgy* 33: 35–48.

Imtiaz-Ul-Islam, M., and T. Langrish. 2008. The Effect of the Salt Content on the Crystallization Behaviour and Sorption Fingerprints of Spray-Dried Lactose. *Food and Bioproducts Processing* 86: 304–11.

Imtiaz-Ul-Islam, M., and T. A. G. Langrish. 2009. Comparing the Crystallization of Sucrose and Lactose in Spray Dryers. *Food and Bioproducts Processing* 87: 87–95.

Jagiello, J., and W. Betz. 2008. Characterization of Pore Structure of Carbon Molecular Sieves Using DFT Analysis of Ar and H2 Adsorption Data. *Microporous and Mesoporous Materials* 108: 117–22.

Keey, R. B. 1992. *Drying of Loose and Particulate Materials*. Bristol, PA: Hemisphere Publishing Corporation, 47.

Kieviet, F. G., J. van Raaij, P. P. E. A. de Moor, and P. J. A. M. Kerkhof. 1997. Measurement and Modelling of the Air Flow Pattern in a Pilot-Plant Spray Dryer. *Chemical Engineering Research & Design, Transactions of the Institute of Chemical Engineers, Part A* 75: 321–28.

Kim, J.-S., M.-S. Kim, H. J. Park, S.-J. Jin, S. Lee, and S.-J. Hwang. 2008. Physicochemical Properties and Oral Bioavailability of Amorphous Atorvastatin Hemi-calcium Using Spray-Drying and SAS Process. *International Journal of Pharmaceutics* 359: 211–19.

Langrish, T., D. Ali, and M. Asplet. 2008. Droplet-Particle Agglomeration of Maltodextrin and Salt in a Small-Scale Spray Dryer. *International Journal of Food Engineering* 4: article 5. http://www.bepress.com/ijfe/vol4/iss6/art5.

Langrish, T. A. G., D. E. Oakley, R. B. Keey, R. E. Bahu, and C. A. Hutchinson. 1993. Time-Dependent Flow Patterns in Spray Dryers. *Chemical Engineering Research & Design* 71: 355–60.

Lehto, V.-P., M. Tenho, K. Vaha-Heikkila, P. Harjunen, M. Paallysaho, J. Valisaari, P. Niemela, and K. Jarvinen. 2006. The Comparison of Seven Different Methods to Quantify the Amorphous Content of Spray Dried Lactose. *Powder Technology* 167: 85–93.

Lekhal, A., B. J. Glasser, and J. G. Khinast. 2001. Impact of Drying on the Catalyst Profile in Supported Impregnation Catalysts. *Chemical Engineering Science* 56: 4473–87.

Liu, Y., B. Bhandari, and W. Zhou. 2007. Study of Glass Transition and Enthalpy Relaxation of Mixtures of Amorphous Sucrose and Amorphous Tapioca Starch Syrup Solid by Differential Scanning Calorimetry (DSC). *Journal of Food Engineering* 81: 599–610.

Masters, K. 1976. *Spray Drying: An Introduction to Principles, Operational Practice and Applications.* London: George Godwin Limited, 284–88.

Masters, K. 2004. Current Market-Driven Spray Drying Development Activities. *Drying Technology: An International Journal* 22: 1351–70.

Migliardo, F., C. Branca, A. Faraone, S. Magazù, and P. Migliardo. 2001. Study of Ascorbic Acid (Vitamin C)/H_2O Mixture across Glass Transition. *Physica B: Condensed Matter* 301: 138–40.

Piorkowska, E., A. Galeski, and J. Haudin. 2006. Critical Assessment of Overall Crystallisation Kinetics Theories and Predictions. *Progress in Polymer Science* 31: 549–75.

Sajkiewicz, P. 2002. Kinetics of Crystallisation of Polymers—A Review. *Progress in Rubber, Plastics and Recycling Technology* 18: 195–215.

Soottitantawat, A., F. Bigeard, H. Yoshii, T. Furuta, M. Ohkawara, and P. Linko. 2005. Influence of Emulsion and Powder Size on the Stability of Encapsulated D-Limonene by Spray Drying. *Innovative Food Science and Emerging Technologies* 6: 7.

Soottitantawat, A., J. Peigney, Y. Uekaji, H. Yoshii, T. Furuta, M. Ohgawara, and P. Linko. 2007. Structural Analysis of Spray-Dried Powders by Confocal Laser Scanning Microscopy. *Asia-Pacific Journal of Chemical Engineering* 2007: 5.

Truong, V., B. R. Bhandari, and T. Howes. 2005. Optimization of Co-current Spray Drying Process of Sugar-Rich Foods. Part I—Moisture and Glass Transition Temperature Profile during Drying. *Journal of Food Engineering* 71: 55–65.

Wang, S., and T. A. G. Langrish. 2007. Measurements of the Crystallization Rates of Amorphous Sucrose and Lactose Powders from Spray Drying. *International Journal of Food Engineering* 3: 1.

Wang, S., and T. A. G. Langrish. 2009. A Review of Process Simulations and the Use of Additives in Spray Drying. *Food Research International* 42: 12.

Wunderlich, B. 2006. The Glass Transition of Polymer Crystals. *Thermochimica Acta* 446: 128–34.

Yoshii, H., T. L. Neoh, T. Furuta, and M. Ohkawara. 2008. Encapsulation of Proteins by Spray Drying and Crystal Transformation Method. *Drying Technology* 26: 5.

Yue, Y., and C. A. Angell. 2004. Clarifying the Glass-Transition Behaviour of Water by Comparison with Hyperquenched Inorganic Glasses. *Nature* 427: 717–20.

CHAPTER 9

Rheology of Complex Fluids Containing Nutraceuticals

Samiul Amin and Krassimir P. Velikov

CONTENTS

9.1 COMPLEX FLUIDS CONTAINING NUTRACEUTICALS

Many food product formats are typical examples of complex fluids. Depending on the product type, nutraceuticals have been formulated in the form of suspensions (solid–liquid dispersions), emulsions (fluid–fluid dispersions), mesophases (e.g., liquid crystals or cubic phases), micellar solutions, and their gels (e.g., flocculated emulsions) or dispersions in the presence of biopolymers.

The rheological properties of complex systems containing nutraceuticals are important for their prereformulation (e.g., in powder, suspension, or emulsion format), processing into products (e.g., ability to disperse or mix with the rest of the products matrix), product stability (e.g., gravity-driven instabilities such as creaming or sedimentation and ability to suspend particles), and sensorial properties (e.g., creaminess, whiteness).

9.2 DESIRED CHARACTERISTICS OF COMPLEX FLUIDS FOR USE AS NUTRACEUTICAL CARRIERS AND COMMON ISSUES WITH FORMULATION

When formulated in food products, nutraceuticals should not affect the product physicochemical stability, appearance, flavor, and texture, and they should have good bioaccessibility. Most of these functionality aspects are also required for preformulated concentrates containing nutraceuticals.

Physical stability is usually a key requirement for any formulation. Complex fluids containing colloidal dispersions should have good stability against gravity-driven phenomena (e.g., creaming or sedimentation), which are particularly important for liquid product formats. Many other unwanted phenomena, such as flocculation and aggregation, can also occur in complex fluids, especially in the presence of biopolymers. In concentrated dispersed phases, which are not common for food products but are encountered in ingredient precursor systems, gelation could also take place as a result of physical arrest or attractive interaction between the particles or droplets. In the presence of polymers, colloidal dispersions are also susceptible to depletion-induced instabilities that cause phase separation and aggregation (Yodh et al. 2001). In general, chemical stabilities reflect the chemical nature of the individual nutraceuticals and their ability to participate in chemical reactions alone (e.g., oxidation, polymerization) or with other ingredients (e.g., complexation). Oxidation and photo-induced oxidation are among the most common problems in food formulations. Many nutraceuticals (e.g., antioxidants like polyphenols) are unstable and undergo oxidation during storage. The strategies for minimizing oxidation in functional foods are not straightforward and can range from the use of antioxidants to special packaging and filling inert atmosphere.

The texture of the complex fluids is determined by the microstructure and interaction between different dispersed phases. Nutraceuticals are typically used in relatively low concentrations (<1% weight); thus, they do not influence much the rheology of the final product. Nevertheless, the ability to formulate nutraceuticals in food products while preserving the desired rheological properties is important. This is also true for the preformulated ingredients in more convenient to use concentrates. In some cases, if the nutraceutical molecule has an amphiphilic character (e.g., saponins), it can act as a surfactant or cosurfactant in the stabilization of dispersed phases, or as part of the mesophases.

Taste and flavor are crucial for the success of any food product. Changes in flavor are often linked to the inherited taste of the nutraceuticals (e.g., polyphenols) or to unwanted chemical interactions (e.g., oxidation). Colloidal dispersions can offer control on solubility and therefore control on taste. Regulation of taste can be achieved by a proper choice of material: water soluble or water insoluble (which can be oil soluble). Chemical stability that can cause off-flavor formation can be addressed with chemical modification. For example, this often can influence the solubility of the ingredient. An example is the fatty esters of ascorbic acid that are more stable in oil solution (or dispersion) (Lim et al. 2006). Application in oil-free products can be achieved by using colloidal dispersions of modified taste and flavor stable chemical derivatives. The strategies for minimizing oxidation of lipid, micronutrients, and nutraceuticals are based on increasing physicochemical stability in the product, but they can vary depending on the chemical nature of the individual compound.

Appearance is an important factor determining the consumer-perceived quality of food products. In general, the product appearance depends on product micro- and nanostructure and composition. Scattering largely determines the turbidity and "lightness" of products, whereas absorption determines their color. The degree of scattering by the complex fluid depends on the concentration of the dispersed phase(s), size, and complex refractive index of any particles or droplets present, whereas the degree of absorption depends on the concentration and type of dyes and absorbing materials present. Transparency often is required when an insoluble material is added to a product that is initially transparent or clear. In many cases, we are looking for the opposite effect—to increase the whiteness of a product. If a color is desired, both material and size properties can be tuned. As is known, depending on the materials, optical properties can be greatly influenced by the particle size. Colloidal delivery systems offer unique possibilities to deliver insoluble pigments both in aqueous and oil phases. The application of carotenoid colloidal nanoparticles is a wonderful illustration on effect of particle size on optical properties (Kristl et al. 2003; Lawless et al. 2003, 2004; Yang et al. 2006).

Bioavailability of nutraceutical ingredients is a rapidly developing area of functional food design (Horn et al. 2001; Kristl et al. 2003; Link et al. 2003). Solubility and membrane permeability are two major factors that contribute to their bioavailability on oral administration (Auwater et al. 1999; Santos-Bauelga et al. 2000; Chu et al. 2007). Often these materials are present as hydrophobic solids and often as crystalline materials with a high melting temperature (e.g., phytosterols). The low absorption rate of poorly water-soluble, and especially lipophilic, substances from the gastrointestinal tract is generally attributed to the poor solubility and wettability in gastrointestinal fluids. There are several common approaches to improve bioavailability. Poorly soluble bioactive agents can in principle be enhanced by the following technological manipulations: chemical modification, reduction of size (FairweatherTait and Hurrell 1996; Horter et al. 1997; Clinton 1998), hydrophilization of particle surfaces to improve the wettability in aqueous media, reduction of the crystallinity of the substance, or changing the polymorphic form. The chemical modification approach is not always applicable, but it has been successfully used for phytosterols. Esterification of sterol with fatty acid to form a sterol ester is currently used to increase the solubility of sterol in oils (Auweter et al. 1999; Chu et al. 2007).

Similarly, in the case of lyophobic micronutrients and nutraceuticals, the solubilization rate in bile micelles, which is important for delivery into the cells, can be a rate-limiting step. As the solubilization process, which is also an important process in many areas such as detergency, emulsion stability, and drug and flavor delivery systems, proceeds in a typical dispersed system, the molecules from the dispersed phase transfer into bile micelles in the aqueous phase. As a consequence of this transfer, a decrease in the emulsion droplet size is expected, as recently confirmed using monodispersed nanosized emulsions (Horter et al. 2001).

Another important process for digestion is enzymatic degradation in the gastrointestinal tract. This is especially important when the micronutrient or nutraceutical is delivered in a digestible matrix (e.g., oil, fat, or protein). In some cases, the active itself undergoes enzymatic degradation (e.g., conversion of sterol ester to pure sterol). Most enzymes are water soluble, and the reaction involves water and a water-insoluble substrate that is part of a large aggregate, like a micelle or an emulsion drop or particle (Hecq et al. 2005). The process of digestion is intricate because of the complex time-dependent compositions and phase behavior (Patravale, Date, and Kulkarni 2004; Hecq et al. 2005; Kesisoglou, Panmai, and Wu 2007). The droplet size was found to have a major effect on lipase activity during lipid digestion (Merisko-Liversidge, Liversidge, and Cooper 2003; Ariyaprakai and Dungan 2007).

9.3 INTRODUCTION TO RHEOLOGY

9.3.1 What Is Rheology?

The formal definition of *rheology* is the study of the flow and deformation of materials (Barnes 2000). Many everyday common words, such as *thickness* or *consistency*, have rheological connotations, and in a gross way rheology can also be described as the study of how "solid-like" or "liquid-like" a material is and how it responds to external stresses that it may encounter during processing and manufacturing or application. Some common examples of such processes can be the spreading of a topical ointment on the skin, spreading of butter on a piece of bread, or spraying an aerosol from a can. All these processes involve application of external stresses that cause the complex fluid to flow. Similarly in the manufacturing or processing environment, there are many processes such as pumping, mixing, and flowing through pipes that subject the fluid to external stresses, and understanding and controlling the fluid's response to such external stresses become of paramount importance in process control and optimization. In addition to the importance of rheology control for manufacturing and application performance, rheology also plays a critical role for various key technical requirements of the product, such as shelf life or stability.

As has been discussed in the introduction to this chapter, most of the nutraceutical-containing systems can be classified as complex fluids or soft matter systems. One of the key characteristics of such materials is that they are organized over an extremely wide range of structural length scales. These materials usually tend to exhibit viscoelastic properties, that is, they exhibit a combination of viscous and elastic responses; therefore, they cannot be described by a simple parameter such as viscosity, but rather require more complex parameters to effectively capture the exhibited rheological response. The complex rheological response of the complex fluid is intricately linked to the underlying microstructure, which in turn is sensitive to formulation conditions. Therefore, effective rheology control entails an accurate establishment of the microstructure–rheology linkages in complex fluids and understanding how formulation conditions affect this.

One of the key objectives of this chapter will be to illustrate the utility of rheology and rheological tools to help control manufacturing and application performance of complex fluids containing nutraceuticals. To achieve this, the key rheological principles, techniques, and methods will first be presented and then discussed in the context of manufacturing and application performance of complex fluids formulated with nutraceuticals.

9.3.2 Steady Shear Flow and Viscosity

The simplest type of flow, and one that has high relevance from a manufacturing and application perspective, is simple shear flow. In shear flow, liquid elements flow past each other. A simple way of thinking about simple shear flow is to think of fluid held between two parallel plates, as illustrated in Figure 9.1. Initially there is no flow because both the upper and lower plates are at rest. If the upper plate is then made to instantaneously achieve a certain velocity, this results in the generation of shear stress in the fluid. The shear strain experienced by the fluid elements will be uniform. The shear rate is a measurement of how fast the sample is flowing and is also known as shear strain rate. In the simplest form, this can be represented as a liquid being sheared between two plates. Mathematically it can be expressed as (Khan, Royer, and Raghavan 1997):

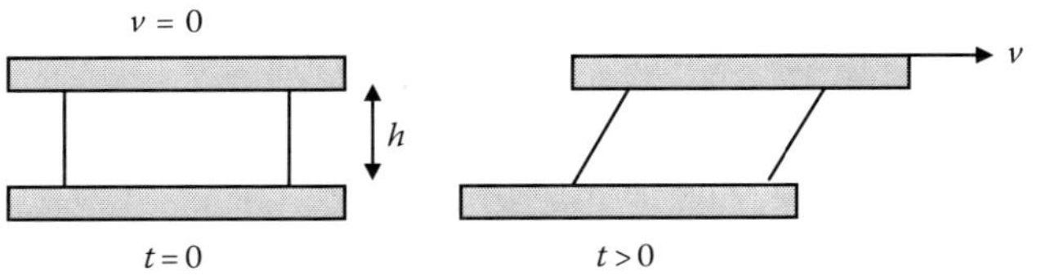

Figure 9.1 Initiation of shear flow in parallel plates.

$$\overset{*}{\gamma} = \frac{v}{h} \tag{9.1}$$

where v = velocity
h = distance between the plates

From a steady shear flow experiment, the key rheological parameter that can be determined is the viscosity.

The shear viscosity is defined as the ratio of shear stress to shear rate, or:

$$\eta = \frac{\tau}{\overset{*}{\gamma}} \tag{9.2}$$

where η = viscosity
τ = shear stress
$\overset{*}{\gamma}$ = corresponding shear rate

The unit of viscosity is the pascal second (Pas). Shear viscosity can be steady state or instantaneous.

Measuring and understanding how the viscosity of a material responds to stress and strain can influence many processes. For example, industrial processes such as pumping, extruding, and blow molding are all stress or strain processes, and their optimization requires an understanding of how the viscosity of the material changes with stress or strain. Many processes in our everyday lives, such as dispensing a shampoo or spreading a skin cream, also require the optimization of the formulation to achieve a desirable response of the viscosity.

Materials can respond to the application of stress or strain in various ways. Some materials do not change on application of stress or strain, and their corresponding viscosities remain constant with stress or shear rate; these are known as *Newtonian fluids.* Other materials undergo significant structural or alignment changes on application of stress or strain; the viscosity of these materials can either increase with increasing stress (shear thickening) or decrease with increasing stress (shear thinning). These fluids are described as exhibiting non-Newtonian behavior. The easiest way to visualize these various material responses to shear is to plot the viscosity versus the shear rate or shear stress. Because for many complex fluids the viscosity can decrease (in the case of shear thinning) over many decades in value as a function of shear rate or shear stress, it is necessary to plot the viscosity and shear rate or stress on a log-log scale. The resulting plot is known as a *flow curve*. Usually from an application perspective, the viscosity versus shear rate or shear stress curve is desired over a large range of shear rate or shear stress, usually spanning several decades. This is necessary because different shear rates or stresses correspond to various different applications. For example, sedimentation and leveling are low shear processes taking place around 10^{-2}–10^{-1} s^{-1}, whereas brushing and rubbing are much higher shear rate processes, typically occurring at 10^3–10^4 s^{-1}.

9.3.2.1 *Modeling Flow Curves*

To mathematically describe the complex fluid response observed in a flow curve, various mathematical models have been proposed, some of which are extensively used. The cross-model is a four-parameter equation that describes the entire flow curve of a structured fluid. The cross-model is given by (Barnes 2000):

$$\frac{\eta - \eta_\infty}{\eta_0 - \eta_\infty} = \frac{1}{1 + \left(K\dot{\gamma}\right)^m} \tag{9.3}$$

where η_0 = zero-shear viscosity (i.e., viscosity at zero shear rate)
K = dimensions of time
m = a dimensionless parameter that describes the extent of shear thinning

Under certain conditions, commonly encountered in many complex fluids, the cross-model reduces to the simpler and more commonly used power law model (Khan, Royer, and Raghavan 1997; Barnes 2000):

$$\eta = K\dot{\gamma}^{n-1} \tag{9.4}$$

where K = consistency
n = power law index

9.3.3 Yield Flow Behavior and Yield Stress

Many materials do not flow until the applied stress exceeds a certain critical stress, known as the *yield stress*. Materials exhibiting this behavior are said to be exhibiting yield flow behavior. Therefore, the yield stress is defined as the stress that must be applied to the sample before it starts to flow. Below the yield stress, the sample will deform elastically (like stretching a spring); above the yield stress, the sample will flow like a liquid and will typically exhibit shear thinning.

A wide range of materials exhibit this behavior, ranging from polymer and surfactant gels to commonly encountered consumer and food products such as toothpaste, mayonnaise, and ketchup. Yield stress fluids are easy to detect. In general, they do not flow by themselves and require the application of stress to begin flowing. There are a range of tests that can be performed to both check the existence of a yield stress and to quantitatively extract the yield stress. These can range from relatively quick measurements, such as a stress ramp, to creep measurements.

Mathematically, yield stress fluids can be described by a range of different models. One of the simplest models to describe yield stress behavior is the Bingham model (Barnes 2000):

$$\sigma = \sigma_0 + \eta_p \dot{\gamma} \quad (9.5)$$

where σ = shear stress
$\dot{\gamma}$ = shear rate
σ_0 = yield stress
η_p = plastic viscosity

9.3.4 Thixotropic Flow Behavior

Thixotropy is a time-dependent effect. Materials exhibiting thixotropy exhibit a decrease in viscosity under shear stress and undergo structural recovery, leading to increasing viscosity with time, after the stress is removed (Mewis 1979). The opposite effect, where viscosity increases with time under stress and undergoes a decomposition of the enhanced structural strength after the stress is removed, is known as *rheopexy*. Rheopexy is also sometimes known as *negative thixotropy* or *antithixotropy*. The variation of viscosity with time is illustrated for a complex fluid exhibiting thixotropy and for one exhibiting rheopexy in Figure 9.2.

Many suspensions and emulsions tend to exhibit thixotropic behavior. Common examples include ketchup, creams, and paints.

Properly characterizing thixotropy requires the right rheological test. Commonly, the thixotropic loop test is used to characterize the extent of thixotropy in a sample. In such a test, the shear rate or shear stress is increased from zero to a certain maximum value and is then returned to zero at the same rate. The area between the up and down curve is then taken as a measure of thixotropy. However, because both shear rate (or shear stress) and time are changing in this experiment, it is not the best

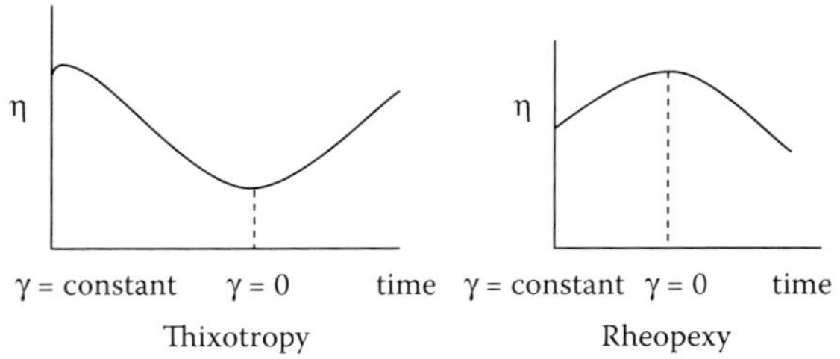

Figure 9.2 Variation of viscosity versus time for a complex fluid exhibiting thixotropy and for a system exhibiting rheopexy.

way to characterize thixotropy. A simpler and better way is to perform a stepwise experiment (Barnes 2000), where the shear rate or shear stress is increased in three steps: a low shear step to establish a reference value, a high shear step to bring about structural breakdown, and a low shear step to allow structural recovery. The shear rate or stress at each of those intervals is maintained for a certain period.

9.3.5 Viscoelasticity and Dynamic Rheology

Viscoelastic materials are materials that simultaneously exhibit both viscous and elastic properties. The extent of viscous or elastic response of the material will depend on the timescale of the experiment in relation to the relaxation time of the material. If the experiment is slow, the material will appear to be viscous; if it is fast, the material will appear to be elastic; and at intermediate timescales, it will exhibit a viscoelastic response. A large majority of materials in a wide range of different industries exhibit viscoelastic behavior.

Viscoelasticity can have major impacts on a number of processes. In consumer and food industries, having high elasticity can bring about problems in dispensing (e.g., generation of long strings on dispensing a shampoo) or in mouthfeel of foods. In the paint industry, high elasticity leads to high tack or stickiness, which can affect coating performance. In the polymer industry, phenomena such as postextrusion die swell can result from viscoelasticity.

A large variety of different complex fluids, such as polymers, surfactant mesophases, and biofluids, exhibit viscoelasticity. This viscoelasticity can affect various industrial and consumer processes; therefore, the proper characterization of the viscoelasticity becomes necessary to optimize process and formulation parameters. Another powerful benefit of viscoelastic measurement is that it provides a possible route into elucidating the microstructure of complex fluids through theoretical relationships developed between viscoelastic parameters and structural parameters.

Some key viscoelastic variables are complex modulus (G^*), elastic modulus (G'), viscous modulus (G''), tan delta ($\tan\delta$), and phase angle (δ).

These and other viscoelastic variables can be obtained from an oscillation experiment, in which a sinusoidally varying stress applied to the sample induces a sinusoidally varying strain (Figure 9.3). The difference in phase between the two waves is defined as phase angle, and the other viscoelastic parameters mentioned above are defined as (Ferry 1980):

$$\begin{aligned} G^* &= G' + iG'' \\ G' &= G^* \, Cos\delta \\ G'' &= G^* \, Sin\delta \end{aligned} \tag{9.6}$$

where δ = phase angle

The following sections highlight some key features of these viscoelastic variables and the insight they provide.

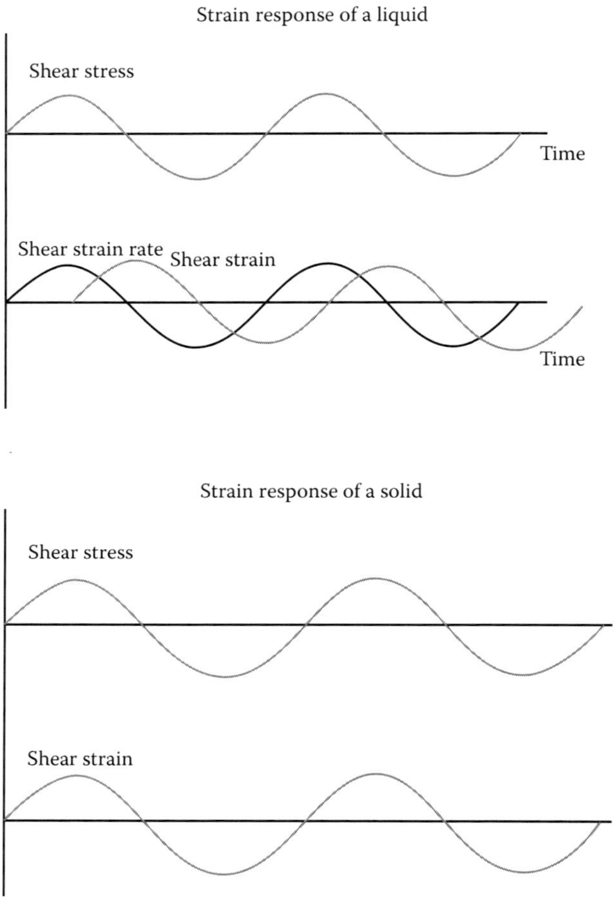

Figure 9.3 An oscillation experiment in which a sinusoidally varying stress or strain is applied.

9.3.5.1 Storage Modulus, G′

The storage modulus G' is a measure of elastic energy stored by a material during the shear process. It is a measure of the elasticity of a sample, with materials in which the energy is completely stored exhibiting pure elastic behavior. It is measured through an oscillation experiment, in which a sinusoidally varying stress (or sinusoidally varying strain) applied to the sample results in a sinusoidally varying strain (or sinusoidally varying stress if a strain was applied).

The storage modulus G' obtained from an oscillation test provides a number of useful insights. When a frequency sweep is carried out, the relative values of $G'G''$ over a range of frequencies provide insight as to over what frequency range or timescales a material's behavior is predominantly going to be gel-like or solid-like. Because elasticity is intimately linked to a material's underlying structure, G' can also be used to understand structure buildup in a system as a function of time. For many processes, G' can be linked to a performance criterion, such as stickiness or tack, with higher G' values being associated with enhanced tack or stickiness.

9.3.5.2 Loss Modulus, G″

The loss modulus G'' is a measure of the viscous dissipation of the energy during shear. Because it is a measure of the energy that is "lost" for the material, it is a measure of the viscous behavior of a material. It is measured through an oscillation experiment, in which a sinusoidally varying stress (or sinusoidally varying strain) applied to the sample results in a sinusoidally varying strain (or sinusoidally varying stress if a strain was applied).

The loss modulus G'' obtained from an oscillation test provides a number of useful insights. When a frequency sweep is carried out, the relative values of $G'G''$ over a range of frequencies provides insight as to over what frequency range or timescales a material's behavior is predominantly going to be liquid-like. Materials that exhibit predominately viscous behavior over a wide range of frequencies are generally not structured liquids or are, at best, weakly structured.

For many processes, G'' can be linked to a performance criterion such as investigating solution–gel phenomena, with $G'' > G'$ indicating a dominance of liquid or solution-like behavior.

9.3.5.3 Tanδ

Tanδ is obtained from an oscillation experiment in which a sinusoidally varying stress (or sinusoidally varying strain) applied to the sample results in a sinusoidally varying strain (or sinusoidally varying stress if a strain was applied).

Mathematically, this can be represented as:

$$\gamma = \gamma_0 \sin(\omega t) \tag{9.7}$$

$$\tau = \tau_0 \sin(\omega t + \delta) \tag{9.8}$$

where ω = frequency
δ = phase angle

The phase shift between the input signal (applied oscillation) and the output signal (sample response) is the phase angle. Tanδ is the tangent of this phase angle. Tanδ can also be represented as $\tan\delta = G''/G'$, that is, the ratio between the viscous and elastic components.

Because tanδ is the ratio between the viscous and elastic components, it provides valuable insight into the balance between these two components and can be widely used to understand transitions between liquid-like or solid (or gel)-like behavior.

Quantitatively $0 \leq \tan\delta \leq \infty$, where pure elastic behavior is characterized by $\tan\delta = 0$ and ideal viscous behavior is characterized by $\tan\delta = \infty$. If they are completely balanced, then $\tan\delta = 1$. Tanδ can be used in any test that involves understanding the transitions between the viscous and elastic components. Typical areas are following solution–gel transitions, glassy to rubbery transitions in polymers, and so on. It is also a useful measure for getting insight into tack or stickiness behavior of materials. In general, the lower the value of tanδ, the more gel-like or solid-like the material is.

9.3.5.4 Modeling Viscoelastic Behavior

One of the widely used models of viscoelastic behavior is the Maxwell model. This is a mechanical model that is used to represent certain viscoelastic fluids. When an instantaneous stress is applied, it will deform elastically and then reach a constant rate of flow. When the stress is removed, it will recover some of its original shape but will not reform totally.

The equation to describe this behavior is (Barnes 2000):

$$\gamma^* = \frac{\tau}{\eta} + \frac{\tau^*}{G} \tag{9.9}$$

where γ^* = shear rate
τ = shear stress
G = shear modulus
η = viscosity
τ^* = change of shear stress over time

Equation 9.9 represents the simplest form of the Maxwell model, which can be used to describe a range of different viscoelastic fluids, which exhibit a single relaxation time. For viscoelastic liquids with a range of relaxation times, such as polymers with a wide molecular weight distribution, it is better to use a generalized Maxwell model, that is, one that contains several Maxwell elements in parallel connection.

For single relaxation time Maxwell fluids, the frequency dependence can be described by (Barnes 2000):

$$G'(\omega) = \frac{G'_\infty \omega^2 t_R^2}{1+\omega^2 t_R^2} \qquad G''(\omega) = \frac{G'_\infty \omega t_R}{1+\omega^2 t_R^2} \tag{9.10}$$

where G'_∞ = plateau modulus
t_R = relaxation time

A range of viscoelastic liquids, polymer melts, emulsions, and surfactant phases in many cases can be described as being a Maxwell model fluid.

For viscoelastic liquids with a single relaxation time, the simplest way to test whether it can be described by a Maxwell model is to carry out an oscillation experiment–frequency sweep, and if the plot of G'' versus G' is a semicircle, then it obeys the mathematical description above and therefore can be described by the simple Maxwell model.

For systems with multiple relaxation times (such as polymers with wide molecular weight distribution) it should be checked whether the generalized Maxwell model describes the data more properly.

Maxwell fluids are viscoelastic fluids, so they exhibit a range of behaviors expected from viscoelastic fluids, such as die swell, Weissenberg effect, and stickiness.

9.4 APPLICATIONS OF RHEOLOGY

Rheology has numerous applications across a wide range of industries, ranging from pharmaceuticals to foods and home and personal care, and is a widely used technique in both research and development, as well as in quality control. Rheology plays a critical role in various stages of complex fluid formulation, manufacturing, and application.

9.4.1 Manufacturing and Processing

Rheology control is an essential requirement in process control and optimization. Many process steps, commonly encountered in the processing of nutraceutical-containing complex fluids, such as pipe flow, pumping, mixing, heat, and mass transfer, are all significantly influenced by the rheological response of the complex fluid. Processing can also be used to affect the microstructure of the complex fluid or soft solid. A brief description of why rheology measurement and control are important in some of these essential processing steps is now further discussed.

9.4.1.1 Pipe Flow and Pumpability

In many manufacturing processes it is often necessary to pump the complex fluid of interest over large distances. Substantial frictional pressure losses in the pipeline may be associated with this flow, and to minimize this and optimize the process, there arises a need to calculate flow rate through the pipe, the pump power requirements, and so on. For a simple Newtonian fluid, the volumetric flow rate is given by the well-known Hagen–Poiseuille relationship:

$$Q = \frac{\pi R^4 \Delta p}{8\eta L} \tag{9.11}$$

where Q = volumetric flow rate
R = radius of a circular pipe
Δp = pressure decrease
η = viscosity

It is clearly seen that the fluid viscosity significantly affects the volumetric flow rate of the fluid. However, this relationship is limited to Newtonian fluids. For complex fluids, typically exhibiting non-Newtonian behavior, the relationship between pipe flow and volumetric flow rate becomes much more complicated. For the case of a shear thinning power law complex fluid, this relationship changes to (Chhabra and Richardson 1999):

$$Q = \pi\left(\frac{n}{3n+1}\right)\left(\frac{-\Delta p}{2K}\right)^{\frac{1}{n}} R^{\frac{(3n+1)}{n}} \tag{9.12}$$

where K and n = the consistency and the power law index from the power law model (Equation 9.4)

This equation illustrates that the volumetric flow rate has a much stronger dependence on R for a shear thinning fluid ($n < 1$) than for a Newtonian fluid, but the pressure decrease, Δp, has a much weaker dependence on Q, $\Delta p \propto Q^n$. Similar relationships for laminar flow of other non-Newtonian fluids, such as Bingham plastics or Herschel–Bulkley fluids, can be derived. Many complex fluids undergo shear-induced changes, and a control of the rheological response can only be brought about by understanding the corresponding microstructure–rheology linkages.

9.4.1.2 Mixing

Mixing of complex fluids with each other or with Newtonian fluids is a common process that takes place in many complex fluid manufacturing processes. Mixing of such fluids and soft solids can give rise to numerous problems, and as will be discussed, it is the rheology of the complex fluid or soft solid that, to a large extent, determines the equipment and the procedures to be adopted for the mixing process.

One of the primary objectives in many mixing processes is to achieve a uniform product by homogeneous mixing of various ingredients. The right level of mixing is an essential requirement in this process. Obviously undermixing is not desired because that would lead to a nonuniform product; overmixing also would not be desired because that would lead to changes or breakdown in the final product microstructure and therefore the final product rheology. Impacting the final product rheology could lead to serious consequences in the product's sensory and textural attributes and therefore affect the consumer acceptability of the product. To ensure proper mixing, it should be ensured that there are no stagnation zones and that there is a zone of high shear mixing that will aid in the breakdown of heterogeneities (Chhabra and Richardson 1999).

Rheology plays a critical role both in terms of the prevalent mixing process that will take place and the required power consumption of the stirred tank (Chhabra and Richardson 1999). There are two types of mixing processes: (1) laminar mixing, which usually occurs in mixing processes involving high viscosity fluids, and (2) turbulent mixing, which occurs in low viscosity fluids. Homogenization is ultimately linked to molecular diffusion, and this occurs much more rapidly for low viscosity fluids than it does for high viscosity fluids. One of the main requirements from a process optimization and design point of view is to accurately determine the power consumption required of stirred tanks used for mixing. The dimensionless version of this, also known as the *power number*, has been shown (Chhabra and Richardson 1999) to be linked to the Reynolds number and the Froude number. In terms of the link with rheology, the dimensionless Reynolds number includes the viscosity; therefore, the viscosity of the fluid has a direct impact on the required power consumption. It should be noted that at high Reynolds numbers, the flow becomes fully turbulent, resulting in rapid mixing.

Although a detailed discussion will not be undertaken, it should be noted that viscoelasticity, which is present in many complex fluids and soft solids, can also affect the mixing process and the power consumption of the stirred tank (Prud'homme and Shaqfeh 1984).

9.4.2 Transportation

Products consisting of complex fluids must be stable to be useful. Many complex fluids have delicate structures that tend to break down easily both by the extent of shear that is applied and by the time of shearing. This will have a significant impact during transportation because it will be subjected to various transportation-related stresses.

A rheological test can identify the mechanical stability of such products. The test involves subjecting the sample to a preshear history similar to that encountered during transport, then following structure rebuild by means of an oscillatory time sweep at low stresses and strains. Because the storage modulus, G', is a measure of elastic structure, the rate of its recovery after a perturbation can be used to assess the product's structural integrity.

9.4.3 Functional and Sensory Performance

Rheology has a direct link with many performance-related aspects of complex soft material matrices that incorporate nutraceuticals. One of the key requirements for food-related nutraceutical delivery is that the basic consumer acceptability criteria in terms of appearance, aroma, taste, and shelf life must be maintained. Incorporation of nutraceuticals into the complex soft matter matrix is in itself highly challenging. For example, incorporation of plant sterols or phytosterols into oil-in-water emulsions is difficult because sterols are poorly soluble in oil. Only through using triglyceride and lecithin as crystallization inhibitors is a high loading of sterols achieved (Appelqvist et al. 2007). Once incorporated, there may be considerable other issues as a result of the nutraceutical's own inherent physicochemical properties and/or because of the interaction with the soft matter matrix. Addition of the nutraceutical into the complex soft matter matrix may cause changes to the soft matter matrix microstructure, thereby affecting a number of the performance criteria, especially texture, appearance, and shelf life. The following is a discussion of some of these key performance criteria and their links with rheology.

9.4.4 Shelf Life and Formulation Stability

A large majority of food, pharmaceutical, and personal care complex fluids that are used to incorporate nutraceuticals are emulsions or dispersions and suspensions. Even in cases in which the complex fluid is a self-assembled surfactant or polymeric mesophase, many times they also incorporate particulates or oil droplets. The technical objective, and sometimes one of the greatest challenges in these systems, is the generation of a formulation that is physically stable. There are many physicochemical processes and mechanisms by which emulsions can break down, the key ones being:

- Creaming
- Sedimentation
- Flocculation
- Coalescence

Creaming and sedimentation are both gravity-driven, and either one can take place depending on the density difference between the droplets and the continuous phase. The rate of creaming or sedimentation is given by Stokes law (Tadros 2004):

$$\nu_0 = \frac{2}{9}\frac{R^2 \Delta\rho g}{\eta_0} \tag{9.13}$$

where η_0 = zero shear viscosity
R = radius of the droplets

From Equation 9.13 it is clearly seen that increasing the zero shear viscosity slows down the process of sedimentation or creaming. Therefore, the usual strategy in eliminating sedimentation or creaming is to add high molecular weight polymers, such as hydroxyl ethyl cellulose, which increase the zero shear viscosity considerably. Alternatively, building in a yield stress through network-forming polymers, such as xanthan gum, can eliminate sedimentation or creaming as the yield stress effectively overcomes the gravitational driving force that leads to sedimentation or creaming. When a particle is dispersed in a complex fluid with a yield stress, τ_y it will exhibit a retaining force:

$$F_R = \tau_y \cdot A \tag{9.14}$$

where A = surface area of the particle

To prevent a particle from creaming or sedimenting, the retaining force needs to be equal or larger than the gravity force, given by:

$$F_B = V \cdot \Delta\rho \cdot g \tag{9.15}$$

where V = particle volume
$\Delta\rho$ = difference between densities of dispersed and continuous phases
g = acceleration resulting from gravity

Many complex fluid mesophases, such as lamellar gel networks, which, for example, are formed by monoglyceride gels, already exhibit a yield stress, and as such there is no requirement to include additional yield stress building materials. It should be noted that many nutraceutical delivery systems can be nanoemulsion based. Nanoemulsions are not thermodynamically stable but do exhibit high kinetic stability because of the small size of the droplets.

Many emulsions also undergo flocculation as a result of attractive van der Waals forces. Again rheology can be used to determine the extent of flocculation. Most emulsions exhibit clear shear thinning characteristics, and the behavior can be modeled using various rheological models, such as the Bingham model, the power law model, or the Casson model. The parameters obtained from the fit of the appropriate model provide insight into the extent of flocculation. For example, if the Bingham model (Equation 9.5) is the most appropriate to describe the data, then σ_0 (yield stress) and η_p (plastic viscosity) can be related to the extent of flocculation. The higher the values of σ_0 and η_p, the greater the extent of flocculation in the emulsion (Tadros 2004). Thixotropy, which has been discussed earlier, is another rheological method that at least qualitatively can also provide insight into the extent of flocculation. Weakly flocculated structures tend to exhibit much more thixotropy than strongly flocculated structures. For a more detailed discussion on these aspects, see Tadros (2004).

9.4.5 Texture

Texture is one of the key criteria that influence consumer acceptability and sensory performance of food-based nutraceutical delivery system. The overall mouthfeel of a food product is not only related to the initial texture of the food but also to changes in texture associated with the breakdown of the product in the mouth. This will influence both taste and flavor perception. The breakdown of the food microstructure in the mouth is influenced by both the applied shear forces during mastication and the dilution caused by the mixing with saliva, which is itself a highly viscoelastic fluid. This microstructural breakdown in the mouth will have two direct consequence-associated changes in the rheology of the product and the impact on the release of activities (may also influence nutraceutical release) and flavors in the mouth.

The mouthfeel of foods is not only related to the bulk rheological properties and the breakdown of the microstructures, but also, as this process progresses, thin-film rheology and the lubrication performance also start to affect the overall perception. Thin-film rheology and tribology require specialized techniques (Clasen, Gearing, and McKinley 2006) but can provide unique insight into sensory perception.

9.5 ENGINEERING THE RHEOLOGICAL RESPONSE

The various important applications of rheology in manufacturing, transportation, and controlling product performance criteria have been highlighted in the previous section.

One of the main areas of focus of research and development in the area of complex fluid formulations containing nutraceuticals is in optimizing their performance. This is best carried out through developing and implementing a knowledge-based approach and strategy. This is based on the fact that understanding and controlling the various performance attributes depends on designing the rheological response, which in turn depends on the complex fluid microstructure and phase behavior, which again depends on the formulation conditions and presence of other additives, such as nutraceuticals. These links between the performance and formulation are illustrated in Figure 9.4.

Commercial formulations are inherently a complex mixture of a number of different complex fluids and additives, such as nutraceuticals, and adopting a knowledge-based approach may seem like a daunting task. However, the key to this approach lies in breaking down the formulation into key components and adopting the approach illustrated in Figure 9.4.

The following are the main steps in this approach:

1. Determine the phase behavior and microstructure of the base formulation. This will primarily be determined by the formulation, but in many cases processing can also affect the micro- and mesostructure. A variety of techniques—light scattering, linear rheology, and imaging—can be used to establish this. A key aspect to this is in characterizing the microstructure at the length scale relevant for rheology control. Linear rheology with associated theories can play a key role in developing this understanding. This base understanding will play a key role in understanding and controlling the performance in a variety of different applications.
2. Determine the links between the microstructure and application-relevant rheology. The specific rheology test and the conditions of the test will be determined by the specific application (in-use processes) and the performance attribute. It should be noted that this is one of the most challenging aspects and difficult steps.
3. Having knowledge of the performance–rheology–microstructure–formulation linkages for the base formulation (main structurant); optimize it through necessary formulation and processing changes.
4. Build up formulation with additives and iterate cycle to enhance performance.

This approach builds toward a modular approach to product design as it builds the knowledge base for the base structurant and how additives affect that

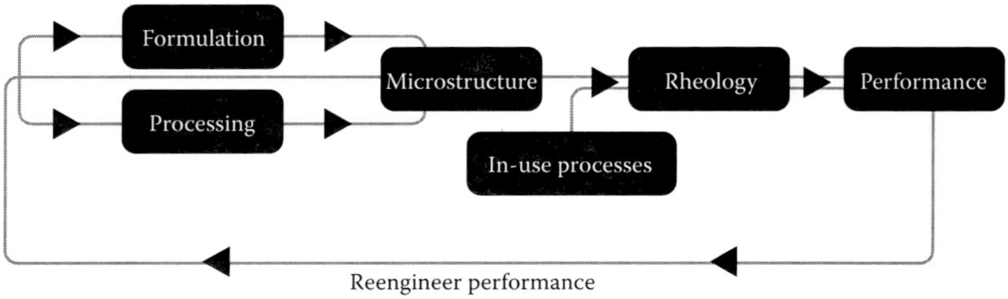

Figure 9.4 Optimizing complex fluid formulations.

performance. The approach also develops an understanding of how the microstructure of the formulation needs to be affected to generate a step change in performance. This can then form the basis of informed discussions with suppliers to tailor synthesis of base structurant and/or additives. This also feeds into the design criteria for generating the next generation of stimuli-responsive or "smart" complex and nanostructured fluids.

9.5.1 Engineering the Response: Emulsions and Suspensions

Emulsion rheology tends to have a strong dependence on the volume fraction of the dispersed phase and the droplet size. The rheological parameters of key interest are the viscosity, normal stress, viscoelasticity, and the yield stress.

The viscosity η of a dilute emulsion with a low capillary number (so that the bubbles do not deform) is given by the following expression (Larson 1999):

$$\eta \equiv \frac{\eta}{\eta_s} = 1 + \frac{1 + \frac{5}{2}M}{1 + M}\phi$$

$$M \equiv \frac{\eta_d}{\eta_s} \tag{9.16}$$

where ηd = dispersed phase viscosity
ηs = viscosity of the suspending fluid
Φ = volume fraction of the dispersed phase

Here it is assumed that the emulsion is not shear thinning; hence the viscosity will be independent of each shear rate.

For higher droplet concentrations ($\phi \geq 0.6$), the system becomes shear thinning, and the relative zero shear viscosity ($\eta_{r,0}$) has to then be linked to the volume fraction of droplets by other relationships (Larson 1999).

On increasing the volume fraction of droplets, a point of phase inversion can be reached. However, if the emulsion droplets are stabilized by surfactant or particles, the droplets can remain stable even as the volume fraction approaches 1. Dense or concentrated emulsions tend to exhibit interesting rheological properties, such as yield stress and high viscoelasticity, as volume fraction of the dispersed phase exceeds that of a close-packed sphere configuration ($\phi = 0.74$ for monodisperse systems, but can be lower as a result of polydispersity, etc.). According to Princen and Kriss (1989), the yield stress (σ_y) generated in such dense emulsions depends on the volume fraction of the droplets and is given by:

$$\sigma_y = \frac{\Gamma}{a_{32}}\phi^{\frac{1}{3}}Y(\phi) \tag{9.17}$$

where ϕ = volume fraction of the droplets
Γ = interfacial tension
a_{32} = volume to surface drop radius

$Y(\phi)$ is a function that is defined as:

$$Y(\phi) = -0.080 - 0.114\log_{10}(1-\phi) \tag{9.18}$$

9.5.2 Engineering the Response: Liquid Crystalline Mesophases

In many cases, the nutraceutical has to be formulated into an existing complex soft matter matrix. The soft matter matrix can be from a wide range of self-assembled surfactant or polymer-based structures, which in turn can exhibit a wide range of rheological responses. A comprehensive review of the various mesophases is beyond the scope of this chapter; however, two liquid crystalline mesophases will be briefly discussed to highlight the links between rheology and the mesostructure. Liquid crystalline mesophases are of high interest because they can be regarded as efficient building blocks of functional foods. This is primarily because of the ability of these mesophases to act as effective carriers for various nutraceuticals, which can be dispersed either in the hydrophobic or hydrophilic phase. Liquid crystalline phases are usually based on the self-assembly of short surfactants such as monoglycerides or phospholipids and water. These mesophases can be used directly in the bulk state or redispersed as colloidal carriers (Ubbink, Burbidge, and Mezzenga 2008). The following describes the rheology–structure linkages in two such systems.

9.5.2.1 Lamellar Gel Networks

Fatty amphiphiles, such as long chain alcohols, acids, and monoglycerides, form lamellar phases when they are dispersed in water in the presence of a high hydrophobic–lipophilic balance surfactant. When the temperature of such a system is decreased, the lamellar liquid crystalline phases of the fatty amphiphiles transform into complex lamellar crystalline gel networks.

These gel network are essentially composed of a network of stiff platelike particles connected by weak links. It is these weak links that play a critical role in the rheology of these mesostructures. However, the rheological response in such systems is highly complicated, with a lack of a strong theoretical model that adequately describes the rheological response, especially the frequency response of G', G''. There have been attempts to link the elastic modulus, which is indicative of the network strength, to the particle volume fraction and the stiffness of the individual layers. This also has been validated experimentally. This relationship is given by (Sein, Verheij, and Agterof 2002):

$$\begin{aligned} G' &\approx \phi^{\mu} \\ \mu &= \frac{\alpha}{(3-D_f)} \end{aligned} \tag{9.19}$$

where ϕ = volume fraction of the network forming particles
μ = an exponential factor of which the value depends on the nature of the particle network

It is obtained from the plot of G' and the volume fraction. Once it is obtained, it can be related to α or the flexibility of the stress-carrying unit or the fractal dimension, D_f. Thus, this rheology-based relationship can be used to determine the flexibility of the stress-carrying unit or the fractal dimension for a particular system. The flexibility of the stress-carrying unit, α, can have values that range from 5 for a flexible chain to approximately 2 for a stiff chain (e.g., a value of 2 has been reported for casein gels).

9.5.2.2 Cubic Phase

Many surfactant systems within a certain composition of oil and water tend to form a highly viscous gel-like phase known as a *cubic phase*. Similar cubic phases also have been seen to form in triblock copolymers such as Pluronics (BASF Corporation, Florham Park, NJ) within certain concentration and temperature ranges. From a rheological standpoint, these phases tend to be extremely strong gels, with the elastic modulus G' exceeding the viscous modulus G'' by a decade or so, especially at the highest frequencies. This is not unexpected because a strong solid-like network is usually formed. From a modeling description, the low frequency behavior can be best modeled by a Maxwell model, whereas the high frequency behavior can be best modeled as a Voigt model (Jones and McLeish 1995).

From a microstructural perspective, much of the rheological response can be best described in terms of "slip planes," which implies that large deformations occur along certain planes in the structure, whereas everywhere else there are relatively small deformations. The density of slip planes is, to a large extent, kinetically determined and is a function of irregularities in the lattice, which is made during the formation stages. The relaxation time in such systems is dependent on the average density of "slip links."

The viscoelastic moduli can be expressed through the following equation (Jones and McLeish 1995):

$$G'(\omega) = AG\frac{\left(\omega^2\tau^2 - \omega^2\tau_1^2\right)}{\left(1+\omega^2\tau^2\right)}$$
$$G''(\omega) = AG\frac{\omega\tau\left(1+\omega^2\tau\tau_1^2\right)}{\left(1+\omega^2\tau^2\right)} \tag{9.20}$$

where $A = \dfrac{\eta_2 N}{\left(\eta + N\eta_2\right)}$
$N - 1$ = slip plane density
τ = bulk relaxation time
τ_1 = local relaxation time

9.6 FUTURE TRENDS IN RHEOMETRY

9.6.1 Expert System Rheometry

The aforementioned examples illustrate a couple of important points regarding a knowledge-based approach to formulation optimization. First, it requires an in-depth knowledge of complex fluid microstructure and rheology; second, even if this is correctly known, setting up the relevant rheological test with the correct set of parameters is not trivial. The majority of errors in rheological measurements result from improper setup of the experimental protocols. A novel development in rheometry based on an Expert System (Amin and Carrington 2009) addresses both of these major issues. It takes the user through every step of this rheological journey from setting up the right test, correct loading of the sample, analysis of the results and extracting key microstructural parameters, to establishing insight on performance. For example, in the examples of the lamellar gel network and the cubic phase insight on the rheology, controlling microstructural parameters, such as the fractal dimension or the slip plane density, can be obtained.

The Expert System includes an extensive library of the most commonly encountered complex fluids from a range of industries—personal care, pharmaceuticals, nanotechnology, drug delivery, coatings, and foods—as well as an extensive library of rheological tests and insight on various material performance aspects in these industries. The software has an easy-to-use search tool that allows material scientists to search for their application problem and/or microstructure type (e.g., wormlike micelle, lamellar gel, or emulsion) or product form (e.g., spreads, pharmaceutical, and personal care skin creams). Within the software are overviews highlighting areas where rheology is important in an industry sector, which then link to individual application notes. These application notes explain the basis behind a particular problem and present the relevant theoretical approach, data, analysis, and interpretation required to provide a measurement solution. Directly linked to each application note (via mouse clicks only) is a test sequence for the rheometer—a sequence is a complete test providing the user with guidance on all aspects of the measurement undertaken in the application note, from optimum sample preparation and loading through to data presentation and analysis. Sequences operate on the rheometer interface with standard operating procedure–type capability and provide relevant instructions for the user at every stage, as well as automatically setting test parameters as appropriate. This Expert System approach is illustrated in Figure 9.5.

9.6.2 High-Throughput Characterization and On-Line Rheometry

One of the major challenges in formulating and processing complex fluids with nutraceuticals is that the addition of the nutraceutical can bring about significant changes to the complex fluid microstructure and rheology, resulting in impacts on key performance criteria such as texture, appearance, and stability. To properly diagnose the problem and engineer an optimized formulation, in addition to the knowledge-based approach, there is sometimes the need for a reliable method to characterize the formulations in a high-throughput way. This helps to identify a formulation space where the nutraceutical could be incorporated effectively and would also help to screen a number of different nutraceuticals in the same formulation space. Once promising formulation space and nutraceuticals have been identified, this subset can be characterized in much more detail using Expert System rheometry.

Currently there are a number of techniques that may hold potential promise as high-throughput rheological characterization devices. Among these, two leading potential techniques are optical microrheology techniques and techniques emerging from the microfluidics and microelectromechanical systems (MEMS) area. Optical microrheological tools, such as those based on dynamic light scattering or imaging-based particle tracking, are rapidly emerging as novel tools for rheological characterization (Waigh 2005). In addition to rapid measurement times compared with mechanical rheological techniques, as well as the accessibility to a wide dynamic range, these techniques require little sample volume and can probe the response of the complex fluid through application of small stresses. These features make optical-based techniques highly appealing for characterizing small quantities of strain-sensitive complex fluids. The main challenge in this area is minimizing probe particle–complex fluid interaction and theoretically robust interpretation of the results. However, as a high-throughput screening tool, these may already prove to be highly promising techniques.

Microfluidic-based devices and devices based on MEMS are also being explored for their potential use in complex fluid characterization (Jakoby et al. 2009; Pipe

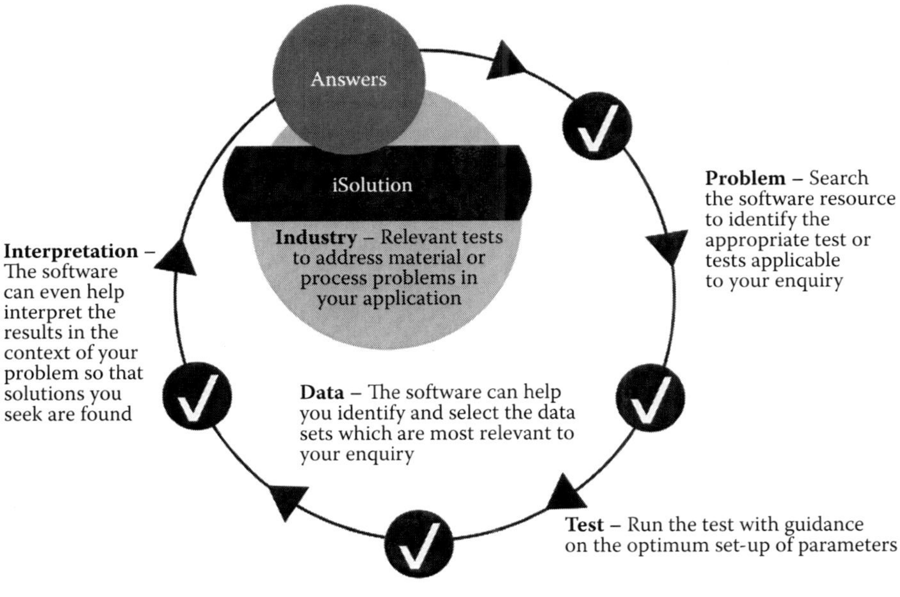

Figure 9.5 The Expert System approach to rheology to allow insight into microstructural characterization of complex and nanostructured fluids.

and McKinley 2009). They also offer the potential to rapidly characterize complex fluids. However, at present they do not cover as wide a dynamic range of frequencies as optical-based methods and are still relatively in their infancy with regard to the theoretical understanding of the links with complex rheological response exhibited by many complex fluids.

Although a detailed discussion will not be undertaken, it should be noted that in addition to high-throughput characterization of complex fluids, there is always a need for on-line process rheology monitoring. As discussed, complex fluids and soft solids can undergo significant changes during the various processing steps, which can have a significant impact on the resulting performance aspects. Effective process optimization and associated process control require measurements to be carried out directly in the process line. This would allow continuous monitoring of a key process parameter and associated enhanced process control, thereby ensuring consistent product quality. Although a number of in-line techniques are being used in the chemical and polymer industries, there are few techniques that would allow proper in-line characterization of complex fluids encountered in the pharmaceutical and food industries. The previously discussed novel optical techniques or techniques based on microfluidics or MEMS devices also hold considerable promise as potential process monitoring devices and sensors. However, work in that particular area is in early stages of research. Roberts (2003) provides a detailed discussion of the area and some of the other potential in-line and on-line rheometry devices.

REFERENCES

Amin, S., and S. Carrington. 2009. Expert System Rheometry. *Materials Today* 12: 44–46.

Appelqvist, I. A. M., M. Golding, R. Vreeker, and N. J. Zuidam. 2007. Emulsions as Delivery Systems in Foods. *Encapsulation and Controlled Release: Technologies in Food Systems*. Oxford, UK: Blackwell Publishing, 41–81.

Ariyaprakai, S., and S. R. Dungan. 2007. Solubilization in Monodisperse Emulsions. *Journal of Colloid and Interface Science* 314: 673.

Armand, M. 2007. Lipases and Lipolysis in the Human Digestive Tract: Where Do We Stand? *Current Opinion in Clinical Nutrition and Metabolic Care* 10: 156.

Armand, M., P. Borel, P. Ythier, G. Dutot, C. Melin, M. Senft, H. Lafont, and D. Lairon. 1992. Effects of Droplet Size, Triacylglycerol Composition, and Calcium on the Hydrolysis of Complex Emulsions by Pancreatic Lipase—An In Vitro Study. *Journal of Nutritional Biochemistry* 3: 333.

Auweter, H., H. Haberkorn, W. Heckmann, D. Horn, E. Luddecke, J. Rieger, and H. Weiss. 1999. Supramolecular Structure of Precipitated Nanosize Beta-Carotene Particles. *Angewandte Chemie International Edition* 38: 2188.

Barnes, H. A. 2000. *A Handbook of Elementary Rheology*. Aberystwyth, UK: University of Wales, Institute of Non-Newtonian Fluid Mechanics.

Chhabra, R. P., and J. F. Richardson. 1999. *Non-Newtonian Flow in the Process Industries*. Oxford, UK: Butterworth, Heinemann.

Chu, B. S., S. Ichikawa, S. Kanafusa, and M. Nakajima. 2007. Preparation of Protein-Stabilized Beta-Carotene Nanodispersions by Emulsification-Evaporation Method. *Journal of the American Oil Chemists' Society* 84: 1053.

Clasen, C., B. Gearing, and G. H. McKinley. 2006. Microrheology Using a Flexure Based Microgap Rheometer (FMR). *Journal of Rheology (New York)* 50: 883–905.

Clinton, S. K. 1998. Lycopene: Chemistry, Biology, and Implications for Human Health and Disease. *Nutrition Reviews* 56: 35.

FairweatherTait, S., and R. F. Hurrell. 1996. Bioavailability of Minerals and Trace Elements. *Nutrition Research Reviews* 9: 295.

Fave, G., T. C. Coste, and M. Armand. 2004. Physicochemical Properties of Lipids: New Strategies to Manage Fatty Acid Bioavailability. *Cellular and Molecular Biology* 50: 815.

Ferry, J. D. 1980. *Viscoelastic Properties of Polymers*. Hoboken, NJ: John Wiley & Sons Inc.

Hecq, J., M. Deleers, D. Fanara, H. Vranckx, and K. Amighi. 2005. Preparation and Characterization of Nanocrystals for Solubility and Dissolution Rate Enhancement of Nifedipine. *International Journal of Pharmaceutics* 299: 167.

Hernell, O., J. E. Staggers, and M. C. Carey. 1990. Physical-Chemical Behavior of Dietary and Biliary Lipids during Intestinal Digestion and Absorption. 2. Phase-Analysis and Aggregation States of Luminal Lipids during Duodenal Fat Digestion in Healthy Adult Human-Beings. *Biochemistry (Moscow Russ Fed)* 29: 2041.

Horn, D., and J. Rieger. 2001. Organic Nanoparticles in the Aqueous Phase—Theory, Experiment, and Use. *Angewandte Chemie International Edition* 40: 4331.

Horter, D., and J. B. Dressman. 1997. Influence of Physicochemical Properties on Dissolution of Drugs in the Gastrointestinal Tract. *Advanced Drug Delivery Reviews* 25: 3.

Horter, D., and J. B. Dressman. 2001. Influence of Physicochemical Properties on Dissolution of Drugs in the Gastrointestinal Tract. *Advanced Drug Delivery Reviews* 6: 75.

Jakoby, B., E. Reichel, C. Riesch, F. Lucklum, B. Weiss, F. Keplinger, et al. 2009. Condition Monitoring of Viscous Liquids Using Microsensors. *Elektrotechnik & Informationtechnik* 126/5: 164–72.

Jones, J. L., and T. C. B. McLeish. 1995. Rheological Response of Surfactant Cubic Phases. *Langmuir* 11: 785–92.

Kesisoglou, F., S. Panmai, and Y. H. Wu. 2007. Nanosizing—Oral Formulation Development and Biopharmaceutical Evaluation. *Advanced Drug Delivery Reviews* 59: 631.

Khan, S. A., J. R. Royer, and S. R. Raghavan. 1997. *Rheology: Tools and Methods. Aviation Fuels with Improved Fire Safety.* Washington, DC: National Academies Press, 31.

Kristl, J., B. Volk, M. Gasperlin, M. Sentjurc, and P. Jurkovic. 2003. Effect of Colloidal Carriers on Ascorbyl Palmitate Stability. *European Journal of Pharmaceutical Sciences* 19: 181.

Larson, R. G. 1999. *The Structure and Rheology of Complex Fluids*. Oxford, UK: Oxford University Press.

Lawless, H. T., F. Rapacki, J. Horne, and A. Hayes. 2003. The Taste of Calcium and Magnesium Salts and Anionic Modifications. *Food Quality and Preference* 14: 319.

Lawless, H. T., F. Rapacki, J. Horne, A. Hayes, and G. Wang. 2004. The Taste of Calcium Chloride in Mixtures with NaCl, Sucrose and Citric Acid. *Food Quality and Preference* 15: 83.

Lim, J., and H. T. Lawless. 2006. Detection Thresholds and Taste Qualities of Iron Salts. *Food Quality and Preference* 17: 337.

Merisko-Liversidge, E., G. G. Liversidge, and E. R. Cooper. 2003. Nanosizing: A Formulation Approach for Poorly-Water-Soluble Compounds. *European Journal of Pharmaceutical Sciences* 18: 113.

Mewis, J. 1979. Thixotropy: A General Review. *Journal of Non-Newtonian Fluid Mechanics* 6: 1–20.

Patravale, V. B., A. A. Date, and R. M. Kulkarni. 2004. Nanosuspensions: A Promising Drug Delivery Strategy. *Journal of Pharmacy and Pharmacology* 56: 827.

Pipe, C. J., and G. H. McKinley. 2009. Microfluidic Rheometry. *Mechanics Research Communications* 36: 110–20.

Princen, H. M., and A. D. Kriss. 1989. Rheology of Foams and Highly Concentrated Emulsions. IV. An Experimental Study of the Shear Viscosity and Yield Stress of Concentrated Emulsions. *Journal of Colloid and Interface Science* 128: 176–87.

Prud'homme, R. K., and E. Shaqfeh. 1984. Effect of Elasticity on the Mixing Torque Requirements for Rushton Turbine Impellers. *AIChE J* 30: 485.

Roberts, I. 2003. In-line and On-line Rheology Measurement of Food. *Texture in Food, Volume 1: Semi-Solid Foods*. Cambridge, UK: Woodhead Publishing.

Santos-Buelga, C., and A. Scalbert. 2000. Proanthocyanidins and Tannin-like Compounds—Nature, Occurrence, Dietary Intake and Effects on Nutrition and Health. *Journal of the Science of Food and Agriculture* 80: 1094.

Sein, A., J. A. Verheij, and W. G. Agterof. 2002. Rheological Characterization, Crystallization and Gelation Behavior of Monoglyceride Gels. *Journal of Colloid and Interface Science* 249: 412–22.

Staggers, J. E., O. Hernell, R. J. Stafford, and M. C. Carey. 1990. Physical-Chemical Behavior of Dietary and Biliary Lipids during Intestinal Digestion and Absorption. 1. Phase-Behavior and Aggregation States of Model Lipid Systems Patterned after Aqueous Duodenal Contents of Healthy Adult Human-Beings. *Biochemistry (Moscow Russ Fed)* 29: 2028.

Tadros, T. 2004. Application of Rheology for Assessment and Prediction of Long Term Physical Stability of Emulsions. *Advances in Colloid and Interface Science* 107–08: 227–58.

Ubbink, J., A. Burbidge, and R. Mezzenga. 2008. Food Structure and Functionality: A Soft Matter Perspective. *Soft Matter* 4: 1569–81.

Waigh, T. A. 2005. Microrheology of Complex Fluids. *Reports on Progress in Physics* 68:685–742.

Yodh, A. G., K. H. Lin, J. C. Crocker, A. D. Dinsmore, R. Verma, and P. D. Kaplan. 2001. Entropically Driven Self-Assembly and Interaction in Suspension Philosophical Transactions of the Royal Society of London Series A. *Mathematical Physical and Engineering Sciences* 359: 921.

CHAPTER 10

Scaling Up and Processing Nutraceutical Dispersions

Emulsions and Suspensions

Jayant Lokhande and Yashwant Pathak

CONTENTS

10.1 DEFINITIONS OF NUTRACEUTICAL DISPERSION FORMULATIONS

Nutraceutical dispersions can be formulated as either nutraceutical emulsions or suspensions. *Nutraceutical emulsions* are the active ingredients dispersed in oil phase, or oil phase itself can be a nutraceutical ingredient, such as castor oil emulsion. The oil phase is further dispersed as globules in the water phase, leading to an emulsion with two immiscible phases in one system. Most of the nutraceutical emulsions are stabilized using an emulsifying agent or a combination of emulsifying agents and interphase stabilizers or surface active agents. Most of the nutraceutical emulsions are turbid milky liquid preparations containing several ingredients, besides the active nutraceutical ingredient to offer better stability to the product. Examples of nutraceutical emulsions available in the US market are dietary omega-3 fatty acid emulsions, dietary omega-3 fatty acids (http://www .croda.com) as functional foods in oil–water emulsions, microencapsulated oil–water emulsion systems, and emulsion droplet engineering to stabilize omega-3 fatty acids. Other examples of emulsified food products are mayonnaise, milk, sauces, and salad dressings (Chen, Weiss, and Shahidi 2006). Nutraceutical dispersions with herbal and mineral ingredients of natural origin pose different kinds of challenges to the formulator.

Nutraceutical emulsions appear as milky-white, opaque solutions when droplets are larger than 0.3 μm because they scatter the incident visible light. Transparent solutions known as *microemulsions* occur when droplets are smaller than 0.1 μm in diameter and are too small to scatter any component of visible light. Emulsion stability is affected by gravitational separation, flocculation, coalescence, and phase separation. Gravitational separation can be minimized by reducing density difference between the oil and water, decreasing droplet size, and increasing continuous-phase viscosity. Flocculation occurs when colloid interaction is out of balance. Droplets can attract each other through van der Waals forces or hydrophobic bonds, which cause depletion or bridging of droplets. Droplets can also be repulsed by electrostatic or steric forces, or hydration. Droplet coalescence happens with aggregation as a result of the fusing together of two or more individual droplets to form a bigger droplet, eventually leading to oiling off. Ostwald ripening refers to the growth of large droplets at the expense of small droplets caused by molecular diffusion of the oil molecules through the aqueous phase driven by a difference in Laplace pressure.

Functional ingredients in beverage emulsions include stabilizers, emulsifiers, wetting agents, and viscosity builders. A modified gum acacia, which is formed by the esterification of octenyl succinic anhydride to the polysaccharide portion of the acacia molecule, reduces surface tension. In flavor emulsion formulation, preparation of the oil phase includes dissolution of the wetting agent in the oil (flavored oil or vegetable oil) to achieve the desired density. This generally takes approximately 1 hour. Preparation of the aqueous phase involves dissolution and hydration of the gum (hydrocolloid) in warm water under medium agitation, avoiding foam, for 4–8 hours. A coarse emulsion formulation requires the addition of the oil phase to the aqueous phase under agitation, followed by shear mixing for 10–15 minutes. Fine emulsion preparation can be achieved via twice homogenizing the coarse emulsion at 3500/500 psi (Mannie 2009).

Wheat flour–lipid and waxy maize starch–lipid composites in emulsion forms have been shown to replace shortening in fortified bread manufacturing. Nutraceutical emulsions are used to combine proteins and lipids to improve film barrier properties in the products. Carbohydrate–lipid composites in an emulsion form optimize tenderness and juiciness of low-fat meat products used as functional foods in some cases. Structured lipids and oils have been replaced by nutritional emulsion systems in functional beverages. Fortified phospholipids in an emulsion form have interesting nutraceutical and functional applications. Pseudoplasticity and elasticity help stabilize citrus drink nutraceutical emulsions; caseinate yields stable oil–water emulsions for many nutraceuticals; and natural proteins inhibit lipid oxidation in emulsions used as nutraceuticals. Mesquite gum, a natural emulsifying agent, improves stability of oil–water nutraceutical emulsions. Carrageenan–wheat emulsifier may partially replace egg yolk in mayonnaise when developing products for people who are egg sensitive. Emulsifying agents affect aroma release in oil–water emulsions, and many optimized carbohydrate systems are useful to encapsulate flavors and stabilize emulsions. Two-phase processing eliminates emulsifiers from ice cream and esterified, subsequently hydrogenated, phytosterol composition for use alone or for incorporation into foods, beverages, pharmaceuticals, nutraceuticals, and the like. The composition has the advantage of enhanced solubility and dispersibility, increased molar potency, and enhanced stability over naturally isolated phytosterol compositions. Methods for the esterification and subsequent hydrogenation of the phytosterols are also provided (Stewart et al. 2000).

10.2 NUTRACEUTICAL SUSPENSIONS

The water-insoluble nutraceutical ingredient is dispersed in water to form a uniform dispersion with the help of suspending agents and a combination of surface active agents that offer stability to the product and keep the water-insoluble ingredients uniformly dispersed in the system. Examples of the nutraceutical suspensions available in the US market are calcium suspensions and nutraceutical supplements for human and veterinary applications, Acigalcane O-Suspension containing Magaldrate, Simethicone, and Oxetacaine (marketed by Shrey's Nutraceuticals & Herbals, New Delhi, India), sucralfate suspensions, β-carotene suspensions in oil, lutein 5% suspensions, lycopene suspensions, and zeaxanthin suspensions. Chen, Weiss, and Shahidi (2006) have recommended incorporation of bioactive compounds—such as vitamins, probiotics, bioactive peptides, and antioxidants—into food systems to provide a simple way to develop novel functional foods that may have physiological benefits or reduce the risks of diseases. As a vital macronutrient in food, proteins possess unique functional properties, including their ability to form gels and emulsions, which allow them to be an ideal material for the encapsulation of bioactive compounds. Based on the knowledge of protein physical–chemistry properties, this review describes the potential role of food proteins as substrates for the development of nutraceutical delivery systems in the form of hydrogels and micro- or nanoparticles. Applications of these food protein matrices to protect delivery-sensitive nutraceutical compounds are illustrated, and the impacts of particle size on release properties are emphasized (Chen, Remondetto, and Subirade 2006). Another interesting patent reported a delivery system for nutraceuticals that uses a low-caloric chocolate base containing one or more nutraceuticals, either blended with the chocolate itself or added as a liquid or cream filling. The chocolate has a relatively high level of oligomeric proanthocyanidins, and preferably also includes a phytosterol and docosahexaenoic acid. It is sweetened with a blend containing tagatose and a secondary low-caloric, high-intensity sweetener, preferably Lo Han Guo extract. Using the inventive system, delivery of nutraceuticals in unit dosage form is facilitated, as the selected dose is carried within individual chocolate product pieces that taste substantially the same as conventional chocolate, although with few calories from carbohydrates or the effects on insulin response encountered with typical chocolate formulations (Mckee and Karwic 2009).

A novel approach has been reported by Semo et al. (2007) using casein micelles (CMs), which are in effect nanocapsules created by nature to deliver nutrients, such as calcium, phosphate, and protein, to the neonate. A novel approach is herein presented to harness CM for nanoencapsulation and stabilization of hydrophobic nutraceutical substances for enrichment of nonfat or low-fat food products. Such nanocapsules may be incorporated in dairy products without modifying their sensory properties. This study introduces new possibilities for encapsulation and delivery of sensitive health-promoting substances using generally regarded as safe natural ingredients. As a model hydrophobic nutraceutical compound, Semo et al. (2007) studied the fat-soluble vitamin D2, which is essential for calcium metabolism. A protocol for incorporation of vitamin D2 into CM was established. The encapsulation process was evaluated, and so was the protective effect of the micelles against photochemical degradation of the vitamin. Semo et al. (2007) have demonstrated, for the first time, the possibility to load a nutraceutical compound into CM, using the natural self-assembly tendency of bovine caseins. The vitamin was approximately 5.5 times more concentrated within the micelles than in the serum, where it was only present bound to residual soluble caseins. Moreover, the morphology and average diameter of the reassembled micelles were similar to those of naturally occurring CMs. Semo et al. (2007) have also demonstrated that the reassembled CMs can provide partial protection against UV light-induced degradation to vitamin D2 contained within them. This study suggests that CM may be useful as nanovehicles for entrapment, protection, and delivery of sensitive hydrophobic nutraceuticals within food products.

10.3 NUTRACEUTICAL DISPERSION FORMULATION: POSSIBLE CHALLENGES

The development of new functional food and nutraceutical ingredients and dosage forms is a contribution of the scientific community toward providing treatment for disease and discomfort suffered by the population at large. Most nutraceutical ingredients are poorly water soluble with discouraging in vitro properties. Few studies have been reported on their in vivo evaluations, of which the results are also disappointing. Possible reasons are:

- Poor solubility of nutraceutical ingredients
- Poor absorption, rapid degradation, and lamination (some natural peptides and protein products) resulting in insufficient concentration
- Nutraceutical ingredient distribution to other tissues where the accumulation of the nutraceutical ingredient occurs and the ingredient is not available for therapeutic action
- Fluctuations in plasma levels owing to unpredictable bioavailability (not much research has been done in this area for nutraceutical products)

The enhancement of oral bioavailability of poorly soluble drugs as well as functional food and nutraceutical ingredients remains one of the most challenging aspects of nutraceutical development (Karanth, Shenoy, and Murthy 2006). Formulation of these drugs and ingredients in dispersions such as emulsions or suspensions with desired release profiles is a solution (Caballero and Duran 2000). The coarser particles range from 10–100 μm. Anything less than 10 μm is considered colloidal dispersion and is out of the purview of this discussion. Dispersions are thermodynamically unstable systems even though the ingredients might be chemically stable (Lu 2007). As the solid particle size decreases in the suspension, the interfacial energy increases, which makes the whole system unstable. This phenomenon further initiates processes of aggregation and particle sedimentation (Nash 1996). Some of the suspension properties—interfacial area associated with suspended particles, polymorphic forms, and growth of large crystals resulting from Oswald ripening—affect physical stability and the bioavailability of suspensions (Lu 2007).

Consumer acceptance and organoleptic properties of nutraceutical dispersions are receiving more attention because these products are developed for pediatric and geriatric medicines (Gallardo, Ruiz, and Delgado 2000; Sohi et al. 2004). Currently the fast developing area in dispersion technology is nanosuspensions, in which the particle size of the dispersed phase is in the nanometer range (Rabinow 2005, 2007; Gruverman 2003, 2004, 2005). Significant process work would be needed to develop and scale up the nanosuspension of nutraceuticals.

To understand the dispersion system technology from the commercial manufacturing perspective with a focus on scale-up and technical transfer from bench scale (1 L) to large scale (up to 5000 L) is critical (Table 10.1). Manufacturing and quality control of nutraceutical oral emulsions and suspensions have presented a dilemma to the industry. There are issues that have led to recalls (Food and Drug Administration [FDA] Document 2006). These include microbiological, potency, and stability problems. As liquid products, emulsions and suspensions are prescribed to pediatric and geriatric patients; therefore, defective dosage forms can pose greater safety risks (Caballero and Duran 2000).

Dispersion stability is the area that has presented a number of problems. For example, there have been a number of recalls of the vitamins with fluoride oral liquid products because of vitamin degradation. Nutraceutical dispersions, including emulsions

Table 10.1 Production Scale-Up for Nutraceutical Dispersions

Serial Number	Stages of Scale-Up and Production	Quantity in Liters
1	Laboratory scale	1–5
2	Phase I study	10–50
3	Phase II study	500–1000
4	Phase III study	500–2000
5	Validation/production batches	2000–5000

and suspensions, are liquid products and tend to interact with closure systems during stability studies; therefore, studies should be designed to determine whether contact of the product with the closure system affects product integrity (FDA Document 1999). Moisture loss in nutraceutical dispersions (which can cause the levels of the nutraceutical ingredients grow per dose level) and microbiological contamination are other problems associated with inadequate closure systems and low density polyethylene plastic containers. Scaling up and manufacturing of nutraceutical dispersions for topical or oral use are challenging (Gallardo, Ruiz, and Delgado 2000).

10.4 NUTRACEUTICAL DISPERSIONS PROCESS DEVELOPMENT: KEY CONCEPTS

The process development of nutraceutical dispersions involves integration of formulator and chemical engineer skills to produce an elegant product with specified properties similar to those of pharmaceutical suspensions (Block 2002; Wibowo and Ng 2004). The scale-up process should determine the operating conditions applicable to large-scale production batches with the goal of obtaining products of the same quality developed on previously optimized laboratory-scale experiments. The scale-up process of a nutraceutical dispersion, which consists of two immiscible phases (solids and liquids for suspensions, and oil and water for emulsions), is a difficult process because of the dynamic behavior of two individual phases. For nutraceutical suspensions in early product development, generally operational variables of powder properties (to be suspended in the liquid phase) are not well established, whereas for emulsions it is comparatively easy because most oil–water dispersion data are readily available.

It is recommended that dimensional analysis, mathematical modeling, and computer simulation (the latter two are less used in nutraceutical dispersion manufacturing) should be performed (Levin 2005). Dimensional analysis (Zlokarnik 2002) is an algebraic treatment of variables affecting a process. The analysis does not result in a numerical equation, but rather the experimental data are fitted to an empirical process equation that results in scale-up (Block 2002). It produces dimensionless numbers and deriving functional relationships among them that completely characterize the process. The analysis can be applied even when the equations governing the process are not known (Pathak and Thassu 2009). Little work has been reported in this area of application in the nutraceutical industry.

Previous development experience is helpful in handling the numerous problems encountered during scale-up in formulation development (Newton et al. 1995). Most work in the field of dispersing small insoluble particles or small oil globules (for emulsions) still depends on the trial and error method and the principles of geometric similarities. The latter describes the interrelationships among system properties on scale-up; thus, the ratio of some variables in small-scale equipment should be equal to that of similar variables in equivalent large-scale equipment (Mehta 1988). However, this rule may not be applied to all dispersions. It is observed that the ratio of solids to liquids in suspension or the ratio of oil phase to water will make the scaling up difficult. As the product viscosity increases, the scaling-up process will need different parameters and different techniques to achieve the desired dispersion product. The next approach, product-oriented process synthesis and development (Wibowo and Ng 2004), involves four steps of process development:

1. Identification of product quality factors: This covers various factors affecting the properties of the dispersion product, such as particle size distribution of the particles; density of the particles, suspending agents, or emulsifying agents; and desired viscosity of the product leading to adequate stability over time.
2. Product formulations: To get a stable formulation and palatable product, various adjuvants must be incorporated in the product formulation, including flavors, colors, suspending or emulsifying agents, and viscosity builders. The properties and characteristics will decide equipment selection and process development.
3. Flow sheet synthesis: The process involves the transfer of materials in a sequential pattern depending on the unit operations involved in the manufacturing of the dispersions; hence, flow sheet describes the flow pattern of the material and equipment used.
4. Product and process evaluation: In process and product controls, parameters are used to ensure the appropriate desired and stable product (Pathak and Thassu 2009).

Specifications to manufacture nutraceutical dispersions can include assay, microbial limits, particle size distribution of the dispersed phase (either solids or oils), viscosity, pH, and dissolution of at least one ingredient in the nutraceutical dispersion formulation. For nutraceutical formulations, one hurdle faced by the formulators is the presence of several active ingredients in different chemical forms as a result of their natural origin. Viscosity can be important from a processing aspect to minimize segregation (Table 10.2). In a few cases, viscosity has been shown to have an impact on bioequivalency (Derkach 2009). An interesting study reported (Logaraj et al. 2008) the rheological properties of emulsions made out of avocado pulp and watermelon seed oils with whey protein concentrate were determined during different storage periods. The oils, as well as the emulsions, behaved like non-Newtonian liquids with shear-thinning characteristics. Both oils showed moderate shear-thinning characteristics as the flow behavior indices were between 0.86 and 0.88. The shear-rate and shear-stress data could be adequately fitted (r = 0.997–0.999) to a common rheological equation, for example, the power law model. Avocado pulp oil

Table 10.2 Test Methods and Checklist for Nutraceutical Dispersion Performance and Stability Testing

Serial Number	Test Methods for Nutraceutical Dispersions	Checklist for Performance and Stability Testing
1	Visual	Uniform appearance
2	Photomicroscopic techniques	Uniform distribution of dispersed phase
3	Coulter counter/laser diffraction methods	Particle size distribution, sedimentation volume, and redispersibility
4	Graduated cylinders for sedimentation studies	Sedimentation rate of the dispersed phase
5	Brookfield viscometers	Viscosity
6	Specific gravity measurements/pH meters	pH, density, and specific gravity
7	Aging tests	Organoleptic properties such as color, taste, and odor
8	Zeta potential	Drug content uniformity/release profiles
9	Aggregation kinetics	Freeze–thaw cycling for stability studies

was markedly more viscous than watermelon seed oil, which was also evident from the higher apparent viscosity and consistency index values.

The rheological parameters during storage did not significantly change flow parameters, for example, flow behavior and consistency indices and apparent viscosity. The values of different droplet sizes and their distribution patterns, as evident from phase contrast microscopy, were considered almost unimodal (i.e., between 2 and 10 μm). Such a narrow range of variations in particle diameters may not markedly influence particle behavior. Cream and serum separation of emulsions were not noticed during storage, indicating that the stability of these two emulsions was not affected during storage for up to 2 months at room temperature.

Another interesting study (Arboleya, Ridout, and Wilde 2008) showed the interactions between emulsion droplets containing solid fat are important for the rheology and functionality of the emulsion as a whole, particularly for aerated emulsion systems where partial coalescence plays a role in the overall structure of the product. In this study, the interactions between emulsion droplets appeared to be sensitive to the relative amounts of solid fat and liquid oil, thus changing the rheology of the whole system. Incorporation of air had a major effect on these interactions as it appeared to force the emulsion to adopt a stronger structure by encouraging partial coalescence. The rheological behavior of a nonaerated emulsion and an aerated emulsion was compared. Nonaerated samples did not show major changes in viscosity with increasing temperature. In contrast, the aerated emulsion seemed to be considerably more temperature sensitive, showing a dramatic increase of viscosity as the temperature was increased above a critical value. The effect of temperature ramp rates was investigated. Higher temperature ramp rates resulted in delayed changes in viscosity. The phase behavior of the fat is both time and temperature dependent; therefore, a faster temperature ramp means that a higher temperature could be reached before critical phase changes in the fat could take place. The rheological behavior of the emulsions was also dependent on the shear rate applied during the experiment. The pH of nutraceutical dispersion formulation affects the effectiveness of preservative systems and may even alter the efficacy of the product in solution.

Particle size distribution is a critical attribute and should be monitored as a quality control parameter for the nutraceutical dispersions. Any change in particle size distribution during the shelf life of nutraceutical dispersion can have a negative impact on the dissolution kinetics and the stability of the product. Particle size, its distribution, and change over shelf life are also critical for extended-release nutraceutical dispersions. Particle size would be a critical parameter for micro- and nano-nutraceutical dispersions.

10.5 PROCESS DEVELOPMENT OF NUTRACEUTICAL DISPERSIONS: IMPORTANT FACTORS TO CONSIDER

Factors that need to be considered while developing and scaling up a manufacturing process for nutraceutical dispersions are:

1. Increase productivity and selectivity through intensification of intelligent operations and multiscale approach to process control.
2. Design novel equipment based on scientific principles and new production methods: process intensification (PI).
3. Extend chemical engineering methodology to product design and product-focused processing using three Ps: (molecular) processes, (nutraceutical dispersion) products, and (manufacturing) processes.
4. Implement multiscale application of computational chemical engineering modeling and simulation to real-life situations from the molecular scale to manufacturing scale (Charpentier 2003).

The major unit operation involved in nutraceutical dispersion manufacturing is the mixing of solids or suspending the solids in liquid phase in the case of suspensions while dispersing the oil phase in water phase in the case of nutraceutical emulsions. Mixing requirements are dictated by the nature of the compounds being mixed (Joseph 2005). Mostly these are structurally and functionally complex compounds, and there is a need to satisfy the controls imposed by the regulatory bodies on nutraceutical products (Paul, Midler, and Sun 2004). Many of the nutraceutical ingredients are natural herbal products, and special care needs to be taken while dispersing such products in nutraceutical dispersion. The majority of mixing operations are carried out in batch or semicontinuous modes rather than continuous mixing modes. Often multiple tasks are carried out in the same vessel (e.g., chemical reaction, heat and mass transfer, concentration, and crystallization), requiring a fine balance in design and operation. Poor mixing can result in loss of product yield and purity, excessive crystalline fines, or oiled out material and/or intractable foam.

One needs to be careful when selecting micro- and macromixing requirements, particularly for dispersion manufacturing where competitive chemical reactions and/or crystallization are involved. Glass-lined mixing equipment is often used in nutraceutical dispersion (suspension and emulsions) manufacturing, especially in chemically unstable or reactive preparations.

An interesting article discusses in detail the mixing phenomena and principles involved in selecting the mixing equipment, mixing of viscous and non-Newtonian fluids, and transport phenomena involved in suspension and semisolid production by Block (2002) in his chapter on nonparenteral liquids and semisolids in pharmaceutical process scale-up. Dispersion systems always need particle size reduction, which is mostly achieved by compression, impact, attrition, or cutting or shear. Equipment for particle size reduction includes crushers, grinders, ultrafine grinders, and knife cutters. In dispersion production, particle size reduction can be an integral part of the processing, or one can reduce the particle size of the particles or globules and then suspend in the vehicle. Block (2002) has given in detail the mathematical equations for particle size reduction processing for suspensions and emulsions.

Freeze drying is another important unit operation involved in the manufacturing of the suspensions and emulsions to provide adequate stability to nutraceutical products. There are not many nutraceutical products in which this technique is used. However, it is widely used in pharmaceutical dispersion production (Pathak and Thassu 2009). The purpose of freeze drying is to remove a solvent (usually water) from dissolved or dispersed solids. Freeze drying involves four stages: freezing of the product, sublimation of the solvent under vacuum, applying heat to accelerate the sublimation, and finally condensation to complete the separation process. Although it appears to be simple, freeze drying can often be complicated in practice because of different properties of the product. The major factors affecting the efficiency of freeze-drying processing are:

1. Suspending solvents
2. Rate of freezing
3. Vacuum control
4. Rate of drying and product temperature
5. Residual moisture content
6. Storage temperatures and atmosphere after drying
7. Rehydration process

Freeze drying provides dry powders for resuspension, offering better stability of the product. The key benefits include retention of morphological, biochemical, and immunological properties; high viability and activity levels of bioactive materials; lower temperature; and oxygen and shear conditions used in the processing, offering better stability for the drug, better retention of structure, better surface area and stoichiometric ratios, high yield, long shelf life, and reduced storage weight, easy for shipping and handling (Pathak and Thassu 2009).

Several patents discuss freeze-drying processing and applying identification marks on the freeze-dried products (Thompson, Yarwood, and Kearney 1995). These inventions comprise various techniques used to emboss the freeze products with an identifying mark, such as a manufacturer's logo, medicinal component strength, or other information related to the product. The desired identification mark is first embossed on the base of the container such as a blister packet, and then the liquid suspension is filled in the container and freeze-dried therein. These techniques are widely used for parenteral suspensions and dry powders to be resuspended as liquid after the addition of water or a suitable solvent. Liu (2006) has provided an excellent overview of the freeze-drying applications. We have few nutraceutical dispersion formulations for parenteral applications. As our knowledge grows, we may also need freeze drying for nutraceutical dispersions.

10.6 PROCESS EQUIPMENT DESIGN, RATING, AND SELECTION

Laboratory testing is the first step in mixing scale-up, primarily because waiting until the last minute for scaling up will be too late. Proper notation of the observations during the laboratory experiments will help significantly during the scaling up. Equipment used for batch mixing of oral solutions, emulsions, and suspensions is relatively basic. Generally these products are formulated on a weight (weight/weight) basis with the batching tank on load cells, so that a final Q.S. can be made by weight. Volumetric (volume/volume) means, such as using a dipstick or line on a tank, have been found to be inaccurate. Equipment should be of a sanitary design. This includes sanitary pumps, valves, flowmeters, and other equipment that can be easily sanitized. Ball valves, packing in pumps, and pockets in flowmeters have been identified as sources of contamination. Mostly steel equipment and connections can prevent much contamination in the products (Pathak and Thassu 2009).

Facilitating the cleaning and sanitization, manufacturing, and filling lines should be identified and detailed in drawings and standard operating procedures (SOPs). In some cases, long delivery lines between manufacturing areas and filling areas have been a source of contamination. Also SOPs, particularly with regard to time limitations between batches and for cleaning, have been found deficient in many manufacturers. Review of cleaning SOPs, including drawings and validation data with regard to cleaning and sanitization, is important.

The design of the tank used for batch production with discharge opening at the bottom has presented fewer problems compared with having the discharge from the top of the tank. However, for thick dispersions sometimes the bottom discharge may also face problems of uneven dispersion as a result of settling of the larger particles. Continuous stirring may help in reducing the problem of uneven dispersions. Ideally, the bottom discharge valve is flush with the bottom of the tank. In some cases, valves, including undesirable ball valves, have been found to be several inches to a foot below the bottom of the tank. In others, natural drug or preservative was not completely dissolved and was lying in the "dead leg" below the tank, with initial samples found to be subpotent. For the manufacture of suspensions and dispersions, valves should be flushed. The batch equipment and transfer lines need to be reviewed and observed constantly.

Transfer lines generally need to be hard piped and can be easily cleaned and sanitized. In some cases, manufacturers have also used flexible hoses to transfer product. It is not unusual to see flexible hoses lying on the floor, thus significantly increasing the potential for both chemical and microbial contamination in the products. Such contamination can occur by operators picking up or handling hoses, and possibly even placing them in transfer or batching tanks after they have been lying on the floor.

It is also a good practice to store hoses in a way that allows them to drain rather than be coiled, which may allow moisture to collect and be a potential source of microbial contamination. Manufacturing areas and operator practices, particularly when flexible hose connection is used, need to be observed carefully. Another common problem occurs when manifold or common connections are used, especially in water supply and premix or raw material supply tanks. Such common connections have been shown to be a source of contamination. Optimization of processes is ensured through multidisciplinary implementation procedures that match mixing and fluid dynamics to the process chemistry and mass transfer requirements (Pathak and Thassu 2009).

Because scale-up is critical to the successful application of mixing technology in industrial processes, a range of geometrically similar reactors 0.2–2.7 m in diameter can be used to establish correct scaling procedures. Novel sensors can be used and developed for this purpose. Computational modeling techniques are used to rationalize experimental programs, describe the fluid dynamics of reactors, and extend to a range of industrially relevant applications.

10.7 NUTRACEUTICAL DISPERSION PROCESSING: BATCH VERSUS CONTINUOUS OPERATIONS

The nutraceutical industry is currently undergoing a radical change. The need to improve process and product quality while controlling overall costs is growing. Continuous processing offers a chance to achieve these goals. The implementation of process analytical technology (PAT) tools may further aid in support of the continuous operations because the tools give real-time indication and assurance of the product quality throughout the process. A variety of batch processes also are currently being evaluated for continuous production in dispersion technology. Presently the batch process is mostly used for the manufacturing of nutraceutical dispersions; it is hoped that within a few years we also may see some continuous manufacturing.

Manufacturers normally assay samples of the bulk solution or suspension before filling. A much greater variability has been found with batches that have been manufactured volumetrically rather than by weight. For example, one manufacturer had to adjust approximately 8% of the batches manufactured after the final Q.S. because of failure to comply with potency specifications. Unfortunately, the manufacturer relied solely on the bulk assay. After readjustment of the potency based on the assay, batches occasionally were found out of specification because of analytical errors. Table 10.2 gives the list of test methods for emulsions and suspensions, as well as the checklist for emulsion and suspension performance and stability testing (Nash 1996).

10.8 TECHNOLOGY SOURCE AND PROCESS INTENSIFICATION

The concept of PI was originally pioneered by Colin Ramshaw and his coworkers at ICI (London, UK). They defined PI as a reduction in plant size by at least a factor of 100. PI is a new approach to process and plant design, development, and implementation that provides a chemical process like mixing with the precise environment it needs to flourish in better products and processes, which are cleaner, smaller, and cheaper (http://www.bhrgroup.com).

Some of the possibilities of PI that can be adapted to dispersion manufacturing are:

1. Converting a batch process into a continuous process
2. Using intensive reactor technologies with high mixing and heat transfer rates (e.g., flex reactors, heat exchanger reactors; http://www.bhrgroup.com) in place of conventional stirred mixing tanks
3. Adopting the multidisciplinary approach when considering opportunities to improve the process technology while underlying the suspension chemistry of the process
4. Plug-and-play process technology to provide flexibility in a multiproduct environment

Some of the established advantages of PI are reduction in reactor volume energy usage, operating costs, and higher yield. Little work has been done and reported applying the PI techniques in dispersion manufacturing; however, because of the competitive prices this approach may also have some applications in the pharmaceutical industry (Pathak and Thassu 2009).

10.9 PROCESSING OF NUTRACEUTICAL DISPERSIONS PROCESS VALIDATION

The FDA (1987) defines *validation* as "written procedures for production and process control designed to assure that the drug products have the identity, strength, quality and purity they purport or represented to possess" (p. 6). The FDA (1987) further states that validation establishes "documented evidence which provides a high degree of assurance that a specific process will consistently produce a product meeting its predetermined specification and quality attributes" (p. 3). The process validation package should provide:

1. Installation qualification verifying that the equipment is installed in a proper manner
2. Operational qualification verifying the performance of equipment for the intended range of applications
3. Process qualification verifying the repeatability and consistency of producing the drug product

Validation is a continuous and evolving process and needs to be given top priority during the production cycle and continuously updated. Maintenance, periodic calibrations, and adjustment of the equipment are continuously done and are always under scrutiny and constant evaluation. There are some maintenance triggers such as:

1. Loss of product quality
2. Replacement of key components in the system
3. Upgrades of software or hardware
4. Change in personnel

These often arise during manufacturing, and there is always a need to address these triggers, preferably with an SOP in place to identify the triggers and solutions.

The FDA guidelines suggest having functional requirements and functional specifications. Functional requirements refer to documents defining what is to be done; the process should be stated succinctly with little or no reference to actual specific devices. The functional specification document usually contains (http://www.fda.gov):

1. An introduction or purpose
2. Intended audience
3. Definitions
4. Reference documentation
5. Mechanical, electrical, computer, and security requirements and specifications

Based on process parameters, the ranges and tolerances of the equipment operation are defined in the functional specifications. There are three important aspects related to the process involved:

1. Identifying the process parameters
2. Identifying the critical process parameters
3. Identifying the tolerances

As with other products, the amount of data needed to support the manufacturing process will vary from product to product. Development (data) should have identified critical phases of the operation, including the predetermined specifications that should be monitored during process validation. For example, for solutions the key aspects that should be addressed during validation include assurance that the drug substance and preservatives are dissolved. Parameters, such as heat and time, should be measured. Also, in-process assay of the bulk solution during and/or after compounding according to predetermined limits is also an important aspect of process validation. For solutions that are sensitive to oxygen and/or light, dissolved oxygen levels would also be an important test. Again, the development data and the protocol should provide limits (Pathak and Thassu 2009).

As discussed, the manufacture of nutraceutical dispersions presents additional problems, particularly in the area of uniformity. Again, development data can address the key compounding and filling steps that ensure uniformity. The protocol should provide for the key in-process and finished product tests, along with their specifications.

10.10 NUTRACEUTICAL DISPERSION COMMERCIAL PRODUCTION: TECHNOLOGY TRANSFER DOCUMENTATION PACKAGE

Technology transfer can be defined as the systemic procedural documentation that results in transfer of the documented knowledge and experience gained during the development and scaling-up process for commercialization of the process leading to manufacturing of the final products. It will always involve a minimum of two stations: the sending (transferring) and the receiving station. Technology transfer will include:

1. Documentation transfer
2. Demonstration of the ability of the transferring station to educate the receiving station properly to perform desired task
3. Demonstration of the ability of the receiving station to duplicate the task performed at desired efficiency
4. Establishment of the evaluation process to measure the success
5. Process at each station should satisfy all concerned parties (departments) and also regulatory bodies.

Technology transfer for suspension manufacturing can happen from laboratory to process development to manufacturing. The manufacturing can involve one location or several locations within the country as well as outsourcing to multiple countries. The requirements for different countries vary, and if the product is sold in the United States and other countries, the local FDA requirements need to be met; hence the technology transfer package will need to take into account all the multiple country FDA requirements (http://www.fda.gov).

10.11 NUTRACEUTICAL DISPERSION MANUFACTURING AND PROCESS ANALYTICAL TECHNOLOGY

The FDA's PAT initiative encourages the pharmaceutical industry to improve process efficiency by developing a thorough understanding of the manufacturing process and by using this knowledge to monitor what are identified as critical quality parameters. These also will be gradually applied to the nutraceutical industry in the near future. The FDA defines PAT as a system for designing, analyzing, and controlling manufacturing through timely measurements of quality and performance attributes of raw and in-process materials and processes with the goal of ensuring the final product quality (http://www.fda.gov). The term analytical PAT is used to describe broadly the chemical, physical, microbiological, mathematical, and risk analyses conducted in an integrated manner. The goal of PAT is to enhance understanding and control the manufacturing process, which is consistent with our current drug quality system. PAT believes that quality cannot be tested into a product; it should be built in or should be by design. There are four dimensions of PAT. When each of these is brought together to optimize the quality, safety, and cost, the real-time process of understanding the process for better risk management can be enhanced. Implementation of PAT in emulsion and suspension development and manufacturing will have a significant effect on the efficiency and quality of production.

Some advantages of implementing PAT are:

1. Flexibility of manufacturing configuration
2. Asset utilization through more effective campaigning of batches
3. Reduction in paperwork
4. Greater productivity and improved quality
5. Lower regulatory scrutiny

Currently batches are processed according to established recipes and slow and wet chemistry-based quality assurance and quality control tests performed at specified points to determine the success of manufacturing steps (Table 10.3). To develop the PAT initiative, an initial investment of both equipment and time is required, but

Table 10.3 Real-Time Monitoring of Nutraceutical Dispersion Manufacturing Processes

Category	Raw Material	Mixing	Formulation, Fill, and Finish
Purpose	Control product quality; ensure product safety	Wet chemistry analysis, hard in case of nutraceutical products because of a large number of ingredients mostly of herbal origin	Particle size distribution, density, pH, viscosity, sedimentation volume, assay, dissolution studies, freeze–thaw cycle, stability studies over time at room temperature and accelerated stability studies
Current monitoring capabilities	Mostly with HPLC and mass analysis for individual ingredients; challenging for nutraceutical products because of large number of ingredients in the products mostly of herbal origin	Use HPLC and other analytical techniques, needs more studies for nutraceutical dispersions as the method development is challenging	Based on individual testing for each parameter
Future and desired monitoring capabilities	Increasing speed of the analytical techniques and using one technique for multiple ingredients to save time and resources	Apply PAT to monitor the mixing efficiency and control the particle size distribution, presently not applied in nutraceutical dispersion manufacturing	In-process PAT application and techniques that can simultaneously determine multiple evaluation parameters

HPLC, high-performance liquid chromatography; PAT, process analytical technology.

at the end the rewards may lead to an efficient, intelligently controlled process, faster time to market, and a significant reduction in waste (Lewis et al. 2005; Kidder et al. 2007). Typical steps one may encounter in a risk assessment process incorporating PAT assessment are (Sekulic 2007):

1. Create a flow chart.
2. Identify the quantity attributes and how these are measured.
3. Identify and prioritize process parameters.
4. Experiment as needed to understand the parameters of evaluation.
5. Perform risk assessment: prioritize experiments.
6. Perform PAT decision analysis.

Several techniques are also proposed for PAT in pharmaceutical dispersion manufacturing that have direct applications in nutraceutical dispersions, but focused beam reflectance measurement (FBRM) is becoming a popular procedure for measuring the particle size in multiphase systems (Blanco et al. 2002; Heath et al. 2002; Laitinen et al. 2003; Kougoulas et al. 2005; Lasinski et al. 2005; Barthe and Rousseau 2006; Saarimaa, Sundberg, and Holmbom 2006; Vaccaro, Sercik, and Morbidelli 2006). This laser probe offered by Mettler Toledo, Inc. (Columbus, OH) provides a precise, sensitive method for tracking changes in both particle dimension and particle populations in suspension. Part of its popularity is because of the simplicity of the method. A laser is rotated in the multiphase flow, and the beam intersects a particle; some of the light is backscattered. The total amount of time the detector experiences backscatter multiplied by the rotation speed of the laser yields a measure of the length of the intersected particles, denoted as a chord of the particle. After numerous intersections, a chord length distribution (CLD) can be generated. This CLD will be strongly dependent on the shape of the particle, with different distributions generated from spheres, ellipsoids, cubes, and other shapes. Altering the shape of the particle will also change the way CLD is converted into a volume distribution, also known as *particle size distribution* (PSD). Modeling techniques such as the population balance method can be used to predict how PSDs evolve in time as a result of breakage, agglomeration, crystal growth, or nucleation of particles, and measurements are necessary to validate those models. Vaccaro, Sercik, and Morbidelli (2006) have described particle mass distribution (PMD) as another parameter while using FBRM for suspensions. They studied the shape-dependent convolution relationships between CLD, PSD, and PMD, and based on these convolution relationships, equations relating moments of the CLD, PSD, and PMD are obtained. Issues related to the definition of particle size of nonspherical objects and its connections are established. Based on the moment relationships, they defined particle size for FBRM in terms of CLD-equivalent spheres for (nonspherical) objects and showed the usefulness of this technique for particles with different shapes (Pathak and Thassu 2009).

An interesting application of FBRM that can be used in suspensions is reported by Saarimaa, Sundberg, and Holmbom (2006). They used the FBRM measurement to continuously assess the aggregation and removal of dissolved and colloidal substances during batchwise dissolved air flotation of process water. They reported the

FBRM results were in agreement with turbidity measurements and chemical analysis by chromatography of pectic acid content. This technique also may be adopted for suspensions in process control.

Kougoulos et al. (2005) have reported the use of FBRM in conjunction with process video imaging in a modified mixed suspension mixed product removal (MSMPR) cooling crystallizer. They used this technique to monitor the steady-state operation in a modified MPMSR crystallizer. These two techniques show potential for application in suspension manufacturing as PAT (Pathak and Thassu 2009). Blanco et al. (2002) have studied the flocculation processes and properties using a nonimaging scanning laser microscope. This methodology allowed them to study the flocculation stability and resistance to shear forces, reflocculation tendency, and reversibility of the flocculations. Furthermore, they recommended this technique for studying the optimal dosage of any polymer and the associated flocculation mechanism. Although most of their study is in the paper industry, the techniques also can be applied in pharmaceutical suspension PAT.

Lasinski et al. (2005) reported error analysis of the focused beam reflectance measurements when used for suspensions. The assumptions using the FBRM techniques are:

1. The assumption that a particle is stationary (zero velocity) while the laser intersects the particle
2. The error associated with intersecting a particle on multiple occurrences
3. The error from converting a CLD into bidisperse (or multidisperse) PSD

These errors were determined as a function of the independent parameters: the rotation speed of the laser, the volume fraction or number density of particles, the ratio of the particle diameter to the laser diameter, the breadth of the particle size distribution, the velocity profile of the particles, and whether there is a preferential

Table 10.4 Summary of Developments and Future Opportunities for Process Monitoring and Real-Time Quality Analytics in Nutraceutical Dispersion Processing

Category	In Development	Opportunities
Improved process monitoring methods	Optical sensors for dissolved gases, pH near-infrared for moisture during lyophilization; rapid microbiology methods; particle size distribution of the suspensions, need to explore the applications in nutraceutical dispersions	Rapid assay for impurities; rapid activity assay for preservatives, challenging to use in nutraceutical dispersions
Real-time quality analytics	Light scattering assessments of the aggregates and particles; high-performance liquid chromatography and mass analysis for the ingredients, may need extensive research in nutraceutical applications	Characterization with modern process analytical technology techniques; rapid assay techniques, may need extensive research for nutraceutical applications

orientation of the particles in the flow. To date, there has been little investigation on these errors; their study reported that these errors and assumptions affect the validity and reliability of the FBRM measurements they have made in suspension formulations. Table 10.4 summarizes the developments and future opportunities for process monitoring and real-time quality analytics (Pathak and Thassu 2009).

Some companies that supply equipment for PAT include Brucker Optics (Billerica, MA), which provides techniques like vibrational spectroscopy, Fourier transform infrared spectroscopy, Fourier transform near-infrared (FT-NIR) spectroscopy, and Ramen spectroscopy (http://www.bruckeroptics.com/pat); Malvern Instruments (Malvern, UK), which provides industrial instruments for particle sizing (http://www.malvern.co.uk); Niro Inc (Columbia, MD), which provides real-time process determination equipment (http://www.niroinc.com); Thermo Fisher Scientific Inc. (Waltham, MA), which supports FT-NIR analyzers, process tools, and integration and informatics solutions (http://www.thermo.com); and Mettler Toledo, which offers equipment for particle system characterization and in situ spectroscopy (http://www.mt.com/pat). All this equipment needs systematic evaluation for application in suspension scale-up and manufacturing (Shaw 2005).

REFERENCES

Arboleya, J.C., M. J. Ridout, and P. J. Wilde. 2008. Rheological Behavior of Aerated Palm Kernel Oil/Water Emulsions *Food Hydrocolloids* 23: 1358–65.

Barthe, S., and R. W. Rousseau. 2006. Utilization of Focused Beam Reflectance Measurement in the Control of Crystal Size Distribution in a Batch Cooled Crystallizer. *Chemical and Engineering Technology* 29: 206–11.

Blanco, A., E. Fuente, C. Negro, and J. Tijero. 2002. Flocculation Monitoring: FBRM as a Measurement Tool. *Canadian Journal of Chemical Engineering* 80: 1–7.

Block, L. H. 2002. Nonparenteral Liquids and Semisolids. *Pharmaceutical Scale Up*. Edited by M. Levin. New York: Marcel Dekker Inc, 57–94.

Caballero, F. G., and J. L. Duran. 2000. Suspension Formulation. *Pharmaceutical Emulsions and Suspensions*. New York: Marcel Dekker Inc, 127–190.

Charpentier, J. C. 2003. Market Demand versus Technological Development: The Future of Chemical Engineering. *International Journal of Chemical Reactor Engineering* 1: A14.

Chen, H., J. Weiss, and F. Shahidi. 2006. Nanotechnology in Nutraceutical and Functional Foods. *Food Technology* 3: 30–36.

Chen, L., G. E. Remondetto, and M. Subirade. 2006. Food Protein-Based Materials as Nutraceutical Delivery Systems. *Trends in Food Science and Technology* 17: 272–83.

Croda. 2002. http://www.croda.com/home.

Derkach, S. R. 2009. Rheology of Emulsions. *Advances in Colloid and Interface Sciences* 151: 1–23.

Food and Drug Administration. 1987. Guideline on General Principles of Process Validation. http://www.fda.gov/Drugs/GuidanceComplianceRegulatoryInformation/Guidances/ucm124720.htm.

Food and Drug Administration. 1999. Container, Closure Systems for Packaging Human Drugs and Biologics, Chemistry, Manufacturing and Control Documentation. http://www.fda.gov/downloads/drugs/guidancecomplianceregulatoryinformation/guidances/ucm070551.pdf.

Food and Drug Administration. 2006. Guide to Inspections Oral Solutions and Suspensions. http://www.fda.gov/iceci/inspections/inspectionguides/ucm074935.htm.

Gallardo, V., M. Ruiz, and A. Delgado. 2000. *Pharmaceutical Suspensions: Their Applications*. New York: Marcel Dekker Inc, 409–64.

Gomez, A. L., and W. A. Strathy. 2002. Engineering Aspects of Process Scale Up and Pilot Plant Design. *Pharmaceutical Scale Up*. Edited by M. Levin. New York: Marcel Dekker Inc, 311–24.

Gruverman, I. J. 2003. Breakthrough Ultraturbulant Reaction Technology Opens Frontier for Developing Life Saving Nanometer Scale Suspensions. *Drug Delivery Technology* 3: 52–55.

Gruverman, I. J. 2004. A Drug Delivery Breakthrough—Nanosuspension Formulations for Intravenous, Oral and Transdermal Administration of Active Pharmaceutical Ingredients. *Drug Delivery Technology* 4: 58–59.

Gruverman, I. J. 2005. Nanosuspension Preparation and Formulation. *Drug Delivery Technology* 5: 71–75.

Heath, A. R., P. D. Fawell, P. A. Bahri, and J. D. Swift. 2002. Estimating Average Particle Size by FBRM. *Particle and Particle Systems Characterization* 19: 84–95.

Joseph, J. A. 2005. Overcoming the Challenges of Blending Multiple Component Dietary Supplements. *Tablets and Capsule* July: 38–42.

Karanth, H., V. S. Shenoy, and R. R. Murthy. 2006. Industrially Feasible Alternative Approaches in the Manufacture of Solid Dispersions: A Technical Report. *AAPS PharmSciTech* 7: E1–E8.

Kidder, L. H., E. Lee, and E. N. Lewis. 2005. NIR Chemical Imaging as a Process Analytical Tool. *Innovations in Pharmaceutical Technology* September: 107–11.

Kougoulas, E., A. G. Jones, K. H. Jennings, and M. W. Wood-Kaczmar. 2005. Use of Focused Beam Reflectance Measurement (FBRM) and Process Video Imaging (PVI) in a Modified Mixed Suspension Mixed Product Removal (MSMPR) Cooling Crystallizer. *Journal of Crystal Growth* 273: 529–34.

Laitinin, N., O. Antikainin, and J. Yliruusi. 2003. Characterization of Particle Size in Bulk Pharmaceutical Solids Using Digital Image Information. *AAPS PharmSciTech* 4: 383–87.

Lasinski, M. E., N. K. Nere, R. A. Hamilton, B. D. James, and J. S. Curtis. 2005. Error Analysis of FBRM. Poster presented at Particle Technology Forum Poster Session. Cincinnati, OH, November 2005, 287B.

Levin, M. 2005. How to Scale Up Scientifically: Scaling Up Manufacturing Process. *Pharmaceutical Technology* 3: S4–S12.

Lewis, E. N., J. W. Schoppelrei, E. Lee, and L. H. Kidder. 2005. NIR Chemical Imaging as a Process Analytical Tool. *Process Analytical Technology*. Edited by K. Bakeev. Oxford, UK: Blackwell Publishing, 187.

Liu, J. 2006. Physical Characterization of Pharmaceutical Formulations in Frozen and Freeze Dried Solid States: Techniques and Applications in Freeze Drying Development. *Pharmaceutical Development and Technology* 11: 3–28.

Logaraj, T. V., S. Bhattacharya, K. Udaya Sankar, and G. Venkateswaran. 2008. Rheological Behavior of Emulsions of Avocado and Watermelon Oils during Storage. *Food Chemistry* 106: 937–43.

Lu, G. W. 2007. Development of Oral Suspensions. *American Pharmaceutical Review* 10: 35–37.

Mannie, E. 2009. Meeting Beverage Formulation Challenges. *Prepared Food Network* http://www.preparedfoods.com/articles/article_rotation/bnp_guid_9-5-2006_a_10000000000000507122.

Mckee, D., and A. Karwic. 2009. Product and Method for Oral Administration of Nutraceuticals. US Patent 7,632,532, filed December 9, 2005, and issued December 15, 2009.

Mehta, A. M. 1988. Scale-Up Considerations in the Fluid Bed Process for Controlled Release Products. *Pharmaceutical Technology* 12: 46–54.

Nash, R. A. 1996. Pharmaceutical Suspensions. *Pharmaceutical Dosage Forms: Disperse Systems,* vol. 2. Edited by H. A. Lieberman, M. M. Rieger, and G. S. Banker. New York: Marcel Dekker, 1–46.

Newton, J. M., S. R. Chapman, and R. C. Rowe. 1995. The Assessment of Scale Up Performance of the Extrusion-Spheronization Process. *International Journal of Pharmaceutics* 120: 95–99.

Pathak, Y. V., and D. Thassu. 2009. Scale Up and Technology Transfer of Pharmaceutical Suspension. *Pharmaceutical Suspensions: From Formulation Development to Manufacturing* Edited by A. K. Kulshreshtha, O. N. Singh, and G. M. Wall. New York: Springer Publishing Company.

Paul, E. L., M. Midler, and Y. Sun. 2004. Mixing in the Fine Chemicals and Pharmaceutical Industries, *Handbook of Industrial Mixing*. Edited by E. L. Paul, V. A. A. Obeng, and S. M. Kresta. New York: John Wiley and Sons.

Rabinow, B. 2005. Pharmacokinetics of Drug Administered in Nanosuspension. *Discovery Medicine* 5: 74–79.

Rabinow, B. 2007. Parenteral Nanosuspensions. *Nanoparticulate Drug Delivery Systems*. Edited by D. Thassu, M. Deleers, and Y. V. Pathak. New York: Informa Healthcare.

Radspinner, D., B. Davies, and Z. Kamal. 2005. Process Analytical Technology and Outsourcing—Impacts on Manufacturing and Process Knowledge. *GOR* 7: 55–58.

Saarimaa, V., A. Sundberg, and B. Holmbom. 2006. Monitoring of Dissolved Air Flotation by FBRM. *Industrial & Engineering Chemistry Research* 45: 7256–63.

Sekulic, S. S. 2007. Is PAT Changing the Product and Process Development? *American Pharmaceutical Review* 10: 30–34.

Semo E., E. Kesselman, D. Danino, and Y. D. Livney. 2007. Casein Micelle as a Natural Nano-capsular Vehicle for Nutraceuticals. *Food Hydrocolloids* 21: 936–42.

Shaw, G. 2005. Micrograms to Kilograms: The Challenge of Scaling. *Drug Discovery and Development* July: 1–4.

Sohi, H., Y. Sultana, and R. Khar. 2004. Taste Masking Technologies in Oral Pharmaceuticals: Recent Developments and Approaches. *Drug Development and Industrial Pharmacy* 30: 429–48.

Stewart, D. J., R. Milanova, J. Zawistowski, and S. H. Wallis. 2000. Phytosterol Compositions and Use Thereof in Foods, Beverages, Pharmaceuticals, Nutraceuticals and the Like. US Patent 6,087,353, filed May 15, 1998, and issued July 11, 2000.

Thompson, A. R., R. J. Yarwood, and P. Kearney. 1995. Method of Identifying Freeze-Dried Dosage Forms. US Patent 5,457,895, filed October 1, 1993, and issued October 17, 1995.

Vaccaro, A., J. Sercik, and M. Morbidelli. 2006. Modeling Focused Beam Reflectance Measurement (FBRM) and Its Application to Sizing of the Particles of Variable Shape. *Particle and Particle Systems Characterization* 23: 360–73.

Wibowo, C., and K. M. Ng. 2004. Product Oriented Process Synthesis: Creams and Pastes. *AIChE J* 47: 2746–67.

Zlokarnik, M. 2002. Dimensional Analysis and Scale Up in Theory and Industrial Application. *Pharmaceutical Scale Up*. Edited by M. Levin. New York: Marcel Dekker Inc, 1–41.

CHAPTER 11

Validation of Nutraceutical Process Equipment

Don Rosendale

CONTENTS

11.1 INTRODUCTION

It is hoped you find this material to be of some assistance during the validation of your particular process equipment. Obviously, one cannot prepare an example for every situation. However, the methods described in this chapter can be modified to reflect the needs of any particular type of equipment. The thought processes and methodology for validation remain the same.

Validation, as it is known today, has developed from the need to maintain quality, consistency, and, above all, public safety. Validation is a rapidly growing and evolving subject. This evolution stems from technology's astonishing growth rate, especially in terms of what is available in computer hardware and software. Since the mid-1980s, machine automation and process control through the use of a computer has caused additional concerns relating the validation of the processing system.

Today the computer is used for everything from controlling the process to automatically providing batch reports and quality control.

The foundation of validation, the methodology behind validation, and the need for validation will likely remain key aspects of the industry in which we work. This chapter reflects the current industry trends and serves as an educational tool in our progressive industry.

11.2 DEFINITIONS AND KEY TERMS

The following key terms are defined for your use in this chapter.

- The Food and Drug Administration's (FDA) definition of *validation* (1987):
  - "Establishing documented evidence which provides a high degree of assurance that a specific process will consistently produce a product meeting its predetermined specifications and quality attributes." (p. 3)

The key idea is to provide a *high level of documented evidence* that the equipment and the process conform to a written standard. The level (or depth) is dictated by the complexity of the system, process, or equipment. The validation "package" must provide the necessary information and test procedures required to prove that the system and the process meet the specified requirements.

- Installation qualification (IQ): Verification that the equipment or system is installed in a proper manner and that all of the devices are placed in an environment suitable for their intended operations.
- Operational qualification (OQ): Verification that the equipment performs as expected throughout the intended range of use.
- Process qualification (PQ): Verification that the system is repeatable and consistently producing a quality product. This is not to be confused with *performance qualification*, another validation term that qualifies the performance of a particular tool, piece of equipment, or software package.
- Protocol: A set of instructions that outline the organization of the qualification documents. The protocol specifies the type of tests required and the order in which the tests should be conducted.

11.3 VALIDATION LIFE CYCLE

Questions to consider:

- What does validation mean to the personnel in your company? What does validation mean to your vendor or equipment supplier?
- Where does the responsibility lie for validation? With the FDA, end user, or vendor?
- Process validation versus equipment validation versus master validation plan? What is the difference between the three?

Validation is a continuing and evolving process. The validation process extends from the basic specifics (how each item works and interacts with another item) to a broad theological and methodical investigation of how the system and processes perform. Its scope encompasses documentation, revision control, training, and maintenance of the system and process. Evidence of validation should be seen at the corporate level and be reflected in the management structure. Validation is not just a set of procedures and rules to satisfy the FDA: validation is a method for building and maintaining ***quality***.

11.3.1 When Does Validation Begin?

- During the chemical development stage?
- When you are currently developing a process?
- When conducting clinical studies?

Ideally validation starts in the beginning, in the laboratory. In the lab, scientists discover exactly how the product reacts, as well as the parameters that are required to produce such a product. They learn under what conditions the product fails or becomes unstable, unusable, and when its quality begins to suffer. Once the boundary processing criteria have been established in the laboratory, this information can then be used for establishing requirements for validation.

11.3.2 When Does Validation End?

- After installation of the equipment?
- After the initial validation?
- After the first successful batch?

The answer to this question is simple: Validation of a system never truly ends. Once a new system and process have been validated, the system still requires maintenance, periodic calibrations, and adjustment. Therefore, the process (and consequently, validation) is always under scrutiny and constant evaluation.

Shown in Figure 11.1 is a simplified flowchart of a validation life cycle. This flowchart has been modified to show just one part of a project rather than the entire master validation plan. The first step involved in validating a new process is to formulate a good definition of it.

This means defining the following:

- What the process is
- Why it is needed
- What it will be used for
- How the elements of the system link together to perform the final process

Once the process has been completely defined, equipment usually will be required to perform the actual processing of the product. This equipment will be collectively called "the system." The system and its operations can then be identified and defined.

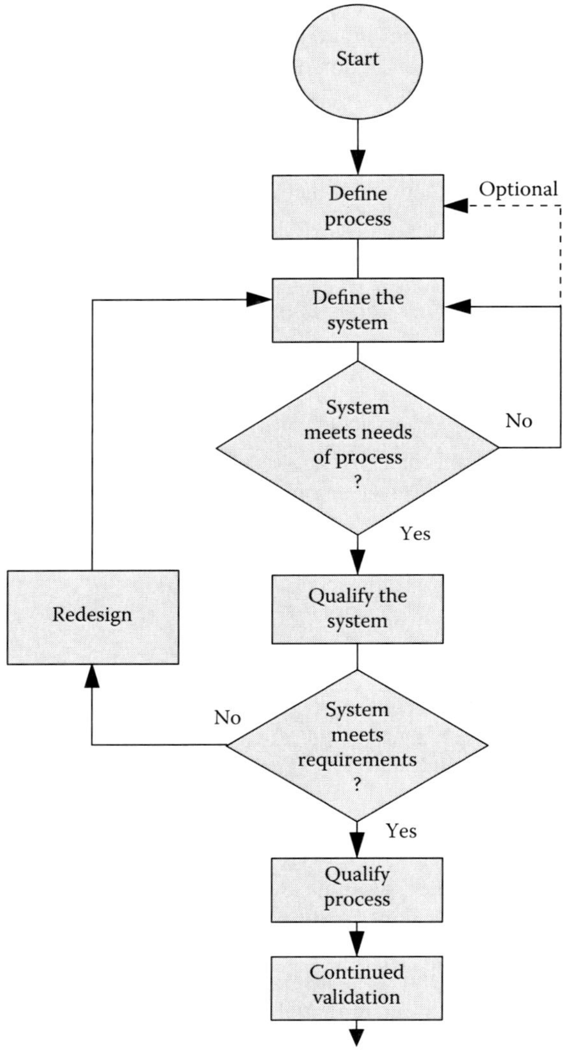

Figure 11.1 The validation life cycle.

11.3.3 Quality Checks: The Beginning

The first check that must occur is to answer the question: "Does the system meet the need of the process?"

A detailed examination of the system required should be conducted. If for some reason the system does not meet the needs of the process, the system may need to be redefined. In some cases, it may become evident that the process should be redefined rather than the system. For example, it may be more cost-effective to execute the process using a different technology or technique. Thus, the option exists for redefining the process (see Figure 11.1).

11.4 DEFINING THE VALIDATION PROCESS

One of the most important parts of the validation life cycle is the definition stages for the process and the system.

11.4.1 Definition Phase

The definition phase starts while a new product is being produced in a laboratory. As the product development progresses, the identification of critical process parameters becomes evident. Critical parameters are those process items that must be maintained to fulfill all the elements required to maintain the quality of the product. To maintain these parameters, the equipment vendors must also be aware of the same critical process parameters for which the quality of the product depends.

Critical process parameters must be transferred to the vendor as early as possible, preferably during the quoting stage of the project. This helps guarantee the purchase of a system that can perform the task as defined by the process as well as holding tolerances within the defined specifications. In communicating and identifying the critical parameter information up front, we help reduce the overall cost of the upscale project, as well as lead times, and ensure a system that will consistently reproduce high quality product.

11.4.2 User Requirement Specification

The International Society of Pharmaceutical Engineers (ISPE) and the European community have introduced the idea of a *user requirement specification* (URS). A URS is simply a specification document ostensibly drawn up by the user group that communicates the operations that the equipment and/or software provide to the user.

Properly implemented, a URS provides a complete document that the production staff can use to obtain exactly what equipment they need to produce their nutraceutical. In fact, there are URS templates available through ISPE and the Joint Equipment Transition Team Consortium that can be used as a starting place for many types of process equipment.

Using a URS as a basis for specifying the equipment helps make sure what you get will be what you need to have to produce the nutraceutical or nutraceutical product that you intend to market.

The URS need not be a strict document that is inflexible. In fact, in the investigation phase of equipment purchases, it may be necessary to adjust the URS requirements to meet commercial needs. Although it may be necessary to list code requirements (e.g., your area may need to have United Laboratories [UL] approval for the controls and electrical systems), it is not necessary to list these requirements in detail; a note will suffice.

The reader is encouraged to visit the ISPE website (http://www.ispe.org) to obtain further information on URS documentation, good automated manufacturing practice (or GAMP, as trademarked by the ISPE) guidelines, and direction toward example URS documentation templates that can assist in the process of defining the machine's requirements.

11.4.3 Importance of Validated and Validatable Software

One of the next key elements to be addressed is system software and its validation. To ensure a successful validation, validated software is a must. Any software that is written, configured, installed, or used in any way to develop the operating process system needs to be validated.

11.4.4 Software Overview

11.4.4.1 Software Definitions

- Application software: Application software is, by definition, custom-written software designed to perform a specific function and/or purpose. Typically this software is written in a high-level language such as C, Pascal, COBOL, or some other language. Thus, application software is not meant to include "off-the-shelf" commercial software.
- Configurable software: Configurable software is off-the-shelf commercially available software that is written to easily customize to the user's process. With configurable software, large sections of code are automatically invoked or created without the need for highly specialized training of the persons performing the configuration.

11.4.4.1.1 Examples of Configurable Software

- GE Fanuc Intellution iFix (GE Intelligent Platforms, Charlottesville, VA)
- Rockwell Software RSView (Rockwell Automation, Milwaukee, WI)
- Allen-Bradley PanelView (Rockwell Automation)
- Wonderware (Invensys, Lake Forest, CA)
- Programmable controller ladder logic

The range of usage of the software will also determine the level of scrutiny required to verify its function. Windows and DOS, programs with wide user bases and large man-hours of mean time between failures, do not need extremely close scrutiny. However, applications and configured software written using these operating systems require a large amount of inspection before placing them into service.

Obviously, the person who knows the most about the software is the programmer. Most validation and quality assurance people do not have a strong software background; therefore, there must be some software guidelines to look for in a validation.

The key questions to ask are:

- Is the software development structured? ***This structure should be published and followed.***
- What is the development cycle of the product?
- What is the development environment?
    - Is it backed up?
    - Is it secure?
    - Is it maintained?
    - Is there revision control of developed software?
    - Is the development methodology comprehensive?
    - Are there higher-level language structures and standards?
    - Are there database structures and verification methods for those databases?
- Are there written test plans and evidence of test cycles?
- Are the testing procedures comprehensive?
- Are there written test programs (stubs) to help with pretesting?
- How comprehensive is factory testing?
- Are the factory tests documented?
- Change control
    - Are there software and hardware change control procedures?
    - How long have they been in place?
    - What level of change control is maintained?
    - Do the people seem aware of the change control methods?
- Documentation
    - Does the vendor have evidence of change control put into use?
    - Does the vendor have evidence of maintenance of the software and hardware?
    - Is there a code management system? Is it followed?
- Code review: The vendor should be able to give an example of operating code and to show adherence to the change control procedures and documentation protocol outlined above.
    - The vendor should have a written support plan.
    - There should be evidence of the support plan in operation.
    - Does the vendor have disaster and contingency plans?

In addition to the basic two types of software presented previously, special considerations should be made concerning the source code and the operating system under which the software is going to run. Arrangements may need to be made for auditing the source code.

- Source code: Source code is the uncompiled version of the software in question. For example, a program written in Pascal that a person can read and interpret is *source code*, whereas the compiled executable is inaccessible and therefore unreadable.
- Operating systems: An operating system is the low-level system code that makes the computer or computer system function. Examples are:
    - Microsoft's DOS
    - IBM's OS/2
    - Microsoft Windows
    - Microsoft Windows NT

An operating system should be chosen that best meets the needs of the process. Figure 11.2 shows a flowchart for a typical validation process.

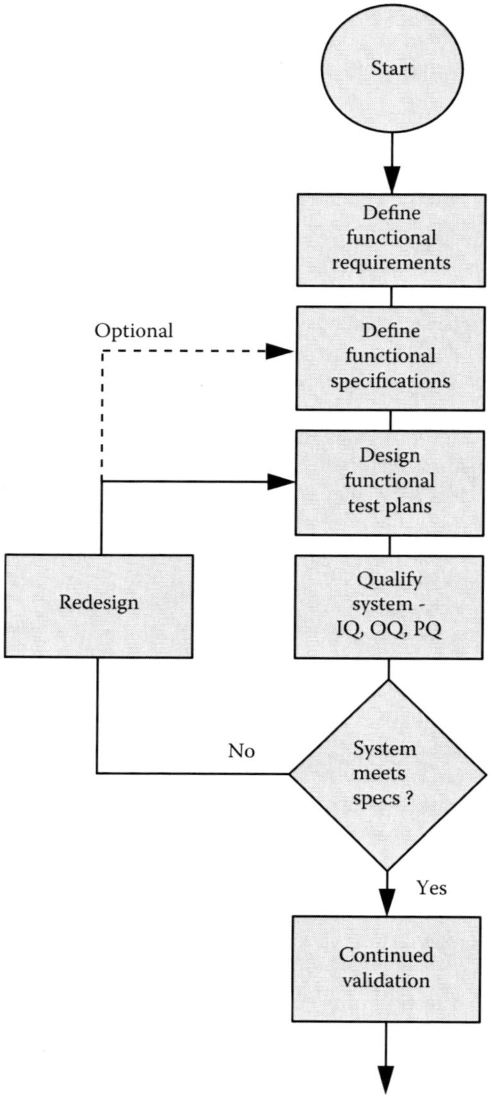

Figure 11.2 The validation process.

11.4.5 Functional Test Plans

The functional test plans lay out a plan to verify, test, and prove the system is installed, configured, and operating according to the requirements set forth by the functional specifications and functional requirements. These test plans are compiled into complete IQ, OQ, and PQ documents. These documents are then tested on the system.

11.4.5.1 If the Tests Fail…

If the system does not pass these procedures, an examination is made of the requirements, specifications, and tests. This helps quality assurance people evaluate the system and narrows the focus of the troubleshooting exercise. A redesign of the system may become necessary to address the system failures. In the worst case, a change in the functional specification may become necessary to address the problem.

11.4.5.2 When the Tests Pass

If the system performs as expected and all the specifications are met, a signoff of the requirements, specifications, tests, and test results indicates to the governing authority that the system has been properly validated. The system then enters a phase of *continuing validation.*

Maintaining validated status should be outlined in the master validation plan (MVP) that governs your plant's validation practices. Continued validation includes periodic checks, calibration, and scheduled maintenance. All the maintenance operations should be documented to prove that the system is still performing within the requirements and specifications as outlined in the functional requirements and functional specifications. Use of the original test is recommended when performing maintenance on the system.

11.4.6 Structure of Test Plans: Installation Qualification and Operational Qualification

The structure of the test plans has been evolving. The current industry trend is to have a distinct IQ section and then proceed to the OQ section after the IQ has been completed.

The FDA recognizes that there will be a certain risk in any process or machinery; thus, there needs to be an evaluation of this risk. This chapter will not go into detail on risk management because there are several good sources on the subject (e.g., GAMP5).

The IQ section handles the obvious installation criteria:

- Is the machine installed correctly (i.e., level, in a room suited to the purpose)?
- Are the utilities satisfactory for the equipment (should be adequate to operate the equipment reliably, without straining any systems)?
- Information should be available:
    - The validation team should be identified in a validation team document.
    - Are the manuals in place and correct?
    - Are the drawings up to date, and do they represent the actual installation?
    - Are all the major pieces of equipment identified in the manual for ease of replacement?

- Configuration of configurable devices should be recorded (e.g., switch settings, programming of recorders).
- Manufacturer, model number, and serial number of critical equipment for any service that may be required should be identified.
- Are all the product contact materials identified and, if necessary, verifiable?

The IQ should have an approval signoff page indicating that the IQ procedure has been thoroughly reviewed and is ready for execution.

There seems to be a push to have a separate IQ signoff that confirms everything is in place and ready for equipment operation. Practically this has not been achievable in most initial installations. Either the drawings are marked up and not finalized, all utilities are not 100% available (such as chiller systems), or one or another major activity is not completed while the machine is being installed and commissioned (e.g., the room finishing and room air validation). Reasonable exceptions can be made and documented for the IQ to be signed off, and the validation engineer can proceed with the OQ testing with these deficiencies in mind.

The OQ may have its own signoff to be completed before conducting the OQ plan, as seems to be the growing trend in Europe. However, one protocol approval covering both the IQ and OQ section approvals may also be used.

OQ testing may well find deficiencies in the system, both from a hardware standpoint and from a software view. These deficiencies can be addressed in the failures, and retesting procedures can be conducted after the fixes have been installed. After successful retesting, the final report can then be written.

A single-page signoff is useful after the OQ to indicate that all matters of the system have been found to be operational.

11.4.7 Revalidation

Revalidation becomes necessary whenever a piece of equipment requires or is given significant upgrades to its capabilities or a key piece of the equipment is replaced. By maintaining a good equipment use and maintenance log, the revalidation of a piece of equipment can be compartmentalized to only the system or device to be repaired or replaced.

Also, there should be a factory standard operating procedure (SOP) in place to identify maintenance trigger items, in addition to the customary recalibration scheduling.

11.4.7.1 Maintenance Triggers

Several "triggers" or indicators should be outlined in the MVP to aid in determining the frequency of retesting. The following triggers are listed as an example:

- Loss of product quality
- Replacement of key components of the system
- Upgrades (software or hardware)
- Change in personnel (e.g., mandating retraining)

Often, the original test plans can act as retest procedures for the equipment after servicing. In others, a special test designed to challenge the installation is best. In the case of replacing a computer system, care should be taken to document that the system replacement has all the elements of hardware and software necessary and in place to run the system in the same manner as it was before replacement. In other words, an examination of the software load (i.e., revision, file names, and file sizes), operability, and retest of critical functions should be approved, test conducted, and test results approved before placing the equipment back into service.

11.5 GENERATING FUNCTIONAL REQUIREMENTS

The functional requirements document is used to define ***what*** is to be done. The process should be stated succinctly with little or no reference to specific devices. The following structure is a useful guideline:

- Purpose or objective: The purpose of the document is to describe the processes necessary to produce quality, marketable product.
- Audience: The intended readership of the document. This may specify that a certain level of knowledge is necessary to adequately understand the information in the document.
- Definitions: An abbreviations and definitions section is helpful.
- Reference documents: Provides supporting evidence to the document, if required. This could be lab test results, charts, or other information that may help explain the process.
- Details: Outlines the functional details of the process and what it does.

To briefly summarize, the functional requirements document describes and defines ***what*** the system or process must do. It does not describe or define ***how*** it is to be accomplished. This shall be accomplished in the *functional specifications.*

An example of a functional requirement technical page is listed below.

THE PROCESS

The product being produced is an agglomeration of product Super X (50%) with ingredient alpha kappa (40%) molded into the form of 10-mg droplets placed on a stainless holding tray. A 10% alcohol solution serves to bind the product together until it is evaporated in the heat chamber during the crystallization phase that is produced in the heat oven system. A heat of 83°C–87°C exposure for 2 minutes will successfully crystallize the material.

A continuous process is desired that will produce product during an infinite production run (i.e., provision should be made for change of shifts). It is desired by marketing to produce 10,000 units/day, with the capability for production of up to 25,000 units/day using three shifts.

Operational Limitations

Heat time over 2 minutes, 15 seconds will degrade the material into a useless structure that does not form the crystalline base combination after cooling. Heat time under 1 minute, 55 seconds will not produce a sufficiently thick shell and will prematurely dissolve in the stomach. (The combination is designed for dissolution in the small intestine.) Heats below 83°C will not properly congeal the outer layer into a smooth case and will exhibit the same dissolution in stomach acid. Heats above 87°C will create a solid that will not dissolve in the human digestive system.

Ideal Conditions

A consistently excellent product was produced in clinical studies at a temperature of 85.2°C for 2 minute, 7-second exposures. Temperatures around this set point within 0.5°C did not seem to degrade this "ideal" product. Hold times within ±5 seconds of this time produced consistently good product.

Records

The resultant process times and temperatures must be recorded and kept on file for at least 5 years after the product has been released.

11.6 GENERATING FUNCTIONAL SPECIFICATIONS

Unlike functional requirements, functional specifications state ***how*** the system is to accomplish the task as described in the functional requirements. It gets specific.

The functional specification document usually contains:

- An introduction, or purpose
- Intended audience
- Definitions
- Reference documentation
- Mechanical requirements and specifications
- Electrical requirements and specifications
- Computer requirements and specifications
- Security requirements and specifications

Its format should be consistent with the functional requirements document. Reference to system drawings, piping and instrumentation drawing (P&ID), and, if necessary, facility layout drawings should also be included. As it is often said, "A picture is worth a thousand words!"

In the following sections, an example of the technical part of a simple functional specifications document is described.

FUNCTIONAL SPECIFICATION

A processing line consisting of a stainless steel conveyor system with a thin insulating layer (to isolate the tablet temperature effects of a large mass of stainless steel) is loaded with 10-mg tablets and passed through an oven that is kept at the ideal 85.2°C temperature required for the reaction.

Loop 1: Temperature Control

The oven temperature is kept constant by a steam valve actuated by a programmable logic controller (PLC)-operated proportional integral and derivative (PID) loop system. Input to the PID loop is via a four-point averaging resistance temperature detector (RTD) probe that spans the length of the oven bed. This RTD input is transmitted to the PLC via an RTD to a 4- to 20-mA conversion module. Control of the steam valve is via a current to pneumatic transducer (I/P) actuated by a 4- to 20-mA signal from the PLC, provided by the PID control loop. Input steam specification is 15 psi saturated. The oven draws 30 lb./h steam when the oven is operational. (There is little load variation in this process.) The operator may adjust the oven temperature from 83°C to 87°C. The machine shall power up with the set point of 85.2°C.

Loop 2: Dwell Control

The time in the oven is set by a conveyor system speed servo that drives the conveyor through a gear set. The speed of the servo motor is monitored by a tachometer and inputted to the PLC by a tachometer input card. Dwell time is converted from the oven length (3 ft.). One revolution of the conveyor servo motor is exactly 1 ft. of travel of the conveyor. The control of the conveyor speed is to read out in dwell time of the product inside the oven. Thus, tachometer speed is converted into dwell time using the following formula:

$$\text{Dwell time (min)} = \frac{\text{3 ft. (oven length)}}{\text{Speed of conveyor (ft./min)}}$$

The operator may adjust the dwell time between 1.916 minutes and 2.25 minutes. The control loop shall power up with the set point of 2.0 minutes. Because of the need to log the information onto the LAN network, a computer-driven PLC system operator interface that has automatic data acquisition capabilities was chosen for this job.

Other Information

The system will operate using 230-VAC, 50-Hz, 3-phase voltage supply. The equipment is to be mounted in a nonhazardous electrical classification. All equipment must be able to be operated at an ambient temperature of 25°C. The equipment is designed to run at 100% duty cycle, 24 hours a day. Equipment must conform to local codes and be certified for European CE (*conformité européenne*) approval. In addition, all validation support documentation will be written in French.

Based on process parameters, the ranges and tolerances of the equipment operation shall be defined in the functional specification. Specific types of equipment, if required, would be outlined here. For example, if a 486 66-MHz PC is required, then it should be specified. If it is not specified, an old 8088 machine could be used, and the system would not meet the specified requirements because it would be too slow. If a component of the system is critical to the system or if a particular type of component is desired, it must be specified.

The functional specification document is useful for describing to your vendor exactly what you want and what will be required. Often if it is not known exactly what should be specified for a given application, the vendor is willing to work with you to develop or finalize the functional specifications. The accuracy of the functional specification is critical to other documents or test plans that will be written around this document. It is important that the functional specification receives proper reviews and acceptance before release.

11.7 METHODS OF IDENTIFYING PROCESS PARAMETERS

Identifying process parameters for a given process is often not difficult; one can simply state all the physical aspects of the system. For example, ambient temperature, internal temperature, external temperature, airflow, air pressure, altitude, relative humidity, compression, power absorption, wavelength, density, and particle size could all be valid process parameters.

Only a limited few process parameters are normally considered process critical. These items should be identified so that they are monitored, controlled, and/or recorded. Typically in the nutraceutical environment, the issues concerning quality are the chemical and physical aspects of the product. These would include release times, shelf life, chemical stability, temperature stability, and appearance. Therefore, any process variables that affect the above would be identified as critical process parameters.

11.7.1 Identifying Critical Process Parameters

Critical process parameters are defined as "those parameters that in any way affect the quality of the product." The term *quality* may be defined to include the physical aspects of the finished product, such as its smoothness, that may not be part of the medicinal quality. If it can be shown or verified that a given change in a process parameter has no effect on the process or product, then it is not considered a critical process parameter.

Many times the process parameters are interrelated, and a change in one parameter directly affects or inversely affects another. The common practice in use today is to provide proof that the process parameters are not changed during the process or are maintained within a specified range. These parameters are documented in the functional specifications. Typically, decision criteria are supported by lab reports and/or historical data on the process. To continue using the example protocol, we can identify the critical process parameters.

IDEAL CONDITIONS

A consistently excellent product was produced in clinical studies at a temperature of 85.2°C for 2 minute, 7-second exposures. Temperatures around this set point within 0.5°C did not seem to degrade this "ideal" product.

Hold times within ±5 seconds of this time produced consistently good product. Thus, ***heat time*** and ***heat temperature*** are the two most important parameters in this process.

For a tablet-coating process, critical process parameters are normally process airflow, temperature, spray rate, humidity, pan speed, and total time.

Airflow, humidity, and temperature are interrelated process variables that directly affect the process by dictating the evaporative capacity of the system. The spray rate is also a critical process parameter. The spray rate would dictate what the required evaporative capacity of the system must be. The spray rate affects the finish of a tablet. If coating of the tablets occurs at too high a rate, the finish may exhibit an orange peel, which is not desirable. Pan speed may not affect the chemical composition of the product, but pan speed can affect the physical appearance of a coated tablet. A fast pan speed could cause chipping, for example.

Identification of the critical process parameters can often be identified by asking the following questions:

- How do the process parameters relate to the chemistry of the product?
- Is there a mathematical process model?
- How do the process parameters affect the physical aspects of the product?
- What change would cause product degradation?

11.7.2 Identifying Tolerances

Simply identifying the critical process parameters is not sufficient. Each critical process parameter will more than likely have an operating tolerance. Both the hardware and the process will have tolerances that need to be identified. For example, a heat exchanger may have a controllable range from 0°F–212°F, but the quality of the product is reduced when the temperature falls below 100°F or increases above 185°F. Thus, from a process standpoint, 100°F–185°F becomes the operating range. This operating range can then be considered to be the worst case. Documentation (such as a batch report) should indicate that the process was maintained between 100°F and 185°F. Alarms and interlocks may also be specified to help maintain this temperature.

When developing and identifying tolerances, keep in mind the following:

- Develop operating tolerances for the ***process***.
- Determine the operating tolerance for the ***equipment***.
- Determine the process limits that are ***required to maintain the quality of the product***.
- ***Verify and document*** this range during the validation process.

11.8 VENDOR AUDITS

One area of concern with the purchase of new equipment is vendor selection. One must ask the question: "Is the vendor prepared to offer systems that have been validated or are validatable?" The primary concern of nutraceutical processors is the ability of the vendor to demonstrate that the software and hardware design has been developed and tested under written guidelines. What to look for in an audit:

- Organization of business—separate QC unit: Look for vendors that have a separate quality control department or quality inspector. This person or department should have the authority to reject components and have direct control over when a product will or will not ship based on its quality.
- Evidence of document revision control: Evidence of document revision control would include a written procedure for how drawings, software, contracts, test plans, and specifications are updated. Looking at the documented history of a particular item should give a clear indication as to the quality of the document revision control system.
- Evidence of compliance with written software procedures: Software is often the most critical point of an audit. The compliance of written software procedures for how software is written, maintained, and updated is critical. Documented history of compliance of these procedures should be reviewed.
- Off-site storage of critical operating software and documentation: A backup and storage procedure for software and documentation should exist. Backup is required in case of a natural catastrophe.
- Evaluate engineering experience and applicability: Qualifications of the technical personnel should be reviewed, as well as an evaluation of the experience of the individuals. The credentials for key positions should also be reviewed.
- Look for and evaluate employee training: Consideration should be given to the continuing education of personnel. For example, if a job description states a requirement for special training, such as how to follow a software procedure, the individual should have in his or her personnel file a record of successfully completing this training.

11.8.1 Conducting an Audit of a Vendor: Initial Considerations

- Publish an agenda: From a vendor standpoint, an audit can be run much more smoothly if the vendor knows your objectives. The agenda can be in an outline form, indicating what the topics of discussion will be.
- Make sure vendor agrees to the agenda: To make sure the required personnel will be available for the audit, it is important that the vendor agrees with the agenda. This is also good business practice.
- Provide a written report after the audit: After the audit is complete, a letter of results should be sent to the vendor. The results should list the items of concern and any unacceptable practices. Also, be sure to include those practices that were done especially well.
- Ask for a company history: In evaluating quality and longevity, often the company's history and growth can be used as an indication. For future service and support, one must have a certain comfort level that the company will be able to support you for the long term.
- Ask for a financial overview: In evaluating the risk of doing business with a particular vendor, a financial statement, yearly report, or other documented evidence should be reviewed. If the company is publicly held, the yearly report can be acquired easily. If the company is privately held, one can look at the history and growth of the company.
- Project overview: The customer should be aware of the current or future projects. Many times an audit can be directed toward a specific project; therefore, a project overview can yield an indication as to what specifically needs to be reviewed.

11.9 VENDOR VALIDATION SUPPORT

Validation support can come from the vendor in several forms:

- Vendor manuals
- Operator training manuals
- Operational documents
- Functional specifications

Receiving validation information from the vendor can present several significant advantages:

- Familiarity with the equipment: Initially the vendor is more familiar with the equipment and its limitations. Therefore, the equipment validation process can usually be accelerated by using vendor protocols.
- Validatable system: Reviewing the vendor's validation documentation up front helps verify that your equipment vendor is prepared to provide a fully documented and validatable system.
- Documentation: Documents that are typically available from the vendor are the IQ, operational qualification, calibration procedures and documents, cleaning procedures, scheduled maintenance documents, functional requirements, and functional

specifications. These documents help define the long-term maintenance that the machine requires.

- System qualification: Receiving validation documentation from the vendor provides a means of qualifying the system with the vendor. For example, when executing the IQ, both the vendor who may be performing the qualification and the purchaser of the equipment would sign off or approve each part of the qualification. This duality provides a good method of checking and verifying that the system is installed as stated.
- Staffing: Pursuing validation documents from the vendor can reduce internal costs by reducing the staffing requirements for creating and writing the validation documents.
- Additional benefits:
  - The equipment can go into production sooner.
  - Problems during the startup of the system can be minimized as a result of the ability to qualify the equipment before shipment.
  - The vendor may be able to provide additional engineering services. Those services provide detailed installation drawings, electrical installation drawings, ductwork drawings, and facility layout drawings. All these documents and services help reduce installation time and can be used as part of the validation package to help qualify the installed system.

11.10 STRUCTURE OF THE VALIDATION DOCUMENTATION

11.10.1 Master Validation Plan

The MVP is the roadmap of the validation exercise. Its principal function is to identify exactly what is being validated and how that validation is to take place. It is also the "catch basin" of all the information necessary to define the machine, software, and hardware terminology, schematics, and diagrams. One of its goals is to act as the index to point out where to look for information regarding the machine and the validation. The other main goal of the MVP is to identify how the tests have been structured and include an explanation of how the test plans fulfill the FDA's requirement for validation. An example of an MVP will be shown in Section 11.15.1, after some other topics have been discussed.

11.10.2 Identifying the Validation Team

The validation team should comprise individuals who are capable of verifying that the test protocol and final machine acceptance testing is done in an acceptable manner. In some organizations, this team is composed totally of in-house personnel; however, this is not always the case.

- In-house and vendor teams: In many cases, the vendor may act as part of the validation team insofar as to provide detailed information on the machine's operation that the end user might not be immediately aware of. This aspect in itself may save the end user months of research and development testing and trial and error that would otherwise be spent learning the system operation.

- In-house and subcontractor teams: The team may also consist of subcontractors and specialists whose job is to perform validation on a professional consulting basis. Care should be taken to qualify the subcontractor properly for the machine validation being requested. Does the subcontractor have previous experience in the field? Is there a list of references to call to verify customer satisfaction? Is the subcontractor's pay based on an hourly rate or at least partially based on protocol acceptance?
- Exclusively in-house teams: The team may also consist of totally in-house employees. In this scenario, care must be taken to have team membership available throughout many disciplines. Be sure that the technical portion of the machine's operation is covered by a knowledgeable person or persons within the group. If several disciplines are involved, identify each area and make sure that all bases are covered. In-house teams are especially attractive when the machine is of proprietary design or the end product is generated under a cloak of confidentiality.

11.10.3 Team Construction: Minimum Constraints

In any case, the team must consist of sufficient team members to guarantee impartiality, usually at least two or three people. The team should also have sufficient technical talent to identify all areas that are to be validated. If an area is lacking, such as a production specialist, the team should recruit an operator or production supervisor to be a part of the approval team. If vendors or subcontractors are part of the team, ask to see the credentials of those individuals on the team. Of course, the credentials of your own organization's people must also be duly recognized and recorded.

11.10.4 Validation Team Document

A necessary conclusion of the creation of the team is a single document that:

- Clearly identifies the validation project.
- Clearly identifies each individual and his or her role on the team. This should include the person's printed name, job title, and role within the validation group if it is not implied within the person's job title.
- Provides a place for a written signature and date for each member of the team.

Attachments may be included or referenced to back up the person's legitimacy for selection by the team (e.g., diploma, transcript, record of training, and résumé). Credentials may also be referenced as "on file" within the organization. In any case, it is important to be able to access this information during an audit.

11.11 INSTALLATION, OPERATIONAL, AND PROCESS QUALIFICATIONS

11.11.1 Installation Qualification

IQ verifies that the equipment has been installed in accordance with the manufacturer's recommendations in a proper manner and that all of the devices are placed in an environment suitable for their intended purpose.

11.11.2 Specification Information

The IQ should build on the list of specification documents as outlined in the MVP. From these lists of specifications and critical components, several checklists are generated. There are several reference documents that help define the IQ parameters; they are:

- An accurate P&ID: The process engineer will be able to identify the critical parts of the process necessary for proper operation of the system.
- An accurate P&ID tag list: Sorted both functionally (alphabetically) and numerically.
- A list of critical components: Critical components are those with which the machine would not operate properly without or would pose a problem if that component were to fail. Also included in this list may be special long-lead components that could jeopardize startup and/or production schedules if they were to fail.
- Complete specifications on the critical components: A list of limitations that the components are designed for (such as operating environment temperature, acceptable humidity range, vibration tolerances, etc.). If no specification is available, suitable judgment should be made about that particular part.
- Engineering design data on critical parts: Most of the time, this information is available on the specification sheets or is part of the equipment data. If the part is custom engineered and affects safety or critical operation, ask to see the engineering calculations.
- Listing of product contact areas: The product contact, solution contact, and air contact areas must be identified and the components listed by the manufacturer, as well as all the materials in contact with any one of these areas. Any area that may come in contact with the product must be identified and approved for use in that manner.
- Surface area calculations: Also important is a calculation of the surface areas involved for large surfaces, so that determination of total active levels of product can be made using sections of swab samples, rather than having to swab the entire machine after a cleaning to determine total residual drug contamination.
- List of lubricants: If lubricants are used in a system, there should be a central place to list those lubricants, as well as where and how often they should be applied. If any lubricants can find their way into the product stream, it must be verified that these lubricants are not harmful to the product or the patient.

11.11.3 Computer and Software Installation Qualification

For software, the IQ data can become somewhat obscure. A list of software components and their record of installation should be available, complete with serial number (when available) and current revision level of each software package. If several sets of software are to work in concert with one another, make sure that each set of software is compatible with one another. For example, a Windows 3.1 platform working on DOS 2.0 would be immediately suspect. If it were running on an 8088 platform, one might also question the integrity of the specification data or the equipment supplier. Personnel who have working knowledge of computers are helpful when evaluating computer systems.

Often software setup documentation specifies minimum hardware and software requirements. Care should be taken to make this criteria part of the checklist and document the findings of the investigation as part of the software IQ. The software IQ is also a good place to record serial numbers of the computer software and hardware.

In a nutshell:

- Software requirements and specifications should be clear, written down, and approved by the appropriate personnel before creating the software package.
- Any development software should be identified and logged.
- Changes to the software specifications and/or requirements should be documented when they are made, suitable explanation given for the changes, and approvals signed for the change.
- Software installation should be recorded, when made, so that the proper model numbers, serial numbers, installation date, and other pertinent facts are kept as a part of an overall disaster recovery plan.
- The record of software installation should be sufficiently complete to reproduce the software from that document.

11.11.4 Development Programming Software

The IQ should also record the *development software* used to program the machine.

- In the case of *executable software*, the language *compiler* and its *version number* should be recorded.
- In the case of *graphic interface packages*, the *manufacturer* and *model number* of all applicable elements should be recorded.
- In the case of *PLCs*, the *make* and *version* of the programming software should be recorded as well as the PLC model and version itself.
- If a package is available on the open market, such as *DOS* or *Windows*, simply writing the version number is all that is required.
- If a *development package* consists of many parts, all should be broken down and identified to *be able to reproduce any software* at any point, if required.

11.11.5 Source Code

Specialized source code written for a *specific function* by an original equipment manufacturer (OEM) may be considered proprietary by the OEM, and source code may not be available to the end user. The source code must be available for auditing purposes, if so desired, by the end user, customer, FDA, or independent source for evidence of testing, structure, and reliability. The OEM should allow audits of its proprietary source codes to ensure that the programming conforms to the OEM's written and approved guidelines.

11.11.5.1 Developing Installation Qualification Challenges

The information gathered above should be used to develop an IQ test protocol. Note that specifications and limitations for all the critical components should be tested wherever possible and practical to ensure that the components are installed according to their designed criteria. Considerations include voltage, loads, noise (e.g., radiofrequency, ground), operating temperatures, humidity, classified environments (e.g., class I, divisions 1 and 2, class II, etc.), pressures, and vibration.

11.11.6 Installation Tests and Startup Protocol

11.11.6.1 Point to Points

Using the electrical schematics, make sure that a point-to-point check of all wires is done, and systematically begin to power up sequences, protecting each branch circuit from power until the branch circuit has been deemed safe. Using the information listed, develop specific tests that are designed to examine the limitations outlined for each instrument.

Here we use a temperature loop process controller, for example:

PROCESS CONTROLLER SPECIFICATIONS

Input voltage: 115 VAC, 50/60 Hz, ±10 % (readings from 103.5–126.5 VAC acceptable), grounded

Power requirements: 2 A at 115 VAC

Input signal: 3-wire RTD, 100 Ω platinum, DIN STD 0.00385 Ω/Ω/°C, shielded at instrument ground

Output signal: 4–20 mA DC, driving into a maximum of 600 Ω resistance

Alarm outputs: High and low alarm contacts provided, with user set point; 2 A at 240 VAC maximum load on alarm contacts

Ambient temperature requirements: 10°C–35°C, 0%–85% humidity, non-condensing

Note: Unit is acceptable for installation in National Electrical Manufacturers Association type 4 enclosures, provided sealing gasket is used. Placement should be horizontal for proper heat dissipation.

An example protocol could take the form of Table 11.1.

11.11.6.2 Software Installation Qualification Tests

Software installation should be verified against the specific software and hardware requirements of the systems. Both hardware and software should be tested in this phase of the project. For the IQ to be successful, the information listed in the functional requirements must have been documented. Also, the software installer should have created a written record of the software installation information, including:

- Software manufacturer's name, address, and telephone number
- Make of software
- Model or version number of software

- Serial number of software (if serial number is given)
- Notes of any problems or special instructions

A form may be created to verify this information using the function requirements and functional specifications as a guideline.

11.11.6.3 Computer Hardware

Because of the complexity of today's modern computer systems, adequate computer hardware diagnostics IQ should also be conducted. Using standard system diagnostics, determine the following:

- Is all random-access memory installed that is required by the operating software?
- Is sufficient hard disk space available for the operating system, and as a buffer for data storage?
- Does the system power up without any error messages?
- Does the power-up sequence test memory, system integrity, etc.?
- Are there any special program diagnostics that can be run? If so, run them and record the results.
- Is there a backup of the entire system in case of equipment failure?
- Is there a record of any firmware installations? If so, obtain and keep a copy.
- Are there any special boards installed on the computer? Is there documentation that can verify that they were installed properly?

11.11.7 Software Backups

There exist many types of software backup systems that "image" the entire hard drive of a computer system. Additionally, there are numerous external hard-drive systems that can automatically and periodically back up the existing system. It would be smart to consider using a backup system such as this to periodically back up the computer to durable media. Data backup can also be made separately. Many times a complete spare computer system may be placed in storage as "insurance" against hardware and/or software crashes.

Experience has shown that these systems can prevent downtime. Is it a validation issue? Not really as much as good business sense. However, the backup and restore process should be tested and verified periodically because computer systems have a way of evolving faster than our industrial needs, and it may become more difficult over time to obtain spare parts for computers in a period as small as 5 years.

11.11.8 Operational Qualifications

OQs require that the following information be generated in the functional specifications:

- Display specifications: If cathode ray tubes (CRTs) are used, or if discrete instrumentation is used, a description of their purpose and detail should be provided of how the instrumentation relates to the process.

Table 11.1 Example Validation Protocol Test Page

Test Number	Specification	Test	Expected Results	Actual Results	(Pass/ Fail)	Initial	Date
1	Controller specification	Measure the input voltage	Input voltage should read between 108 and 123 VAC	______ VAC ______ HZ			
2		Examine the input to the controller	The input shall be a 100-Ω platinum 3-wire RTD, with DIN STD curve 0.00385 Ω/Ω/°C with ground				
3		Examine the calibration certificates	The RTD and display calibration shall be documented				
4		Examine the alarm output connections and loads	High and low alarm output loads shall not exceed 2 A (AC or DC) or 240 VAC	______ VAC ______AMPS			
5		Examine the ambient temperatures, humidity, and operating environment	Ambient temperature of the inside of the control panel shall be between 10°C and 35°C, 0%–85% humidity, noncondensing	Temperature of control panel interior during operation: ___°C@__%relative humidity			
6		Observe the orientation of the instrument	Controller should be mounted horizontally to account for heat dissipation	Record orientation:			

- Security specifications: ***All*** nutraceutical equipment shall show evidence of security systems, be it passive or active. In the event 21 Part 11 Compliance is required, the system must be shown to be secure against tampering. (21 Part 11 Compliance and the European counterpart, Electronic Records, Electronic Signatures [ERES], involve a separate topic that we will not discuss in detail here.)
- Fundamental sequence of operations (FSO; also known as the standard operating procedure): Exactly how is this machine or process to function and interface with the real world, its inputs and outputs, and how is it to be operated on? How is it cleaned? How is it set up for production? How is it torn down after production is complete? How is it disposed of, should that become necessary? (Note: The last item is a European CE requirement.)

Also, the following questions should be asked, and the appropriate information should be gathered:

- Does the machine have several modes of operation? If so, each mode should be discussed in detail by the functional specifications.
- Does the system have any diagnostics? If so, can some of these be used in the OQ process to speed testing?
- Does the system have interlocks? If so, a comprehensive interlock list should be provided in the functional specifications.
- Does the system have alarms? If so, a comprehensive alarm list should be provided in the functional specifications.
- Are there any other specifications unique to this equipment? If so, they should be listed in the MVP, and some sort of testing should be included to verify that the machine operates according to the functional requirements and functional specifications.

11.11.9 Writing the Operational Qualifications

OQs should be conducted in two stages:

- Component operational qualifications (of which calibration can be considered a large part)
- System operational qualifications (does the entire system operate as an integrated whole)

For some field devices, the calibration exercise may be all that is needed to consider the instrument operationally qualified. Input and output are examined and checked for proper operation. For other devices that are not "calibrated" or "calibrateable," such as CRTs, software, PLCs, and computers, special tests must be designed from the functional specifications so that the range of interlocks, alarms, displays, and functional operations are tested adequately to ensure consistent operations.

11.11.10 Use the P&ID

The P&ID is instrumental in verifying system operations. If possible and practical, separate each component of input and output (I/O) signals of the PLC and each critical device, and conduct tests of each item to verify that it operates according to the designated function on the P&ID diagram. If loop diagrams are provided, these are an excellent way to categorize checklists. References to P&ID loops and/or loop diagrams may be used in the OQ tests. After the P&ID has been run through, it can usually be said that the individual field components have been thoroughly tested for operability. Use of a database program may help the validation group conduct an audit. Ask whether the manufacturer uses databases to organize the data, and, if practical, obtain a version of that database to help organize all the components to be tested.

11.11.11 Take Advantage of Diagnostics

Some manufacturers provide diagnostic programs and systems. PLCs normally have force instructions to manually force I/O on and off. Other systems provide maintenance items that allow diagnostic exercises to be run on the various substructures of the equipment. Used properly, the diagnostics are invaluable in saving time and providing assurance that the machine and all its parts are operating properly.

11.11.12 Thoroughly Examine Interlock and Alarms

The interlock and alarm listings are the most comprehensive and organized functional descriptions of the system. Develop tests that exercise these two lists to the specifications. If there are any changes in these lists during startup, make sure that the most current lists are given to the validation team before conducting testing.

11.11.13 Other Operational Tests

If the machine has configurable recipes, tests should be conducted to challenge the recipe engine, boundary conditions of the recipes or sequences, and its operability. An FSO is helpful in providing a test guideline. Also, a method of operation or SOP of various product runs to be used on this machine may help to provide suitable automatic and manual tests on the total functional system.

11.11.14 Use of a Fundamental Sequence of Operations or Standard Operating Procedure

The FSO is a detailed write-up or flowchart that explains the operation of the entire system or machine. It includes sequencing operations and detailed explanations of how the equipment is to perform its function. Taken to its most complete form, the FSO becomes the basis of communication between the equipment designer, programmers, and operators. If one does not exist, you may consider constructing one. It provides insight into the details of the equipment operation that might not surface until the startup phase of the project, when it might be much more difficult and expensive to make modifications to software and/or equipment.

An FSO or SOP is designed to convey the proper operating sequences to the programmer, and serves as a doublecheck of alarms, interlocks, and system functionality. Usually this step is missed during the OQ testing. Some will argue that it is not to be included in the IQ/OQ testing protocol, but rather should be included only in the PQ. In any case, *a comprehensive written sequence is invaluable in understanding and conveying the operation of the entire sequence to the validation team, operators, and maintenance personnel.*

In generating a test, run through a typical operation sequence, preferably using a placebo batch. Several modes may be tested, such as automatic and manual mode. If it is practical, this testing of the FSO may also be a part of the PQ. However, keep in mind that the purpose of testing the *equipment operation* to the FSO is to test for proper *equipment functionality* and not necessarily to test the *process.* Although testing to the FSO in the OQ stage is optional, it should be useful as a tool to verify proper operation of the entire system before the equipment is placed into service.

11.11.15 Process and Performance Qualification

11.11.15.1 Process Scale-Up Testing

Process scale-up testing should be part of the overall PQ. A properly run project should already have scale-up considerations taken under advisement. The laboratory runs of the process should also be well documented and reliable before the system has been placed up for bid. From the laboratory data and scale-up tests, the important items needed to be scrutinized will be evident from the test data. The resultant PQ should exercise the limits of the operational criteria for proper product. In some cases, this may be a multidimensional phase array of several criteria. If possible, equations generated from lab testing should be used to evaluate where the limit points should be. Practical real-world knowledge should also help narrow down the variables for testing and criteria. *It is crucial during this phase of the test qualification construction to have trained personnel involved in creating and approving the PQ procedures.*

PQs at a minimum should include the following:

- Processing at normal (nominal) processing conditions
- Processing at several boundary conditions that would normally be encountered during yearly production
- Processing at less than optimal processing conditions to verify boundary criteria
- Retesting within selected boundary criteria to verify that the proper boundary criteria are acceptable to use in normal production runs
- Testing of alarm and interlock set points resulting from the process boundary testing above

The system should be able to protect itself (within reason) from abnormal processing conditions and alarm, modify, or halt the process accordingly.

11.12 CALIBRATION

Calibration should be conducted in a sequenced manner after the IQ has been completed. Calibrations should at a minimum cover the measuring instrumentation, transmitters, PLC or other processor input and outputs, display readouts, and recording devices.

11.12.1 Problem Areas

Calibration should be performed at the instrument level when possible. Some devices are difficult, if not impossible, to fully calibrate on site either because of their nature or as a result of excessively high costs.

Some examples of these are:

- High-volume airflow stations 1000 cfm and higher
- Infrared temperature sensors
- Dew-point sensors
- Vibration sensors

For the above item classifications, strive to have the vendor obtain the best information possible. In several cases, the sensor supplier may be able to provide a National Institute of Standards and Technology (NIST) certificate or certificate of compliance. If possible, a check should be made on the device by another source. For example, high-volume airflow stations could be checked by an anemometer for approximate compliance. Because this is a mature technology, a lot is known about Pitot tubes and their properties. Therefore, a consistent and proven design is beneficial.

In the case of mirrored dew-point sensors, they are, by the laws of physics, calibration standards in themselves. Verification and certification from the factory that the instrument is reading properly may be all that is required.

Modern dew-point sensors that are not based on mirror technology also require verification and calibration on a regular basis.

Infrared sensors generally work well when what they are measuring has the same albedo. When several products or products that change properties are used, care must be taken to investigate the properties of the devices and compare them with known standards. In many cases, the manufacturer of the sensor may be able to provide NIST certification to a given product for a fee.

Vibration sensors usually have a set method of "calibration" that relies on empirical field tests. For calibration of these items, it is best to refer to the manufacturer's instructions on setup. A NIST traceable vibration source may not be practical to place in the field.

11.12.2 Other Devices

Some measuring elements are somewhat time-consuming to calibrate. In those cases, calibration sheets from the sensor manufacturer may prove useful. It may be more cost-effective to have an outside source calibrate them.

11.12.3 Calibration Tolerances

Calibration tolerances should be governed by the *production process* and are cumulative. For example, we may have an overall need to be ±7°C on a particular product before measureable degradation of quality occurs. Therefore, a suitable guideline for calibration could be relaxed from a stiffer standard that may be required in another part of the process. This may save considerable amounts of money and later calibration hassles. If the tolerance is too strict, it will be difficult to maintain conformance in the devices.

11.12.3.1 Cumulative Errors

The entire sensor loop must be considered to fully evaluate a reading. For example:

- A temperature element has a 0.25°C accuracy.
- Its transmitter (range 0°C–100°C) has a 0.05°C accuracy.
- The PLC input card has a 0.25 % accuracy (translating in this case to ±0.25°C).
  - Therefore, if we take the extreme ranges into account, we get a total accuracy of ±0.55°C (0.25 + 0.05 + 0.25).
  - For the above application, ±7°C variation is possible, but likely a ±3°C variation is possible for the control loop.
  - Thus, leaving a safety margin of 1°C, we can conclude that ±3.0°C overall calibration accuracy is acceptable. That means that an acceptable calibration range for each section would be:
    - ±1.0°C for the element
    - ±0.5°C for the transmitter
    - ±1% for the PLC input (or ±1.0°C)
    - With a 0.5°C margin for error.

11.12.3.2 Conduct a Total Loop Test

A verification check should be made after unit calibration has been completed to verify that the entire system is functioning properly and measures within limitations. This tolerance should take into account the process tolerance of the product, and should fall within the specifications of those tolerances, if the entire system is functioning properly.

11.12.3.3 Calibration Summary

- Specify the product tolerances to the OEM up front for bid.

- If calibration of a device is not available, obtain certificates of compliance and devise other methods for checking the instrument's accuracy.
- Take the sum of the errors into account when creating calibration tolerance specifications.
- Conduct a final instrument loop check of the entire calibrated sensor loop to verify that the entire circuit is operating within specifications.

11.13 CONDUCTING THE TESTS

11.13.1 Proper Up-Front Protocol Work

When the system specifications have been nailed down, the vendor has supplied the equipment; the riggers are finished moving the equipment in place; the electrician's, bricklayer's, and plumber's work have been completed, and it is time to conduct the tests.

11.13.2 Reviews

The following reviews are recommended throughout the life of the startup project:

- Initial audit: Qualify your vendors.
- Initial review: Discuss scope of work, responsibilities, and level of validation required.
- Functional specification review: Audit the functional specifications for correctness, appropriate information, and considerations for real-world problems and inaccuracies.
- In-process review and company audit: A review should be conducted of the vendor during construction of the software so that adherence to accepted programming practices is ensured. Is the vendor using "stubs?" Is the vendor correcting mistakes? Is the vendor keeping revision control on the software and other documentation?
- Factory acceptance test (FAT): An acceptance test of the equipment is recommended whenever possible, so that problem items may be resolved before shipment of the equipment to your facility. This should be part of the original purchase order to conduct an acceptance test before shipment. Make sure that suitable testing has been made on all aspects of the equipment, and that a record of the test results is available for inspection. Ask for a copy of the FAT results. If aspects of the equipment are simulated, ask to see the stub programs or simulation setup.
- Unpacking inspection: An in-depth, detailed inspection of the equipment should occur before acceptance of the equipment on site. A copy of this report should appear in the project file. A shipping list is also helpful.
- IQ review: Review the IQ with the startup crew before installation of the site. Explain to the contractors why the IQ is being conducted, and be aware of any construction timetables and methods used during the construction phase. This is crucial because some equipment may be difficult to access to gain serial number information after the equipment has been installed.

The test protocol should be compared with the functional specifications for test accuracy. If any discrepancies are present, they should be discussed immediately and resolved before validation of the system can occur.

11.13.3 Consider the "Real World"

With few exceptions, startups usually involve adjustments to alarms, interlocks, and operational specifications. Marked-up revisions of the changes should be available to the validation test crew. If possible, the changes to the specifications should be reviewed and typed before execution of the tests. A final test protocol revision should be made before conducting the validation tests. This is usually done in the last stages of startup, so allotment should be made up front to allow for time to conduct a review. This requirement is less important if a machine has been installed before and the unknowns are minimized.

11.13.4 Do Not "Validate Yourself into a Corner"

Sometimes protocol is written such that any change or modification is not allowed and is grounds for revalidation of the entire equipment set. Although this issue is important, consideration of circumstances needs to be weighed when changes to functional specifications occur. Often a few practical adjustments to test procedures, specifications, or protocol are required to address these changes.

In the case where changes to the *functional requirements* are necessary to adjust real-world conditions, an evaluation must be made to the validation protocols so that the full impact of the changes is incorporated into the revised specifications and validation protocols. *Documentation should be provided and should remain in the file to explain these deviations when they occur.*

11.13.5 The Right Tools

Validation and calibration require measuring tools. A proper calibration set of instruments should include NIST-traceable RTD simulator, thermocouple (T/C) simulator, 4- to 20-mA simulator and meter, multimeter, ammeter, differential pressure meters of appropriate ranges, calipers, and level. The validation crew should have NIST-traceable multi-meter, 4- to 20-mA simulator, stopwatch (NIST-certified), and a black or blue, nonsmearing, medium-point pen.

11.13.6 Proper Ways to Complete the Forms

Validation and calibration should be done in a clear and concise manner, neatly, in a patient and orderly fashion. The following rules are useful:

- Always use a pen; ***never use a pencil!***

- When signing a test, write ***Pass*** or ***Fail***. Check marks are not considered acceptable, and unless they are properly documented and allowed for on the form, they are not sufficient evidence that a device has been checked.
- Note ALL deviations from expected outputs.
- Sign each sheet of the validation test procedure using your full signature.
- You should first print clearly and then sign your name on at least one document, so that any auditor can clearly identify your signature with your proper name.
- Review of the completed test procedure is recommended before final acceptance of the final test procedure to verify that all tests were conducted thoroughly and properly.

11.13.7 Proper Ways to Make Corrections to the Forms

As much as we would like to think we are perfect, we are all human, and sometimes mistakes and oversights make it through the most thorough examination. Other times, fate intervenes and the test procedure needs to be adjusted (rewritten) to reflect the actual test conditions and test results.

Making corrections to approved documents is acceptable by the FDA, provided you follow the rules below:

- Make corrections using one line through the mistake, and write the correction clearly and legibly near the line. Initial and date the correction on that piece of paper, as close as possible to the correction. Red is appropriate for corrections, although black ink is also acceptable for corrections.
- Do not "black out" any section of the validation test plan.
- Attachments are acceptable to the FDA, as long as they are written in pen and permanently attached to the paper in question, by stapling or some other means.

11.13.8 What to Do When the System Does Not Work Exactly to the Written Specification

When the scale-up process or system does not work according to planned specification, it is time to regroup and formulate a game plan. The validation personnel should act in a cautious but cooperative manner to make adjustments to the test plans and work with the technical staff to develop upgraded specifications and solutions that match the original functional requirement. If the functional requirement cannot be met by the equipment or process, there are several questions to ask:

- Can the system reliably produce quality product using alternative methods of production? If so, a revision of the functional requirements and specifications may save the day. Revisions to the test plans may be made from the newly revised specifications, approvals signed, and retesting done to verify proper consistency and operation of the equipment.
- Can the system be modified to meet the functional requirements and specifications? Again, this may be expensive but will save the day as well. The validation team only has to adjust its tests to the modifications once they have been properly defined by approved revisions to the functional requirements and specifications.

- Can the requirements be "downgraded" to levels the machine currently can produce? Although unattractive, this alternative is sometimes used by companies and is acceptable as long as safety, integrity, and reliability are maintained to the new standards. In any case, these standards must be approved by the appropriate parties before writing revised test procedures.

The key idea in all three scenarios is to *clearly define* revised procedures and specifications *before* revalidation is allowed to occur. The FDA greatly frowns on adjusting validation procedures to match data without properly authorized approval of the change in process.

11.14 REVALIDATION AND RETESTING PROCEDURES

Revalidation is required when the operating equipment or system has been changed in some way. The following scenarios are good "triggers," such as when:

- System upgrades have been added
- Major mechanical equipment has been replaced
- Computer systems have been replaced
- Additions to the functionality of the system have been made
- New products have to be run on the system (requiring additional PQ testing)
- Critical items have been replaced or repaired

When major revisions are performed on the system, revalidation may have to encompass the entire machine. As a rule, only the portions of the machine that have been modified need to be revalidated. When only a transmitter or element has been replaced, usually a record of calibration of that element, plus a record of calibration for the entire readout loop, is required. For software upgrades and computer system replacements, this effort might be much more involved.

11.14.1 Software and Computer Hardware Upgrades and Fixes

If possible, get a comprehensive list of the changes made to the software that are to be added to the system. If this list is available, specific test protocol may be written to address the changes. However, spot checks of critical functions should also be made to ensure that regular operations have not been affected. If the nature of the software or hardware upgrade is profound or massive (such as a new operating system upgrade), an entire computer system check should be done to verify that the computer system is operating properly. Follow the common-sense plan below when conducting these upgrades.

1. Back up the system before you start. (It is amazing how many people forget this simple step!)
2. Obtain all upgrade specifications before installation, and have the responsible parties approve all aspects of the upgrade.

3. Before conducting the upgrade, ask to see test results of module testing or a complete list of upgrades included for your system.
4. Divide and conquer. Concentrate on those items that are the most meaningful to the operation of the system. Divide the tests into functional segments.
5. Write test plans specific to the items that have been changed.
6. Include some random checks on existing systems that supposedly have not been changed in your revalidation protocol. Especially concentrate on those operations that are "mission critical."
7. Have the testers write down anything abnormal that happens, not just the results of the tests. Investigate any anomalies found. Prepare a final report.
8. Make sure the system is backed up daily during the test and debug phases.

11.14.2 Repair and Replacement of Components

Component repair and replacement is a relatively simple process. A form should exist when doing maintenance or replacement of an item that addresses:

1. The nature of the problem.
2. What has been replaced or repaired (the technician should be specific, using model and serial numbers whenever they are provided).
3. A supposition of why the part failed or needed to be replaced (for total quality management). If the reason is vague, this should trigger some investigation as to exactly why the part failed.
    - If recalibration is required, it should not only be done on the component replaced but also on the entire loop that the component was in, to verify full functionality of the system.
    - Any other systems that the component was connected to that might be affected by incorrect operation of the failed part or addition of the new part should be tested using the same protocol that was used when initially validating the system.

A copy of the maintenance records should be kept for each major repair made.

11.14.3 Calibration and Recalibration

Calibration data often contain pertinent information on the quality of the part being calibrated. If possible, calibration data should be in a database so that trends may be made of the devices. Logging device historical information may yield important information that could help troubleshoot trends of failures. Record:

- The calibration device used, as well as its manufacturer, model number, serial number, and calibration date and calibration reference number, should be clearly marked on the calibration sheet.
- The existing measurements made before calibration of the device has been done. (As found results)—this documents any instrument drift.
- The resulting calibration data of the device after it has been calibrated.

- A description of how the calibration was performed.
- Listing of all data points calibrated.
- A place for comments should be made on the form.
- A "Pass/Fail" criterion (normally tolerances) should appear on the calibration form.
- A place for the calibrator to put "Pass or Fail" is required.
- A place for the calibrator to sign and date the calibration and a place for the calibrator to print his or her name should also be included. (A copy of a printed name and signature should be on file for comparison inspection at any time because often personal signatures may be illegible.)

11.14.4 Recalibration Frequency

Recalibration should be done on a semiannual (6-month) or annual (yearly) basis according to FDA guidelines. However, some items may require recalibration more frequently than others. Those items should be identified, and calibrations should be conducted on a basis suitable to their physical properties.

11.14.5 Retroactive Validations

Retroactive validation must be done on equipment for which records have been lost or validation was not provided for in the initial life of the machine. Retroactive validations become more difficult the older the machine is. Several questions surface when retroactive validations are planned:

- What is the level of validation the machine requires? If the machine is used for a tertiary operation, the level of validation required may not need to be extensive because the machine has a historical record of operation. This should reveal any, if not all, of the problems associated with the machine.
- What is the past performance history of the machine? Past performance of the machine will yield valuable information when planning a validation strategy. Where are its weak points? Talk to the operators and gather informational data on the operation of the machine.
- Is the manufacturer of the machine available? Is the manufacturer cooperative with information about the machine?
- Does the machine have product contact surfaces? If so, are those surfaces kept clean? Do they show any signs of contamination? The manufacturer should be able to provide the design specifications of the materials. Chances are the manufacturer will not have kept material certifications of the parts. Evidence should be gathered from the daily operations. (Basically, if surfaces show no signs of corrosion, they are probably suitable for the purpose.) If a part is in serious doubt, an analysis should be done.
- Is there a written sequence of operation? Is programming well documented? Is a current copy of the programming available for inspection?
- If there is software source code, is the manufacturer willing to have an audit conducted on that source code? Did the manufacturer have programming guidelines in

place at the time of software programming? If so, was the guideline followed for the programming of that machine?
- Does the manufacturer have other machines of that type and model operational in other facilities? What is the history regarding the machine?
- Did the manufacturer keep factory test data on the machine? Did ***you*** conduct an FAT? Is there record of an FAT? Does the manufacturer have a copy of its quality test procedure that was in place to inspect the machine before shipment? After it was installed? After it was started up?
- Is the manual up to date? If not, what has changed?
- Have there been modifications to the machine after it was initially installed? Were there records kept of the modifications?
- Is there a P&ID of the system? Is it up to date? Schematics? Other?

In the modern nutraceutical world, retroactive validation should not be an issue—the systems in place should be in a validated state and maintained that way throughout the life of the machine. However, there are situations when a system that has already been in service needs to be validated (e.g., used equipment). In these cases, a retroactive validation should be performed.

11.14.6 What to Do When Faced with Retroactive Validation

11.14.6.1 Step 1: Size up the Job, Identify Your Team, Write a Master Validation Plan, Gather Data

Size up the job, and follow a written MVP. Your team should be aware of this plan. Gather all the information on the machine that you can. Ask operators for a copy of the manuals and drawings. The most important aspect of the validation is obtaining a written procedure of how the machine operates and its history. A good manual helps. A P&ID is ***invaluable***.

11.14.6.2 Step 2: Create a Functional Requirement if One Has Not Been Made for the System

A functional requirement is the basis for all the testing you are about to do on the machine. Although most times the operating manual can serve as a functional requirement, it may not be sufficiently specific for your product. The functional requirement starts back at the creation of the product. ***Why*** was the machine necessary? ***What*** does it do?

11.14.6.3 Step 3: Create a P&ID if There Is Not One

If there is a P&ID, make sure it is updated. The P&ID is your best roadmap of the system. If one does not exist, take the time to make one. Label all the components, and if possible, place tags on the critical parts. The P&ID is your most useful checklist of functionality and should encompass all aspects of machine functionality.

11.14.6.4 Step 4: Make a Database

Using the P&ID, a database listing will help you organize your plan of attack and identify what is important and what is not. A database need not be in a computer. It can simply be a list of the items in the system, ordered conveniently for reference; however, a computerized database is helpful!

11.14.6.5 Step 5: Gather All Information to Form the Functional Specification

How does the machine perform its function? What I/O does it need to operate? What interlocks does it have? Any alarms? What are its displays? Does it have any diagnostics? What are the operational limitations of its individual components? Of its critical components? All this information should already be provided for you in an organized fashion. If it is not, make sure it becomes available before a test plan is written. The functional specifications are the basis of any good test plan.

11.14.6.6 Step 6: Make a List of All Product Contact Areas, Possible Contamination Sites, and Critical Equipment Items

This list should be done with the aid of suitable process expertise. This identifies all important areas. These should be areas on which you concentrate.

11.14.6.7 Step 7: Write a Fundamental Sequence of Operations

This one is usually tough to accomplish retroactively but is useful if done correctly. Again, the manufacturer or operations people should have written sequences or SOPs, and you can build on those. If an FSO does not exist, generate one and make sure that the appropriate technical personnel sign off on its validity.

11.14.6.8 Step 8: Prioritize

With all the data in front of the team, decide where the most critical areas are, and concentrate on them. Decide where to stop by looking at the priority of the item. When the priorities get trivial, stop. Document the priorities.

11.14.6.9 Step 9: Write the Test Plans

Using the priorities established above, as well as all the information that has been gathered, have the team write the test plans.

11.14.6.10 Step 10: Review the Test Plans

Have the test plans reviewed by at least one other knowledgeable source, and have them approve the final test protocol documentation.

11.14.6.11 Step 11: Conduct the Tests

The reviewed tests may then be run on the equipment. Follow established change procedures, and take any corrective action necessary. From here onward, the validation should proceed as though this were a machine that has just been commissioned.

11.15 MASTER VALIDATION PROTOCOL

11.15.1 Master Validation Protocol: Structure of a Validation Program

The MVP is the outline of the validation for a facility. Its purpose is to provide the users, coordinators, managers, and auditors with a clear guide to validate the equipment and processes in a plant. Simply put, a good MVP will allow an untrained person to work with and understand the entire system, its purpose for being, how it functions, how it is to be tested, who tests it, what the order of the tests is, and the results of those tests. All this should provide a clear path for the FDA to evaluate the machine and/or process in question.

All documentation pertinent to validation and/or qualification should be referenced here and clearly explained to eliminate any ambiguity for their purpose and function. It also provides an overview of the procedures and the recommended workflow for examination of the equipment. It should outline the planning stages of definition, review, tests, and final qualification.

11.15.2 Example Outline of a Master Validation Protocol

I. Introduction
Purpose of plant and scope of validation procedures to be used
II. Reference documentation
Reference P&IDs, laboratory test results, I/O lists, engineering definitions, etc.
III. Procedures
- A. Identification of validation team
- B. Reviews and audits
- C. Test plan creation methodology
- D. Installation qualification
- E. Calibration
- F. Operational qualification
- G. Security
- H. Process qualification

I. Errata: provisions for corrections, change control, new procedures, changes, etc.
J. Maintenance Revalidation: how much and how often?*

FURTHER READING

Chamberlain, R. 1991. *Computer Systems Validation for the Nutraceutical and Medical Device Industries.* Libertyville, IL: Alaren Press.

Chapman, K. 1994. A History of Validation in the United States. Paper presented at the ISPE Annual Meeting, 1994.

Food and Drug Administration. 1987. *Guideline on General Principles of Process Validation.* http://www.fda.gov/drugs/guidancecomplianceregulatoryinformation/guidances/ucm124720.htm.

Food and Drug Administration. 1990. *Code of Federal Regulations, CFR 21: Parts 210 and 211.* http://www.accessdata.fda.gov/scripts/cdrh/cfdocs/cfcfr/cfrsearch.cfm.

* **Disclaimer:** All the terms mentioned in this chapter that are known to be trademarks or service marks have been appropriately capitalized. Use of a term in this chapter should not be regarded as affecting the validity of any trademark or service mark. Microsoft, Microsoft Excel, Access, DOS, and Windows are registered trademarks of Microsoft Corporation. PLC is a registered trademark of Allen-Bradley Corporation. DMACS is a registered trademark of Intellution, Inc. OS/2 is a registered trademark of IBM. GAMP is a registered trademark of ISPE.

CHAPTER 12

Bioprocessing of Marine Products for Nutraceuticals and Functional Food Products

Se-Kwon Kim, Isuru Wijesekara, and Dai-Hung Ngo

CONTENTS

12.1 INTRODUCTION

Marine organisms are rich sources of structurally diverse bioactive compounds with various beneficial biological activities that promote good health. The importance of marine organisms as a source of novel bioactive substances is growing rapidly. With marine species comprising approximately one half of the total global biodiversity, the sea offers an enormous resource for novel compounds (Barrow and Shahidi 2008). Moreover, different kinds of substances have been procured from marine organisms because they live in an exigent, competitive, and aggressive surrounding—different in many aspects from the terrestrial environment, a situation that demands the production of specific and potent active molecules.

Nowadays, with the renewed industrial interest for deriving products from the sea, there is increased need to find different ways to develop more suitable and sustainable processes. It is evident that marine bioprocess engineering will play an important role in this development. The bioprocessing of marine food products to develop novel functional ingredients to be used in the food industry has been previously reviewed (van der Wielen and Cabatingan 1999; Kim and Mendis 2006). Recent studies have identified a number of bioactive compounds from marine foods and marine food processing by-products using various bioprocessing techniques, mainly the ultrafiltration membrane bioreactor system (Shahidi, Vidana Arachchi, and Jeon 1999). Marine food products can be consumed to play a vital role in the growth and development of the body's structural integrity and regulation, as well as other functional properties. Apart from that, marine food proteins have been used as essential raw materials in most industries, but most marine food sources are underused. Recently, both marine-derived food proteins and biopeptides have potential in novel commercial trends because they are widely commercialized in the food, beverage, pharmaceutical, and cosmetic industries, as well as other fields such as photography, textiles, leather, electronic, medicine, and biotechnology. Industrialists are eager to embrace a novel product if it can deliver what consumers want, and at the same time industry needs to balance its involvement against the perceived market potential for a new trend.

Moreover, consumers' demand for food products with functional ingredients has increased recently, and this situation has underlined the need to guarantee the safety, traceability, authenticity, and health benefits of such products. Therefore, commercially available marine food products have also prompted newer challenges to face the consumers demand for bioactive functional foods. Bioactive marine-derived biopeptides, collagen, gelatin, chitooligosaccharide (COS) derivatives, sulfated polysaccharides (SPs), phlorotannins, sterols, carotenoids, and lectins have been isolated by bioprocessing various marine sources, including marine invertebrates, fish, and algae by-products (Table 12.1). They are potential candidates as functional ingredients for new commercial trends in several industries. This chapter presents an overview of membrane bioprocessing of marine food products, as well as their current status and future perspectives in the functional food industry.

Table 12.1 Potential Functional Ingredients from Bioprocessing of Marine Products

Functional Ingredient	Marine Product
Bioactive peptides	Fish, invertebrates, seaweeds
Chitooligosaccharides	Crustaceans
Sulfated polysaccharides	Seaweeds, invertebrates
Phlorotannins	Seaweeds
Lectins	Fish
Fucoxanthin	Seaweeds

12.2 BIOPROCESSING OF MARINE PRODUCTS BY MEMBRANE BIOREACTOR

There is a great potential in the marine bioprocessing industry to convert and use most marine food products and marine food by-products as valuable functional products. Apparently there has been an increasing interest on the use of marine products, and novel bioprocessing technologies are developing for the isolation of bioactive substances from marine food products for use as functional foods and nutraceuticals. Development of these functional ingredients involves certain biotransformation processes through enzyme-mediated hydrolysis in batch reactors. Membrane bioreactor technology equipped with ultrafiltration membranes is recently emerging for the bioprocessing and development of functional ingredients and is considered a potential method to use marine food products efficiently (Nagai and Suzuki 2000; Kim and Rajapakse 2005; Kim and Mendis 2006).

This system has the main advantage that the molecular weight distribution of the desired functional ingredient can be controlled by the adoption of an appropriate ultrafiltration membrane (Cheryan and Mehaia 1990; Kim et al. 1993; Jeon et al. 1999). Enzymatic hydrolysis of marine food products allows preparation of functional ingredients such as bioactive peptides and COS. The physicochemical conditions of the reaction medium, such as the temperature and pH of the reactant solution, must then be adjusted to optimize the activity of the enzyme used (Table 12.2). Proteolytic enzymes from microbes, plants, and animals can be used for the hydrolysis process of marine food products to develop bioactive peptides and COS derivatives. Moreover, one of the most important factors in producing bioactive functional ingredients with desired functional properties is the molecular weight of the bioactive compound. Therefore, for efficient recovery and to obtain bioactive functional ingredients with both a desired molecular size and functional property, an ultrafiltration membrane system is used. To obtain functionally active peptides, use a three-enzyme system for sequential enzymatic digestion. Moreover, it is possible to obtain serial enzymatic digestions in a system using a multistep recycling membrane reactor combined with an ultrafiltration membrane system to bioprocess and develop marine food–derived bioactive peptides and COS derivatives (Jeon et al. 1999; Byun and Kim 2001). Membrane bioreactor technology equipped with ultrafiltration membranes recently has emerged for the development of bioactive compounds and is

Table 12.2 Optimum Conditions of Enzymes to Bioprocessing of Some Marine Products by Membrane Bioreactor

Enzyme	Buffer	pH	Temperature (°C)
Alcalase	0.1 M Na_2HPO_4-NaH_2PO_4	7.0	50
α-Chymotrypsin	0.1 M Na_2HPO_4-NaH_2PO_4	8.0	37
Papain	0.1 M Na_2HPO_4-NaH_2PO_4	6.0	37
Pepsin	0.1 M Glycine-HCl	2.0	37
Neutrase	0.1 M Na_2HPO_4-NaH_2PO_4	8.0	50
Trypsin	0.1 M Na_2HPO_4-NaH_2PO_4	8.0	37

considered a potential method to use marine food products as value-added nutraceuticals with beneficial health effects.

12.3 POTENTIAL FUNCTIONAL INGREDIENTS FROM MARINE PRODUCTS

The bioprocessing of marine food products to improve the functional characteristics of marine food ingredients could be a way for novel food products to be used as functional foods and nutraceuticals. Potential functional products from the bioprocessing of marine food sources are discussed below.

12.3.1 Bioactive Peptides

Components of proteins in marine foods contain sequences of bioactive peptides, which could exert a physiological effect in the body. These short chains of amino acids are inactive within the sequence of the parent protein but can be released during gastrointestinal digestion, food processing, or fermentation. Marine-derived bioactive peptides have been obtained widely by enzymatic hydrolysis of marine proteins (Kim and Wijesekara 2010). In fermented marine food sauces such as blue mussel sauce and oyster sauce, enzymatic hydrolysis has already been done by microorganisms, and bioactive peptides can be purified without further hydrolysis. In addition, several bioactive peptides have been isolated from marine processing by-products or wastes (Kim and Mendis 2006). Marine-derived bioactive peptides have been shown to possess many physiological functions, including antihypertensive or angiotensin-converting enzyme (ACE) inhibition (Yokoyama et al. 1992), anticoagulant (Rajapakse et al. 2005), and antimicrobial (Liu et al. 2008) activities. Moreover, some of these bioactive peptides have been identified to possess nutraceutical potentials that are beneficial in human health promotion (Defelice 1995), and recently the possible roles of food-derived bioactive peptides in reducing the risk of cardiovascular diseases have been reported (Erdmann, Cheung, and Schroder 2008). The use of marine-derived antioxidative peptides (Table 12.3) as natural antioxidants in functional foods is promising.

Table 12.3 Some Antioxidative Bioactive Peptides from Bioprocessing of Marine Products

Amino Acid Sequence of the Peptide	Marine Product
Phe-Asp-Ser-Gly-Pro-Ala-Gly-Val-Leu	Jumbo squid
Leu-Lys-Glu-Glu-Leu-Glu-Asp-Leu-Leu-Glu-Lys-Glu-Glu	Oyster
Phe-Gly-His-Pro-Tyr	Blue mussel
Val-Lys-Ala-Gly-Phe-Ala-Trp-Thr-Ala-Asn-Glu-Glu-Leu-Ser	Tuna
Leu-Leu-Gly-Pro-Gly-Leu-Thr-Asn-His-Ala	Rotifer
Val-Glu-Cys-Tyr-Gly-Pro-Asn-Arg-Pro-Glu-Phe	Microalgae

12.3.2 Chitooligosaccharide Derivatives

Chitin is the second most abundant biopolymer on earth after cellulose and one of the most abundant polysaccharides. It is a glycan of β(1→4)-linked *N*-acetylglucosamine units, and it is widely distributed in crustaceans and insects as the protective exoskeleton and in the cell walls of most fungi. Chitin is usually prepared from the shells of crabs and shrimps. Chitosan, a partially deacetylated polymer of *N*-acetylglucosamine, is prepared by alkaline deacetylation of chitin (Kim, Nghiep, and Rajapakse 2006). COSs are chitosan derivatives (polycationic polymers comprised principally of glucosamine units) and can be generated via either chemical or enzymatic hydrolysis of chitosan (Jeon and Kim 2000a, 2000b). Recently COSs have been the subject of increased attention in terms of their pharmaceutical and medicinal applications (Kim and Rajapakse 2005) because of their lack of toxicity and high solubility, as well as their positive physiological effects such as ACE enzyme inhibition (Hong et al. 1998), antioxidant (Park, Je, and Kim 2003), antimicrobial (Park, Lee, and Kim 2004a), anticancer (Jeon and Kim 2002), antidiabetic (Liu et al. 2007), hypocholesterolemic (Kim et al. 2005), hypoglycemic (Miura et al. 1995), anti–Alzheimer's disease (Yoon, Ngo, and Kim 2009), and anticoagulant (Park et al. 2004a) properties and adipogenesis inhibition (Cho et al. 2008).

12.3.3 Sulfated Polysaccharides

Edible marine algae, sometimes referred as seaweeds, have attracted special interest as good sources of nutrients, and one particular interesting feature is their richness in SPs, the uses of which span the food, cosmetic, and pharmaceutical as well as microbiology and biotechnology industries (Ren 1997). These chemically anionic SP polymers are widespread not only in marine algae but also occur in animals such as mammals and invertebrates. Marine algae are the most important source of nonanimal SPs, and the chemical structure of these polymers varies according to the species of algae. The amount of SPs present is found to differ according to the three major divisions of marine algae: *Chlorophyceae* (green algae), *Rhodophyceae* (red algae), and *Phaeophyceae* (brown algae). The major SPs found in marine algae include fucoidan and laminarans of brown algae, carrageenan of red algae, and ulvan of green algae. In recent years, various SPs isolated from marine algae have attracted much attention in the fields of food, cosmetics, and pharmacology. Carrageenans, a family of SPs isolated from marine red algae, are widely used as food additives, such as emulsifiers, stabilizers, and thickeners (Chen et al. 2007). Ulvan displays several physiochemical and biological features of potential interest for food, pharmaceutical, agricultural, and chemical applications. Compared with other SPs, fucoidans are widely available commercially from various cheap sources; hence fucoidans have increasingly been investigated in recent years to develop drugs and functional foods.

Table 12.4 Phlorotannins and Potential Health Effects

Phlorotannin	Health Effect
6,6′-Bieckol	Anti-HIV
Dioxinodehydroeckol	Antiproliferative
Eckol	Radioprotective
Dieckol	Antidiabetic
7-Phloroeckol	Skin protection
Fucodiphloroethol G	Antiallergic
Phlorofucofuroeckol A	Antiallergic

12.3.4 Phlorotannins

Phlorotannins are phenolic compounds formed by the polymerization of phloroglucinol, or defined as 1,3,5-trihydroxybenzene monomer units and biosynthesized through the acetate–malonate pathway. They are highly hydrophilic components with a wide range of molecular sizes ranging between 126 and 650,000 Da (Ragan and Glombitza 1986). Marine brown algae and red algae accumulate a variety of phloroglucinol-based polyphenols, and phlorotannins could be used in functional foods with potential health effects (Table 12.4). Among marine algae, *Ecklonia cava*, edible brown algae, is a richer source of phlorotannins than others (Heo et al. 2005). Phlorotannins have several biological activities that are beneficial to health, including antioxidant (Li et al. 2009), anti-HIV (Artan et al. 2008), antiproliferative (Kong et al. 2009), anti-inflammatory (Jung et al. 2009), radioprotective (Zhang et al. 2008), antidiabetic (Lee et al. 2009), anti–Alzheimer's disease (acetyl- and butyryl-cholinesterase inhibitory) (Yoon et al. 2009), antimicrobial (Nagayama et al. 2002), and antihypertensive (Jung et al. 2006) activities. Hence phlorotannins are potential functional ingredients.

12.3.5 Lectins

Lectins are carbohydrate-binding proteins or glycoproteins that are playing a role in the valuable commercial trends in the food and biomedical industries. In the food industry, calcium-binding marine food–derived lectins can be incorporated in nutraceuticals or functional food to prevent calcium deficiency (Jung et al. 2003). Moreover, some marine algae lectins can be developed as antibiotics against marine vibrios.

12.4 CONCLUDING REMARKS

Recent studies have provided evidence that marine-derived functional ingredients play a vital role in human health and nutrition. The possibilities of designing new functional foods and nutraceuticals from marine products by ultrafiltration membrane bioreactor are promising. In addition, marine food processing by-products like food proteins can be easily used for producing nutraceuticals and functional foods via membrane bioprocessing. This evidence suggests that because of their valuable biological functions with beneficial health effects, bioprocessed marine-derived foods have potential as active ingredients for the preparation of various functional foods and nutraceutical products. Furthermore, the wide range of biological activities associated with the natural compounds derived from marine food sources has potential to expand their beneficial health value not only in the food industry but also in the pharmaceutical and cosmeceutical industries.

REFERENCES

Artan, M., Y. Li, F. Karadeniz, S. H. Lee, M. M. Kim, and S. K. Kim. 2008. Anti-HIV-1 Activity of Phloroglucinol Derivative, 6,6′-Bieckol, from *Ecklonia cava*. *Bioorganic and Medicinal Chemistry* 16: 7921–26.

Barrow, C., and F. Shahidi. 2008. *Marine Nutraceuticals and Functional Foods*. Boca Raton, FL: CRC Press.

Byun, H. G., and S. K. Kim. 2001. Purification and Characterization of Angiotensin I Converting Enzyme (ACE) Inhibitory Peptides from Alaska Pollack (*Theragra chalcogramma*) Skin. *Process Biochemistry* 36: 1155–62.

Chen, H., X. Yan, J. Lin, F. Wang, and W. Xu. 2007. Depolymerized Products of λ-Carrageenan as a Potent Angiogenesis Inhibitor. *Journal of Agricultural and Food Chemistry* 55: 6910–17.

Cheryan, M., and M. A. Mehaia. 1990. Membrane Bioreactors: Enzyme Process. *Biotechnology and Food Process Engineering*. Edited by H. Schwartzberg and M. A. Rao. New York: Marcel Dekker.

Cho, E. J., A. Rahman, S. W. Kim, Y. M. Baek, H. J. Hwang, J. Y. Oh, H. S. Hwang, S. K. Lee, and J. W. Yun. 2008. Chitosan Oligosaccharides Inhibit Adipogenesis in 3T3-L1 Adipocytes. *Journal of Microbiology and Biotechnology* 18: 80–87.

Defelice, S. L. 1995. The Nutritional Revolution: Its Impact on Food Industry R&D. *Trends Food Science & Technology* 6: 59–61.

Erdmann, K., B. W. Y. Cheung, and H. Schroder. 2008. The Possible Roles of Food-Derived Bioactive Peptides in Reducing the Risk of Cardiovascular Disease. *Journal of Nutritional Biochemistry* 19: 643–54.

Heo, S. J., E. U. Park, K. W. Lee, and Y. J. Jeon. 2005. Antioxidant Activities of Enzymatic Extracts from Brown Seaweeds. *Bioresource Technology* 96: 1613–23.

Hong, S. P., M. H. Kim, S. W. Oh, C. H. Han, and Y. H. Kim. 1998. ACE Inhibitory and Antihypertensive Effect of Chitosan Oligosaccharides in SHR. *Korean Journal of Food Science and Technology* 30: 1476–79.

Jeon, Y. J., H. G. Byun, and S. K. Kim. 1999. Improvement of Functional Properties of Cod Frame Protein Hydrolysates Using Ultrafiltration Membranes. *Process Biochemistry* 35: 471–78.

Jeon, Y. J., and S. K. Kim. 2000a. Continuous Production of Chitooligosaccharides Using a Dual Reactor System. *Process Biochemistry* 35: 623–32.

Jeon, Y. J., and S. K. Kim. 2000b. Production of Chitooligosaccharides Using Ultrafiltration Membrane Reactor and Their Antibacterial Activity. *Carbohydrate Polymers* 41: 133–41.

Jeon, Y. J., and S. K. Kim. 2002. Antitumor Activity of Chitosan Oligosaccharides Produced in Ultrafiltration Membrane Reactor System. *Journal of Microbiology and Biotechnology* 12: 503–07.

Jung, H. A., S. K. Hyun, H. R. Kim, and J. S. Choi. 2006. Angiotensin-Converting Enzyme I Inhibitory Activity of Phlorotannins from *Ecklonia stolonifera*. *Fisheries Science* 72: 1292–99.

Jung, W. K., Y. W. Ahn, S. H. Lee, Y. H. Choi, S. K. Kim, S. S. Yea, I. Choi, S. G. Park, S. K. Seo, S. W. Lee, and I. W. Choi. 2009. *Ecklonia cava* Ethanolic Extracts Inhibit Lipopolysaccharide-Induced Cyclooxygenase-2 and Inducible Nitric Oxide Synthase Expression in BV2 Microglia *via* the MAP Kinase and NF-kB Pathways. *Food Chemical Toxicology* 47: 410–17.

Kim, K. N., E. S. Joo, K. I. Kim, S. K. Kim, H. P. Yang, and Y. Jeon. 2005. Effect of Chitosan Oligosaccharides on Cholesterol Level and Antioxidant Enzyme Activities in Hypercholesterolemic Rat. *Journal of Korean Society of Food Science and Nutrition* 34: 36–41.

Kim, S. K., H. G. Byun, T. J. Kang, and D. J. Song. 1993. Enzymatic Hydrolysis of Yellowfin Sole Skin Gelatin in a Continuous Hollow Fiber Membrane Reactor. *Bulletin of Korean Fisheries Society* 26: 120–32.

Kim, S. K., and E. Mendis. 2006. Bioactive Compounds from Marine Processing Byproducts—A Review. *Food Research International* 39: 383–93.

Kim, S. K., N. D. Nghiep, and N. Rajapakse. 2006. Therapeutic Prospectives of Chitin, Chitosan and Their Derivatives. *Journal of Chitin Chitosan* 11: 1–10.

Kim, S. K., and N. Rajapakse. 2005. Enzymatic Production and Biological Activities of Chitosan Oligosaccharides (COS): A Review. *Carbohydrate Polymers* 62: 357–68.

Kim, S. K., and I. Wijesekara. 2010. Development and Biological Activities of Marine-Derived Bioactive Peptides: A Review. *Journal of Functional Foods* 2: 1–9.

Kong, C. S., J. A. Kim, N. Y. Yoon, and S. K. Kim. 2009. Induction of Apoptosis by Phloroglucinol Derivative from *Ecklonia cava* in MCF-7 Human Breast Cancer Cells. *Food Chemical Toxicology* 47: 1653–58.

Lee, S. H., Y. Li, F. Karadeniz, M. M. Kim, and S. K. Kim. 2009. α–Glycosidase and α–Amylase Inhibitory Activities of Phloroglucinal Derivatives from Edible Marine Brown Alga, *Ecklonia cava*. *Journal of Science of Food Agriculture* 89: 1552–58.

Li, Y., Z. J. Qian, B. M. Ryu, S. H. Lee, M. M. Kim, and S. K. Kim. 2009. Chemical Components and Its Antioxidant Properties in Vitro: An Edible Marine Brown Alga, *Ecklonia cava*. *Bioorganic and Medicinal Chemistry* 17: 1963–73.

Liu, B., W. S. Liu, B. Q. Han, and Y. Y. Sun. 2007. Antidiabetic Effects of Chito-oligosaccharides on Pancreatic Islet Cells in Streptozotocin-Induced Diabetic Rats. *World Journal of Gastroenterology* 13: 725–31.

Liu, Z., S. Dong, J. Xu, M. Zeng, H. Song, and Y. Zhao. 2008. Production of Cysteine-Rich Antimicrobial Peptide by Digestion of Oyster (*Crassostrea gigas*) with Alcalase and Bromelin. *Food Control* 19: 231–35.

Miura, T., M. Usami, Y. Tsuura, H. Ishida, and Y. Seino. 1995. Hypoglycemic and Hypolipidemic Effect of Chitosan in Normal and Neonatal Streptozotocin-Induced Diabetic Mice. *Biological Pharmaceutical Bulletin* 18: 1623–25.

Nagai, T., and N. Suzuki. 2000. Isolation of Collagen from Fish Waste Material—Skin, Bone, and Fins. *Food Chemistry* 68: 277–81.

Nagayama, K., Y. Iwamura, T. Shibata, I. Hirayama, and T. Nakamura. 2002. Bactericidal Activity of Phlorotannins from the Brown Alga *Ecklonia kurome*. *Journal of Antimicrobial Chemotherapy* 50: 889–93.

Park, P. J., J. Y. Je, W. K. Jung, and S. K. Kim. 2004a. Anticoagulant Activity of Heterochitosans and Their Oligosaccharide Sulfates. *European Food Research and Technology* 219: 529–33.

Park, P. J., J. Y. Je, and S. K. Kim. 2003. Free Radical Scavenging Activity of Chitooligosaccharides by Electron Spin Resonance Spectrometry. *Journal of Agricultural and Food Chemistry* 51: 4624–27.

Park, P. J., H. K. Lee, and S. K. Kim. 2004b. Preparation of Hetero-chitooligosaccharides and Their Antimicrobial Activity on *Vibrio parahaemolyticus*. *Journal of Microbiology and Biotechnology* 14: 41–47.

Ragan, M.A., and K. W. Glombitza. 1986. *Handbook of Physiological Methods*. Cambridge, UK: Cambridge University Press, 129–241.

Rajapakse, N., W. K. Jung, E. Mendis, S. H. Moon, and S. K. Kim. 2005. A Novel Anticoagulant Purified from Fish Protein Hydrolysate Inhibits Factor XIIa and Platelet Aggregation. *Life Sciences* 76: 2607–19.

Ren, D. 1997. Biotechnology and the Red Seaweed Polysaccharide Industry: Status, Needs and Prospects. *Trends in Biotechnology* 15: 9–14.

Shahidi, F., J. K. Vidana Arachchi, and Y. J. Jeon. 1999. Food Applications of Chitin and Chitosans. *Trends in Food Science and Technology* 10: 37–51.

van der Weilen, L. A. M., and L. K. Cabatingan. 1999. Fishing Products from the Sea—Rational Downstream Processing of Marine Bioproducts. *Journal of Biotechnology* 70: 363–71.

Yokoyama, K. H., H. Chiba, and M. Yoshikawa. 1992. Peptide Inhibitors for Angiotensin-I-Converting Enzyme from Thermolysin Digest of Dried Bonito. *Bioscience, Biotechnology and Biochemistry* 56: 1541–45.

Yoon, N. Y., S. H. Lee, Y. Li, and S. K. Kim. 2009. Phlorotannins from *Ishige okamurae* and Their Acetyl- and Butyry-Lcholinesterase Inhibitory Effects. *Journal of Functional Foods* 1: 331–35.

Yoon, N. Y., D. N. Ngo, and S. K. Kim. 2009. Acetylcholinesterase Inhibitory Activity of Novel Chitooligosaccharide Derivatives. *Carbohydrates Polymers* 78: 869–72.

Zhang, R., K. A. Kang, M. J. Piao, D. O. Ko, Z. H. Wang, I. K. Lee, B. J. Kim, Y. I. Jeong, T. Shin, J. W. Park, N. H. Lee, and J. W. Hyun. 2008. Eckol Protects V79-4 Lung Fibroblast Cells against ύ-Ray Radiation-Induced Apoptosis *via* the Scavenging of Reactive Oxygen Species and Inhibiting of the c-Jun NH_2-Terminal Kinase Pathway. *European Journal of Pharmacology* 591: 114–23.

CHAPTER 13

The Packaging of Nutraceuticals Derived from Plants

Melvin A. Pascall

CONTENTS

13.1 INTRODUCTION

Bioactive compounds in plants exist in relatively small quantities and are known to exhibit nutritional and healing characteristics. These compounds exist as a wide variety of chemical structures, and the functions they accomplish in the human body vary according to the health effect they induce. These include improved performance of targeted organs and/or bodily functions. Examples of these chemicals include phenolic and organosulfur compounds. Examples of phenolic compounds include flavonoids, phytoestrogens, hydroxytyrosols, resveratrol, and lycopene. These compounds can be found in a variety of plant products such as cereal grains, fruits, vegetables, tea, nuts, legumes, herbs, spices, red wine, and edible oils such as olive, flax, and soy (Kris-Etherton et al. 2002). The plant structures from which these compounds are found vary from roots, stems (including tubers), leaves, flowers, barks, bulbs, fruits, seeds, buds, stalks, and clusters. In some plants, the bioactive compounds can be found in secretions such as resin, latex, and sap.

In many cases, the health benefits of bioactive compounds from plants are obtained without performing any processing. For example, vegetables, fruits, and seeds can be eaten raw. In some cases, the plants are consumed after minimal processing, such as cooking at home. However, if the bioactive compounds are to be extracted and preserved for long-term use or in the case of commercial operations, the plants must be harvested, cleaned, preserved, packaged, and stored so that the maximum benefits are not compromised and the bioactive compounds are available when needed. If these processing steps are not properly done, the plants can be a vehicle for bacterial, fungal, or infestation cross-contamination; the bioactive compounds can be deactivated by chemical or radiant action; or they can be consumed by microorganisms or insect pests.

After processing, the packaging of plant products is a proven method to protect them during long-term storage and for future use. This is important because environmental factors (e.g., temperature, relative humidity, and oxygen) are known to affect the quality of these compounds (Lusina et al. 2005). In addition to this, packaging offers many other useful benefits. These include labeling for easy identification, warnings, instructions, and nutritional information; ease of transportation;

security from pilferage; protection from contamination; convenience; and marketing. Packaging comes in a variety of shapes, sizes, and material types, and packaging should be carefully chosen to maximize the benefits that it is expected to impart to the bioactive compounds. Thus, if a bioactive compound is susceptible to rancidity, the package should protect the plant product from light, oxygen, and moisture. Because bacteria, fungi, and insects live and strive in moist environments, most plant products are dried and then packaged for long-term storage. In such cases, the packaging should prevent the ingression of moisture into the product from the storage environment. This dictates that a material such as glass, metal, or ceramic, which has a high barrier to moisture, should be considered. For protection against light, for example, a dark storage environment, a colored package, or a container wrapped by a label should be considered.

In some cases, the bioactive compound may be packaged, stored, and transported in a liquid form. In such cases, the package may be required to prevent the loss of moisture from the product. In addition to this, consideration should be given to the absorption (sorption) of the bioactive compound by the packaging material. Because many plant bioactive compounds are organic in nature, the choice of plastic packaging should be carefully considered so that the one chosen has a low partitioning for the bioactive compounds. If a high partitioning exists between the packaging material and the compounds in a liquid product, there may be either high levels of chemical migration from the package to the product or a high level of sorption of the compounds by the packaging material. In the case of migration, the safety of the product can be compromised by the uptake of toxic moieties from the packaging (Jeon et al. 2007). In the case of sorption, some of the bioactive compounds can be removed by the packaging material.

In the following subsections of this chapter, the types and characteristics of packaging and the materials used to make packaging will be discussed. This will be considered in relation to the different forms in which bioactive compounds are stored and marketed. Factors responsible for spoilage of these products will be examined, and how packaging plays a role in preserving bioactive compounds will be discussed. Because many bioactive compounds are marketed as food, drugs, dietary supplements, cosmetics, and other pharmaceutical products, the legal requirements of labeling will be discussed. This includes the need to inform the public about the nutritional content, instructions on usage, dosage, warnings, claims, and manufacturers' contact information, for example. The subject of tamper-evident packaging will also be discussed.

Packaging can be divided into three main categories depending on its function. A primary package is one that is in direct contact with the product. It is designed to protect the product from factors that can influence a reduction in its shelf life. A secondary package contains a number of primary packages, and in addition to providing protection to the primary package, it facilitates retail marketing, advertising, and transportation. Tertiary packaging is used mainly for cushioning, storage, and transportation and can contain a number of secondary packages.

13.2 MARKETING OF BIOACTIVE PLANT PRODUCTS

Bioactive compounds are marketed in a variety of forms. These include liquids, granules (including tablets), powders, creams, and gels.

13.2.1 Liquids

Bioactive compounds in liquid products come in several forms, including teas, juices, extracts, oils, syrups, and red wines. Although many of them are ingested orally or can be used as enemas, some are for external or pseudoexternal uses, such as eye or ear drops and mouthwashes. Depending on the method of administration, liquid bioactive compounds should be packaged for adequate protection, convenience of use, accuracy of dosage, and the provision of pertinent information to the consuming public. If the liquid product is sensitive to light, then the container should be impermeable to light and colored or wrapped in an opaque label. To minimize the chances of overdose, some manufacturers include a suitable measure cup within the secondary package. This secondary package can also serve to advertise the product and act as a source of cushioning. In the case of glass bottles, the packaging helps to reduce breakage because it prevents them from physically contacting each other. It also prevents bruising of metal containers and the resultant removal or damage to the label.

Depending on the intended use of the product, consideration should be given to the type of closure used on the package. Eye and ear drops may be required to be packaged in bottles fitted with a dropper within the cap. Because some of these products are measured by droplets, it is essential to ensure that the material used to make the pipette is inert to the product because it may be expected to be in contact with the compound for a prolonged period. Because glass is inert to most liquids, it may be the material of first choice. Factors that should be considered when choosing glass would be the cost and increased weight when compared with plastic.

In general, because liquids are in more intimate contact with the container that is used to package them, care must be taken to ensure that the liquids are compatible with the packaging material. Some metals will corrode if the liquid is too acidic, but coated cans may be considered to overcome this problem. In the case of plastic containers, the mass transfer of chemical from or to the polymeric material or the contents can occur depending on the formulation of the product. In almost all cases, wines are not packaged in plastic containers because the alcohol they contain can be absorbed by the material, and this may act to plasticize the polymeric chains and result in a breakdown of the polymer. Limonene, for example, is a flavor compound that is found in citrus fruits. Polyolefins are susceptible to attack by limonene and can break down when in contact with this chemical additive. Because paper and linen offer limited containment, they are not suitable for the packaging of liquid products, except when the paper is part of a composite material. Composites are made from different types of materials that are laminated together. Each sublayer in the composite is chosen to impart a specific advantage to the material in its entirety. Figure 13.1 shows an example of composite packaging.

Figure 13.1 Various "brick-type" packages made from composite material. A gable-top package is shown on the left.

13.2.2 Granules (Pills, Tablets) and Powders

When bioactive plants are harvested and dried, a convenient method of storage is in the form of powders, pills, granules, or in a partially crushed state. This reduces volume and makes them more convenient for transportation and storage. The parts of a plant that can be crushed, chopped, or ground include the leaves, stems, barks, roots, seeds, fruits, and flowers. Before any form of size reduction, the plant product is usually dried. In some cases, drying is done after the size reduction. Drying is a progressive removal of moisture from the tissue of the plant. When properly dried, the potential for enzymatic activities and bacterial and fungal growth is significantly reduced, especially if the crushed product has a water activity of less than 0.85. To maintain the plant product in this dried state, the packaging should act to eliminate or significantly reduce the ingression of moisture toward the product. This means that the packaging material should have a high barrier to moisture. Even though this might be the case in glass and metal containers, leaks can develop if the closure is defective, not properly applied, or is incompatible with the sealing surface of the main package. In some cases, a consumer may not consume the entire product after opening the package. This demands that the closure system permits the container to be adequately resealed by the consumer. In an attempt to limit the presence of high moisture contents in the headspace of the package either before or after opening by a consumer, some manufacturers include a sachet with a desiccant inside the package. This method of using additional remedies to enable the package to maintain a desired atmosphere within the product is referred to as *modified atmosphere packaging* and will be discussed in more detail later in Section 13.6. If too much moisture enters the package, it can cause caking of powders, swelling and cracking of pills, loss in texture to baked products, reoccurrence of enzymatic activities within the plant tissue, and growth of microorganisms.

When used to market a plant bioactive compound, small pellets, tablets, or pills are often coated with an edible additive, especially if the compound is expected to be swallowed whole or chewed. This coating is sometimes referred to as *edible packaging* and is designed to make the consumption of the product more palatable (Gil et al. 2008). Edible packaging is usually made of starches or proteins. When two-piece capsules are used, the outer coat is made from edible plastics. When over-the-counter capsules are used to package plant bioactive compounds, Food and Drug Administration (FDA) regulations mandate the use of a colored tamper-evident strip between the two halves of the capsules. This is designed to alert consumers to any attempt of intentional tampering with the product. This regulation arose as a result of the Tylenol poisoning incident that took place in 1982 (Greenberg 1996).

13.2.3 Creams and Gels

Creams containing plant bioactive compounds are made by combining and mixing vegetable oils and glycerin with beeswax and tincture of benzoin as a preservative, for example (Pamplona-Roger 1998). In some formulations, vitamin E is used as an antioxidant. Because a small amount of sterile water is added, lecithin is used as an emulsifying agent to make the product homogeneous in nature and prevent separation of the water from the oils. The bioactive ingredient from the plant is usually added and blended with this mixture. Examples of these bioactive compounds include aloe vera, cocoa butter, and anhydrous lanolin. The main storage problems with creams are oxidation and resultant rancidity of the product and microbial contamination. If oxidation occurs, it can cause the development of off-odors and a loss in the consistency of the product. The products of oxidation can also cause degradation of bioactive compounds. Because creams are mainly for external use, flavor is not an issue that warrants consideration. To minimize the potential for oxidation to develop within the packaged product, an inert gas blanket can be injected into the headspace that lies between the product and the closure or cap of the container (Mahanjan et al. 2005). This inert gas may be nitrogen or a mixture of nitrogen and carbon dioxide. The use of a chosen headspace gas that replaces air within a package is also referred to as *modified atmosphere packaging.*

To prevent bacterial contamination, bottles and jars used to package these products should not have pinholes and should be made with adequate resealable devices. To prevent pilferage in stores, consideration should be given to the use of tamper-evident devices. These include the addition of shrink bands, pull-tabs, or other suitable devices. This is necessary because some consumers are in the habit of opening and looking at or sampling items, then deciding whether they want to purchase them. This habit has the potential to contaminate the product with bacteria or harmful chemicals while it is still on the store shelf.

Many bioactive compounds can be used in the form of gels. Examples of these are seen in hair products, lip balms, and deodorants. Gels are liquids, emulsions, and suspensions into which gelling agents are dissolved. These agents thicken and stabilize the liquid phase and form a colloidal mixture. This results in a relatively stabilized structure that looks solid in appearance although mostly composed of a liquid. Figure 13.2 shows containers used to package creams and gels.

Figure 13.2 A collapsible tube and plastic bottles with various types of dispensers used to package gels and creams.

Gels are susceptible to oxidation, microbial contamination, and deactivation of the bioactive ingredients they contain. These products can also lose consistency if stored in conditions of high temperature or if subjected to severe vibrations over time. Because gels are high moisture products, they can lose moisture and become hard. Thus, the packaging types to be used for these products should protect them against light, heat, moisture loss, oxygen ingression, and microbial contamination. For convenience and ease of dispensing, gels are usually packaged in jars, squeezable bottles, disposable pouches, or seamless collapsible tubes. Pouches are usually made of plastic films, whereas tubes are mainly made from aluminum or plastic. In most cases, the packaging offers little protection from temperature abuse.

13.3 FACTORS INFLUENCING SPOILAGE AND LOSS OF SHELF LIFE

Shelf life is defined as the length of time that a bioactive compound is given before it is declared unfit for consumption and unsuitable for sale to the consuming public. In the United States, this time is usually printed on the packaging as a *best before*, *use by*, or *sell by* date. The natural shelf life of plant-based products depends on the species, maturity, and the part of the plant that is used to obtain the bioactive compound. In an unprocessed form, most seeds have a longer shelf life if kept dried and at relatively cool temperatures. Fruits, leaves, and flowers, on the other hand, have a relatively shorter shelf life. Barks, stems, and roots can remain usable for a fairly long period if they are taken from mature parts of the plant or, if needed, dried to a significantly lower moisture content. They may not need expensive packaging if stored in a dry environment that is free from insects.

The shelf life of bioactive plant compounds can be considerably increased if moisture is removed from the tissue and the product is stored in a dry environment (Farag Badawy, Gawronski, and Alverez 2001). The removal of moisture reduces enzymatic activity and microbial growth within the product. Without drying, fruits, flowers, and leaves will have a fairly short shelf life. Moisture within the tissues of plants exists in two forms: bound and free water. Bound water can be seen as H-O-H molecules that form hydrogen bonds between the molecules that make up the plant tissues. This water is unavailable for enzymatic action or for bacterial growth. Free water, on the other hand, is not chemically bound to the molecules in the tissue of the plant and can be used by microorganisms for growth and survival. Thus, to extend the shelf life of plant tissue, the free water must be removed during the drying process. Once this occurs, packaging and storage conditions must prevent its return if an extended shelf life is required. Pills exposed to high moisture environments can absorb water, swell, and eventually crack. If the bioactive compound is extracted and stored as a liquid, cream, gel, syrup, latex, or sap, it may have a tendency to lose moisture. In this case, the product would experience a decrease in its free water content, especially if the product is stored at a high temperature. In such cases, the packaging would be required to keep a desirable level of moisture within the product and limit its evaporation to the external environment. The loss of moisture in such cases can lead to caking and a progressive hardening of the product. If the bioactive compound is a bakery product, this can cause undesirable texture changes and an increase in the susceptibility to staling.

In most plant-based products, seeds generally have the highest lipid content. This makes them highly susceptible to lipid oxidation. Fruits, such as avocadoes, are also known to have fairly high oil content. Typically lipids from any source are the most chemically unstable components of foods, especially those from plant origins and which have fatty acids with one or more unsaturated double bonds. Fats and oils are susceptible to free-radical chain reactions that not only deteriorate the lipids but also produce oxidative degradation products that produce off-flavors; degrade proteins, vitamins, and pigments; and can produce toxic compounds in some cases. Many bioactive compounds are not spared from the destructive effect of lipid oxidation. Lycopene, for example, is the pigment principally responsible for the characteristic deep-red color of ripe tomato fruits and tomato products. Several researchers have reported about its natural antioxidant properties and its ability to provide protection against a broad range of epithelial cancers. However, the main causes of tomato lycopene degradation during processing are isomerization and oxidation. Undesirable degradation of lycopene not only affects the sensory quality of the final products but also the health benefit of tomato-based foods for the human body (Shi and Le Maguer 2000).

The main factors responsible for the oxidation of plant-based products include oxygen, moisture, and irradiation and photo-oxidation caused by light. In some cases, various pro-oxidants in the tissue of the plant can act to accelerate the oxidative process once it begins. Fortunately, plants also have several antioxidants that work to hinder the oxidative process. Examples of these antioxidants include α-tocopherol, β-carotene, and ascorbic acid. A detailed discussion of the oxidation process is beyond the scope of this book. Several packaging interventions can be used to reduce the incidence of oxidation in processed plant bioactive compounds. These include the use of packaging made with materials that are high barriers to oxygen and moisture, such as glass, metal, and plastic films with laminated sublayers that are good barriers. Composite materials, with aluminum foil as one of the components, are also known to act as high barriers. Another approach is the use of modified atmosphere or active packaging. These techniques are known to work synergistically with the packaging material to limit the ingression of oxygen, or they can remove oxygen trapped within the product after packaging. Coloring or the use of a label barrier is also an effective method of limiting the harmful effect of light in causing photo-oxidation.

13.3.1 Microbial Deterioration

Cuticles and barks are external protective coverings found on plants. Cuticle is made of a waxy compound and can be found as a covering for leaves, fruits, and the stems of some shrubs, especially in younger plants. On the other hand, bark is the outer covering of the main trunk of trees, woody branches, stems, roots, and other woody plants. If the cuticle of a fresh-cut leaf or fruit is broken, the soft tissues of the plant can be attacked by bacteria and fungi. However, if the harvested plant is dried, the potential for microorganisms to grow on crushed and broken leaves and stem tissue is significantly reduced. If the dried plant tissue is subsequently exposed to high levels of moisture, it can be easily attacked by bacteria and fungi, which can alter its bioactive compounds. Pamplona-Roger (1998) reported that bacteria need more than 40% humidity to reproduce, and fungi need 15%–20%. A well-dried plant will not contain more than 10% moisture; thus, those microorganisms cannot reproduce. Therefore, the role of packaging in protecting bioactive compounds in dried plants would be to keep the moisture content below the critical limit of 10%.

13.3.2 Insect Infestation

Insect infestation refers to an attack by one or more different species of insects. In the case of dried plant parts, it may serve as a source of food for the insects. The attack may not be by adult insects but can be from insects in various juvenile development stages. This may not have been initiated as a problem with packaging, but rather can be related to the cleanliness of the plant when it was harvested or the condition in which it was stored. If the plant parts were infested with insect eggs before harvesting and the drying conditions were not sufficiently harsh to kill them, the eggs may hatch at a latter time and cause the growth of more mature insect subtypes. FDA regulations limit the quantity of insect parts that can be found in plant-based products that are consumed by humans and/or animals. This is documented in the FDA's *Defect Levels Handbook* and is referred to as the "defect action level" for unavoidable additives (FDA 2009).

13.3.3 Chemical Mass Transfer to and from Plastic Material

The mass transfer of chemical moieties from packaging materials to the product occurs mainly from plastic containers or cans coated with polymers. This is referred to as *chemical migration*. The converse is also possible and is called *sorption*. This occurs when certain chemicals from the packaged product migrate toward the plastic container. Plastic is made of polymers that are organic in nature. Although the polymers themselves do not migrate, the same cannot be said for the many additives that they contain. These additives include plasticizers, residual monomers, solvents from printing inks, flame retardants, antislip and static agents, antioxidants, ultraviolet absorbents, and curing agents, for example. These additives are organic compounds that are compatible with the organic nature of the polymer. However, because many food and pharmaceutical products that contain bioactive compounds are either organic or may have additives that are organic, a partitioning between these additives for the packaged product can develop. If this partitioning is stronger than that for the polymer, large concentrations of these additives can migrate out of the plastic and into the packaged product. FDA regulations limit the given threshold levels of these chemicals in food, drug, or dietary supplements or pharmaceutical products. This is referred to as the *threshold of regulations*, and details about it can be found on the FDA's website related to food contact materials (FDA 2007). If the concentrations of migrating species are excessive, they can be toxic if consumed with the packaged product. Because liquid products are in more intimate contact with the containers that package them, a greater level of mass transfer would be expected to occur compared with solid, powdery, or semisolid products. Mass transfer is also increased when the package is stored at higher temperatures and the design of the container provides more surface area for contact.

In the case of sorption, if the additives and ingredients in the food, drug, or pharmaceutical products have a high partitioning for the plastic, these compounds can be absorbed into the polymer. Examples of sorption can be seen in published reports from several researchers on the mass transfer of limonene, lycopene, and hexanal into polyolefins and other polymers (Hernandez-Muñoz, Catala, and Gavara 2001; Wong 2003; Sajilata et al. 2007). If the concentrations of these sorbed compounds are sufficiently high, they can plasticize the polymeric chains within the plastic, disrupt its morphology, and severely affect its chemical, physical, and thermal properties, which in turn can compromise the quality of bioactive compounds. If the sorbed compound is a flavor, the product can rapidly lose quality.

13.4 PACKAGING MATERIALS USED TO FABRICATE CONTAINER TYPES

13.4.1 Polymers

Plastics are made from polymers that are long-chain high molecular weight compounds made of repeating monomeric units. Polymers are classified according to the functional groups in their hydrocarbon chains and the method of their synthesis. They are further classified according to the arrangement of the molecules in the chains. Table 13.1 shows a list of the main polymers used in food, drug, and

Table 13.1 Main Types of Polymers Used in Food, Drug, and Pharmaceutical Packaging

Plastic Name	Repeating Monomeric Units	Packaging Applications
Polyethylene (PE)	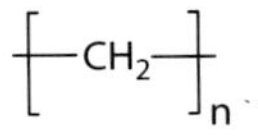	HDPE: bottles, jugs, bags, sacks, drums LDPE: films, pouches, bags, heat-sealing layer LLDPE: film, heat seal
Polypropylene (PP)	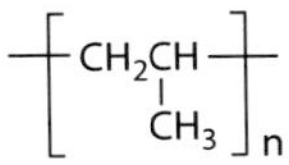	Films, bottles, caps, hinges for dispensing closures
Polyvinyl chloride (PVC)	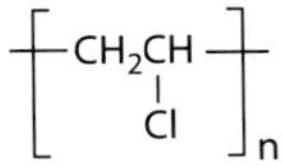	Plasticized PVC: films, blister packs, water bottles, stretch wrap
Polyvinylidene chloride (PVDC)	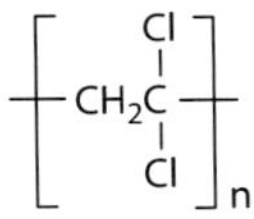	Laminated with polyolefins for high oxygen and water vapor barrier use in films, cups, bowls, etc.
Polystyrene (PS)	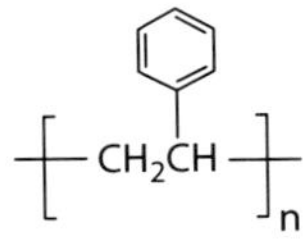	Crystal (glassy) PS: drinking glasses, cups, eating utensils, films Foam PS: cushioning material, cups, plates, bowls, and insulation
Nylon-6	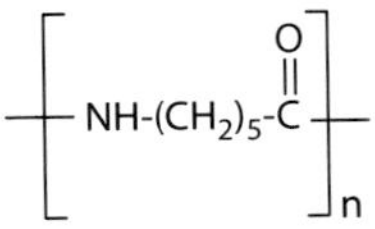	Low temperature flexibility, good flex strength, films, bags, boil-in-bag applications, vacuum packaging, medical packaging; nylon MXD-6 has good oxygen barrier
Polyethylene terephthalate (PET)	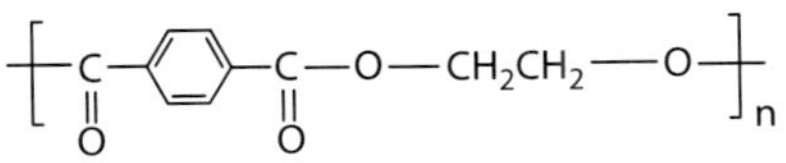	APET: bottles, films CPET: ovenable trays PETG: tubing, films, sheets, blister packs, bottles, medical devices
Ethylene vinyl alcohol (EVOH)	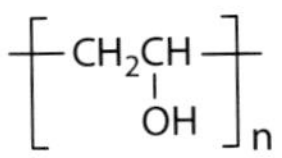	Laminated with polyolefins for high oxygen barrier use in films, cups, bowls, etc.; must be protected from moisture

APET, amorphous polyethylene terephthalate; CPET, crystalline polyethylene terephthalate; HDPE, high-density polyethylene; LDPE, low-density polyethylene; LLDPE, linear low-density polyethylene; PETG, glycol polyethylene terephthalate.

pharmaceutical packaging. These polymers are synthesized by either the addition or condensation polymerization methods. Polyolefins, polystyrene, polyvinyl chloride (PVC), and polyvinylidene chloride (PVDC) are produced by the addition method. Polymers synthesized by condensation polymerization are polyamides, polyesters, and polyurethanes. Because of the presence of the carbonyl functional groups in the condensation polymers, they can cross-link with adjacent chains under certain conditions. This ability to cross-link allows them to form crystals and thus be better barriers to diffusing gases and vapors. Polymers that do not cross-link are less likely to form large percentages of crystals and would have more amorphous regions in their polymeric matrix. The amorphous regions of polymers are characterized by having large numbers of void spaces and are highly susceptible to diffusing gases and vapors. Polymers with polar functional groups such as hydroxyls, carbonyls, esters, and amides, for example, tend to have a higher partitioning for polar solvents, including water molecules.

These polymers tend to have a low barrier to water vapor, although they may have low oxygen transmission rates. If the percentage of crystallinity is high in these polymers, it will serve to increase the barrier to moisture. Polymers that do not have polar groups, on the other hand, have a higher partitioning for nonpolar solvents, oils, and greases. Although these polymers are usually good barriers to moisture and polar compounds, they may be poor barriers to oxygen because of the lack of cross-linking. Ethylene vinyl alcohol, shown in Table 13.1, is a polymer formed by hydrolyzed copolymerization of ethylene and vinyl alcohol. It forms cross-linkages between adjacent chains and is highly crystalline because of the presence of hydroxyl groups. This makes it an excellent barrier to oxygen, but it is easily attacked by moisture.

Recently the use of polymers made from nonpetroleum (bio-based) sources has been increasingly used for food and nutraceutical packaging. Examples of these include polylactic acid (PLA), cellophane, polyhydroxyalkanoates (PHAs), polyhydroxybutyrate (PHB), polyhydroxyvalerate (PHV), chitosan, and pullulan. Cellophane and PLA are derived from starches; chitosan is made from the exoskeleton of certain crustaceans and insects; whereas PHA, PHB, PHV, and pullulan are produced by microorganisms. Most of these materials have fairly good oxygen-barrier properties, but they tend to slowly diminish in the presence of moisture. Thus, these materials are mainly used to package dried products when extended shelf life is required.

13.4.2 Metal

The three main types of metals used in food, drug, and pharmaceutical packaging are steel, tin, and aluminum. These are used to make cans, tubes, films, pails, bowls, and cups, for example. Depending on the nature of the product to be packaged, some metal containers may be coated with a suitable inert compound. Coatings include various lacquers, epoxy, polymers, and even chromium. In some cases, steel cans may be coated with tin as a protective layer. Metals are excellent barriers to gases and vapors of all types. However, they can be attacked by corrosive products if not protected with an adequate coating. Metal containers are generally more expensive compared with plastic. However, they are capable of providing the product with a longer shelf life because of their superior light- and gas-barrier properties. Metal containers are generally heavier in weight and thus would increase freight cost.

13.4.3 Paper

Paper can be made from wood, rags (linen), straw, cotton, certain plant stems such as bagasse from sugar cane processing, banana, hemp, jute, and flax, for example. Currently, wood is the main source of paper raw material. Wood is mainly composed of carbohydrates, lignin, and cellulose. Paper fibers are made from the cellulose. In the United States, examples of hardwood trees use for papermaking include poplar, aspen, and maple. Softwood trees include pine, spruce, and hemlock. A large percentage of recycled paper is also used in the paper-making industry. Long fibers are used for paper with great mechanical strength (sometimes referred to as *coarse paper*), whereas shorter fibers produce paper with smoother texture and more consistency (i.e., *fine paper*).

The main types of paper that are used in packaging include natural kraft, bleached kraft, and linerboard. Tissue paper is sometimes used for protective packaging (cushioning). Other types of paper include glassine and greaseproof paper. Glassine paper is made by excessive calendaring, which gives it a glasslike appearance. Greaseproof paper is glassine paper that has been coated with waxes, lacquers, or other types of additives that give it a high barrier to moisture, grease, fats, and oils (Hanlon 1984). Other names given to selected types of paper include parchment, waxed, and coated paper. Types of packages made from paper products include bags, folding cartons, setup boxes, tubes, trays, drums, and shipping cartons. Paperboard is also combined with metal and plastic films to produce composite packaging materials. These are usually used to make gable-top and brick-type packages (see Figure 13.1). These are mainly used for aseptically processed juices and soups. Dual-ovenable paperboard trays are laminated with polyethylene terephthalate (PET) and can be used when high temperatures are required. At the other extreme, waxed-coated paperboard is used to make containers for frozen products such as ice cream. Corrugated fiberboard is mainly used to make shipping cartons.

These are usually fabricated with two facings bonded to a fluted (corrugated) medium (Figure 13.3). These are sometimes referred to as *regular slotted containers*. Depending on the desired strength of the shipping carton, they can be made of single, double, or triple lining with flutes and linerboard facings. Paper packaging provides little barrier to gases and vapors and is not used by itself for packaged products that are highly susceptible to oxygen, moisture loss or gain, or have high levels of volatile flavors.

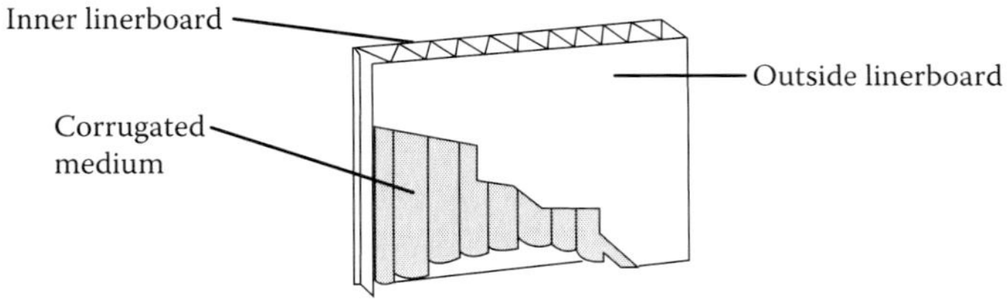

Figure 13.3 Corrugated fiberboard material used for the making of shipping cartons.

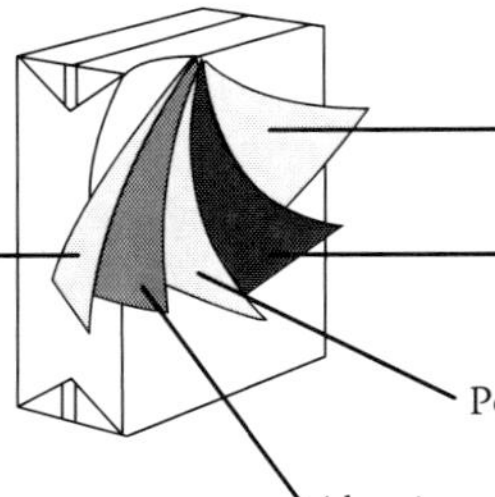

Figure 13.4 The sublayers in a composite material used to fabricate brick-type packages.

13.4.4 Composite

As mentioned previously, composite materials are composed of fabricated structures that have more than one type of homogeneous material. Each material type used in the composite is chosen to impart a specific property to the composite. It is not unusual to see aluminum foil being used as a laminate together with plastic and/or paperboard materials. For example, the composite material used to make brick-type packages is made of polyolefin, aluminum foil, paperboard, and selected adhesives to bind the individual sublayers (Figure 13.4).

Depending on the processing, handling, or storage needs that the package is to experience, one type of material may be substituted for another. For retortable brick-type packages, the polyolefin food contact layer will be polypropylene instead of polyethylene. Composites are also used to make various types of cans, for example, with a combination of metal, paper, and plastic parts. In some cases, these cans may have bodies made of aluminum foil and paperboard, whereas the ends are made of metal.

13.5 PACKAGING TYPES USED TO MARKET BIOACTIVE COMPOUNDS

13.5.1 Bottles

Bottles and jars are extensively used to package a variety of foods, drugs, and pharmaceutical products, many of which have bioactive plant compounds. Bottles are made mainly from plastic and glass. Current trends have seen a large swing from glass toward plastic bottles. Factors influencing this trend include light weight, lower cost, unbreakable nature, recyclable, convenience, more easily manufactured, lower capital cost in plant and equipment, and the fact that plastics are inert to many products. Disadvantages include slower production line filling speeds, mass transfer between certain chemicals in the polymer and in the packaged product, and a negative impact on the environment in some cases.

13.5.1.1 Plastic Bottles, Cups, and Trays

Polymers mainly used to make bottles include high-density polyethylene (HDPE), polypropylene, PET, and polycarbonate. Many carbonated beverages are packaged in PET bottles because PET shows good structural stability at a wide range of temperatures and is a fairly good barrier to moisture and carbon dioxide. PET bottles are also widely used for noncarbonated beverages such as teas, juices, and certain pharmaceutical preparations. Bottles made from HDPE are widely used to package dairy products, water, and various juices and beverages. Recent advances in extrusion blow molding have made it possible to produce coextruded multilayered bottles. As a result, high barrier polymers such as ethylene vinyl alcohol (EVOH) can now be coextruded with polypropylene or PET to product bottles with lower gas transmission rates. Injection blow molding has allowed manufacturers to produce preform bottles (parisons) in one location and sell them to end users at another location. These preform bottles can then be heated and blown into fully shaped bottles at the filling location. The process of reheating and blowing preform bottles requires simple machines and significantly reduces space because the preform bottles are relatively small. Figure 13.5a shows a photograph of fully blown bottles (with and without labels) and preforms, while Figure 13.5b shows performs alone.

Other types of plastic containers used to package bioactive compounds include cups, tubes, tubs, drums, and pouches. When polymers are synthesized, they are usually sold to converters in the form of a resin. These resins can be converted into sheets, films, tubes, and bottles by heating them in an extruder and then performing a rapid cooling process once the desired shape is achieved. Bottles can be formed by extrusion blow molding or injection stretch blow molding. The preform bottles mentioned earlier are converted into bottles by the injection stretch blow molding process. In the bottle-forming process, the molten plastic is placed inside a mold; air is then blown into the plastic; and it causes the plastic to press against the sides of the mold. This causes the plastic to take the shape of the mold. Once it is sufficiently cooled, the mold opens, and the bottle is extracted. Packages such as cups and bowls

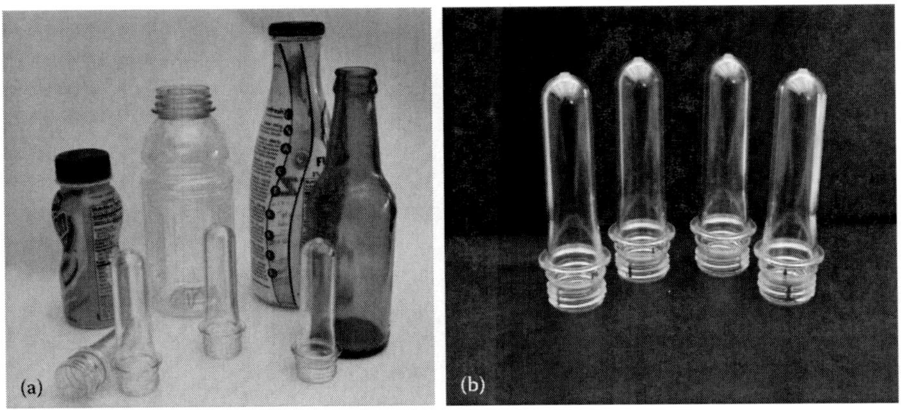

Figure 13.5 (a) Fully blown bottles (with and without labels) and preforms; (b) performs alone.

can be made from plastic sheets that were previously extruded. These sheets can be used to make the cups by a process called *compression molding*. In this process, the sheet is heated and then forced into a mold that forms the container. Collapsible plastic tubes are manufactured by a process called press blowing. This is a combination of injection molding and blow molding. In this process, the molten plastic is injected into a mold that forms the tube in a three-step process. Tubes are used to package a variety of creams, gels, and over-the-counter and prescription drugs, in addition to other products (Soroka 2009).

13.5.1.2 Glass Bottles and Jars

Almost all glass used in food, drug, and pharmaceutical packaging is referred to as *soda lime glass*. This is because the main ingredients used to make the glass are 70% sand (silica), 12% sodium carbonate (soda), and 13% calcium carbonate (lime). Approximately 20% of the raw material mix is made of recycled glass, often referred to as *cullet*. Other minor ingredients are also added to either add or remove color where necessary or to assist in the smelting and purification process of the molten glass. Colorless glass is called *flint*, but selected additives can give it a brown, green, or blue color. After mixing and smelting at approximately 1250°C, molten soda lime glass can be blown into bottles or jars. Before being ready for normal use, the glassware is annealed in an oven, coated with a lubricant to reduce friction, inspected for quality characteristics, and then packaged for shipment.

Bottles are distinguished from jars by the forming method and the diameter of their mouths. Bottles are formed by a blow and blow method after a gob of molten glass falls into the bottle-forming machine from the furnace. In some operations the molten glass is pulled against the walls of the mold by the assistance of a vacuum. Bottles tend to have a relatively narrow mouth and slender neck compared with the wide-mouth, short-neck jars. Jars are made in the forming machine by a press and blow method. Figure 13.6 shows jars used to package plant-based bioactive compounds.

Figure 13.6 Jars used to package plant-based bioactive compounds.

Glass is inert to most food, drugs, and pharmaceutical products, which is its big advantage over plastic and metal containers. As a result, glass is compatible with a wider array of products, and it offers the longest shelf life compared with packages made from other types of materials. There is also virtually no chemical migration or sorption between glass and packaged food products. However, glass is relatively heavy, breakable, and more costly. Soda lime glass is less heat tolerant than the metals used in packaging.

13.5.2 Cans

Cans are made and marketed in a variety of sizes and shapes. Most beverage cans are cylindrical and made of aluminum, although some are made of steel. Two-piece cans are made by repeated hammering of a metal disc until the body of the can is formed. Two-piece cans are made of aluminum or steel. Three-piece cans are made from a flat piece of steel sheet that is rolled into a cylinder and welded along the length of the joint where the material meets. Three-piece cans are sealed with can ends at both openings of the cylinder. Two-piece cans are only sealed at the open end of the cylinder. When the ends are joined to the body of the can, the seal is called a *double seam*. Can manufacturers sell cans with one end open to food, drug, and pharmaceutical processors (Figure 13.7).

After filling the product into the open-ended cans, they are sealed and then processed by heat treatment, if that is desired. If this sealing process is not done properly, the cans may become a potential health hazard if leaks develop before the product is ingested by consumers. This is because liquid low-acid canned products are sterilized after sealing the can, and the product is expected to be shelf stable. Such products do not require an additional preservative step for extended shelf life. However, if these cans are underprocessed or if they leak after processing, spoilage bacteria can enter and contaminate the product. If the spoilage organism is not

Figure 13.7 An assortment of cans used to package products with plant bioactive compounds.

gas forming, it can become difficult for consumers to recognize that the product is contaminated before consuming it. This is referred to as *flat sour* because no swelling of the can occurs. This highlights the importance of proper sealing of cans and other packaging types that are used for shelf-stable foods, drugs, and pharmaceutical products. If properly sealed and not abused after processing, cans are capable of withstanding the stresses of processing, handling, and transportation. This includes both pressurized and nonpressurized cans. Cans (metal and composite) are also used to package powdered products. These may not be heat treated after filling. Some of these cans may be fitted with a metal membrane as a tamper-evident, pilfer-proof, or freshness-preservation device. These cans also carry a metal pressure plug or plastic snap-on outer cover and can be resealed if the entire can content is not used after the initial opening.

13.5.3 Pouches

These are made mainly from plastic, paper, or a combination of plastic and metal films. In many cases, more than one type of plastic is used in the material from which the pouch is made. When more than one polymer is used in the film, it can be in the form of a laminate. In such cases, single films of materials are bonded together by an adhesive or when one molten film is cast on top of another and the laminate is formed after the structure cools (Figure 13.8).

Materials can also be made from coextrusion and/or polymer blends. Copolymers are defined as polymers having more than one type of repeating monomeric units. Examples of these include styrene-butadiene and acrylonitrile. Glassy polystyrene is tough but brittle and shatters easily on impact. By copolymerizing it with butadiene, it maintains its toughness and transparency and gains good shatter resistance. Polystyrene can also be blended with polypropylene and polycarbonate to give a tough resilient material that is not transparent. This can be used as an outer layer in

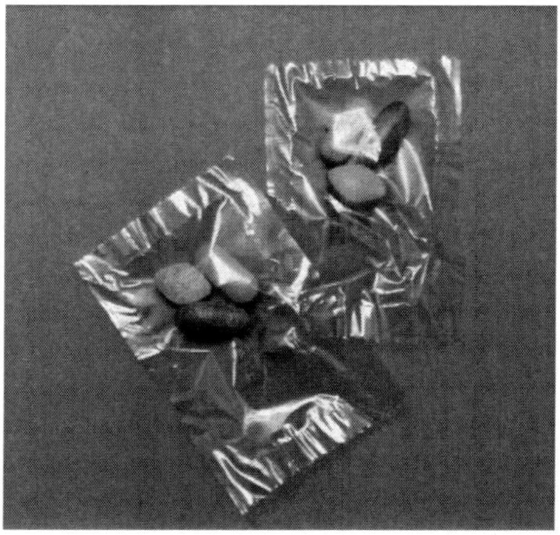

Figure 13.8 Plastic pouches used to package tablets.

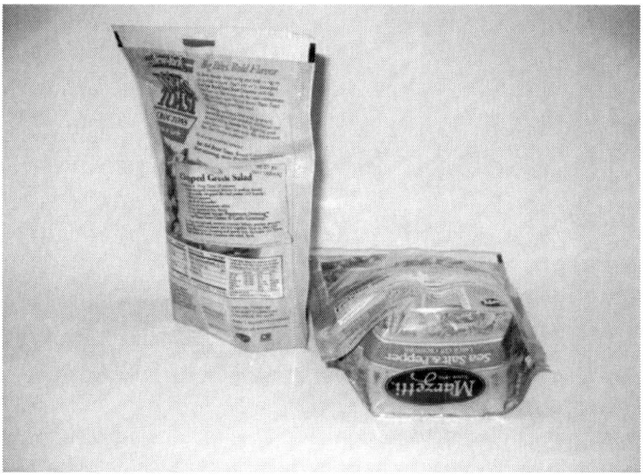

Figure 13.9 Standup pouches used to package a variety of foods.

cups and bowls to give them shatter resistance. Polyblends are made by mixing two or more compatible polymers and extruding them as a single film (Figure 13.9).

Pouches are made by folding, cutting, and sealing films that were previously extruded. Some pouches can be fabricated by one manufacturer and sold to an end user, whereas most processors have machines that form, fill, and then seal pouches in one continuous operation. The equipment in such a case is called a *form–fill–seal machine*. These can either be referred to horizontal or vertical machines. Pouches are categorized based on their geometry. As the name implies, pillow pouches are shaped like pillows and are usually sealed along their backs, top, and bottom ends. Three-side seal pouches are made by folding a length of film into two halves and then sealing the two sides and the top. Four-side seal pouches are sealed at their four quadrants (Selke 1997). Standup pouches are in another category, and one is shown in Figure 13.9. The advantage of this type of pouch is that it can be better displayed on the store shelves. They are more expensive than pillow pouches and use more material. However, because these types of pouches are made with sides and/or bottom gussets, they can stand up without falling over when filled. This makes them a good alternative to cans, bottles, jars, and other types of rigid packaging that can be used for liquids, powders, and granules. They are also lower in cost than cans, glass, and plastic bottles because they use fewer materials. When fitted with caps or zippers, they can be resealed and are convenient. They can be made on high-speed form–fill–seal machines, and when high barrier laminates are used to make them, they can be retorted or aseptically processed and used to package shelf-stable foods and pharmaceutical products with bioactive compounds.

13.5.4 Blister Packs

These are made by heat sealing printed paperboard cards to preformed plastic blisters. The plastic blisters are mainly made of thermoformed PVC or PETG sheets,

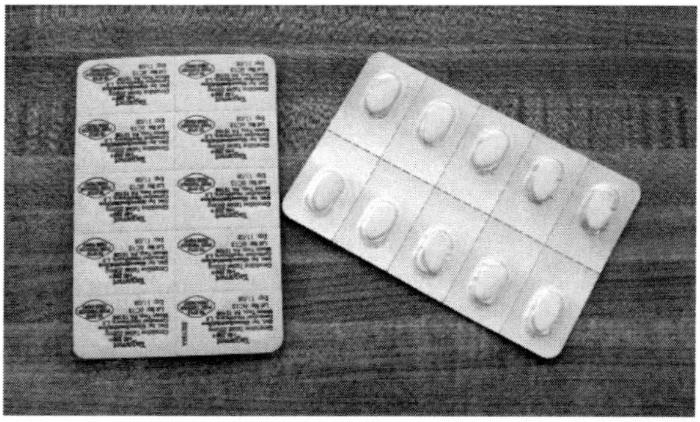

Figure 13.10 Blister packs used to package tablets, capsules, and softgels.

but other polymers with suitable properties can be substituted. Some packs use foil, plastic, or laminated materials to bond to the preformed blisters. Blister packs are widely used in the pharmaceutical industry to package unit-dose tablets and capsules. They can also be used for packaging food items such as chewing gum and selected candies, for example. Blister packs are referred to by other names such as *blister cards* or *blister strips* in places outside the United States (Chan 2000). Figure 13.10 shows a typical blister pack used to package tablets. Skin packaging is a type similar to blister packs. However, the difference is in the blisters. Skin packaging involves the use of a similar paperboard card, but instead of a heat-sealed blister, the product is draped over by a plastic film that is drawn to the card by a vacuum.

13.6 HOW PACKAGING INCREASES SHELF LIFE

13.6.1 Barrier to Oxygen

As was mentioned earlier, several factors can cause a food, drug, or pharmaceutical product to compromise its shelf life. At this point, a few examples will be discussed, and how packaging can be used to limit the progress of these unwanted changes will be explained. Oxygen is well known as the main cause of oxidation in many products, as well as the resultant rancidity with objectionable odors, flavors, colors, and the lost of potency of bioactive compounds (German 1999; Miller et al. 2003; Rhee and Watts 2006). Thus, the reduction or elimination of oxygen from the product is essential. Apart from oxidation, oxygen also supports the growth of various aerobic microorganisms that can also induce spoilage in packaged products. Many products such as powders, granules, creams, gels, and, to a lesser extent, liquids do have dissolved air (oxygen, nitrogen, and carbon dioxide) within their matrix. Thus, the use of vacuum packaging is one method to remove this oxygen. Another technique is to flush the headspace of the package with an inert gas such as nitrogen or helium. This technique is referred to as *modified atmosphere packaging* (Mahanjan et al. 2005). Because the removal or reduction of oxygen within the package creates a large concentration gradient between that of the oxygen in the outer atmosphere, it is essential that the package itself maintains this gradient as long as possible, so that the benefit of the modified atmosphere can be realized (Pascall et al. 2008).

Packaging materials made of glass or metal offer the best protection in such cases. Plastic packaging, on the other hand, is not as efficient in limiting the return of oxygen to the package. As a result, a tremendous amount of research is currently focused on the development of high gas-barrier plastics for extended shelf life of products susceptible to oxidation or microbial spoilage. Plastics that are known to form cross-linkages between polymeric chains tend to have less void spaces and a higher percentage of crystallinity. The permeation of gases and vapors through a plastic medium occurs within the void spaces. Because crystals have regular geometric shapes, the crystalline regions of polymers are more tightly packed and thus limit the diffusion of gases and vapors. In addition to cross-linking, examples of other factors known to increase crystallinity are orientation (stretching of the plastic while still at a high temperature), the nature of the repeating monomeric functional unit in the chain, the absence of side branches, and the stereochemistry of the polymer. Examples of high oxygen-barrier plastics include EVOH, PVDC known by its Saran trade name (Dow Chemistry Co., Midland, MI), and nylon MXD6. As was mentioned earlier, the lamination of aluminum foil with other types of lower barrier plastics makes an excellent high barrier structure. Current research efforts are also focused on the use of nanoparticles to increase the barrier, thermal, and mechanical properties of polypropylene, PET, and nylons (Halim et al. 2009).

As an additional step in limiting the ingression of oxygen into a packaged bioactive plant-based product, some manufacturers use a technique called *active packaging* (Miller et al. 2003). This can take the form of a device or chemical within a sachet in the headspace of the package, or a chemical coated onto the wall of the material or dispersed within the matrix of the polymer. Irrespective of the location of the chemical, it is designed to absorb or block the presence of oxygen from reaching the product. These devices are called *oxygen scavengers* or *absorbers*. To be called active packaging, these devices must be activated only in the presence of oxygen. Several other types of active packaging devices are known to exist. These include antimicrobials, desiccants, odor removers, aroma emitting, gas permeability controllers, and ethylene controlling devices (Brody, Strupinsky, and Kline 2001).

13.6.2 Barrier to Moisture

Glass and metals are excellent barriers to moisture, but polymers exhibit varying moisture-barrier properties. The intensity of the barrier of polymers to moisture depends on the nature of the functional groups in the repeating monomeric units. For example, EVOH is a good barrier to oxygen because of the hydroxyl groups in the chain. This allows the formation of hydrogen bonds between adjacent chains, which increases cross-linking. However, because the hydroxyl groups are hydrophilic, when exposed to moisture, the polymeric chains of EVOH are plasticized, and the polymer loses its barrier properties to oxygen. Thus, when

EVOH is used in packaging, it is usually laminated between hydrophobic polymers such as polyolefins. This serves to protect it from moisture and thus allows it to keep its good barrier to oxygen.

A similar interaction is also seen with nylon in the presence of moisture. Because nylon is a polyamide, it has hydrophilic characteristics. However, when exposed to a small concentration of moisture, hydrogen bonds are formed between adjacent chains. This causes cross-linking to occur, and it increases the oxygen-barrier properties of the polymer. However, if the exposure to moisture exceeds a certain threshold, water molecules will eventually plasticize the nylon chains, and this also leads to the loss of oxygen-barrier properties.

Polymers that are commonly used to make pouches for food, drug, and pharmaceutical products and that are good barriers to moisture are polyolefins and PVDC. In the case of the polyolefins, the repeating monomeric groups on the chain are hydrophobic in nature and have no affinity for polar compounds. Both PVC and PVDC have chlorine molecules in their repeating units (see Table 13.1). The presence of chlorine in the chemical structure adds dipole-induced dipole interactions to the polymeric chains. This is exerted by the polar C-Cl bonds that PVC imparts to the polymeric chains. This increases the attractions between adjacent chains, and thus cross-linking. This is the reason why unplasticized PVC is tough and rigid. These dipole interactions are reduced by the addition of plasticizers, and it allows the polymer to be easily drawn into films and sheets. However, when this occurs, cross-linking and the permeability of the material are affected negatively.

13.6.3 Barrier to Light

Because metals are not transparent, they are natural barriers to light. Flint glass is transparent, but amber and green glass provide adequate light barrier to many light-sensitive products. Examples of this can be seen in the packaging of beer, stout, and milk in colored or translucent cans, bottles, and jugs. When plastic containers are chosen to package light-sensitive products, they can be colored or wrapped in light-impermeable labels. When light-sensitive products are exposed to the fluorescent lighting in display cases, for example, the process of oxidation can be hastened compared with exposure to natural light. The problem of light-sensitive products being packaged in transparent plastic and exposed to fluorescent light can be solved by the addition of ultraviolet absorbers to the polymer. An example of this can be seen by the use of the hinder amine Tinuvin-326 (BASF, Florham Park, NJ) (Pascall et al. 1995). Paper as a packaging material also provides an excellent barrier to light. However, paper is a poor barrier to oxygen and moisture if not coated with wax or laminated with other higher barrier materials.

13.6.4 Barrier to Filth

One of the functions of a package is the protection of the product. This protection includes microorganisms, filth, infestation, moisture, oxygen, and/or any other additive that is not wanted in the product. In most cases, this protection is a physical one. Filth is legally defined as any contaminant in food, drugs, or pharmaceutical products that are consumed by humans or animals (Federal Food, Drug and Cosmetic Act of 1938). This is not limited to excrement but includes things such as dirt, hair, lubricants from machinery, and scraps of packaging material.

13.7 PACKAGING INTEGRITY AND SHELF LIFE

To maximize the functions of a package during the storage of a processed product, the integrity of the container must be maintained until a consumer opens it and consumes the contents. This is much more crucial for shelf-stable products that depend solely on the packaging for the prevention of spoilage (Dagel 1999). In the case of glass containers, the major site of possible leaks and integrity loss is the contact points between the closure (cap) and the sealing rim of the packaging. Examples of testing methods that are done to ensure the quality of cap-sealing procedures include application and removal torque measurements, security and pull-up tests for jars with lug-type closures, and cap and glassware finish dimensional measurements. Figure 13.11 shows a diagram of the different parts of a glass or plastic container. Quality control measurements on the bottle or jar finishes include internal and external bore diameters, thread pitch, the finished height, perpendicularity, and the absence of cracks (checks). Measurements on the cap are also done to provide assurance that they are within the required specifications.

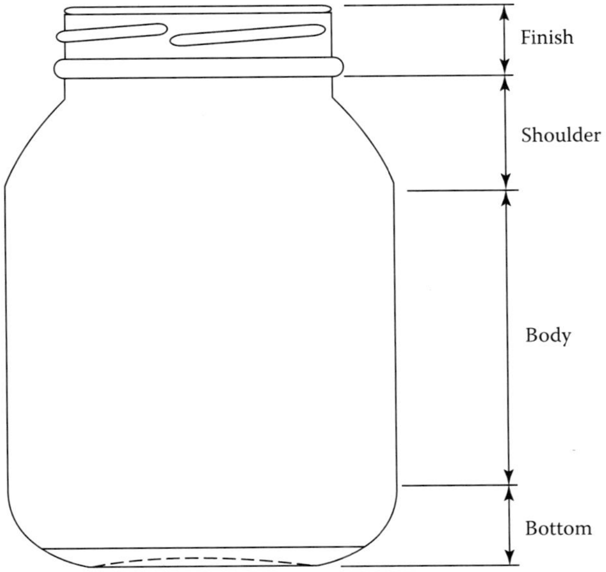

Figure 13.11 Different parts of a glass or plastic bottle.

With metal containers, it is important that corrosion does not occur and results in leakage in the walls or side seam of three-piece cans or other types of metal packages. In the double seams of cans, many types of defects are possible. Some of these defects can result from problems with the can-seaming equipment, from defects in the design of the parts of the can, and from damage to the container after processing. Even though a defective double seam may not leak when a can is sealed, it may do so during or after retorting, if that is the processing method of choice. Retorting is similar to pressure cooking, and the product and container are subjected to intense heat and pressures during the process. If the pressures within the retort are not properly regulated during the cooking process, paneling or buckling of the cans can occur, and this has the potential to cause leaks to develop. To test the quality of the double seam on cans, it is necessary to perform a tear-down analysis and a subsequent measurement of the body hook, cover hook, and the overlap of the cover with that of the body hook. Other measurements that are done on cans include seam width, height, and the countersink of the can ends. Usually the can manufacturer provides the processor with the upper and lower limits of these measurements.

As was mentioned earlier, brick-type packages are made from composite materials. However, in fabricating these packages, it is necessary to make a number of folds in the material to achieve the rectangular shape. To accomplish this, it is necessary to insert a number of creases in the material. If these are not done properly, cracks can develop along these folds and can cause leaks to develop. Because most of these types of packages are used for shelf-stable products, leaks can cause the product to spoil. These types of packages also have a longitudinal and two transverse seals. To perform integrity dye testing of these types of packages, it is necessary to strip away the nonsealing layers of the laminate and then expose the heat-sealing polyolefin material for leak testing. This is necessary because the sealing layer is blocked from direct view by the aluminum and paperboard layers (Sivaramakrishna et al. 2007). Current research efforts are focused on the development of automatic techniques to measure seal integrity of these types of packaging.

Because of its flexible and semirigid nature, plastic containers are subjected to a wider array of defects compared with packages made from the other types of materials. Some of these defects can be difficult to detect with the naked eye. The increased use of plastic bottles for food packaging has created a need for more sophisticated and sensitive automatic leak detection devices compared with manual inspection. This need has spurred research and development in many types of packaging integrity testing devices (Hurme et al. 1998 ; Pascall 2002). Leaks have been a continued source of concern to food, drug, and pharmaceutical processors, especially in situations where packages are filled and sealed at high speeds. Papadakis and Yam (2003) reported that this is because of the potential for leaking packages to become contaminated by microbial entry, lose flavors, or become spoiled as a result of oxygen ingression through the leaks (Donghwan, Papadakis, and Yam 2003). Because most plastic pouches and cups are heat sealed, anything that decreases or increases the heat, sealing time, or the pressure applied to the seal during the process can cause defects to occur. When sealing these structures, care must be taken to prevent extraneous matter from getting between the two sealing surfaces.

If this occurs, the seal can develop leaks. Also, if wrinkles are not removed from the material, channel leaks can be formed between the seals. Flexible packages are also susceptible to punctures by other packages or external objects. Defects in the materials caused during the extrusion process can result in the development of pinholes in thin films. Various types of testing devices for leaks in plastic packaging include pressure differential, ultrasound, helium gas leak, dye testing, electrolytic, x-ray, and electrical conductivity testing, for example. Plastic bottles and jars with caps are subjected to integrity testing similar to that for glass containers. In the case of plastic closures, allowances must be given to the phenomenon called *creep*. This is a property of plastics that causes them to relax under the load of a sustain force. The application torque used to seal caps on bottles and jars is an example of such a force. This is less of a problem if the plastic used is a thermoset polymer versus a thermoplastic one.

During the shipping and handling of consumable products, the package can be subjected to various shock and vibration stresses. These include stresses from conveying, palletization, warehouse stacking with fork trucks, road, rail, sea, and/or airfreight transportation and subsequent handling. To reduce the impact of these stresses, it is necessary to cushion the packages. Corrugated cartons are the usual tertiary packaging of choice for the shipping of filled packages. To reduce contact between adjacent individual packages, some shipping containers are fitted with corrugated dividers. These work well for glass containers, and they prevent breakage that can be caused by one package impacting another one. Foam polystyrene is widely used as the cushioning material of choice. Currently, foam materials made from bio-based polymers are being used increasingly. Irrespective of the cushioning method used to protect packaged products, the intent is to reduce shock and vibration to safe levels so that damage to the package and/or product is avoided. In some cases, high amplitude vibrations can cause segregation to occur in certain emulsified products. This is one example of how the product can be affected by these forces.

13.8 LEGAL ISSUES INVOLVING PACKAGING AND BIOACTIVE COMPOUNDS

In the United States, regulations that impact packaging are grouped into various categories. Examples of these and references to the regulation are (1) packaging integrity and low-acid canned foods (Title 21 Code of Federal Regulations Part 113, subpart D; FDA 2010a); (2) migration of chemical additives from polymers to packaged foods (Title 21 Code of Federal Regulations Part 177; FDA 2010b); (3) labeling and food, drug, supplement, and pharmaceutical packaging (Title 21 Code of Federal Regulations Part 101; FDA 2010c); (4) tamper-evident packaging (Anti-tampering Act of 1983); (5) Fair Packaging and Labeling Act of 1967 (Title 21 Code of Federal Regulations Part 1); and (6) recycling laws (Safe Drinking Water and Toxic Enforcement Act of 1986). This chapter will not go into a detailed discussion about these regulations, but the first five mentioned directly impact day-to-day production and packaging of products, and these will be discussed briefly.

The regulations relating to low-acid canned food processing and packaging also include the testing of cans, glassware, and plastic containers for sealing quality. The regulations specify the limits for specific quality control measurements, the frequency of these measurements, and who is qualified to monitor them. Although the specification on quality measurements for seal quality of plastic containers is not mentioned in the regulation, it does specify the frequency of measurements and who is qualified to do so. Defects are classified according to their severity as Class I (critical), Class II (major), and Class III (minor). All containers with Class I defect are expected to be discarded because they are susceptible to microbial entry into the package.

FDA regulations provide detailed directives on the names of polymers approved for food contact. It also states which materials are approved for packaging foods that are to be irradiated (Sadler, Chappas, and Pierce 2001). New or modified food contact polymers must get FDA approval before they can be used. To obtain approval, manufacturers must provide the FDA with information showing that the migration of objectionable chemicals from the polymer to food simulants is below threshold limits that are specified in the regulations. The petition to the FDA must also state the test method used.

Both the FDA and the US Department of Agriculture have provided regulations for the labeling of foods, drugs, supplements, and pharmaceuticals. These regulations state in detail the design of various categories of labels and the information that must be presented. These include the principal display panel, the ingredient statement, the nutritional fact panel, signature line, warning statements if applicable, name and address of the manufacturer, handling statements if applicable, dosage instructions, and net weight, for example (Keller 2007). Labeling laws also include different categories of additives and the details of how they are to be used in foods, drugs, dietary supplements, and cosmetics. Additives addressed in the regulations include those that are referred to as direct, indirect, secondary, color, and unintentional additives. Labeling laws also include wordings on unavoidable contaminants. Violations of these regulations can render a packaged product as misbranded. A package can also be misbranded if a claim is made about a food, drug, or dietary supplement that does not conform to FDA guidelines or if the claim is not substantiated by fact. These include issues such as health, disease, and structure–function claims. The Fair Packaging Act requires manufacturers of food, drugs, cosmetics, and medical devices to label their products in such a manner that they are not misbranded. The law also outlines what constitutes legible and conspicuous writing on labels and what is unfair and deceptive packaging. If packaged products are found to have contaminants in excess of what is allowed, it can be deemed adulterated. Recent laws relating to allergenic compounds obligate manufacturers to label products that are potential allergen suspects. The recent Homeland Security Act of 2002 requires the registration of all food plants within the United States and foreign plants that export to the United States. The act also mandates that manufacturers inform the FDA about imports in advance of landing in the United States and that the country of origin must be printed on the label.

13.8.1 Tamper-Evident Packaging

The FDA's (2010d) definition of a tamper-evident device states that it is a device "having an indicator or barrier to entry which, if breached or missing, can reasonably be expected to provide visible evidence to consumers that tampering has occurred." Tamper-evident devices are required on all over-the-counter drugs except dermatological, dentifrice, insulin, or lozenge products. The law also requires manufacturers to put a statement on the package alerting consumers to the device and information on its proper function. For certain products, this requirement is exempted. Tamper-evident devices can be divided into three categories: overt (can be easily seen with the naked eye), covert (requiring the use of a specially designed detector), and reserved (devices that are only known to exist by a selected group of individuals). Currently, many forms of electronic detection devices are being tested or have been added to various packages (Bix et al. 2004). In some cases, the lack of a covert or reserved tamper-event device on a package can be used to identify counterfeit products (Laven 2006).

The success of any tamper-evident device hinges on the effectiveness of the device to be different in appearance when tampered (or opened) compared with its appearance when it was initially applied to a package. This success is also conditional on the absence of false-negative and false-positive findings in the operation of a tamper-evident device, the ability of consumers to recognize that the device was tampered with, consumer awareness of the presence and purpose of the device, and consumers' understanding of the tamper-evident signals (or lack of signals if untampered) to be expected by the device. FDA regulations also require that two-piece, hard gelatin capsules must be sealed using an acceptable tamper-evident technology. This is usually a colored band located at the junction between the two halves of the outer coat of the capsule. Although tamper-evident devices are widely seen on many food items, they are voluntary and not required by law within the United States.

13.9 PACKAGING MACHINERY

The modern packaging industry has evolved to the stage where highly specialized equipment is essential for economical operations. The main factors influencing this trend are high output volumes and product consistency. These factors are in turn driven by marketing and government regulatory demands. Because of the high diversity in food products and the myriad methods of processing them, most packaging machines are custom built to meet the specifications of the processor. If a processor decides to acquire prefabricated machines, it then becomes necessary to purchase and/or design ancillary equipment that is compatible with the initially obtained packaging machines. When deciding to purchase packaging machines, processors need to consider several important factors, including power requirements, available space, installation necessities, and specialized requirements. The power requirements may require considerations such as the nature of the electricity supply, compressed air, vacuum, potable water, steam, fuel, and compressed and/or liquefied gases of various types, for example. Specialized requirements can include consideration for things such as ventilation, cleaning, drainage, waste removal, raw material or product in feed, product outflow, quality control monitoring, troubleshooting, parts replacement, retooling, maintenance, and product changeover. Other factors that must be considered before acquiring packaging and/or processing machinery include any government regulations associated with the use of the equipment. The manufacturers of infant formulation and/or medication require special approval from the FDA if the product will be manufactured within the United States. Depending on the needs of the equipment, approval from the US Environmental Protection Agency may be required. In addition to these, before purchasing a machine, consideration should be given to the capital cost, the nature of the product, the type of packaging, the market needs, the machine capacity and speed, availability of spare parts, warranty, after-sale service, machine durability and versatility, compatibility with existing equipment, and flexibility to subsequent planned equipment add-ons.

The main functions of packaging machines in bioactive compound processing are empty container handling, product filling, package sealing, labeling, case packing, and palletization. Depending on the nature of the packaging, its cleaning may be a necessary function. In addition to these, some systems may have secondary functions relating to quality control. Examples of these include weight checking, headspace monitoring, metal detection, leak testing, and other needs for defects testing and elimination that may arise during production.

13.9.1 Empty Container Handling

Two main types of packaging approaches are used in bioactive product processing. These are the use of preformed packages or form–fill–seal operations. Almost all metal and most glass packages are preformed containers when used in production. In the case of polymeric packages, they can be used as preformed packaging or the final package can be formed from roll-stock material, filled with the product, and sealed by the same packaging machine at the production site. Some paper and composite packages can be similarly used in production.

Form–fill–seal machines are usually described as horizontal or vertical in orientation when in use. For polymeric materials, horizontal machines can be used to fabricate pouches, trays, cups, blister packs, and clam shells, for example. Vertical machines can be used to fabricate polymeric pouches and bags, brick-type composite packages, and glass ampoules in pharmaceutical packaging. Polymeric trays, cups, blisters, and clam shells are usually made by a process called *compression molding*. Glass ampoules are made by the process of *blow molding*.

When preformed packages are used, the initial concern is their removal from shipping containers if they are not manufactured on site. If no cleaning and sanitization are required, the packages are then fed to the filler by a series of conveying systems, including unscramblers, conveyor belts or chains, accumulators, time screws, star wheels, motion sensors, and robots of various types. Some operations may have a system for the inspection of empty containers for defects.

13.9.2 Filling Machines

The nature of the product and the selected packaging has a direct influence on the choice of filler. For liquid products, the filler can be adjusted for constant-volume or constant-level filling. Constant-level filling is used when the product is relatively inexpensive. Constant-volume filling is used for more expensive products, such as dietary supplements, drugs, and other pharmaceuticals, for example. The dispensing of the product into the package is usually facilitated by gravity, vacuum, pressure, or a combination of the three. An appropriate metering device is usually used when volume control is required. In such cases, piston, diaphragm, or metering pump fillers are the techniques of choice. Depending on the viscosity of the product, the rate of filling and the temperature are chosen to reduce splashing, increase the filling speed, and/or control the nature and volume of the headspace gas.

For dry products, the filling method is controlled by volume, weight, or count. The main types of volume filler are volumetric cup or flask, vacuum, and auger filling. In volumetric flask filling, the product is first filled into a measuring container, and then it is dispensed into the package. Vacuum volumetric fillers are used for powders that are light in weight, low in density, and have a high volume of entrapped air. In such situations, the vacuum removes air from the powder, increases its density, and compacts the product for easy deposition into the package. Auger fillers are generally used for products that are not easily free flowing. The device can be adjusted to dispense a given volume of the product per revolution. Auger fillers are sometimes used for viscous liquids such as gels, creams, and ointments. Net weight fillers are designed with a weighing device that measures and then dispenses the product in the package before sealing. Number-counting fillers are used extensively for products such as tablets and capsules. The filler operates by allowing a specified number of the product to fall into the package. Some filling machines are fitted with an electronic photo cell or some other sensor that counts the number of items as they fall into the package. Other types of filling devices are made with a revolving disc with perforations compatible with the size and shape of the product. Each perforation is fitted with a hollow tube that collects the product until the desired quantity is obtained. Once this is reached, it discharges the product into the package previously situated at its other open end. Some systems are designed with chutes or grooves instead of perforations in the disc. The chutes are designed to collect the product, and once a linear measurement is collected, it discharges them into the package. This arrangement works best for products that are uniform in size. Depending on the size of the product and its nature, some processors use robot arms to fill packages. Examples of these are pick-and-place systems, which work well for delicate products. It is possible that variations of these counting filling devices are currently used in the industry. However, the ones mentioned here are examples of number-counting fillers.

13.9.3 Closing Machines

Hermetic sealing is designed to prevent the entry of gases (including air) and microorganisms into the package. The closure used on a package depends on the nature of the package and that of the product. Bottles are usually fitted with screw-type, snap-fit, pressure plug (friction), sports cap, crimp-on (crown), roll-on, or press-on/twist-off caps. Some bottle caps are modified to accommodate a dispensing arrangement (e.g., dropper, pump dispensing head, flip-spout), a child-resistant device, reclosing capability, or tamper-evident features. Irrespective of the type of cap, a capping machine is usually situated in close proximity to the filler. Some bottles are also closed with an induction sealed liner before the application of an outer cap. In such cases, an induction sealing machine is required. In most cases, the liner is attached to the underside of the cap and is affixed to the bottle concurrently with the cap. Cans are closed by the application of a lid. The joint between the lid and the body of the can is called the *double seam*. This is usually applied by a can seamer that is positioned adjacent to the filler. Packages made from polymeric materials are closed by any of several methods that clamp the sealing surfaces and cause them to melt into a single layer. For proper sealing to occur, four important conditions must be met. These are adequate pressure, time, heat, and material compatibility. Currently, several heating methods are used in the industry. Examples of these include conduction, ultrasound, induction, radiofrequency, spin-welding, and impulse sealing. Composite containers with polymeric heat sealing layers are usually closed with methods such as ultrasonic or laser sealing. This is because the heat needs to be focused at the polymeric layers and not on the paperboard, which will be susceptible to burning if other sealing methods are used.

Nonhermetic sealing is generally not considered to be air tight, including devices such as twisters, clamps, fasteners, and various types of applied adhesives, for example. Irrespective of the type of sealing device used on a package, the quality of the closure should be ensured. Defective seals can cause the package to lose integrity, and this can compromise the safety of the product, reduce its shelf life, or be the cause of product losses. In addition, the seal should be attractive, user friendly, fairly easy for the average consumer (including geriatric persons) to open, child resistant if required, pilfer proof, difficult to counterfeit, resealable if desired, and cost-effective.

13.9.4 Aseptic Packaging Machines

These are designed to fill and seal containers for shelf-stable storage. To successfully accomplish this, these machines are designed to create a sterile zone where a sterile product is loaded into a sterile container and sealed under sterile conditions. Depending on the design of the machine, glass, metal cans, plastic trays, pouches, cups, and composite brick-type containers can be filled and sealed successfully. The sterile zone in an aseptic packaging machine is usually created before the beginning of production and must be maintained throughout the production cycle. If sterility is lost during production, filling of the packages must cease until the sterile zone is recreated. The main advantage of aseptic processing and packaging is the minimal heat required to sterilize liquids or semisolids approved for that type of technique. This works well for products that are susceptible to quality loss if exposed to excessive heat. The drawback with aseptic processing and packaging is the high capital cost for the equipment.

13.10 CONCLUSION

Plant bioactive compounds are found in leaves, stems, bark, roots, flowers, seeds, and fruits. When harvested for commercial processing, they are consumed when freshly cut and as extracts in juices, syrups, teas, wines, and oils. They can also be processed and marketed as tablets, capsules, creams, gels, powders, and granules. To increase the shelf life of these compounds, they can be packaged in small quantities in bottles, jars, cans, pouches, blister packs, cups, and trays or in bags, drums, and pails for bulk storage. These packages can be made from metal, glass, plastic, and paper products. Factors that cause these plant-based compounds to deteriorate include oxidation, moisture, light, microorganisms, heat, enzymatic activity, and infestation. To increase the shelf life of these products, the method of processing, storage conditions, maturity of the harvested plant, cleanliness, and the packaging selection must be optimized. Because most commercial operations that process plant bioactive compounds may be subjected to federal regulations, knowledge of good manufacturing practices, labeling regulations, and other relevant laws, including state and municipal ordinances, is essential. Depending on the package and product choice, an appropriate closure must be selected. These should be convenient to use, enhance the marketing of the product, and should not compromise the safety of the product.

REFERENCES

Bix, L., S. Sansgiry, R. Clarke, F. Cardoso, and G. S. Shringarpure. 2004. Retailers' Tagging Practices: A Potential Liability? *Packaging Technology and Science* 17: 3–11.

Brody, A. L., E. R. Strupinsky, and L. R. Kline. 2001. Oxygen Scavenger Systems. *Active Packaging for Food Applications*. Washington, DC: CRC Press, 31–60.

Chan, T. Y. K. 2000. Improvements in the Packaging of Drugs and Chemicals May Reduce the Likelihood of Severe Intentional Poisoning in Adults. *Human & Experimental Toxicology* 19: 387–91.

Dagel, Y. 1999. Pharmaceutical and Medical Packaging. *Packaging Technology and Science* 12: 203–05.

Donghwan, C., S. E. Papadakis, and K. T. Yam. 2003. Simple Models for Evaluating Effects of Small Leaks on the Gas Barrier Properties of Food Packages. *Packaging Technology and Science* 16: 77–86.

Fair Packaging and Labeling Act. U.S. Code 15, Chapter 39 §1451 (1967). http://www.fda.gov/RegulatoryInformation/Legislation/ucm148722.htm.

Farag Badawy, S. I., A. J. Gawronski, and F. J. Alverez. 2001. Application of Sorption-Desorption Moisture Transfer Modeling to the Study of Chemical Stability of a Moisture Sensitive Drug Product in Different Packaging Configurations. *International Journal of Pharmaceutics* 223: 1–13.

Federal Anti-Tampering Act. U.S. Code 18, Chapter 65 § 1365 (1983). http://www.fda.gov/RegulatoryInformation/Legislation/ucm148785.htm.

Federal Food, Drug, and Cosmetic Act. U.S. Code 21 (1938).

Food and Drug Administration. 2007. *Preparation of Premarket Submissions for Food Contact Substances: Chemistry Recommendations*. http://www.fda.gov/food/guidancecomplianceregulatoryinformation/guidancedocuments/foodingredientsandpackaging/ucm081818.htm.

Food and Drug Administration. 2009. *Defect Levels Handbook*. http://www.fda.gov/Food/GuidanceComplianceRegulatoryInformation/GuidanceDocuments/Sanitation/ucm056174.htm#CHPT3.

Food and Drug Administration. 2010a. Part 113, Subpart D. *Code of Federal Regulations Title 21*. Washington, DC: U.S. Government Printing Office. http://www.accessdata.fda.gov/scripts/cdrh/cfdocs/cfCFR/CFRSearch.cfm?CFRPart=113&showFR=1&subpartNode=21:2.0.1.1.12.4.

Food and Drug Administration. 2010b. Part 117. *Code of Federal Regulations Title 21*. Washington, DC: U.S. Government Printing Office. http://www.accessdata.fda.gov/scripts/cdrh/cfdocs/cfCFR/CFRSearch.cfm?CFRPart=177.

Food and Drug Administration. 2010c. Part 101. *Code of Federal Regulations Title 21*. Washington, DC: U.S. Government Printing Office. http://www.accessdata.fda.gov/scripts/cdrh/cfdocs/cfCFR/CFRSearch.cfm?CFRPart=101.

Food and Drug Administration. 2010d. Part 211, Subpart G. *Code of Federal Regulations Title 21*. Washington, DC: U.S. Government Printing Office. http://www.accessdata.fda.gov/scripts/cdrh/cfdocs/cfcfr/CFRSearch.cfm?fr=211.132.

German, J. B. 1999. Food Processing and Lipid Oxidation. *Advances in Experimental Medicine and Biology* 456: 23–50.

Gil, E. C., A. I. Colarte, J. L. L. Sampedro, and B. Bataille. 2008. Subcoating with Kollidon VA 64 as Water Barrier in a New Combined Native Dextran/HPMC-Cetyl Alcohol Controlled Release Tablet. *European Journal of Pharmaceutics and Biopharmaceutics* 69: 303–11.

Greenberg, E. F. 1996. Packaging Term with Legal Commentary. *Guide to Packaging Law: A Primer for Packaging Professionals*. Herndon, VA: Institute of Packaging Professionals, 78–79.

Halim, L., M. A. Pascall, J. Lee, and B. Finnigan. 2009. The Effect of Pasteurization, High Pressure Processing and Retorting on the Barrier Properties of Nylon 6, Nylon 6/Ethylene Vinyl Alcohol, and Nylon 6/Nanocomposites Films. *Journal of Food Science* 74: 9–15.

Hanlon, J. F. 1984. Paper and Paperboard. *Handbook of Package Engineering*, 2nd ed. New York: McGraw-Hill, 2-1–2-23.

Hernandez-Muñoz, P., R. Catala, and R. Gavara. 2001. Food Aroma Partition between Packaging Materials and Fatty Food Simulants. *Food Additives & Contaminants* 18: 673–82.

Hurme, E., G. Wirtanen, L. Axelson-Larsson, and R. Ahvenainen. 1998. Testing Reliability of Non-destructive Pressure Differential Leakage Tester with Semi-rigid Aseptic Cups. *Food Control* 9: 49–55.

Jeon, D. H., G. Y. Park, I. S. Kwak, K. H. Lee, and H. H. Park. 2007. Antioxidants and Their Migration into Food Simulants on Irradiated LLDPE Film. *LWT Food Science and Technology* 40: 151–56.

Keller, J. J. 2007. Food Labeling. *Compliance Manual for Food Quality and Safety*. Neenah, WI: J. J. Keller and Associates, Inc., 1–113.

Kris-Etherton, P. M., K. D. Hecker, A. Bonanome, S. M. Coval, A. E. Binkoski, K. F. Hilpert, A. E. Griel, and T. D. Etherton. 2002. Bioactive Compounds in Foods: Their Role in the Prevention of Cardiovascular Disease and Cancer. *American Journal of Medicine* 113: 71S–88S.

Laven, D. L. 2006. Prescription Drug Wholesalers: Drug Distribution and Inspection Process (A Florida Perspective). *Journal of Pharmacy Practice* 19: 196–214.

Lusina, M., T. Cindric, J. Tomaic, M. Peko, L. Pozaic, and N. Musulin. 2005. Stability Study of Losartan/Hydrochlorothiazide Tables. *International Journal of Pharmaceutics* 291: 127–37.

Mahanjan, R., A. Templeton, A. Harman, R. Reed, and R. T. Chern. 2005. The Effect of Inert Atmospheric Packaging on Oxidative Degradation in Formulated Granules. *Pharmaceutical Research* 22: 128–40.

Miller, C. W., M. H. Nguyen, M. Rooney, and K. Kailasapathy. 2003. The Control of Dissolved Oxygen Content in Probiotic Yoghurts by Alternative Packaging Materials. *Packaging Technology and Science* 16: 61–67.

Pamplona-Roger, G. D. 1998. *Encyclopedia of Medicinal Plants*. Edited by G. X. Francesc. Madrid: Editorial Safeliz, S.L., vol. 1, 27–51.

Pascall, M. A. 2002. Evaluation of a Non-destructive Bench-Top Pressure Differential (Force/Decay) System for Leak Detection in Polymeric Trays Used for Food Packaging. *Packaging Technology and Science* 15: 197–208.

Pascall, M. A., U. Fernandez, R. Gavara, and A. Allafi. 2008. Mathematical Modeling, Non-destructive Analysis and Gas Chromatographic Method for Headspace Oxygen Measurement of Modified Atmosphere Packaged Soy Bread. *Journal of Food Engineering* 86: 501–07.

Pascall, M. A., J. R. Giacin, I. Gray, and B. R. Harte. 1995. Decreasing Lipid Oxidation in Soybean Oil by a UV Absorber in the Packaging Material. *Journal of Food Science* 60: 1116–19.

Rhee, K. S., and B. M. Watts. 2006. Evaluation of Lipid Oxidation in Plant Tissues. *Journal of Food Science* 31: 664–68.

Sadler, G., W. Chappas, and D. E. Pierce. 2001. Evaluation of E-Beam γ and X-ray Treatment on the Chemistry and Safety of Polymers Used with Pre-packaged Irradiated Foods: A Review. *Food Additives and Contaminants* 18: 475–501.

Safe Drinking Water and Toxic Enforcement Act (Proposition 65). *California Code of Regulations, Title 27* (1986).

Sajilata M. G., K. Savitha, R. S. Singhal, and V. R. Kanetkar. 2007. Scalping of Flavors in Packaged Foods. *Comprehensive Reviews in Food Sciences and Food Safety* 6: 17–35.

Selke, S. E. M. 1997. Major Packaging Polymers. *Understanding Plastic Packaging Technology*. Cincinnati, OH: Hanser/Gardner Publications, Inc., 15–49.

Shi, J., and M. Le Maguer. 2000. Lycopene in Tomatoes: Chemical and Physical Properties Affected by Food Processing. *Critical Reviews in Biotechnology* 20: 293–334.

Sivaramakrishna, V., F. Raspante, S. Palaniappan, and M. A. Pascall. 2007. Development of a Timesaving Leak Detection Method for Brick-Type Packages. *Journal of Food Engineering* 82: 324–32.

Soroka, W. 2009. Plastic Applications. *Fundamentals of Packaging Technology*, 4th ed. Naperville, IL: Institute of Packaging Professional, 271.

Wong, M., R. Marion, K. Reed, and Y. Wang. 2006. Sorption of Unoprostone Isopropyl to Packaging Materials. *International Journal of Pharmaceutics* 307: 163–67.

CHAPTER 14

Addressing Powder Flow and Segregation Concerns during Scale-Up of Nutraceutical Solid Formulations

Thomas Baxter, James Prescott, Roger Barnum, and Jayant Khambekar

CONTENTS

14.1 INTRODUCTION

This chapter provides guidance in designing bulk solid ("powder") handling equipment to provide consistent, reliable flow and to minimize segregation. The principles discussed in this chapter can be applied to analyzing new or existing equipment designs. The principles can also be used to compare different powders using the various test methods discussed.

The chapter will focus on the equipment used from the final blend step to the inlet of the packaging equipment (e.g., press, encapsulator, vial filling, and sachets). This chapter is divided into the following primary topics:

1. Introduction: A review of introductory concepts, such as flowability, blending, and segregation
2. Typical powder flow and segregation concerns: A review of common flow and segregation concerns, and the two primary flow patterns (mass flow vs. funnel flow)
3. Measurement of flow properties and segregation potential: A summary of the flow properties that need to be measured to obtain the equipment design parameters required for consistent, reliable flow
4. Basic equipment design techniques: A review of the basic design techniques for the blender-to-packaging equipment to provide reliable flow and minimize the adverse effects of segregation, including troubleshooting methods for assessing segregation and product nonuniformity concerns

Many nutraceutical processes include powder handling, such as blending, transfer, storage, and feeding. A full understanding of powder flow behavior is essential when developing, optimizing, or scaling up a process. This may include designing new equipment or developing corrective actions for existing equipment. There are several instances where the robustness of a process is adversely affected by flow or segregation problems that develop.

Common powder flow problems can have an adverse effect on:

1. *Production costs* as a result of reduced production rates, restrictions on raw ingredient selection, method of manufacturing, equipment selection, and overall yield
2. *Product quality* as a result of segregation, nonuniform feed density, weight variation, etc.
3. *Time to market* as a result of delays in product and process development because flow and segregation problems may not occur until the process has been scaled up

14.1.1 Introduction to Flowability

A *bulk solid* is defined as a collection of solid particles. The term *powder* is often used to describe a fine bulk solid. This common term will be used predominantly throughout this chapter. The concepts discussed in this chapter apply to many types of powders with different particle sizes, shapes, and distributions. The powders may include raw ingredients, granulations, and final blends. The powder could be either a single substance, such as an ingredient, or a multicomponent blend. The principles outlined in this chapter can be used to design for all these different types of powders.

A simple definition of *flowability* is the ability of a powder to flow through equipment reliably. By this definition, there is often a tendency to define flowability as a single parameter of a powder ranked on a scale from "free-flowing" to "nonflowing." Unfortunately, a single parameter is not sufficient to define a powder's complete handling characteristics. In addition, a single parameter is not sufficient to fully address common handling concerns encountered by the formulator and equipment designer. In fact, several design parameters may need to be known for a successful design. The behavior of a powder will depend on several different parameters or "flow properties." Flow properties are the specific properties of a powder that can be measured and affect flow behavior in equipment. Therefore, a full range of *flow properties* will need to be measured to fully characterize the powder.

In addition, the flowability of a powder is a function of the powder's flow properties *AND* the design parameters of the handling equipment. For example, "poor flowing" powders can be handled reliably in properly designed equipment. Conversely, "good flowing" powders may develop flow problems in improperly designed equipment. Therefore, our definition of flowability is the ability of powder to flow in the desired manner in a specific piece of equipment.

The flow properties of the powder should be *quantitative* and *scalable* design parameters. The term *flow properties* often refers to the physical characteristics of the powder that were measured. The term *powder flow* often refers to an observation of how the powder flowed through a given piece of equipment. In discussing or reporting flowability, both the powder flow properties and the handling equipment must be included. Therefore, the measurement of the powder flow properties can be used during scale-up to predict powder flow behavior in specific equipment.

14.1.2 Introduction to Blending

Powder blending processes are used during the manufacture of products for a wide range of industries, including nutraceuticals. In the nutraceutical industry, a wide range of ingredients may be blended together to create the final blend used to manufacture the consumer product. The range of ingredients that may be blended presents a number of variables that can affect blend uniformity. These variables may include the particle size distribution (including aggregates or lumps of material), particle shape (spheres, rods, cubes, plates, and irregular), presence of moisture (or other volatile compounds), particle surface properties (roughness, cohesiveness), among others.

It is critical to understand the physical properties of the final blend ingredients and how they affect the flowability and segregation of the final blend when selecting and designing the powder handling equipment. The process steps and equipment parameters before the final blend step are often critical to the flowability and segregation of the final blend, especially if they affect the particle size distribution (e.g., milling, screening, and granulation of select ingredients).

The quality of the consumer product is dependent on the adequacy of the blend. Producing a uniform mixture is paramount in being able to deliver the required product quality. Once an adequate blend is obtained, it is also critical to ensure that it does not segregate in the postblending handling steps. Formulation components and process parameters involved with blending operations should be carefully selected and validated to ensure uniform blends and dosage units are produced.

The complexity of the blending process can vary substantially, but most blenders use one or more of the following mechanisms to induce powder blending: convective blending, shear blending, and diffusive blending (Rippie et al. 1980; Venables and Wells 2001; Prescott and Garcia 2008). Each of these mechanisms has its role in affecting a uniform blend. Depending on the ingredients used, a blender may need to be selected that puts more emphasis on one mechanism over the others. Large-scale production batches often use equipment capable of blending hundreds of kilograms of material. Depending on the product, commercial-scale blending processes can be complex and may require the preparation of preblends or the inclusion of milling and screening operations to achieve acceptable content uniformity. Regardless of the scale of manufacture, the goal remains the same: to prepare a blend that is adequately blended and can be further processed into uniform consumer products.

Blending should not be seen as an independent unit operation, but rather as an integral part of the overall manufacturing process. Blending includes producing an adequate blend, maintaining that blend through additional handling steps, and verifying that both the blend and the finished product are sufficiently homogeneous. Therefore, a complete approach should be used to assess the uniformity of blends and the subsequent dosage forms produced from them.

14.1.3 Introduction to Segregation

Segregation can be defined as having particles of similar properties (e.g., size, composition, density, resiliency, and static charge) preferentially distributed into different zones within given equipment or processes (Prescott and Garcia 2008). Segregation can result in blend and content uniformity problems in the consumer product. In addition, the segregation of other blend components can be responsible for variations in properties such as taste, appearance, and color. Even if the blend remains *chemically* homogeneous, variations in particle size can affect flowability, bulk density, weight uniformity, tablet hardness (for compacted products), and dissolution.

Segregation can occur any time there is powder transfer, such as discharging the final blend from the blender into a bin. Segregation can also occur when forces acting on the particles in the blend, such as air flow or vibration, are sufficient to induce particle movement. This may include handling steps upstream of a blender, including segregation of raw materials at a supplier's plant or during shipment, as well as movement within the blender, during its discharge, or in downstream equipment. Of all these potential instances where segregation can occur, the most common area for problems is post–blender discharge. Therefore, this chapter will focus on segregation of the final blend in the postblending handling steps.

The current state of understanding segregation is limited to having empirical descriptors of segregation mechanisms and previous experiences with diagnosing and addressing specific segregation behaviors, which are addressed in this chapter. Unlike the flow property tests used to assess how a powder flows through equipment, there are no "first principle" models that describe segregation. Therefore, one cannot currently input the particle properties of the blend components (e.g., particle size and chemical composition of the ingredients) into a mathematical model and obtain a prediction of segregation potential. At best, computational models such as discrete element modeling (DEM) are evolving and can be tuned to match specific segregation behaviors that are created in physical models. As these models evolve they will become more powerful and have fewer assumptions and limitations. At present, the average nutraceutical scientist may not be able to use DEM to predict or solve the segregation problems he or she will likely encounter. Therefore, when assessing segregation concerns, it is critical to use as many resources as possible such as lab-scale tests to assess different segregation mechanisms and stratified blend and content uniformity data to assess potential segregation trends. The empirical tests that should be run to assess the segregation potential of a final blend, as well as how to apply the results to a reliable equipment design, are discussed in Section 14.3.

14.2 TYPICAL FLOW AND SEGREGATION CONCERNS

14.2.1 Common Flow Problems

A number of problems can develop as powder flows through handling equipment such as bins (also commonly referred to as a *tote* or *intermediate bulk container* [IBC]; see Figure 14.1), transfer chutes (Figure 14.2), and feed hoppers. If the powder is cohesive, an arch or rathole may form, resulting in "no flow" or erratic flow. In addition, flooding or uncontrolled discharge may occur if a rathole spontaneously collapses. A deaerated bed of fine powder may experience flow rate limitations or no-flow conditions as a result of the two-phase flow effects between the powder and the interstitial air. Each of these flow problems is discussed in more detail in the following sections.

14.2.1.1 No Flow

No flow from a bin or hopper is a common and significant handling problem when working with solids. In production, it can result in problems such as starving downstream equipment, production delays, and the requirement for frequent operator intervention to reinitiate flow. No flow can be caused by either arching (sometimes referred to as *bridging* or *plugging*) or ratholing (also referred to as *piping*).

Arching occurs when an obstruction in the shape of an arch or bridge forms above the bin outlet and prevents any further material discharge. It can be an interlocking arch, where the particles mechanically lock to form the obstruction, although this is less common with fine nutraceutical powders. An interlocking arch occurs when the particles are large compared with the outlet size of the hopper. The arch could also be a cohesive arch where the particles pack together to form an obstruction (Figure 14.3). Both of these problems are strongly influenced by the outlet size of

Figure 14.1 Intermediate bulk container (IBC; also referred to as a bin or tote).

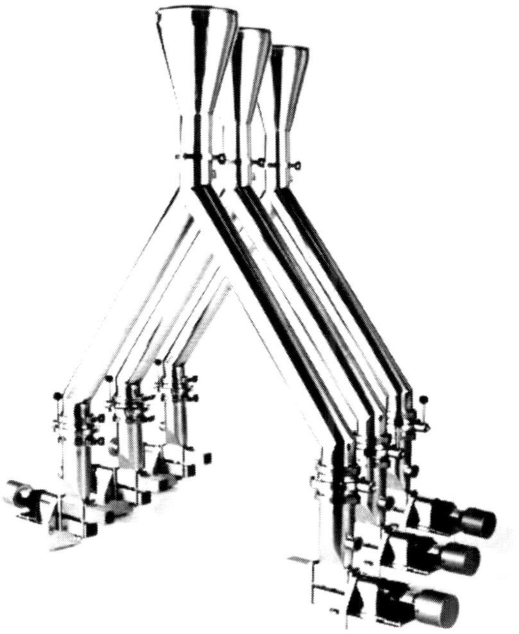

Figure 14.2 Bifurcated press transfer feed chute.

the hopper through which the material is fed. Powder flow properties, discussed in Section 14.3, can be used to determine whether these problems will occur and address them during scale-up. In particular, the cohesive strength of a powder will dictate what size outlet over which it can arch (the greater the cohesive strength, the higher the likelihood of arching).

Figure 14.3 Cohesive arch.

Figure 14.4 Rathole.

Ratholing can occur in a bin when the powder empties through a central flow channel, but the material at the bin walls remains stagnant and leaves an empty hole ("rathole") through the material starting at the bin outlet (Figure 14.4). Ratholing is influenced by the bin or hopper geometry, material level, and outlet size through which the material is fed. Similar to the problem of arching, this problem will arise if the material has sufficient cohesive strength. In this case, the material discharge will stop once the flow channel empties.

14.2.1.2 Erratic Flow

Erratic flow is the result of obstructions alternating between an arch and a rathole. A rathole may collapse as a result of an external force, such as vibrations created by surrounding equipment or a flow-aid device such as an external vibrator. Although some material may discharge, falling material may impact over the outlet and form an arch. An arch may break as a result of a similar external force, and material flow may resume until the flow channel is emptied and a rathole is formed again. This not only results in erratic feed to the downstream equipment but also can result in a nonuniform feed density as well.

14.2.1.3 Fine Powder Flow Concerns: Two-Phase Flow Effects

Additional flow concerns can arise when handling *fine powders*, generally in the range less than 100 μm in average particle size. These concerns are the result of the interaction of the material with entrained air or gas, which becomes significant in describing the behavior of the material. This interaction can result in two-phase (powder or interstitial gas) flow effects. There are three modes that can occur when handling fine powders that are susceptible to two-phase flow effects: *steady flow*, *flooding* (or flushing), and *a flow rate limitation* (Royal and Carson 2000).

Steady flow will occur with fine powders if the target flow rate (feed rate through the system) is less than the critical flow rate that occurs when the solid's stress is balanced by the air pressure at the outlet. The target flow rate is often controlled by a feeder, such as at the inlet to a compression machine (press feed frame). The critical flow rate and the flow property tests used to determine it are described in Section 14.3. At target flow rates exceeding the critical flow rate, unsteady flow can occur by two different modes.

Flooding (or "flushing") is an unsteady two-phase flow mode that can occur as falling particles entrain air and become fluidized. Because powder handling equipment often cannot contain a fluid-like powder, powder can flood through the equipment (feeders, seals) uncontrollably. Flooding can also occur when handling fine powders in small hoppers with high fill and discharge rates. In such situations, the powder does not have sufficient residence time to deaerate, resulting in flooding through the feeder. One adverse effect of flooding or flushing may be high variation in the consumer product weight (e.g., tablet weight).

A flow rate limitation is another unsteady two-phase flow mode that can occur with fine powders. Fine powders have low permeabilities and are affected by any movement of the interstitial air (air between the particles). This air movement will occur as a result of the natural consolidation and dilation of the powder bed that takes place as it flows through the cylindrical (straight-sided) and hopper (converging) geometries. As the material is consolidated in the cylinder, the air is squeezed out. As the powder flows through the hopper and outlet, it dilates, and additional air must be drawn in. The air pressure gradients caused as a result of this air movement can retard discharge from a hopper, significantly limiting the maximum achievable rates. This limit may be observed when the speed of the packaging equipment or press is increased and the product weight variation increases.

During unsteady two-phase flow modes, the material's bulk density can undergo dramatic variations. This can negatively affect downstream packaging or processing operations. Two-phase flow problems can result in excessive product weight variations, reduced filling speeds, and segregation. Equipment and process parameters that affect two-phase flow behavior include hopper geometry and outlet size, applied vacuum and other sources of air pressure differences (such as dust collection systems), material level, time since filling, and the target feed rate. Material properties such as permeability and compressibility (discussed later in Section 14.3) will also affect two-phase flow behavior, as will variations in the material's state of aeration that can occur based on its residence time or degree of compaction from external forces and handling. One of the most important factors in determining whether a powder will discharge reliably from a hopper is establishing what flow pattern will develop, which is discussed in the following section.

14.2.2 Flow Patterns

Two flow patterns can develop in a bin or hopper: *funnel flow* and *mass flow*. In *funnel flow* (Figure 14.5), an active flow channel forms above the outlet, which is surrounded by stagnant material. This results in a first-in, last-out flow sequence. It generally occurs in equipment with relatively shallow hoppers. Common examples

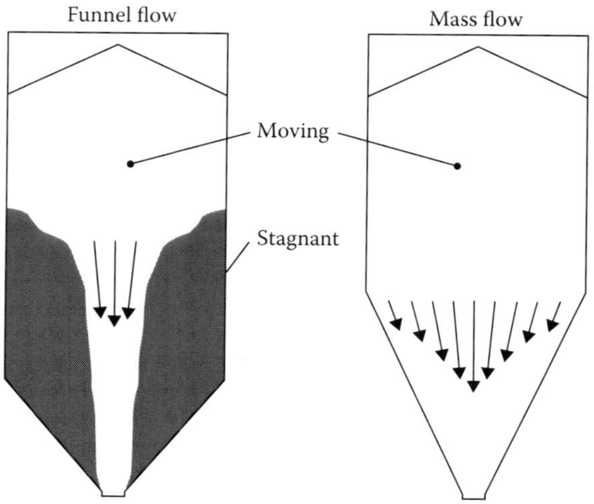

Figure 14.5 Funnel flow versus mass flow discharge pattern.

of funnel flow bins include hopper geometries such as asymmetric cones and rectangular-to-round transitions (Figure 14.6).

As the level of powder decreases in funnel flow, stagnant powder may fall into the flow channel if the material is sufficiently free-flowing. If the powder is cohesive, a stable rathole may develop. Funnel flow occurs if the powder is unable to flow along the hopper walls as a result of the combination of friction against the walls and hopper angle. Funnel flow's first-in/last-out flow sequence can also have an adverse effect on segregation and result in a nonuniform feed density to the downstream equipment (product weight variation).

In *mass flow* (see Figure 14.5), all the powder is in motion whenever any is withdrawn. Powder flow occurs throughout the bin, including at the hopper walls. Mass

Figure 14.6 Examples of funnel flow bins.

flow provides a first-in/first-out flow sequence, eliminates stagnant powder, provides a steady discharge with a consistent bulk density, and provides a flow that is uniform and well controlled. Ratholing will not occur in mass flow because all the material is in motion. A mass flow bin design may also be beneficial with respect to reducing segregation (further discussed in Section 14.4).

The requirements for achieving mass flow are:

1. Sizing the outlet large enough to prevent arch formation
2. Ensuring the hopper walls are steep and smooth enough to allow the powder to flow along them

14.2.3 Common Segregation Mechanisms

The term *segregation mechanism* refers to the mode by which the components of the blend separate. There are many different segregation mechanisms that can adversely affect the uniformity (Bates 1997), but three primary segregation mechanisms are of interest in typical nutraceutical blend handling operations. These three primary segregation mechanisms are:

1. *Sifting segregation* (sometimes referred to as *percolation segregation*)
2. *Fluidization segregation* (sometimes referred to as *air entrainment*)
3. *Dusting segregation* (sometimes referred to as *particle entrainment in an air stream*)

The three segregation mechanisms are described in more detail below. These terms are not universally defined, so one must use caution when using them. Segregation may occur as a result of just one of these mechanisms, or a combination of several mechanisms.

14.2.3.1 Material Properties That Affect Segregation

Whether segregation occurs, to what degree, and which mechanism or mechanisms are involved depends on a combination of the properties of the blend and the process conditions encountered. Some of the primary material properties that influence segregation tendencies include:

1. The mean particle size and particle size distribution of the blend components and final blend: Segregation can occur with blends of any mean size, but different mechanisms become more pronounced at different particle sizes, as further discussed in Sections 14.2.3.2 to 14.2.3.4.
2. Particle density: The particle density will affect how the blend components fluidize, or get carried along a pile or in an airstream.
3. Particle shape: Rounded particles may have increased mobility than irregularly shaped particles, which can allow more segregation.
4. Particle resilience: This property influences collisions between particles and surfaces, which can lead to differences in where components accumulate during the filling of equipment.

5. Cohesive strength of the blend: As a general rule, more cohesive blends are less likely to segregate. However, if sufficient energy is added to dilate the blend and/or separate particles from one another, even a cohesive material can segregate.
6. Electrostatic effects: The ability of components to develop and hold an electrostatic charge, and their affinity for other ingredients or processing surfaces can also contribute to segregation tendencies.

Of all these, segregation based on particle size is by far the most common (Williams 1976). In fact, particle size is the most important factor in the three primary segregation mechanisms considered here, as further described in the following sections.

14.2.3.2 Sifting Segregation

Sifting segregation is the most common form of segregation for many industrial processes. Under appropriate conditions, fine particles tend to sift or percolate through coarse particles. For segregation to occur by this mechanism, four conditions must exist:

1. There must be a range of particle sizes. A minimum difference in mean particle diameters between components of 1.3:1 is often more than sufficient (Williams 1996).
2. The mean particle size of the mixture must be sufficiently large, typically greater than approximately 100 μm (Williams and Khan 1973).
3. The mixture must be relatively free-flowing to allow particle mobility.
4. There must be relative motion between particles (interparticle motion).

The last requirement is important because without it, even highly segregating blends of ingredients that meet the first three tests will not segregate. Relative motion can be induced in a variety of ways, such as when a pile is formed when filling a bin, vibration from surrounding equipment (such as a tablet press), or as particles tumble and slide down a chute.

If any one of these conditions does not exist, the mix will not segregate by this mechanism. However, all these are typically present in nutraceutical production processes to at least some extent.

The result of sifting segregation in a bin is usually a side-to-side variation in the particle size distribution. The smaller particles will generally concentrate under the fill point, with the coarse particles concentrating at the perimeter of the pile (Figure 14.7).

14.2.3.3 Fluidization Segregation

Variations in particle size or density often result in vertically segregated material when handling powders that can be fluidized. Finer or lighter particles often will be concentrated above larger or denser particles. This can occur during filling of a bin or other vessel, or within a blending vessel once the blending action has ceased.

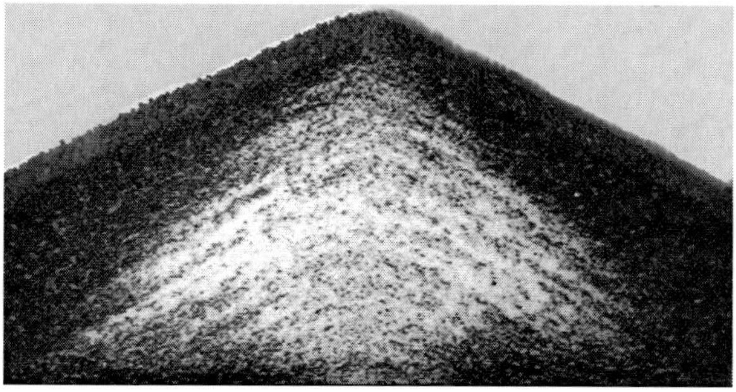

Figure 14.7 Example of sifting segregation.

Fluidization segregation often results in horizontal layers of fine and coarse material. A fine powder can remain fluidized for an extended period after filling or blending. In this fluidized state, larger and/or denser particles tend to settle to the bottom. Fine particles may be carried to the surface with escaping air as the bed of material deaerates. For example, when a bin is filled quickly, the coarse particles move downward through the aerated bed, whereas the fine particles remain fluidized near the surface. This can also occur after blending if the material is fluidized during blending.

Fluidization is common in materials that contain a significant percentage of particles smaller than 100 μm (Pittenger, Purutyan, and Barnum 2000). Fluidization segregation is likely to occur when fine materials are pneumatically conveyed, when they are filled or discharged at high rates, or if gas counterflow occurs. As with most segregation mechanisms, the more cohesive the material, the less likely it will segregate by this mechanism.

Fluidization via gas counterflow can occur as a result of insufficient venting during material transfer. For example, consider a tumble blender discharging material to a bin, with an airtight seal between the two (Figure 14.8). As the blend transfers from the blender to the bin, air in the bin is displaced, and a slight vacuum is created in the blender. If both are properly vented, air moves out of the bin and separately into the blender. If both are not vented, the air must move from the bin to the blender through the blender discharge. In doing so, the fine particles may be stripped off the blend and carried to the surface of the material still within the blender.

14.2.3.4 Dusting Segregation

Like fluidization segregation, dusting is most likely to be a problem when handling fine, free-flowing powders that have particles smaller than approximately 50 μm (Pittenger, Purutyan, and Barnum 2000), as well as a range of other particle sizes. If dust is created on filling a bin, air currents created by the falling stream will carry particles away from the fill point (Figure 14.9). The rate at which the dust settles is governed by the particle's settling velocity. The particle diameter is much more significant than particle density in determining settling velocity.

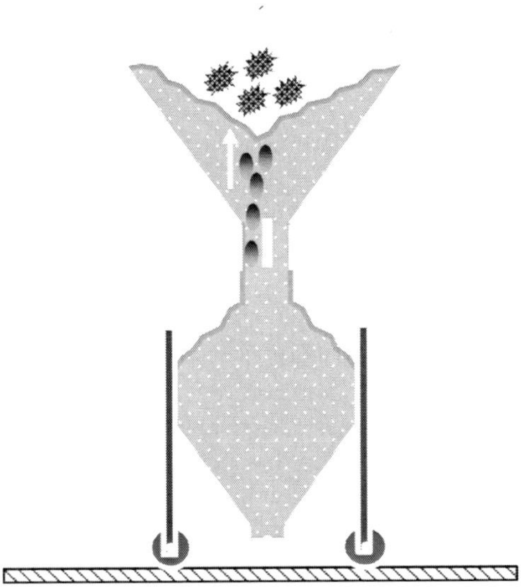

Figure 14.8 Example of fluidization segregation from blender to bin.

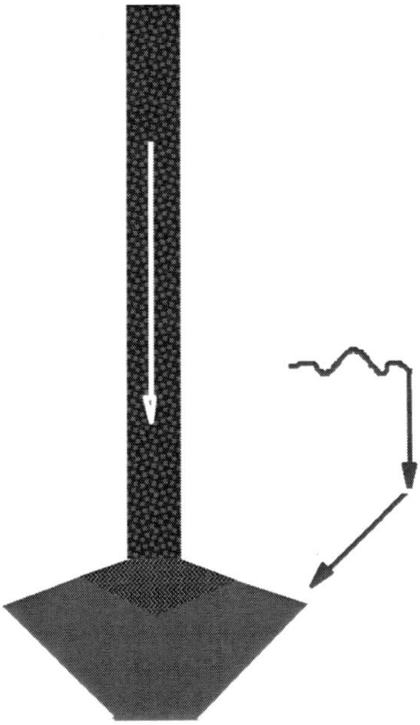

Figure 14.9 Example of dusting segregation example during filling of bin.

As an example of this mechanism, consider a mix of fine and large particles that is allowed to fall into the center of a bin. When the stream hits the pile of material in the bin, the column of air moving with it is deflected. The air then sweeps off the pile toward the perimeter of the bin, where the air becomes highly disturbed. After this, the air generally moves back up the bin walls in a swirling pattern. At this point, the gas velocity is much lower, allowing many particles to fall out of suspension. Because settling velocity is a strong function of particle diameter, the finest particles (with low settling velocities) will be carried to the perimeter of the bin. The larger particles will concentrate closer to the fill point, where the air currents are strong enough to prevent the fine particles from settling. Dusting segregation can also result in less predictable segregation patterns, depending on how the bin is loaded, venting in the bin, and dust collection use and location.

The empirical tests used to assess these segregation mechanisms are reviewed in Section 14.3, along with applying the results to design equipment to minimize segregation.

14.3 MEASUREMENT OF FLOW PROPERTIES AND SEGREGATION POTENTIAL

14.3.1 Cohesive Strength Tests: Preventing Arching and Ratholing

Jenike (1964, revised 1980) developed his mathematical model for the flow of powders by modeling the powder as a plastic (not a viscoelastic) continuum of solid particles. This approach included the postulation of a "flow–no flow" criterion that states the powder would *flow* (e.g., from a bin or hopper) when the *stresses* applied to the powder *exceed* the *strength* of the powder. The strength of a material will vary, depending on how consolidated it is. For example, the strength of lactose increases as the consolidation pressure is increased (e.g., the more one packs it). Therefore, it is critical to be able to measure the cohesive strength of a powder as a function of the consolidation pressure.

14.3.1.1 Test Methods

One of the primary flow problems that can develop in powder handling equipment is a "no flow" obstruction resulting from the formation of a cohesive arch or rathole. The required outlet size to prevent a stable cohesive arch or rathole from forming is determined by applying the flow–no flow criterion and using the results of a cohesive strength test. To apply the flow–no flow criterion, we need to determine:

1. The cohesive *strength* of the material as a function of the *major consolidation pressure* acting on the material. The consolidation pressure acting on the powder changes throughout the bin height as a result of the weight of material above it. Therefore, the cohesive strength must be measured over a range of consolidation pressures. The cohesive strength can be measured as a function of major consolidating pressure using the test methods described in this section.

2. The *stresses* acting on the material to induce flow. Gravity pulls downward on a potential arch that may form. The stresses acting on the powder can be determined using mathematical models (Jenike 1964, revised 1980).

Several different test methods have been developed to measure cohesive strength. The respective strengths and weaknesses of these different cohesive strength tests are reviewed in the literature (Schulze 1996a, 1996b). Although many different test methods can be used to measure cohesive strength, this chapter focuses specifically on the Jenike Direct Shear Test method. The Jenike Direct Shear Test method is the most universally accepted method and is described in ASTM International standard D6128 (Standard Shear Testing 2006). It is important that these tests be conducted at representative handling conditions such as temperature, relative humidity, and storage at rest because all these factors can affect the cohesive strength. An arrangement of a cell used for the Jenike Direct Shear Test is shown in Figure 14.10. The details of the Jenike Direct Test Method are provided in *Storage and Flow of Solids* (Jenike 1964, revised 1980), including the generation of:

1. Mohr's circle to plot the shear stress (τ) versus the consolidation pressure (σ)
2. Effective yield locus
3. Flow function

The data generated *experimentally* from the Jenike Direct Shear Test can be used to determine the following derived parameters:

1. The *flow function* that describes the cohesive strength ("unconfined yield strength," F_c) of the powder as a function of the major consolidating pressure (σ_1). The flow function is one of the primary parameters used to calculate the minimum outlet diameter or width for bins, packaging or press feed hoppers, blender outlets, etc. to prevent arching and ratholing. The calculation of the minimum outlet diameter or width is discussed in Section 14.3.1.2.

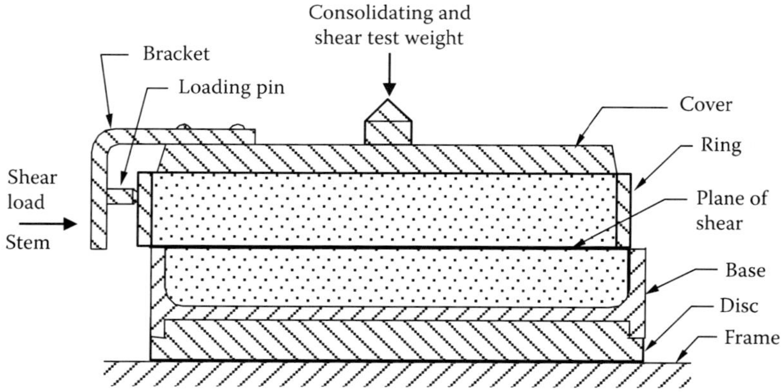

Figure 14.10 Jenike Direct Shear Test, cohesive strength test setup.

2. The *effective angle of internal friction* (δ) that is used to calculate the minimum outlet to prevent arching and the required hopper angles for mass flow (described in Section 14.3.2.2).
3. The *angle of internal friction for continuous flow or after storage at rest* (ϕ and ϕ_t). The static angle of internal friction (ϕ_t) is used to calculate the minimum outlet to prevent ratholing (described in Section 14.3.1.3).

Other testing methods exist that use the same principles of consolidation and shearing to determine the cohesive strength of a bulk powder. Annular (ring) shear testers produce rotational displacement between cell halves containing material, rather than a lateral displacement. The loading and shearing operations are more readily adapted to automation because of the unlimited travel that can be achieved with this type of test cell. The successful use of an annular ring shear tester to measure cohesive strength is well documented (Nyquist 1984; Bausch et al. 1998; Hausmann et al. 1998; Ramachandruni and Hoag 1998).

14.3.1.2 Calculation of the Minimum Required Outlet Dimensions to Prevent Arching: Mass Flow Bin

The flow behavior of powders through bins and hoppers can be predicted by a complete mathematical relationship. This is beneficial when scaling up a process and designing powder handling equipment. If gravity discharge is used, the outlet size required to prevent a cohesive arch or rathole from forming over a bin outlet can be calculated. The term B_c refers to the *minimum outlet diameter for a circular outlet* to prevent a cohesive arch from forming in a mass flow bin. The term B_p refers to the *minimum outlet width for a slotted outlet* ("B_p"), in which the length/width ratio exceeds 3:1, to prevent arching in a mass flow bin.

The majority of bins used in nutraceutical processes use hoppers with circular outlets. Therefore, we will focus our discussion on the calculation of the B_c parameter. It is worth noting that the outlet diameter required to prevent arching over a circular outlet (B_c) will typically be approximately 2 times greater than the required outlet width of a slotted outlet (B_p). The calculation of B_p is provided in Jenike (1964, revised 1980).

For mass flow, the required minimum outlet diameter to prevent arching is calculated in Equation 14.1:

$$B_c = H(\theta')\, f_{crit}/\gamma \tag{14.1}$$

1. $H(\theta')$ is a dimensionless function of the hopper angle (θ') derived from first principles (mathematical model). The complete derivation of $H(\theta')$ is beyond the scope of this chapter but is provided in Jenike (1964, revised 1980).
2. f_{crit} (units of force/area) is the unconfined yield strength at the intersection of the hopper "flow factor" (ff) and the experimentally derived Flow Function (FF), as shown in Figure 14.11. The flow factor (ff) is a mathematically determined value that represents the minimum available stress to break an arch. The calculation of the flow factor (ff) is also beyond the scope of this chapter but is provided in

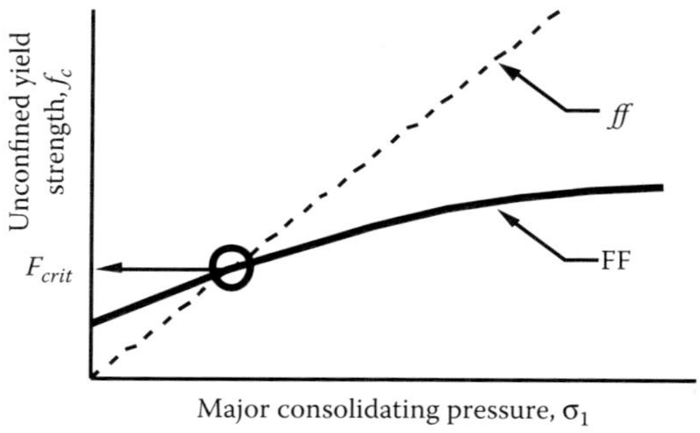

Figure 14.11 Example of Flow Function (FF) and flow factor (*ff*) intersection, showing f_{crit} at their intersection.

Jenike (1964, revised 1980) and is a function of the flow properties and the hopper angle (θ').

3. γ (units of weight/volume) is the bulk density of the powder at the outlet.

This calculation yields a dimensional value of B_c in units of length, which is *scale independent.* Therefore, for a mass flow bin, the opening size required to prevent arching is not a function of the diameter of the bin, height of the bin, or the height/diameter ratio.

The determination of B_c is especially valuable in making decisions during process scale-up. As a formulation is developed, a cohesive strength test can be conducted early in the development process to determine the cohesive strength (flow function). This material-dependent flow function, in conjunction with Equation 14.1, will yield a minimum opening (outlet) size to prevent arching in a mass flow bin. For example, this opening size may be calculated to be 200 mm. This 200-mm diameter will be required whether the bin holds 10 or 1000 kg of powder and is scale independent. In this example, because a 200-mm-diameter opening is required, feeding this material through a packaging or press feed hopper or similarly small openings would pose problems with an arch developing over the outlet. This information could then be used early in the development process to consider reformulating the product to reduce the cohesive strength and improve flowability.

14.3.1.3 Calculation of Minimum Required Outlet Dimensions to Prevent Ratholing: Funnel Flow Bin

If the bin discharges in funnel flow, the bin outlet diameter should be sized to be larger than the *critical rathole diameter* (D_f) to prevent a stable rathole from forming over the outlet. For a funnel flow bin with a *circular* outlet, sizing the outlet diameter to exceed the D_f will also ensure that a stable arch will not form (because a rathole is inherently stronger than an arch). The D_f value is calculated in Equation 14.2, and additional details of the calculation are provided in Jenike (1964, revised 1980).

$$D_f = G(\phi_t)\, f_c(\sigma_1)/\gamma \tag{14.2}$$

1. $G(\phi_t)$ is a mathematically derived function from first principles and is provided in Jenike (1964, revised 1980).
2. $f_c(\sigma_1)$ is the unconfined yield strength of the material. This value is determined by the flow function at the actual consolidating pressure σ_1 (see Equation 14.3).

The relationship in Equation 14.2 cannot be reduced further (e.g., to a dimensionless ratio) because the function $f_c(\sigma_1)$ is highly material dependent.

The consolidation pressure σ_1 is a function of the "head" or height of powder above the outlet of the bin and takes into account the load taken up by friction along the walls, as derived by Janssen (1895) in Equation 14.3:

$$\sigma_1 = (\gamma\, R/\mu k)\,(1 - e^{(-\mu k h/R)}) \tag{14.3}$$

where γ = average bulk density of the powder in the bin
R = hydraulic radius (area/perimeter)
μ = coefficient of friction, μ (μ = tangent ϕ'); the value of ϕ' is determined from the wall friction test (discussed below)
k = ratio of horizontal to vertical pressures; a k value of 0.4 is typically used for a straight-sided section)
h = depth of the bed of powder within the bin

14.3.2 Wall Friction: Determining Hopper Angles for Mass Flow

14.3.2.1 Test Method

The wall friction test is crucial in determining whether a given bin will discharge in mass flow or funnel flow. Wall friction is caused by the powder particles sliding along the wall surface. In Jenike's continuum model (1964, revised 1980), wall friction is expressed as the *wall friction angle* (ϕ'). The higher the wall friction angle, the steeper the hopper or chute walls need to be for powder to flow along them. This coefficient of friction can be measured by shearing a sample of powder in a test cell across a stationary wall surface using a Jenike direct shear tester (Jenike 1964, revised 1980; Standard Shear Testing 2006). One arrangement of a cell used for the wall friction test is shown in Figure 14.12. In this case, a coupon of the wall material being evaluated is held in place on the frame of the tester, and a cell of powder is placed above. The *coefficient of sliding friction* (μ, μ = tangent ϕ') is the ratio of the shear force required for sliding (τ) to the normal force applied perpendicular to the wall material coupon (σ_n). A plot of the measured shear force (τ) as a function of the applied normal pressure (σ_n) generates a relationship known as the *wall yield locus* (Figure 14.13).

The wall friction measured is a function of the powder handled and the wall surface (type, finish, orientation) in contact with it. Variations in the powder, handling conditions (e.g., temperature, relative humidity), and/or the wall surface

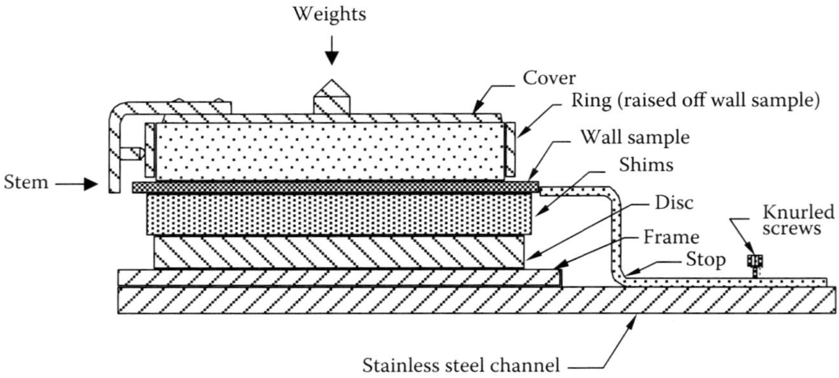

Figure 14.12 Jenike Direct Shear Test, wall friction test setup.

finish can have a dramatic effect on the resulting wall friction coefficient (Prescott, Ploof, and Carson 1999). The results of the wall friction test are used to determine the hopper angles required to achieve mass flow, as discussed in the following section.

14.3.2.2 Calculation of Recommended Mass Flow Hopper Angles

Based on mathematical models, design charts have been developed to determine which flow pattern is expected during gravity discharge from a bin (Jenike 1964, revised 1980). The design charts use the following inputs for the powder being handled and the bin design being considered:

1. The hopper angle ("θ_c" for a conical hopper or "θ_p" for a planar hopper), as measured from *vertical*
2. The wall friction angle (ϕ'), as measured from the wall friction tests
3. The effective angle of internal friction (δ), as measured from the cohesive strength tests

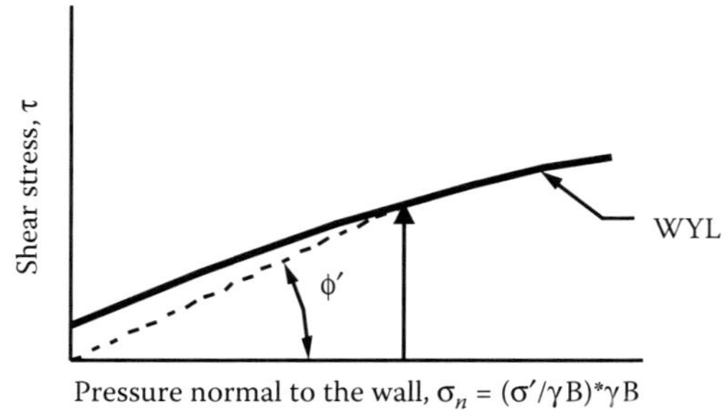

Figure 14.13 Example of wall yield locus generated from wall friction test data.

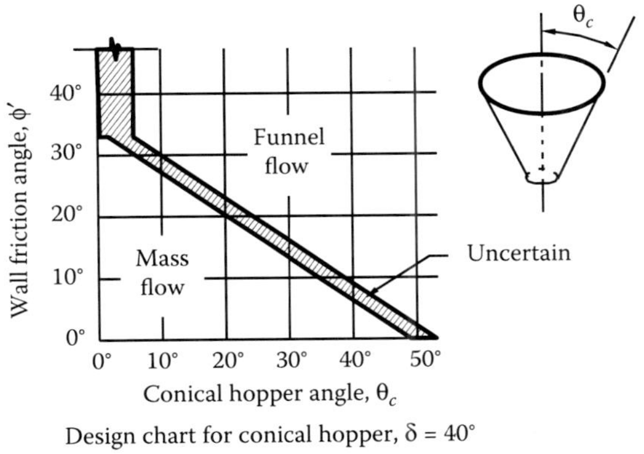

Figure 14.14 Mass flow/funnel flow design chart for conical hopper handling a bulk solid with an effective angle of internal friction (δ) of 40°.

This chapter will focus on the calculation of the *recommended mass hopper angles for a conical hopper* (θ_c) because the majority of nutraceutical processes use bins with a conical hopper. The methods to calculate the *recommended mass hopper angles for a planar hopper* (θ_p) with a slotted outlet are similar in approach and are outlined in Jenike (1964, revised 1980). It is worth noting that the recommended mass flow angles for planar hopper walls (θ_p) can often be 8°–12° shallower than those for a conical hopper (θ_c) for the same-sized opening.

An example of such a design chart for a conical hopper is shown in Figure 14.14. The design chart shown is specifically for a powder with an effective angle of internal friction (δ) of 40°; the design charts will be different for different values of δ (Jenike 1964, revised 1980). Hopper angles required for mass flow are a function of δ because flow along converging hopper walls involves interparticle motion of the powder. For any combination of ϕ′ and θ_c that lies in the mass flow region, mass flow is expected to occur. If the combination lies in the funnel flow region, funnel flow is expected. The "uncertain" region is an area where mass flow is expected to occur based on theory but represents a 4° margin of safety on the design to account for inevitable variations in test results and surface finish.

As an example of using the design chart, a bin with a conical hopper angle (θ_c) of 30° from vertical is being used. Wall friction tests are conducted on the hopper wall surface, and a wall friction angle of 35° is measured for the normal pressure calculated at the outlet. Based on the design chart, this bin would be expected to discharge in funnel flow. In that case, the designer would need to find another wall surface with a wall friction angle that is less than 13° to ensure mass flow discharge from a hopper with a 30° (from vertical) wall angle .

The wall friction angle (ϕ′) is determined by the wall friction tests described previously. The value of ϕ′ to use for the hopper design charts will be selected for the expected normal pressure (σ_n) against the surface at the location of interest in the bin (e.g., the hopper outlet). For many combinations of wall surfaces and powders, the wall friction angle changes depending on the normal pressure. When mass flow develops, the solids' pressure normal to the wall surface is given by the following relationship:

$$\sigma_n = (\sigma'/\gamma B) \times \gamma B \tag{14.4}$$

where the $(\sigma'/\gamma B)$ term = a dimensionless parameter that can be found in Jenike (1964, revised 1980)
B (units of length) = span of the outlet: the diameter of a circular outlet or the width of a slotted outlet
γ = bulk density at the outlet

Generally ϕ' *increases* with *decreasing* normal pressure (σ_n). The corresponding normal pressure to the wall (σ_n) is the lowest at the outlet where the span (B) is the smallest. Therefore, it is at the outlet where the wall friction angle (ϕ') is the highest for a given design, provided the hopper interior surface-finish and angle remain constant above the outlet. As a result, if the walls of the hopper are sufficiently steep to provide mass flow at the outlet, mass flow is to be expected at the walls above the outlet (regardless of total bin size).

The hopper angle required for mass flow is principally dependent on the outlet size selected for the hopper under consideration. The hopper angle required for mass flow is not a function of the flow rate, the level of powder within the bin, or the diameter or height of the bin. Because the wall friction angle generally increases with lower normal pressures, a steeper hopper is often required to achieve mass flow for a bin with a smaller outlet.

14.3.3 Bulk Density

The bulk density of a given powder is not a single or even a dual value, but varies as a function of the consolidating pressure applied to it. There are various methods used in industry to measure bulk density. One prominent method is using different-sized containers that are measured for volume after being loosely filled with a known mass of material ("loose" density) and after vibration or tapping ("tapped density"), such as the US Pharmacopeia method (A Bulk and Tapped Density 2002). These methods can offer some repeatability with respect to the conditions under which measurements are taken. However, they do not necessarily represent the actual compaction behavior of a powder being handled in a bin, chute, or packaging or press feed hopper. Therefore, it is necessary to measure the bulk density over a range of consolidation pressures via a compressibility test (Jenike 1964, revised 1980; Carson and Marinelli 1994) for design purposes. The results of the compressibility test can often be plotted as a straight line on a log-log plot (Figure 14.15). In powder handling literature, the slope of this line is typically called the *compressibility* of the powder.

The resulting data can be used to determine capacities for storage and transfer equipment and to evaluate wall friction and feeder operation requirements. For example, when estimating the capacity of a bin, the bulk density based on the

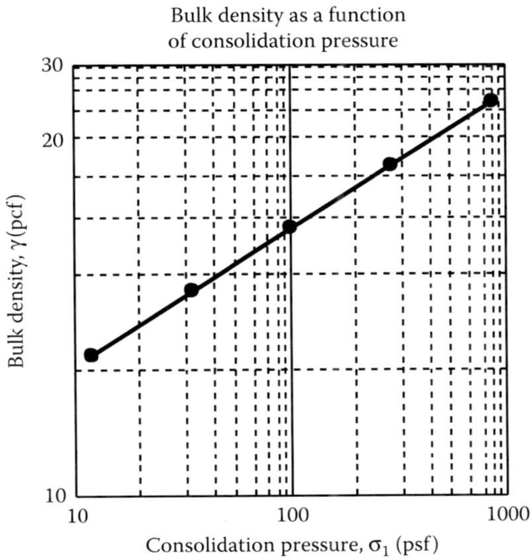

Figure 14.15 Example of bulk density versus consolidation pressure plot from compressibility test data.

average major consolidation pressure in the bin can be used. For the calculation of the arching dimensions (B_c) and recommended mass flow hopper angles (θ_c), the bulk density based on the major consolidation pressure at the bin outlet can be used.

14.3.4 Permeability

The flow problems that can occur as a result of adverse two-phase (powder/interstitial gas) flow effects were reviewed previously in this chapter. These problems are more likely to occur when the target feed rate (e.g., tableting rate) exceeds the critical flow rate based on the powder's physical properties. The results of the permeability test are one of the primary flow properties used to determine the critical flow rate. The *permeability* of a powder is a measurement of how readily gas can pass through it. The permeability will have a controlling effect on the discharge rate that can be achieved from a bin or hopper with a given outlet size. Sizing the outlet of a piece of equipment or choosing the diameter of a transfer chute should take into consideration the target feed rate.

Permeability is measured as a function of bulk density (Carson and Marinelli 1994). A schematic of the permeability tests is provided in Figure 14.16. In this test setup, gas is injected at the bottom of the test cell through a permeable membrane. During the test, the pressure decrease and flow rate across the powder are measured. The method involves measuring the flow rate of air at a predetermined pressure decrease through a sample of known density and height. The permeability is then calculated using Darcy's law. The permeability of a powder

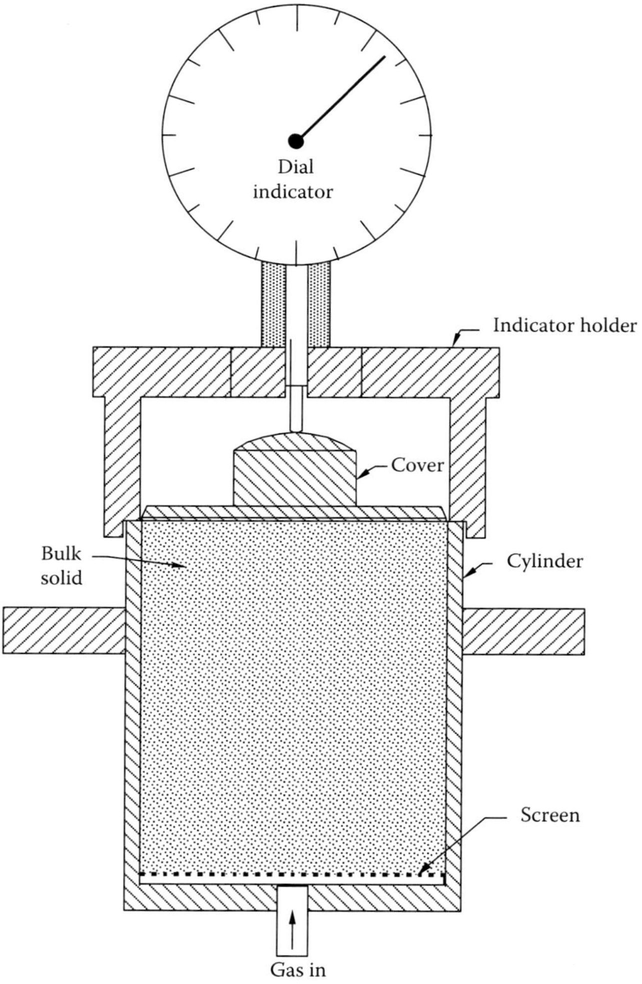

Figure 14.16 Schematic of permeability test setup.

typically *decreases* as the bulk density *increases*, so the test is conducted over a range of bulk densities.

Once the permeability–bulk density relationship is determined (Figure 14.17), it can be used to calculate the critical flow rates that will be achieved for steady flow conditions through various outlet sizes. The critical flow rate is dependent on the permeability, compressibility, bin geometry, outlet size, and consolidation pressure (fill level). The details of calculating critical flow rates are outside the scope of this chapter, but mathematical models have been developed for these calculations.

Higher flow rates than the calculated critical flow rate may occur, but may result in nonsteady or erratic feed and the resulting adverse effects. Permeability values

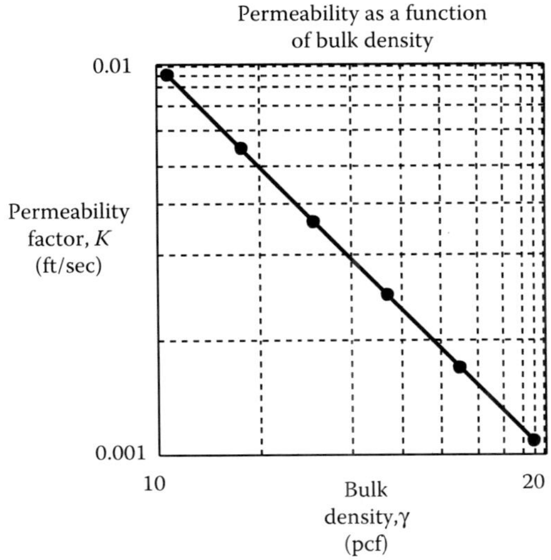

Figure 14.17 Example of permeability versus bulk density plot from permeability test data.

can also be used to calculate the time required for fine powders to settle or deaerate in equipment.

14.3.5 Segregation Tests

When developing a product or designing a process, it is beneficial to know if the blend will be prone to segregation. If the blend is prone to segregation, it is beneficial to know what *segregation mechanism*(s) will occur because this information can be used to modify the material properties (e.g., ingredient selection, ingredient particle size distribution) to minimize the potential for segregation. An understanding of the potential for segregation can alert the equipment or process designer to potential risks that may then be avoided during the scale-up process. In some cases, significant process steps, such as granulation, may be required to avoid potential segregation problems.

There are two ASTM standard practices for sifting and fluidization segregation test methods (ASTM International 2003a, 2003b). These testers are designed to isolate specific segregation mechanisms and test a material's tendency to segregate by that mechanism. A brief description of these test methods follows.

14.3.5.1 Sifting Segregation Test Method

The sifting segregation test (Figure 14.18a) is performed by center-filling a small funnel flow bin and then discharging it while collecting sequential samples. If sifting segregation occurs either during filling or discharge, the fine particle content of the discharging material will vary from beginning to end, with the fine particles

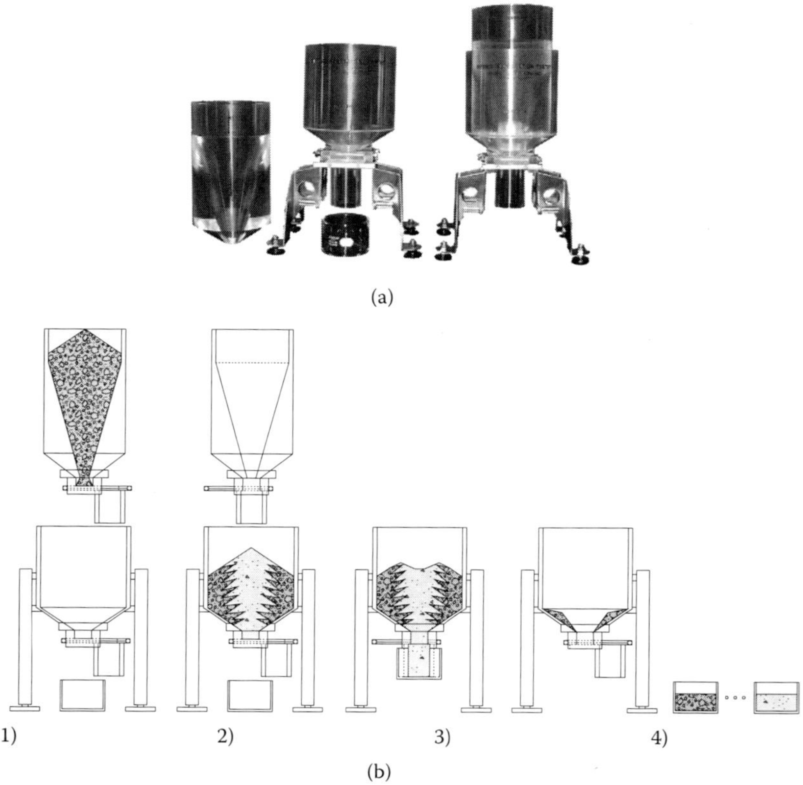

Figure 14.18 (a) Sifting segregation testers and (b) test sequence.

discharged initially, followed by coarse-rich material discharged at the end of the test. Samples are collected from the various cups (i.e., the beginning, middle, and end of the discharge). These collected samples can then be measured for segregation by particle size analyses, assays, and/or other variables of interest.

The sequence for performing the sifting segregation test is depicted in Figure 14.18b and is as follows:

1. The blend is placed in mass flow bin.
2. The material is discharged from a fixed height, dropping into a funnel flow bin. This transfer of material will promote segregation if the material is prone to segregate as a result of sifting.
3. The material is discharged from the funnel flow bin. The discharge pattern will cause material from the center to discharge first and material from near the walls to discharge last.
4. The collected samples are then measured for segregation.

A material-sparing version of the sifting segregation tester has also been developed (Prescott et al. 2008).

14.3.5.2 Fluidization Segregation Test Method

The fluidization segregation test (Figure 14.19) is run by first fluidizing a column of material by injecting air at its base. After the column is thoroughly fluidized, it is held near a minimum fluidization velocity for a predetermined period. The air is then turned off, and the material is allowed to deaerate. The column is then split into three sections (top, middle, and bottom), and the resulting samples are measured for segregation. A material-sparing version of the fluidization segregation tester has also been developed (Prescott et al. 2006).

Segregation tests are useful for identifying the following:

1. Segregation mechanism(s) that might be active for a given blend
2. General segregation trend that may be observed in the process
3. Comparisons between different formulations, variations of the same formulation, etc.

However, the test results have limitations. Most notably, the segregation results are *not* scalable and cannot be tied quantifiably to the process. The segregation tests do not necessarily mean that a highly segregating material cannot be handled in a successful manner. Therefore, the segregation test methods are primarily used as a "stress test" to identify the dominant segregation mechanism(s) expected to occur. This information enables the equipment designer to take the appropriate precautionary measures during scale-up or to make corrective actions to existing equipment. Design techniques to minimize the potential for segregation are outlined in the following section.

Figure 14.19 Fluidization segregation tester (controls not shown).

14.4 BASIC EQUIPMENT DESIGN TECHNIQUES

There are several basic design techniques for the bin-to-packaging equipment to provide consistent, reliable gravity flow and minimize segregation, including:

1. Reliable mass flow designs for bins, transfer chutes, and feed hoppers
2. Minimizing adverse two-phase flow effects (e.g., feed rate limitations, flooding)
3. Minimizing segregation during postblending transfers steps

For each of these different design concerns, we will review the key equipment design parameters. Regardless of whether the equipment being designed is a bin, transfer chute, or feed hopper, a crucial first step in designing a reliable feed system is determining the flow pattern and designing the equipment accordingly. The wall friction tests and design charts used to determine whether a hopper will discharge in mass flow or funnel flow were discussed previously.

14.4.1 Reliable Mass Flow Designs for the Bin, Chute, and Feed Hopper

Mass flow discharge from a bin occurs when the following two design criteria are met:

1. The bin walls are smooth and/or steep enough to promote flow at the walls.
2. The bin outlet is large enough to prevent an arch.

The wall friction tests and design charts used to determine whether a bin will discharge in mass flow or funnel flow were discussed previously. This section focuses on design techniques for mass flow bins, but these techniques may be extended to obtaining mass flow in a transfer chute and feed hoppers (e.g., a press hopper) as well. These techniques may be applied to designing new equipment or modifying existing equipment to provide mass flow.

When designing the bin to provide mass flow, the following general steps should be taken:

1. *Size the outlet to prevent a cohesive arch*: The bin designer should ensure that an arch will not form by making the outlet diameter equal to or larger than the minimum required outlet diameter ("B_c"; Figure 14.20). If a slotted outlet is used (maintaining a 3:1 length/width ratio for the outlet), the outlet width should be sized to be equal to or larger than the minimum required outlet width ("B_p"; see Figure 14.20). The outlet may also need to be sized based on the feed rate and two-phase flow considerations, as discussed in Section 14.4.2. If the outlet cannot be sized to prevent an arch (e.g., press hopper outlet that must mate with a fixed feed frame inlet), an internal mechanical agitator or external vibrator could be considered.
2. *Make the hopper walls steep enough for mass flow*: Once the outlet is sized, the hopper wall slope (angle) should be designed to be equal to or steeper than the recommended hopper angle for the given outlet size and selected wall surface. For a conical hopper, the walls should be equal to or steeper than the recommended mass

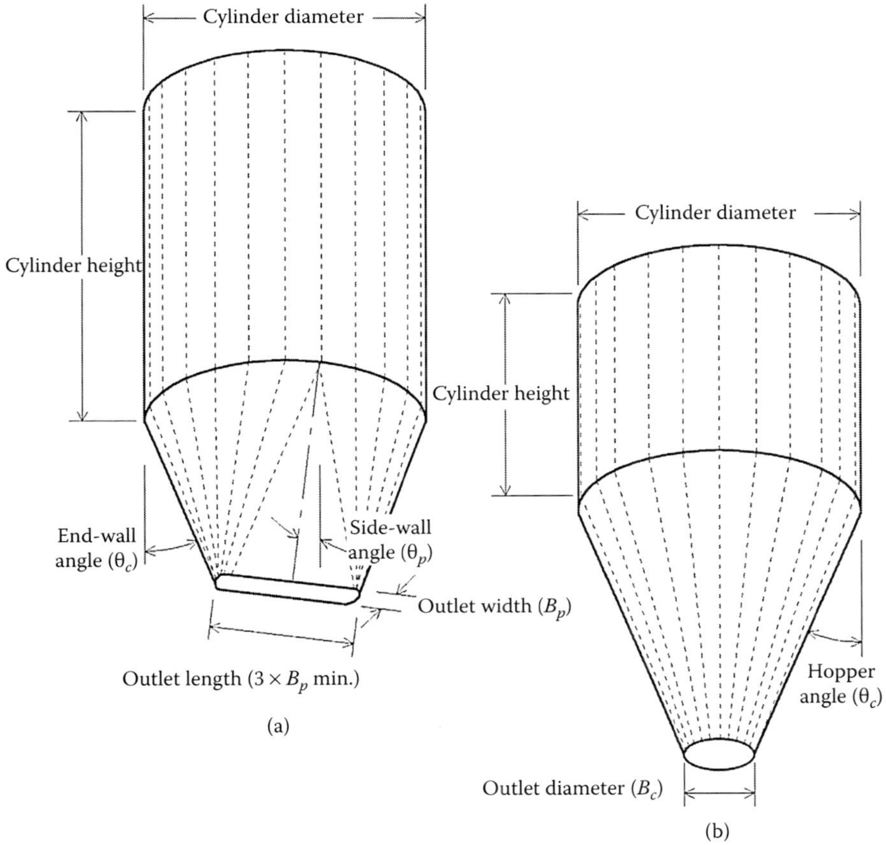

Figure 14.20 Mass flow design parameters: (a) transition hopper; (b) conical hopper.

flow angle for a conical hopper ("θ_c"; see Figure 14.20). If the bin has a rectangular-to-round hopper, the valley angles should be equal to or steeper than θ_c. For planar walls, the walls should be equal to or steeper than the recommended mass flow angle for a planar hopper ("θ_p"; see Figure 14.20).

3. *Pay careful attention to the interior wall surface finish*: When conducting the wall friction tests, it is beneficial to conduct tests on several different finishes (e.g., 320 grit finish, 2B cold-rolled finish, or 2B electropolished finish). This will provide the bin designer with a range of design options for the bin. Testing multiple wall surfaces will also enable the designer to assess the sensitivity of the wall friction results to different finishes. It is not sufficient to simply test a 304 or 316 stainless steel with no regard to the interior finish. The wall friction of the blend may vary significantly from finish to finish. The orientation of directional finishes such as a mechanical polish is also critical to assess and control during fabrication. In addition, it cannot be assumed that an interior surface finish with a lower average roughness (R_a) will provide the best wall friction properties.

4. *Consider velocity gradients*: Even when a bin is designed for mass flow, there still may be a velocity gradient between the material discharging at the hopper walls (moving slower) versus the center of the hopper (moving faster), assuming a symmetric bin with a single outlet in the center. Depending on the application, the bin designer may want to increase the velocity gradient to enhance blending between vertical layers of material in the bin. Alternatively, the bin designer may want to reduce the velocity gradient to enhance blending on a side-to-side basis. The decision to increase or decrease the velocity gradient will depend on the segregation that occurs on filling the bin and its effect on content uniformity. The velocity gradient is *reduced* by making the hopper slope *steeper* with respect to the recommended mass flow hopper angle (θ_c). The velocity *gradient* is *increased* by making the hopper slope closer to (but still steeper than) the recommended mass flow hopper angle. Changing the interior surface to reduce friction and using an insert (discussed more in the next paragraph) are other methods used to control the velocity gradient. Asymmetric hoppers, which are common for press hoppers, are especially prone to velocity gradients because the material will move faster at the steeper hopper wall. In addition, a velocity gradient cannot be completely eliminated, especially as the material level in the hopper empties. Velocity profiles, and their effect on blending material, can be calculated a priori given the geometry of the bin (θ_c) and measured flow properties discussed previously.
5. *Avoid upward-facing lips and ledges resulting from mismatched flanges*, level probes, view ports, or partially opened valves, especially in the hopper section. Ideally, interior protruding devices should be located in the straight-sided (nonconverging) section of a bin or feed hopper if possible, where they will be less detrimental in upsetting a mass flow pattern.

If the bin designer is modifying an existing funnel flow bin to provide mass flow, several different options can be considered, including:

1. *Using a different interior surface finish* with better (lower) wall friction properties. The bin designer should conduct wall friction tests on alternate wall surfaces to assess whether changing the surface finish (e.g., electropolishing an existing 2B finish) will convert the bin from funnel flow to mass flow. This is often one of the most cost-effective modifications to obtain mass flow.
2. *Using a flow-controlling insert* such as a Binsert (Jenike & Johanson, Inc., Tyngsboro, MA; Figure 14.21) to obtain mass flow within the same bin. A properly designed insert can change the stresses that develop in the bin during discharge so that mass flow can be obtained at a wall where the material was previously stagnant.
3. *Modifying the hopper geometry*. Use a different geometry that is more likely to provide mass flow (e.g., conical instead of a rectangular-to-round hopper with shallower valley angles). If the hopper is modified to have a slotted outlet, it is crucial that the feeder to which the hopper mates withdraws material across the entire outlet.

In addition to these design techniques for bins, there are several additional design techniques for designing transfer chutes for reliable mass flow, including:

1. For *converging* sections that are flood loaded to have a full cross section (i.e., "hoppers"), use the same design criteria discussed previously for a mass flow bin.

Figure 14.21 Example of a Binsert (Jenike & Johanson, Inc., Tyngsboro, MA) design.

2. For *nonconverging* sections of the chute, the chute should be sloped to exceed the wall friction angle (ϕ') by at least a 10° margin of safety (preferably more). For example, if the measured wall friction angle for the given wall surface (from the wall friction test results) is 40° from horizontal, the recommended chute angle for the nonconverging portion of the chute would be at least 50° from horizontal.
3. If a bifurcated chute is used, the sloping chute legs should be symmetric to prevent velocity gradients and the possibility of stagnant material in the shallower leg.
4. Use mitered joints between sloping and vertical sections.

14.4.2 Minimizing Adverse Two-Phase Flow Effects

The primary focus in preventing adverse two-phase flow effects is to ensure that the powder handling equipment is designed so that the *critical flow rate* through a given outlet is greater than the *target feed rate*. The critical flow rate is determined using mathematical models with the permeability and compressibility test results as primary inputs. The critical flow rate is also a strong function of the outlet size and increases as the outlet size increases. Adverse two-phase flow effects are typically most pronounced at the packaging or press feed hopper because it often has the smallest outlet in the entire feed system. Therefore, the feed hopper will typically have the lowest critical flow rate.

When designing the powder handling equipment to minimize adverse two-phase flow effects, the following general design techniques are beneficial:

1. *Design the equipment for mass flow*: Mass flow will provide consistent feed and a more uniform consolidation pressure acting on the powder. In addition, having a first-in/first-out flow sequence will allow the material more time to deaerate before being discharged through the outlet. This will reduce the likelihood of flooding. Mass flow will also prevent collapsing ratholes that can result in the powder

aerating and flooding as it falls into the central flow channel. It is worth noting that mass flow can result in a lower critical flow rate than funnel flow but will be more stable. Therefore, simply using a mass flow bin design may not be the only corrective action required if a flow rate limitation occurs. However, designing the equipment for mass flow is often the first step in addressing adverse two-phase flow effects.

2. *Use larger outlets for the handling equipment*: The critical flow rate is a strong function of the cross-sectional area of the outlet. Therefore, increasing the outlet can often be highly beneficial in reducing two-phase flow effects. The goal would be to increase the outlet size until the critical flow rate for the selected outlet size exceeds the target flow rate. Because this may not be feasible for a feed hopper (e.g., press feeder hopper) in which the outlet size is fixed, additional design techniques are discussed below. Computer software can be used to model the two-phase flow behavior to assess the effect of changing the outlet diameter.
3. *Reduce the fill height in the handling equipment*: The critical flow rate through a given outlet *increases* as the major consolidation pressure (σ_1) *decreases*. Therefore, reducing the fill height will be beneficial but much less effective than increasing the outlet size.
4. *Reduce the target feed rate*: If possible, reducing the target feed rate (e.g., tableting rate) to be less than the critical flow rate will be beneficial, but it is often impractical because it will result in a decreased production rate.
5. *Consider gas pressure differentials*: A gas pressure differential can have a beneficial or adverse effect on two-phase flow effects. A positive gas pressure differential at the outlet (i.e., bin at a higher gas pressure than the equipment downstream) may be beneficial in overcoming a feed rate limitation. In this case, the air pressure is forcing the material in the direction of flow. Conversely, a negative gas pressure differential at the outlet can further reduce the critical flow rate because the negative gas pressure acts to further retard the flow rate.
6. *Add air permeation*: Air permeation may be added to the system actively via an air injection system or passively through a vent. In particular, adding judicious (often small) amounts of air at the location in the packaging or press feed system where the interstitial gas pressure is lowest can often be beneficial. However, this can be unstable for small systems and/or low permeability materials.
7. *Changing the particle size distribution of the powder*: The permeability of a powder is a strong function of its particle size distribution. Powders with a finer particle size distribution are often less permeable and therefore more prone to adverse two-phase flow effects. Even a reduction in the percentage of fine particles can often be beneficial in increasing the permeability of a powder and decreasing the likelihood of adverse two-phase flow effects (Barnum, Baxter, and Prescott 2007).

The key to implementing any corrective actions designed to reduce adverse flow effects will be using a mathematical two-phase flow analysis to assess the effects on the bulk solid stresses and interstitial gas pressure (Baxter 2009). This analysis would need to use inputs such as key flow properties (permeability, compressibility) and equipment or process parameters (tableting rate, bin or hopper geometry, and gas pressure gradients) to assess the effect of the potential corrective actions outlined previously.

14.4.3 Minimizing Segregation in the Postblending Transfer Steps

It may be a challenging process for a designer to determine which segregation mechanism(s) is dominant and to develop appropriate corrective actions. This requires knowledge of the material's physical and chemical characteristics, as well as an understanding of the segregation mechanisms that can be active. One must identify the process conditions that can serve as a driving force to cause segregation. Flow property measurements (e.g., wall friction, cohesive strength, compressibility, and permeability) can help to provide understanding of the behavior of the material in storage and transfer equipment. Consideration should be given to the fill and discharge sequence, including flow pattern and inventory management, which gives rise to the observed segregation. Testing for segregation potential can provide additional insight about the mechanisms that may be causing segregation. Sufficient sampling is required to support the hypothesis of segregation (e.g., blend samples and final product samples, samples from the center vs. periphery of the bin). Finally, one must consider the impact of analytical and sampling errors specific to the blend under consideration, as well as the statistical significance of the results, when drawing conclusions from the data.

From the previous discussion about segregation mechanisms, it can be concluded that certain *material properties* and *process conditions* must exist for segregation to occur. Elimination of one of these will prevent segregation. It stands to reason then that if segregation is a problem in a process, one should look for opportunities to either (1) change the material or (2) modify the process equipment or conditions. This chapter focuses on the *equipment and process* design techniques to minimize segregation.

Some generalizations can be made when designing equipment to minimize segregation. The complete details on how to implement these changes correctly are beyond the scope of this chapter. However, all equipment must be designed based on the flow properties and segregation potential of the blends being handled.

Primary equipment and process design techniques to minimize segregation during the postblending transfer steps include:

1. *Minimize the number of postblending transfer steps.* The tendency for segregation increases with each transfer step and movement of the bin. In-bin blending is as close to this as most firms can practically obtain. This assumes that a uniform blend can be obtained within the bin blender in the first place.
2. *Storage bins, feed hoppers, and chutes should be designed for mass flow.* In mass flow, the entire contents of the bin are in motion during discharge. In funnel flow, stagnant regions exist.
3. *Minimize transfer chute volumes* to reduce the volume of displaced air and the volume of potentially segregated material. However, the chute must remain large enough to provide the required throughput rates.
4. *Bins and blenders should be vented to avoid gas counterflow.* Air in an otherwise "empty" bin must be displaced out of the bin as powder fills it. If this air is forced through material in the V-blender, it can induce fluidization segregation within the blender. A separate pathway or vent line to allow the air to escape without moving through the bed of material can reduce segregation.

5. *Velocity gradients within bins should also be minimized.* The hopper must be significantly steeper than the mass flow limit to achieve this. A steeper hopper section may result in an impractically tall bin. Alternate approaches include the use of inserts (discussed previously). If an insert is used, it must be properly designed and positioned to be effective. Asymmetric bins and hoppers should be avoided if possible, and symmetric ones should be used whenever possible.
6. *Dust generation and fluidization of the material should be minimized during material movement.* Dust can be controlled by way of socks or sleeves to contain the material as it drops from the blender to the bin, for example. There are many commercially available "let down" chutes that may be beneficial.
7. *Drop heights should be minimized where possible.* Drop heights may aerate the material, induce dust, and increase momentum of the material as it hits the pile. This will increase the tendency for each of the three segregation mechanisms to occur.
8. *Valves should be operated correctly.* Butterfly valves should be operated in the full open position, not throttled to restrict flow. Restricting flow will virtually ensure a funnel flow pattern, which is usually detrimental to uniformity.
9. *Use a symmetric split whenever a process stream is divided.* A symmetric split, such as a bifurcated chute to feed two sides of a packaging machine or press (a "Y-branch"), will eliminate potential differences in the flow between the two streams. Consideration must be given to any potential for segregation upstream of the split. Even seemingly minor details, such as the orientation of a butterfly valve before a split, can affect segregation. Proper designs should be used for Y-branches to avoid stagnant material and air counterflow.

Other specific solutions may be apparent once the segregation mechanism has been identified. For example, mass flow is usually beneficial when handling segregation-prone materials, especially materials that exhibit a side-to-side (or center-to-periphery) segregation pattern. Sifting and dusting segregation mechanisms fit this description.

It is important to remember that mass flow is not a universal solution because it will not address a top-to-bottom segregation pattern. For example, consider the situation in a portable bin where fluidization on filling the bin has caused the fine fraction of a blend to be driven to the top surface. Mass flow discharge of this bin would effectively transfer this segregated material to the downstream process, delivering the coarser blend first, followed by the fine particles. In summary, when addressing segregation concerns, it is crucial to know your process and how the blend will segregate before implementing equipment designs or corrective actions.

It is also critical to distinguish the underlying root causes of the blend or product variation, as discussed in the following sections.

14.4.4 Troubleshooting Methods for Addressing Segregation and Variation in Product Uniformity

The blend uniformity (BU) and content uniformity (CU; i.e., consumer product uniformity, tablet uniformity) of nutraceutical products are not currently regulated by the Food and Drug Administration (FDA). However, FDA guidance for conducting and analyzing the BU and CU of pharmaceutical products can provide a basis for addressing segregation and product uniformity concerns with nutraceutical products. A 2003 FDA draft guidance document on BU and CU outlines a science-based approach for analyzing uniformity using stratified in-process dosage unit sampling (FDA 2003). Although this FDA draft guidance document has not been finalized, its methodology has been successfully applied to numerous products by multiple pharmaceutical companies, and it provides a guiding methodology for identifying BU and CU concerns.

The troubleshooting methods outlined in this section are not intended to serve as the basis for regulatory policy or as acceptance criteria for manufacturing. Both BU and CU data should be analyzed together to fully assess the capability of the process because CU variation may be adversely affected by BU variation. This section focuses on CU with respect to compression processes (i.e., the uniformity of the tablets produced), but the methodology outlined can also be used to address CU concerns with other dosage units (e.g., capsule, vial, or pouch).

14.4.4.1 In-Process Stratified Sampling and Analysis

In-process stratified sampling and data analysis is a primary diagnostic tool for addressing CU variation. Stratified sampling is the collection of multiple samples from multiple "locations" in a process and is required to measure the variability inherent at a single location. For the purposes of this discussion, the term *sampling location* refers to a specific time point (percentage of batch, number of tablets compressed) during compression, but it could also be a physical location in a blender if analyzing BU. Although the term *stratified sampling* generally has referred to samples collected in-process and then analyzed off-line, process analytical technologies such as near-infrared sensors can be used to determine blend or tablet composition in-line or at-line.

Analyzing the normality, overall mean, relative standard deviation (RSD), and location means of the stratified sampling data is a recommended first step in beginning to assess CU concerns. An analysis of variance (ANOVA) should also be conducted to determine the within-location and between-location (time) variation (Davies 1960). An ANOVA provides insight into potential root causes of CU variation because the root causes of between-location versus within-location variation are often distinct. For example, consider two sets of stratified sampling data with the same overall variation (RSD) but different sources of variance. For the data in Figure 14.23, the between-location variation is the primary variance component (i.e., the variance of the three tablets at a given location is comparatively low). In contrast, within-location variation is the primary variance component for the data in Figure 14.26.

The within-location and between-location variation of the stratified BU and CU data should also be compared because there are significant implications based on differences in their respective values (Prescott and Garcia 2008).

14.4.4.2 *Identifying and Addressing Variation in Content Uniformity Data*

On completion of the statistical analysis discussed previously, a scientist can next assess whether the CU data fit a common category or pattern. Identifying the category of CU variation can provide further insight into the root cause(s) of the variance. These categories are based on tendencies of the overall mean, location means, between-location variance, and within-location variance. Identifying these categories is subjective (especially when analyzing only a few batches) because the complete product history, acceptance criteria, process and equipment parameters, and other details must also be considered when assessing CU variation. The categories describe general behaviors only and are not necessarily correlated to any specific regulatory specifications. In this chapter, these categories are treated as separate (decoupled) groups, but in practice, the CU variation may fit two or more of these categories. Additional categories of CU variation are identified elsewhere in the literature (Prescott and Garcia 2008), but the categories discussed in this chapter identify the majority of data sets the authors have observed.

Identifying the category the CU variation fits is a crucial first step in a root cause analysis and development of corrective actions. These general categories of CU variation are discussed in the following sections, along with potential next steps in a root cause analysis and corrective actions to reduce the variation.

14.4.4.3 *Satisfactory Content Uniformity Data*

Satisfactory CU data (see Figure 14.22) meet the acceptance criteria being applied and are reproducible for all batches. Satisfactory CU data often appear as a flat line (minimal between-location variation) with minimal within-location variation.

14.4.4.4 *Trending Content Uniformity Data*

Trending CU data can be identified by a distinct directional pattern in the variation of the location means (between-location variation) as a function of time (see Figure 14.23). Trending may result in a high overall RSD, but the overall mean is often close to 100% target and the within-location variation is often comparatively low.

Trending is typically caused by "macro" segregation that occurs during a process transfer step. These segregation mechanisms often occur on a large-scale basis (with respect to the scale of the equipment) during transfer between process steps, resulting in a trending pattern across the batch. Trending can exhibit several different forms with low or high location means occurring at the beginning, middle, or end of the batch. The different trends can often be better understood by identifying the flow pattern that develops in the postblending handling equipment, mass flow versus funnel flow.

Assuming the blend data are uniform (low between- and within-location variation), potential next steps in determining the root causes of the CU trending as a result of postblending segregation include:

- Conduct bench-scale segregation tests (discussed previously) to confirm the dominant segregation mechanism(s) that may occur during the postblending transfer steps.

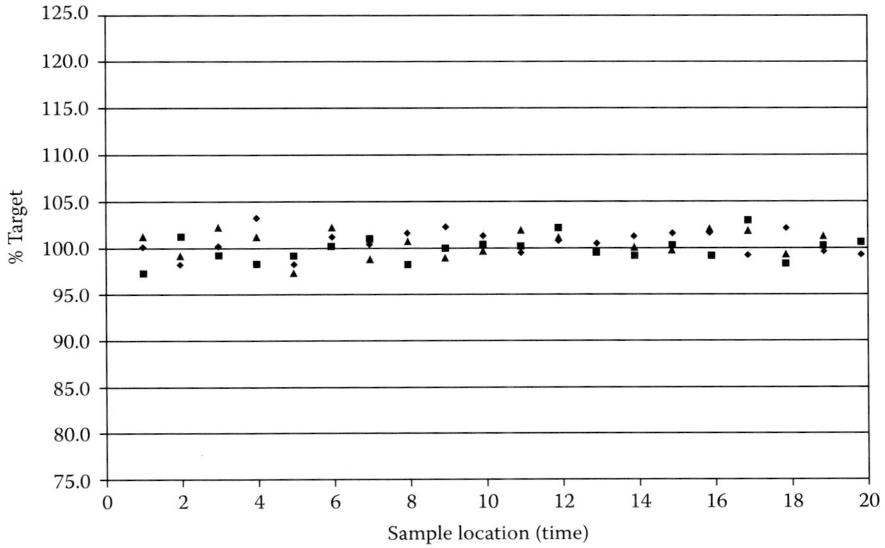

Figure 14.22 Satisfactory content uniformity data.

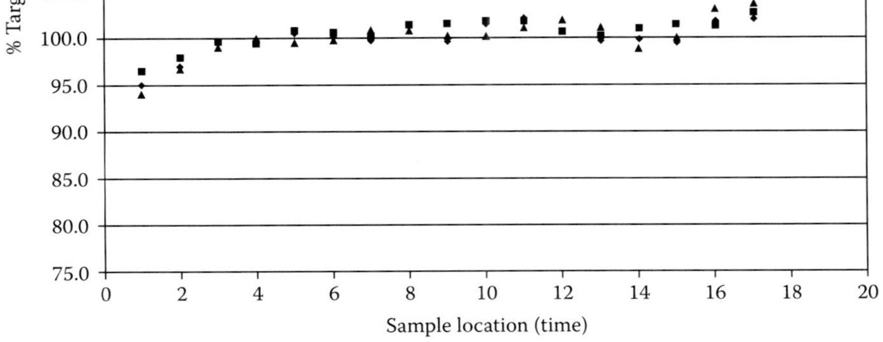

Figure 14.23 Trending content uniformity data.

- Conduct flow property tests (discussed previously) to identify the flow pattern in the postblending handling equipment.
- Confirm the fill and discharge sequence of the process equipment.
- Add additional sampling points to further confirm a trend (e.g., increased sampling frequency at the beginning or end of a batch).
- Document key equipment parameters that may affect segregation (e.g., transfer rate, drop heights, batch size, bin volume, and venting arrangements).
- Document key material properties that may affect segregation (e.g., particle size and density of the key ingredients in the blend).

This additional information can be used to confirm the root causes of the CU trending and to determine what corrective actions should be implemented. Potential corrective actions may include:

- Redesign the process to eliminate a transfer step (e.g., eliminating a blender-to-bin transfer step by using an in-bin blender).
- Redesign the transfer step and equipment to minimize the motive forces that cause interparticle motion and segregation (e.g., a let-down transfer chute that drops the blend in stages).
- Redesign the handling equipment to minimize the effects of segregation (i.e., designing for mass flow if beneficial).
- If necessary, reformulate the blend so it is less prone to segregation.

A nutraceutical scientist should consider the ease of implementation, cost implications, and likelihood of success in determining which corrective actions to implement.

14.4.4.5 Wandering Content Uniformity Data

Wandering CU data can be identified by high between-location variation that does not exhibit a distinct upward or downward trend across the batch (Figure 14.24). Although wandering CU data result in high between-location variation, the overall mean is often close to 100% target and the within-location variation is relatively low. Identifying and addressing CU wandering may be challenging because a distinct pattern in the data may not be apparent within a single batch and between multiple batches of product. Wandering is often caused by high between-location variation in the blender, but it could also be the result of segregation in the postblending transfer steps (discussed previously) or poor product or tablet weight control.

If high between-location variation in the blender is a potential root cause of the wandering CU data, potential next steps for further investigation include:

- Analyze the blend data (ANOVA of the stratified data), including intensified sampling if needed, to calculate the between-location variation in the blender.
- Review the blender sampling procedures (e.g., type of thief, thief sample size, and location of samples).

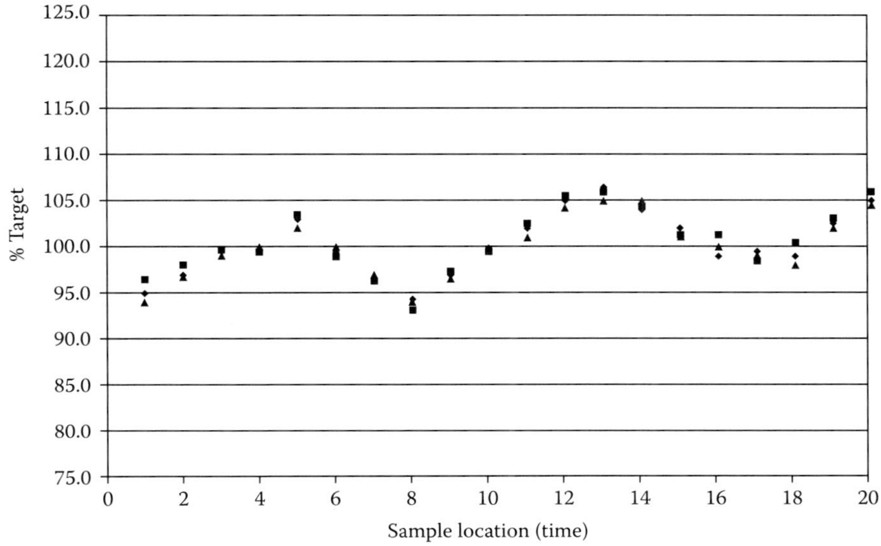

Figure 14.24 Wandering content uniformity data.

- Review key parameters that can affect the blend performance, such as the scale-up techniques (lab-scale to full-scale), loading percentage, order of addition, and number of revolutions (blend time).

If high between-location variation in the blender is identified as the primary root cause of the CU variation, the following potential corrective actions can be considered:

- Optimize the number of rotations via a blend study in which multiple blend times are analyzed to determine the number of revolutions that provides the lowest variation (RSD).
- Change the percentage fill of the blender.
- Change the order of addition and/or consider preblending.
- Change the blender type.
- If needed, reformulate the blend.

All these potential corrective actions would need to be analyzed via a series of controlled blend studies before implementing for production.

14.4.4.6 Hot Spots in Content Uniformity Data

CU data with a "hot spot" can be identified by high between-location variation as a result of an isolated location that has a significantly higher or lower mean than the other locations (Figure 14.25). Although a hot spot may result in a location

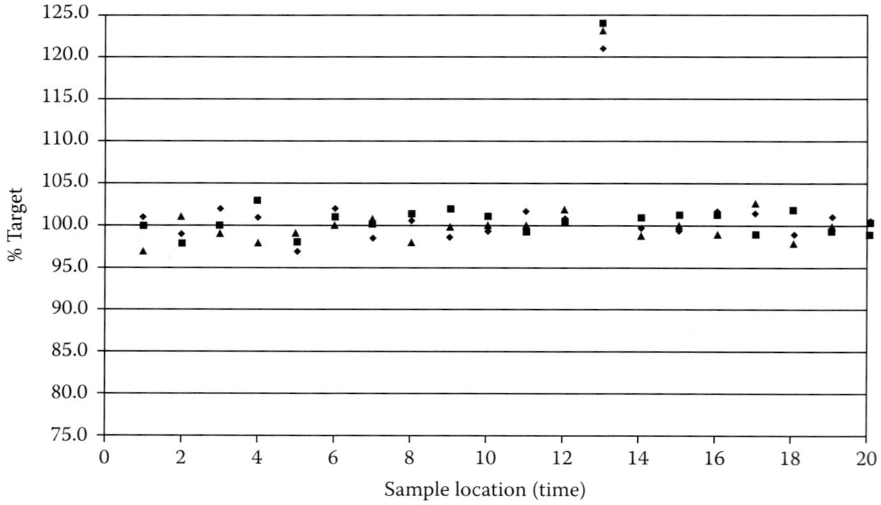

Figure 14.25 Hot spot in content uniformity data.

mean that is not within specification, the overall mean is often close to 100% target and the within-location variation is comparatively low. A hot spot can occur at any point during compression (beginning, middle, and end) as a seemingly random event. Identifying the root cause of a hot spot can be challenging, especially if it is not observed in subsequent batches. Therefore, the appearance of a single hot spot should trigger further investigation (even if it complies with specifications) because the worst-case hot spot may not always be captured via stratified sampling.

A common root cause of a hot spot is high between-location variation in the blender as a result of a specific location that is not well blended (i.e., a dead spot). Therefore, the potential next steps and corrective actions outlined previously for addressing high between-location variation in the blender can be followed for this case as well. The potential corrective actions would need to be assessed via a controlled set of experiments and additional batches.

14.4.4.7 Scatter in Content Uniformity Data

Scatter in CU data is identified by high within-location variation for the majority of the locations across the batch (Figure 14.26), even if the concentration of the blend (and location means) remains relatively uniform across the batch and the between-location variation is comparatively low. CU scatter may be the result of high within-location variation in the blend data, agglomerates in the blend, or high tablet weight variation.

CU scatter caused by high within-location variation in the blender can be further investigated and addressed by following the potential next steps and corrective

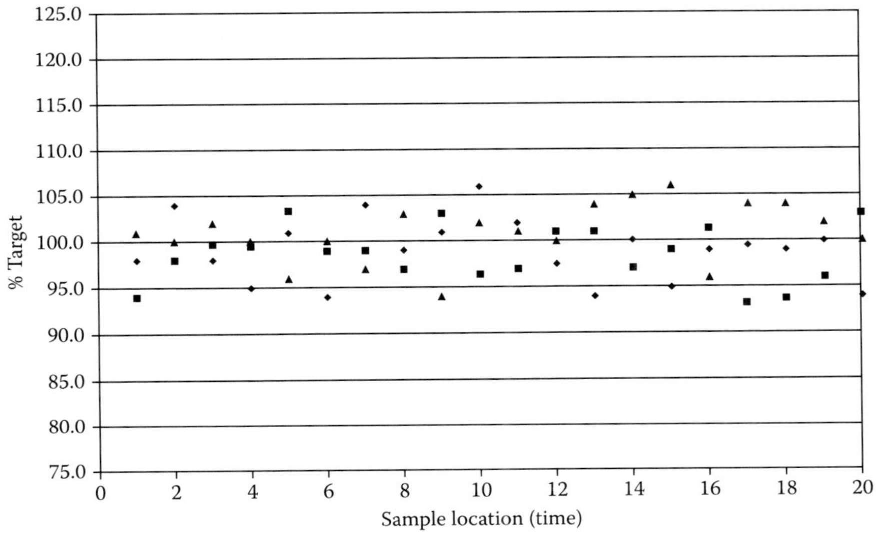

Figure 14.26 Scatter in content uniformity data.

actions outlined previously for addressing high between-location variation in a blender.

Another potential root cause of scatter is the formation of agglomerates that are high (or potentially low) in a key or active ingredient. These agglomerates may be present in the raw materials (and survive dispensing) or form during the blending process or storage in a bin. If agglomeration is a likely root cause of CU scatter, potential next steps in an investigation include:

- Analyze the agglomerates, including conducting sieve tests to isolate and capture the agglomerates, obtaining micrograph images, and reviewing the particle size of the key ingredient(s).
- Review the role of key blender parameters on the agglomerate formation, such as the order of addition, blend time, use of preblending, and use of an intensifier.
- Review the effect of the storage at rest conditions in a bin on agglomerate formation, such as the amount of material being stored (bin capacity), the temperature, and relative humidity conditions during dispensing, blending, storage, and compression.

If agglomeration is identified as a root cause of a CU scatter, potential corrective actions to reduce agglomeration include:

- Mill different components in the blend.
- Use a blender with an agitator (intensifier bar).
- Assess the effect of a different order of addition or preblending key ingredients.
- Confirm that a long blend time is not increasing the risk of agglomerate formation.

- Use a blender with additional shear such as an intensifier bar.
- If necessary, reformulate the blend.

If high product weight variation (e.g., tablet, vial, or pouch weight) is a suspected root cause of CU scatter, potential next steps for further investigation include:

- Normalize (weight-correct) the CU data to see whether the within-location variation remains high; if the within-location variation remains high, poor microblending and an insufficient distribution of key ingredients are likely root causes.
- Conduct flow property tests on the blend to assess whether no flow or erratic flow is a concern.
- Review the postblending handling equipment to identify the flow pattern (mass flow vs. funnel flow).
- Conduct a two-phase (powder/interstitial air) flow rate analysis to determine whether there is a concern with the feed rate stability to the packaging equipment or press at the target production rate.
- Review the process and equipment parameters that affect product weight variation (e.g., packaging and press equipment details, feeder speeds, and agitator designs).

Depending on the results of this root cause assessment, the nutraceutical scientist may reduce product weight variation by improving the consistency of powder feed to the packaging equipment and press (e.g., modifications to the bin, transfer chute, and press hopper), changing the production speed, modifying the feed equipment, using a different packager or press, modifying the formulation, or changing the handling conditions.

The test methods, basic equipment design techniques, and troubleshooting methodology outlined in this chapter should provide a nutraceutical scientist with guidance for designing and scaling-up processes to provide reliable flow and reduce the risk of segregation.

REFERENCES

A Bulk and Tapped Density. 2002. The United States Pharmacopeia. *Pharmacopeial Forum* 28: 3.

ASTM International. 2003a. *Standard Practice for Measuring Fluidization Segregation Tendencies of Powders, D6941-03.*West Conshohocken, PA: ASTM International.

ASTM International. 2003b. *Standard Practice Guide for Measuring Sifting Segregation Tendencies of Bulk Solids, D6940-03.* West Conshohocken, PA: ASTM International.

Barnum, R., T. Baxter, and J. Prescott. 2007. The Effect of Particle Size Changes on Two-Phase Flow Behaviors. Presented at the Annual Meeting of the American Association of Pharmaceutical Scientists, San Diego, CA, November 2007.

Bates, L. 1997. *User Guide to Segregation.* Hartford, UK: British Materials Handling Board.

Bausch, A., R. Hausmann, C. Bongartz, and T. Zinn. 1998. Measurement of Flowability with a Ring Shear Cell, Evaluation and Adaptation of the Method for Use in Pharmaceutical Technology. Proceedings of the 2nd World Meeting of the APGI/APV, Paris, May 1998, 135–36.

Baxter, T. J. 2009. When Powders Flow Like Water: Addressing Two-Phase Flow Effects in Tablet Press Feed System. *Tablets & Capsules* 7: 26–32.

Carson, J. W., and J. Marinelli. 1994. Characterize Powders to Ensure Smooth Flow. *Chemical Engineering* 101: 78–90.

Davies, O. L., ed. 1960. *Design and Analysis of Industrial Experiments*. New York: Hafner Publishing Co. (Macmillan).

Food and Drug Administration. 2003. Guidance for Industry (draft). *Powder Blends and Finished Dosage Units—Stratified In-Process Dosage Unit Sampling and Assessment*. Washington, DC: US Department of Health and Human Services, Food and Drug Administration, Center for Drug Evaluation and Research (CDER), Pharmaceutical CGMPs, October 2003; *Revised Attachments*, November 2003.

Hausmann, R., A. Bausch, C. Bongartz, and T. Zinn. 1998. Pharmaceutical Applications of a New Ring Shear Tester for Flowability Measurement of Granules and Powders. Proceedings of the 2nd World Meeting of the APGI/APV, Paris, May 1998, 137–38.

Janssen, H. A. 1895. Versuche uber Getreidedruck in Silozellen. *Verein Deutcher Igenieure, Zeitschrift* 39: 1045–49.

Jenike, A. W. 1964 (revised 1980). Storage and Flow of Solids. *Bulletin 123 of the Utah Engineering Experimental Station* 53: 26.

Nyquist, H. 1984. Measurement of Flow Properties in Large Scale Tablet Production. *International Journal of Pharmaceutical Technology and Product Manufacturing* 5: 21–24.

Pittenger B. H., H. Purutyan, and R. A. Barnum. 2000. Reducing/Eliminating Segregation Problems in Powdered Metal Processing. Part I: Segregation Mechanisms. *P/M Science Technology Briefs*, 2: 5–9.

Prescott, J., D. L. Brone, S. A. Clement, B. C. Hancock, D. B. Hedden, M. A. McCall, T. G. Troxel, and Assignee Jenike & Johanson, Inc. 2008. Segregation Testing Apparatus for Powders and Granular Materials. US Patent 7,347,111 B2, filed December 15, 2005, issued March 25, 2008.

Prescott, J., M. McCall, and S. Clement (Jenike & Johanson, Inc.), D. B. Hedden, D. Brone, A. Olsofsky, P. J. Patel, and B. C. Hancock. 2006. Development of an Improved Fluidization Segregation Tester for Use with Pharmaceutical Powders. *Pharmaceutical Technology* http://pharmtech.findpharma.com/pharmtech/mixing+and+blending/development-of-an-improved-fluidization-segregatio/articlestandard/article/detail/390984.

Prescott, J. K., and T. P. Garcia. 2008. Blending and Blend Uniformity. *Pharmaceutical Dosage Forms: Tablets*, 3rd ed. Zug, Switzerland: Informa Healthcare, 111–74.

Prescott, J. K., D. A. Ploof, and J. W. Carson. 1999. Developing a Better Understanding of Wall Friction. *Powder Handing and Processing* 11: 27–35.

Ramachandruni, H., and S. Hoag. 1998. Application of a Modified Annular Shear Cell Measuring Lubrication of Pharmaceutical Powders. Poster presented at the American Association of Pharmaceutical Scientists Annual Meeting, San Francisco, CA, November 1998.

Rippie, E. G. 1980. Powders. *Remington's Pharmaceutical Sciences*, 16th ed. Edited by A. Osol, G. D. Chase, A. R. Gennaro, M. R. Gibson, and C. B. Granberg. Easton PA: Mack Publishing Company, 1535–52.

Royal, T. A., and J. W. Carson. 1991. Fine Powder Flow Phenomena in Bins, Hoppers and Processing Vessels. Presented at Bulk 2000, London, October 29–31.

Schulze, D. 1996a. Measuring Powder Flowability: A Comparison of Test Methods Part I. *Powder and Bulk Engineering* 10: 45–61.

Schulze, D. 1996b. Measuring Powder Flowability: A Comparison of Test Methods Part II. *Powder and Bulk Engineering* 10: 17–28.

Standard Shear Testing Method for Powders Using the Jenike Shear Cell, ASTM Standard D6128-06. 2006. West Conshohocken, PA: American Society for Testing and Materials.
Venables, H. J., and J. I. Wells. 2001. Powder Mixing. *Drug Development and Industrial Pharmacy* 27: 599–612.
Williams, J. C. 1976. The Segregation of Particulate Materials: A Review. *Powder Technology* 15: 245–51.
Williams J. C., and M. I. Khan. 1973. The Mixing and Segregation of Particulate Solids of Different Particle Size. *Chemical Engineer* January: 19–25.

CHAPTER 15

Complying with FDA Regulations for the Manufacture and Quality Control of Nutraceuticals

Do What You Say, Say What You Do

Mike Witt, Girish J. Kotwal, and Yashwant Pathak

CONTENTS

15.1 INTRODUCTION

With regulatory bodies around the world increasing the regulatory demands placed on nutraceutical companies, the need to understand what is required becomes a larger part of the daily function of those companies. Even though regulations vary greatly depending on where you manufacture or ship your products, one thing is clear: regulations are becoming more consistent between all the governing bodies.

As the total world revenue increases with the sale of nutraceuticals, not all the companies doing business are reputable, and some may introduce unsafe products to capitalize on the current demand. There have been many examples of such companies in the news, such as the use of melamine as protein filler in dog food to keep cost lower and increase profit. The result of the change was the loss of many pets to kidney failure and the increased demand on companies to prove there is no melamine in their products.

In the United States, the control of nutraceuticals falls under the Food and Drug Administration (FDA). The FDA has led the way to increased control and quality assurance for nutraceuticals sold within the United States. The implementation of testing and quality control within the industry is following a path similar to the pharmaceutical industry. Many of the restrictions in the pharmaceutical industry were implemented by the FDA after a company released an unsafe or poorly produced (adulterated) product.

Through the actions of the FDA, the public has access to safe and effective drugs that they know have been reviewed and approved by the FDA for consumption. Therefore, the natural process has been for the FDA to enact restrictions and requirements on the nutraceutical industry that have been successful within the pharmaceutical industry.

The regulations set forth by most of the governing bodies are general guidelines, and how a company implements them within its facility will differ between companies. Because there is no one correct way to implement a regulation, the FDA will usually start by looking at the standard operating procedures (SOPs) in place within a facility. When a company writes an SOP, it should keep one simple fact in mind: Explain what employees need to do in the SOP, and then make sure the employees are doing what the SOP states they need to do—Do what you say, say what you do. This does not mean a company will not get into trouble if employees just follow the SOP; however, if procedures are designed to follow the regulations and the employees are following the SOP, then the auditors will be more willing to help a company make corrections if a problem exists.

In preparation for ensuring that the product being manufactured complies with the regulation, some principles with regard to development of novel nutraceuticals should be considered so that the product that is ultimately manufactured is safe, stable, and optimally efficacious and nontoxic at the recommended dose. There are no clear guidelines developed by the FDA for the development process leading to nutraceuticals with health benefits, and generating a formulation from known and widely administered or accepted products does not require mandated major additional research work. However, researching, developing, and adapting protocols to determine and ensure the safety and efficacy of the final formulation, as well as to ensure the quality and batch-to-batch consistency of the manufactured formulations, will contribute to consumer confidence as has been suggested previously (Crowley and Fitzgerald 2006). In addition, clear understanding of the mechanism through systematic research by which the formulation provides benefit will enable the information to be transmitted to the consumer, when included as a product insert.

The sterility of the formulation can first be evaluated by testing the formulation for contamination with bacterial pathogens like *Escherichia coli* or *Salmonella*, *Staphylococcus/Streptococcus* contamination, or fungal contamination by plating the formulation on MacConkey agar, blood agar, and Sabouraud agar and incubating for the requisite time. If there are colonies on any of the plates, the formulation should be sterilized and retested. Viral testing to determine whether there are any enteric viruses that could be transmitted will have to be done either in-house or by sending a sample to ViroLogic Inc. (South San Francisco, CA). Endotoxin levels will be determined by endotoxin detection kits from GenScript (Piscataway, NJ), and if the levels exceed the permissible levels, then the GenScript removal system should be used. The levels of arsenic, iron, and lead need to be determined by the standard procedures; if the levels exceed those considered safe, then either those metals have to be removed or the ingredients tested for the source of contamination.

To evaluate hepatotoxicity in HepG2 cells, neurotoxicity in neuroblastoma cells and in live rodents would be beneficial. Established short-term cytotoxic and genotoxic effects on HepG2 cells can be assessed by pretreatment of cells with saline as a negative control, or with cadmium chloride, an inducer of reactive oxygen species levels, as a positive control, or test formulation alone at multiple concentrations. To measure cell survival, MTT [3-(4,5-dimethylthiazol-2yl)-2,5-diphenyltetrazolium-bromide, a tetrazole], clonogenic, and apoptotic assays could be used as described previously (Rao et al. 2009). Short-term genotoxic effects can be evaluated by a micronucleus and comet assay (Rao et al. 2009), whereas mutagenic activity can be evaluated by the Ames test (Ballardin et al. 2005). To determine the neurotoxicity of the formulation, we will treat neuroblastoma cells with either saline as a negative control or infection with a neurotropic but attenuated vaccine strain of vaccinia virus, which would generate free radicals, as a positive control, or the formulation and assess the membrane fluidity and cell death by using the lactate dehydrogenase, reactive oxygen species measurement, and mitochondrial membrane potential measurement as described previously (Mishra et al. 2009). If the toxicity shows no adverse cytotoxicity or apoptosis, then one can proceed to determine the efficacy. To test the efficacy of the formulation, the assays can be modified; for example, the HepG2 cells can be treated with cadmium chloride with and without the formulation and using mangiferin as a positive control for antioxidant activity. Similarly, in the case of determining antioxidant properties, minocycline, a known antioxidant, can be used as a positive control; a panel of in vitro radical scavenging assays as described previously (Kraus et al. 2005) can be used to compare the antioxidant properties of the nutraceutical formulation. One can also determine the ability of human cultured neuroblastoma cells to take up choline and measure the choline acetyltransferase activity in the presence and absence of the formulation as described recently by Ray et al. (2009). This will shed light on whether there is an enhanced uptake of choline, an indication of the proportion of cholinergic neurons, a major constituent of the normal mammalian central nervous system that is diminished during neurodegeneration.

To test the toxicity, distribution, bioavailability, and efficacy, transgenic mice predisposed to a disease could serve to prepare SOPs and a dossier for obtaining FDA approval to manufacture a nutraceutical.

15.2 STANDARD OPERATING PROCEDURES—THE HEART AND SOUL OF THE BUSINESS

SOPs tell the employees and the regulating bodies who is responsible for what process, how the process is performed, and how to record the process. Even though it may sound redundant, one of the most important SOPs, and usually the first one created, is an SOP on how to write an SOP. This document is needed to create consistency within the company and make it easier for employees and auditors to determine what the proper procedures are for a particular process or function.

The SOP on SOPs should outline the standard format and contain the following sections:

1. Objective: A clear, concise purpose of the procedure.
2. Areas affected: Functional groups affected by the new or revised SOP.
3. Responsibilities: An explanation as to what each functional group is expected to do with respect to its area of expertise. The Responsibility section does not provide instruction on how to carry out the responsibility.
4. Forms and attachments: Will be used in the performance or execution of the SOP.
5. Definitions: Words not generally recognized as "terms of art" and not included in the XYZ Company Glossary of Terms will be defined in the SOP.
6. Procedure: Provide sufficient instruction on how the SOP is to be executed. The level of detail will be commensurate with departmental requirements.
7. Process flow diagram: Optional; the flow diagram will summarize the activity called out by the SOP.
8. Current revision changes: Will be listed on the SOP for information and tracking purposes.
9. Signature approvals

The following is an example of an SOP on SOPs.

1. Objective
   1.1. This standard operating procedure (SOP) describes the general requirements for style, content, writing, controlling, revising, and issuing SOPs. This SOP defines the following:
      1.1.1. Processes that require SOPs
      1.1.2. SOP format
      1.1.3. Identifying/numbering/indexing SOPs by assigning a unique alphanumeric designation
      1.1.4. SOP approvals
      1.1.5. SOP master files
      1.1.6. Issuing SOPs
      1.1.7. Effective dates
      1.1.8. History and obsolete SOPs
      1.1.9. SOP review
      1.1.10. Archival of SOPs
      1.1.11. Use of forms and attachments
2. Areas affected

Unit Area	Definition	Affected
GEN	General: affects all departments	X
QS	Quality systems	
DS	Documentation systems	
LAB	Chemistry laboratory	
	Microbiology laboratory	
FEQ	Plant utilities	
MTL	Materials	
OPS	Operations	
LBL	Packaging/labeling	
WHS	Warehouse	
IT	Information technology	

3. Responsibilities
   3.1. Documentation systems (DS) is responsible for the following:
      3.1.1. Issuance and effective dates of all SOPs
      3.1.2. Issuance of SOP numbers using the SOP issuance log
      3.1.3. Maintaining an index of current SOPs
      3.1.4. Maintaining SOP history files
      3.1.5. Archival of all SOPs
      3.1.6. Verification that training has been completed before making an SOP effective
   3.2. The Area Manager is responsible for development and approval of all SOPs affecting his or her area of responsibility, and for ensuring that all new or revised SOPs comply with this SOP.
   3.3. Area managers are responsible for the biennial review of SOPs for their respective areas.
   3.4. The Director of Quality Systems or designee is responsible for the approval of the SOP and ensuring compliance.
   3.5. The SOP Preparer is responsible for assembling a cross-functional group of individuals affected by the SOP (new or revised), and for gathering comments and resolving any issues before submitting to Quality Systems for approval.

4. Forms and attachments
   Attachment 1: SOP Assignment Log
   Attachment 2: Biennial SOP Review Sheet
5. Definitions
   Please refer to the XYZ Company Glossary for a list of technical terms and acronyms that may be found in this procedure.
6. Procedure
   6.1. Processes requiring SOPs
      6.1.1. Written procedures shall be established and approved by the Director of Quality Systems or designee and the affected Area Manager for all operational processes involved in manufacturing, processing, packaging, labeling, storing, testing, or control of product or materials that are governed by requirements of the US Pharmacopeia or current good manufacturing practice regulations.
   6.2. Format
      6.2.1. Format details not specified explicitly in this section are not considered critical. However, a standard template document, modeled on the format of this SOP, shall be used to develop all SOPs.
      6.2.2. The font will be Arial with 11-point size.
      6.2.3. The right and left hand margin will be 1.00″.
      6.2.4. The table header information (e.g., logo, address and location, SOP category designation, Document Number, Revision Number, Page Numbering, Title, and space for the Effective Date stamp) will be included as illustrated in this SOP. Note: SOP attachments are required to have the same header information as the related SOP.
      6.2.5. Section headings (e.g., 6. is bold, 6.1. or more is not) will be bold typeface.
      6.2.6. SOPs will be in an outline format and will contain the following sections:
         6.2.6.1. Objective: A clear, concise purpose of the procedure.
         6.2.6.2. Areas affected: Functional groups affected by the new or revised SOP.
         6.2.6.3. Responsibilities: An explanation as to what each functional group is expected to do with respect to its area of expertise. The Responsibility section does not provide instruction on how to carry out the responsibility.
         6.2.6.4. Forms/attachments: Will be used in the performance or execution of the SOP.
         6.2.6.5. Definitions: Words not generally recognized as "terms of art" and not included in the XYZ Company Glossary of Terms will be defined in the SOP.
         6.2.6.6. Procedure: Provide sufficient instruction on how the SOP is to be executed. The level of detail will be commensurate with departmental requirements.
         6.2.6.7. Process flow diagram: Optional, the flow diagram will summarize the activity called out by the SOP.
         6.2.6.8. Current revision changes: Will be listed on the SOP for information and tracking purposes.
         6.2.6.9. Signature approvals
   6.3. Identification/numbering
      6.3.1. SOPs will be identified by a combination of the unit area (Section 2), a two-digit sequential SOP number, and a two-digit sequential revision number. New SOPs will begin with REV# 00. Example: QS-01 REV# 00, the first SOP and revision issued for the Quality Systems unit area.

6.3.2. Check the affected area from the table below. The "GEN" designation is reserved for procedures that affect all functions within the organization. Otherwise, SOPs will be identified by the department that will be responsible for following the procedure.

TABLE OF UNIT DESIGNATIONS

Unit Area	Definition	Affected
GEN	General: affects all departments	
QS	Quality systems	
DS	Documentation systems	
LAB	Chemistry laboratory	
	Microbiology laboratory	
FEQ	Plant utilities	
MTL	Materials	
OPS	Operations	
LBL	Packaging/labeling	
WHS	Warehouse	
IT	Information technology	

6.4. SOP approvals
- 6.4.1. At a minimum, SOPs are required to have at least two approval signatures:
  - 6.4.1.1. Preparer
  - 6.4.1.2. Area manager approval
  - 6.4.1.3. Quality systems approval
- 6.4.2. General procedures will require additional approvals by affected departments as necessary and by the Vice President.
- 6.4.3. Some procedures will require approval from the Chief Executive Officer (CEO) as designated on the approval area. One example is this SOP.

6.5. SOP master files
- 6.5.1. DS will maintain all current effective SOPs and will update all numbered SOP binders (sets) distributed as soon as possible after the procedure's posttraining approval.
- 6.5.2. All obsolete documents will be removed from the distributed SOP binder sets simultaneously with the addition of the revised SOP.
- 6.5.3. DS will maintain a history file for each SOP through its initial (new), revision phase (updates), and obsolescence.

6.6. Issuance of new SOPs
- 6.6.1. Once a new SOP is in draft form and has been reviewed by the signers, the Preparer will submit the SOP to Document Systems for the assignment of a number.
- 6.6.2. The assignment of the number will also be documented in the SOP Assignment Log maintained by Quality Systems. See Attachment 1.
- 6.6.3. The Preparer will obtain final signatures for the new SOP and submit the document to DS. If during the development process an SOP number is assigned and then cancelled, this shall be noted in the SOP Assignment Log. Voided or cancelled numbers may not be reassigned. SOP numbers may not be cancelled after they have been approved. In these cases, the process for making an SOP obsolete shall be followed.
- 6.6.4. Training will be performed and documented before the effective date of the SOP.

6.6.5. DS will verify training has been completed before the issuance of the SOP.

6.7. Effective date

6.7.1. The effective date of a new or revised SOP will be assigned by DS.

6.7.2. The SOP is put into practice for an area once training on the SOP has been completed and the effective date has been established.

6.7.3. DS will update SOP binders with the new or revised procedure and will remove the previous version from the binder. The previous version of the SOP removed from the binders will be destroyed by DS.

6.8. History or obsolete SOP files

6.8.1. DS will be responsible for the handling and maintenance of all history and obsolete SOP files.

6.8.2. When a current SOP is revised, the new version will receive the next sequential revision number, and the previous version will be moved to the appropriate history file.

6.8.3. When an SOP is obsolete, the master original will be removed from the current file, stamped OBSOLETE, and moved to the appropriate history file.

6.9. SOP changes/review

6.9.1. All new and/or revised SOPs will have a change control initiated on them before submission to Quality Systems (see SOP GEN-002 for Change Control forms). SOP change requests can be initiated anytime after the effective date of an SOP. The initiator will be responsible, in conjunction with the Area Manager, to prepare the revised document, make changes to the procedure, and complete a change control form.

6.9.2. A document revision number will be assigned by the preparer and will increment the current revision by one number from the current SOP.

6.9.3. SOPs will be reviewed no later than every 2 years by the Area Manager for the affected department to determine whether revisions are necessary. The review process will be documented using the SOP Review Log Sheet. See Attachment 2.

6.9.4. If a revision is deemed not necessary, the review sheet will indicate that no changes are required and will be sent to DS to be maintained in the appropriate Master SOP file. DS will then stamp all SOPs in current use binders that were reviewed but needed no changes. The stamp will indicate that the initial 2-year SOP review was completed and list the due date of the next 2-year review.

6.10. Archival of SOPs

6.10.1. SOP history files will be stored in DS files for a minimum of 2 years from the calendar date of inactivation. Once a year, the history file documents 2 years or older will move into an inactive archive file for a period of 5 years.

6.10.2. All inactive archived SOPs that are obsolete may be destroyed 5 years from the date of creation of the inactive archive file. A list/database of archived SOPs will be maintained by DS to document the archival of all SOPs.

6.11. Archival of SOP Issuance Log pages

6.11.1. DS will periodically archive SOP Assignment Logs.

6.11.2. On archival, all pages will be inclusively paginated.

6.12. Use of forms and attachments

6.12.1. Forms and attachments will be attached to the SOPs.

6.12.2. If a form or attachment must be revised, then the related SOP in which it is included must also be revised.

7. Process flow diagram
   7.1. Process flow diagrams are optional (but encouraged) and may be used to summarize or demonstrate the key elements of the procedure described in the previous section.

Example:

Step One

↓

Step Two

↓

Step Three   Current Revision Changes

Current revision changes will be identified in a completed change control form.

15.3 DESIGNATED AUTHORITY AND RESPONSIBILITIES

This SOP should detail the authority and responsibilities of the upper management. The SOP should detail who has what responsibility within the facility and how that responsibility can be passed on to designates. It should also explain who has the responsibility in the case of the head person being out or in the case of an emergency. For example, the Director of Quality has final say on the disposition of any product manufactured within the facility.

15.4 ADVERSE EFFECTS REPORTING AND PRODUCT COMPLAINTS

This SOP details how a company will deal with adverse effects reporting and any product complaints. This SOP should detail how a company will receive a report or complaint, who the report or complaints will be assigned to, how it will be recorded, and what the procedure will be in case of a product recall.

All nutraceutical products must have a company contact number and address so the public can submit a complaint or concern about the product. Pay special attention to this SOP because it will be scrutinized by the FDA.

15.5 RESPONSIBILITIES OF THE QUALITY UNIT

This SOP details how a quality unit is organized and what its responsibilities are. This SOP should detail all the areas that are governed by quality. Examples of quality responsibilities are receipt of incoming materials, release of raw materials, sampling and inspection of final product, disposition, and release of final product.

15.6 RECEIPT AND HANDLING OF MATERIALS

This SOP details how the company will receive and handle incoming raw materials. This SOP should state who can receive materials, what the disposition of the material is once received, and what steps are taken to release the material for use within the facility.

Items to consider when setting up the receipt of material are the vendor qualified by the company and what type of testing is required to ensure the product was received at the strength stated by the vendor.

15.7 PERSONNEL ORGANIZATION AND RESPONSIBILITIES

This SOP details the authority and responsibilities of the employees within the company. This SOP should have an organization chart and detail what departments are responsible for and who is responsible within the departments. For example, the quality department is responsible for the final disposition of all products manufactured within the facility.

15.8 TRAINING PROGRAM

This SOP will explain how new employees are trained, how employees are updated on changes to the SOPs, and how the training will be recorded and stored. Employees must be continually reminded of the procedures and notified when changes are taking place to the existing procedures.

15.9 GOOD MANUFACTURING PRACTICE ATTIRE

This SOP will explain what is appropriate attire for company employees and the areas associated with manufacturing and testing. Examples of proper attire would be no street clothes may be worn in the manufacturing area to prevent contamination of the product; steel-toed shoes must be worn in the manufacturing area; and lab coats must be worn in the lab at all times.

15.10 PROPER GOOD MANUFACTURING PRACTICE DOCUMENTATION

This SOP details how procedures will be recorded and how corrections will be made to the records. The SOP should explain to employees and auditors how the different steps of a process will be documented, the acceptable way to document the steps, at what points the documentation will be reviewed, and how to make changes to the document if an error is found. This SOP should also explain what is not an acceptable way to make corrections; for example, using white correction fluid to cover the original mistake, blacking out the original mistake, or simply throwing out the recorded mistake.

For many of the important steps of a procedure, it would be advisable to have a second person verify that the first person completes the step or double checks the values that are recorded in the record. When a record is reviewed, the reviewer should sign and date that the record was reviewed.

Auditors understand that errors can happen during the process. One of the most import items detailed in the SOP is how to correct errors. The best way to correct an error is to line out the mistake with a single line through the error; write the corrected answer as close to the error as possible; initial and date the correction; and if the error is not obvious (e.g., the wrong year recorded in the date), provide a short explanation of the error.

The documentation process is time consuming, and employees may believe it is a waste of time, but the importance of good documentation is as important as the process itself. No matter how good the process is or how reliable the employee, the rule of thumb "if it wasn't recorded, it wasn't done" always applies (FDA 2004).

15.11 SPECIFICATIONS AND TESTING

Companies have different requirements for the same products that are used in many different areas. Therefore, it is the responsibility of the company to determine what is a reasonable amount of testing and what is an acceptable specification for the product. Specifications should be set to help the company ensure it is using a quality product without greatly increasing the final cost of the product.

15.12 VENDOR QUALIFICATION

Vendor qualification is a way to reduce the amount of testing needed once a material has been received. By doing more testing of the product for the first few times it is received within the facility, a material can be accepted with a minimal amount of testing after that point. Once the vendor's certificate of analysis can be verified for a set number of times (usually three times in a row), the vendor becomes validated, and at that point only a test for identity needs to be performed. In the future, a full verification of the product is completed at a reasonable period. The amount of time between full testing needs to be stated in the SOP.

15.13 INVESTIGATION OF OUT-OF-SPECIFICATION TEST RESULTS

This SOP will detail what steps need to be taken if an out-of-specification result is generated. An out-of-specification result may be caused by a lab error, a sampling error, a manufacturing error, or any number of reasons. This SOP should lay out an order in which the result is investigated. All the steps taken during the investigation need to be documented. The ultimate goal of a good investigation is to determine the underlying problem that caused the error initially and to help prevent it in the future. If an investigation determines that an error occurred during the manufacturing process, what caused the error and a correction to the procedure to help prevent the problem in the future need to be resolved.

15.14 RETAINED SAMPLE REQUIREMENTS

This SOP details what samples should be retained by the manufacturer in the case of product complaints or problems. Retained samples can be used by the manufacturer to determine whether there is a problem with the product or to show that there is not a problem. The SOP should describe how much of the final product should be retained after manufacturing, when the samples will be taken, and how the samples will be stored.

15.15 STABILITY PROGRAM

This SOP details what testing will be completed to determine the expiration dating of a product. Depending on the product, this testing could be as simple as a visual inspection, or more complicated, such as microtesting and high-performance liquid chromatography assay. This SOP should detail how expiration dating will be determined (e.g., through concurrent testing or accelerated testing).

15.16 FINISHED PRODUCT DISPOSITION AND RELEASE

This SOP details how all finished products will be released from the company for sale. This SOP will determine how samples will be collected for final release testing and how the disposition of the product will be determined throughout this process. This SOP should also detail what steps will be taken if there is a problem with the manufacturing or testing process.

15.17 MATERIAL SPECIFICATIONS

This SOP details how specifications will be determined for all raw materials, components, and final products. Depending on the company and the material, the complexity of this SOP will vary. Keep in mind that by implementing the right testing, the company can save time and money in the long run. This does not always mean doing the least amount of testing or the easiest, but rather the testing that determines that the company has a quality product (e.g., testing for melamine in protein).

15.18 QUALIFYING THE EQUIPMENT

This SOP details how equipment will be qualified for use within the company. Depending on the equipment and how it is used, this SOP should detail what process will be taken to determine that the equipment will perform as stated.

15.19 STANDARD OPERATING PROCEDURES FOR MANUFACTURING

15.19.1 Periodic Checking and Maintenance Records for the Equipment

This SOP details when equipment records will be reviewed by personnel. Equipment use logs and maintenance logs are used to record when products are made and when maintenance and repairs take place. These documents need to be reviewed regularly to ensure proper cleaning, use, and maintenance. The SOP will also determine who is responsible for the review.

15.19.2 Batch Records and Process Evaluation

This SOP will detail what is required for manufactured batch records. Much like the SOP on SOPs, this SOP will set up a consistency for all batch records that will allow for easier use and review. This SOP will also detail who is responsible for batch record review, how batch records will be issued, and how long batch records will be retained.

15.19.3 Cleaning

This SOP details what steps are required for cleaning before and after a production run. The SOP should determine when cleaning must be completed after a production run and how long a cleaned piece of equipment can sit before use.

15.19.4 Packaging

Your facility should have a set of SOPs that details all aspects of packaging within the facility. This set of SOPs will include how labeling for products will be approved and who is responsible for the approval. Packaging material may be handled differently from raw materials if it does not come in direct contact with the product. Packaging material can include the container in which the product is packaged, containers into which the final packaged products are packed, cardboard boxes for shipping, and insert material placed in the final package. Each of the components can have different needs for testing, but the same component may have a different need for testing depending on which product is being packaged at the time (e.g., a box may be used for shipping in one case and used as the primary package in another). It is dependent on the company to determine at what level the component needs to be tested for all products and to test accordingly.

15.19.5 Labeling

This SOP should detail how labeling is approved, who is responsible for label approval, and how labeling will be issued for use. Labels need to be controlled to ensure only approved labels are used for the final product. Label review and approval should be an important step in the process to ensure correct and necessary information is printed on the label. The SOP should also detail how lot numbers will be issued and printed on the labels (FDA 2007).

15.20 INSPECTION

15.20.1 Internal Audits and Inspections

This SOP will detail how often internal audits are conducted within the facility and who is responsible for conducting the inspections. The SOP will set up timelines for when areas within the company will undergo inspection. The International Organization for Standardization certification system does an excellent job in detailing how to create an acceptable program for inspections.

15.20.2 Speaking with Inspectors

This SOP details who is responsible for interacting with external auditors and how to handle correspondence with governing bodies. This SOP should detail the primary contact within the company for a governing body and how all correspondence is handled from those agencies. This SOP will also explain who accompanies auditors during the visit.

15.20.3 Fix What They Tell You to Fix

This is not an SOP but rather a good business plan. Once the auditors find a problem that needs to be fixed, the way to fix it is left to the company. Recommendations may be made by the auditors, but they are only recommendations. The correction to the problem needs to be acceptable with the governing body that cited the problem. Use the experience of the auditors to implement a strong change, but make sure that the correction suggested is something that can be accomplished. Once a correction is put into place, make sure that it accomplishes what it is designed to do. One of the first things a good auditor will do when he or she comes back to audit the next time is to check on any problems found the previous visit.

15.20.4 Honesty Is the Best Policy

When dealing with auditors, keep in mind that the best policy is to be honest. Every company makes mistakes, but it is the way the company deals with the mistakes that leaves a lasting impression. Toyota is an example of how not being honest with a regulator body can cause severe damage to a company's reputation and pocketbook. By not reporting customer complaints about sticking gas pedals, Toyota hid a problem, and in turn the company was fined more severely by the US government, on top of money spent to make the necessary corrections. Thus, be honest with auditors, and in return, if the company makes mistakes that need correction, auditors may be more helpful in finding the easiest correction.

There are many guidelines to keep in mind while implementing an SOP system, but the most important to remember when putting a system into place is to "Say What You Do, Do What You Say."

REFERENCES

Ballardin, M., R. Scarpato, G. J. Kotwal, and R. Barale. 2005. In Vitro Mutagenicity Studies of the Antiretrovirals AZT, Didanosine, 3TC, and Fulvic Acid. *Annals of the New York Academy of Sciences* 1056: 303–10.

Crowley R., and L. H. Fitzgerald. 2006. The Impact of cGMP Compliance on Consumer Confidence in Dietary Supplement Products. *Toxicology* 221: 9–16.

Food and Drug Administration. 2004. FDA 101: Dietary Supplements. http://www.fda.gov/forconsumers/consumerupdates/ucm050803.htm.

Food and Drug Administration. 2007. Dietary Supplement Current Good Manufacturing Practices (CGMPs) and Interim Final Rule (IFR) Facts. http://www.fda.gov/food/dietarysupplements/guidancecomplianceregulatoryinformation/regulationslaws/ucm110858.htm.

Kraus, R. L., R. Pasieczny, K. Lariosa-Willingham, M. S. Turner, A. Jiang, and J. W. Trauger. 2005. Antioxidant Properties of Minocycline: Neuroprotection in an Oxidative Stress Assay and Direct Radical-Scavenging Activity. *Journal of Neurochemistry* 94: 819–27.

Mishra, M. K., D. Ghosh, R. Duseja, and A. Basu. 2009. Antioxidant Potential of Minocycycline in Japanese Encephalitis Virus Infection in Murine Neuroblastoma Cells: Correlation with Membrane Fluidity and Cell Death. *Neurochemistry International* 54: 464–70.

Rao, S., M. V. Sreedevi, and N. Rao. 2009. Cytoprotective and Antigenotoxic Potential of Mangiferin, a Glucosylxanthone against Cadmium Chloride Induced Cytotoxicity in HepG2 Cells. *Food and Chemical Toxicology* 47: 592–600.

Ray, B., J. R. Simons, and D. K. Lahiri. 2009. Determination of High Affinity Choline Uptake (HACU) and Choline Acetyltransferase (ChAT) Activity in the Same Population of Cultured Cells. *Brain Research* 1297: 160–68.

CHAPTER 16

Fortification and Value Enhancement of Food Products during Nutraceutical Processing Using Microencapsulation and Nanotechnology

Xiaoqing Yang, Yuping Huang, and Qingrong Huang

CONTENTS

16.1 INTRODUCTION

Health- and wellness-promoting food products containing micronutrients and nutraceuticals have led to increasing attention in the food industry (DeFelice 1995). Micronutrients mainly contain vitamins and minerals that are essential for human health, whereas nutraceuticals are not essential for human life, but are widely used in functional foods, which are defined as "any food or ingredient that has a positive impact on an individual's health, physical performance, or state of mind, in addition to its nutritive value" (Goldberg 1994, pp. xv–xvi). Examples of common micronutrients and nutraceuticals are listed in Table 16.1 (Velikov and Pelan 2008).

Researchers have already proved that the intake of bioactive compounds including nutraceuticals can bring health benefits to humans (Weiss et al. 2008), although the complete physiological roles of those compounds are not fully understood. Most nutraceuticals are water insoluble, which leads to low dissolution levels in the gastrointestinal tract and thus poor absorption and limited bioavailability (Spernath and Aserin 2006). More importantly, nutraceuticals extracted from the food matrix are easily subjected to quality losses because of their sensitivity to environmental conditions, such as oxygen, humidity, and light (Wildman 2007). Both poor bioavailability and stability of nutraceutical compounds have become two major issues during processing in the food industry.

Table 16.1 Examples of Common Micronutrients and Nutraceuticals*

Micronutrients	Nutraceuticals
Water-soluble vitamins	*Polyphenols*
Vitamin B_1 (thiamine)	Flavonoids
Vitamin B_2 (riboflavin)	Isoflavone
Vitamin B_3 (niacin)	Anthocyanins
Vitamin B_5 (pantothenic acid)	*Conjugated linoleic acid*
Vitamin B_6 (pyridoxine)	*Omega-3 polyunsaturated fatty acids*
Vitamin B_7 (biotin)	*Terpenoids*
Vitamin B_{12} (cyanocobalamin)	*Alkaloids*
Vitamin C (ascorbic acid)	Caffeine
Oil-soluble vitamins	Theobromine
Vitamin A (retinol, retinoids, carotenoids)	
Vitamin D (ergocalciferol and cholecalciferol)	
Vitamin E (tocopherol, tocotrienol)	
Vitamin K (phylloquinone, menaquinone)	
Minerals	
Ca, Mg, Zn, K, Fe, Mn, Cu, Se	
Organic and inorganic salts of these minerals	

Source: Velikov, K. P. and E. Pelan, *Soft Matter*, 4, 1964, 2008. Reproduced with permission of The Royal Society of Chemistry.
*Some micronutrients also have nutraceutical action.

This chapter reviews the carrier systems for active compounds fabricated through nanotechnology, which is an emerging multidisciplinary area that combines knowledge from chemistry, physics, and biotechnology. These carrier systems include nanoemulsions, solid lipid nanoparticles, coacervates, micelles, and so on, which can protect sensitive compounds from degradation during storage, as well as increase their absorption and bioavailability.

16.2 NANOEMULSIONS

Emulsions with droplet sizes smaller than 200 nm are usually referred to as *nanoemulsions* (Solans et al. 2005). Small droplet size provides the colloidal system novel characteristics that have attracted increasingly more attention in the food, drug, and cosmetic industries. For example, nanoemulsions always have long-term kinetic stability to avoid sedimentation or creaming during storage because their small particle size reduces gravity such that brownian motion can overcome gravity to avoid creaming or sedimentation. Other advantages include enhanced gastrointestinal absorption, improved release profile because of the large interfacial area, and reduced inter- and intraindividual variability of the active compounds (Brüsewitz et al. 2007).

16.2.1 Preparation

16.2.1.1 High-Energy Input Method

Emulsion formation is a nonspontaneous process; therefore, energy is usually required to produce small droplets based on the equation:

$$\Delta G = \Delta A\gamma - T\Delta S$$

where ΔA = increase of the interfacial area
γ = interfacial tension

Both ΔA and γ have positive values. The entropy term $T\Delta S$ is also positive, but its value is too small to compensate the term of $\Delta A\gamma$. Thus, energy input usually from mechanical devices is required to produce emulsion, especially nanoemulsion with its small droplet size and large interfacial area.

16.2.1.1.1 High-Speed and High-Pressure Homogenizers

High-speed and high-pressure homogenizers are most commonly used to prepare nanoemulsions. Small droplets can be obtained because of the high shear stress. A high-speed homogenizer is often used to prepare the so-called “pre-emulsion” with relatively large particle sizes, whereas a high-pressure homogenizer is used to further reduce the particle size because it provides sufficient energy to produce homogeneous

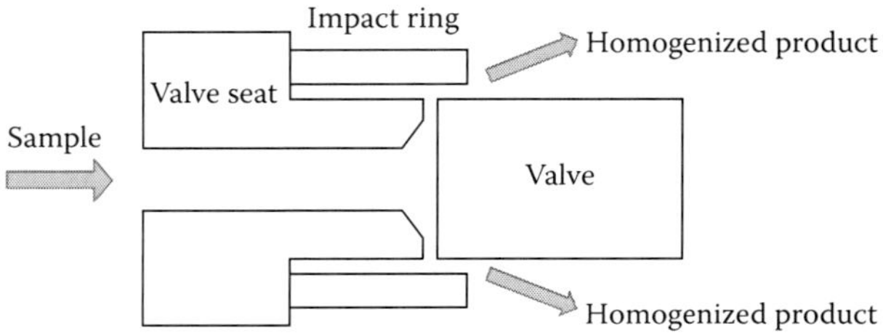

Figure 16.1 Schematic diagram of high-pressure homogenizer.

dispersion flow through a restrictive valve (Solans et al. 2005). Figure 16.1 shows the schematic diagram of a high-pressure homogenizer. Wang, Jiang et al. (2007) used a high-pressure homogenizer to reduce the oil–water emulsion particle size from 683.2 nm (high-speed homogenization only) to 82.1 nm (high-pressure homogenization at 1500 bar for 6 cycles); in addition, emulsions loaded with curcumin showed improved inhibition on the edema of mouse ear. Such anti-inflammation activity was further enhanced when the emulsion droplet sizes were reduced below 100 nm after applying a high-pressure homogenizer to the emulsion system (Wang, Jiang et al. 2007).

16.2.1.1.2 Microfluidizer

Microfluidizer is also a mechanical device that is used to produce high energy to reduce particle sizes of emulsions. The difference between microfluidizers and traditional high-pressure homogenizers is the presence of interaction and auxiliary chambers with microchannels in microfluidizers. These extra chambers can provide cavitations and shear forces to facilitate emulsion formation in microfluidizers (Sanguansri and Augustin 2006). Compared with the traditional emulsifying devices, the interaction and auxiliary chambers help to produce emulsions with narrower particle size distribution (Robin et al. 1992; Pinnamaneni, Das, and Das 2003; Jafari, He, and Bhandari 2007). Jafari, He, and Bhandari (2006) used both microfluidization and sonication to produce nanoemulsions and evaluated the efficiencies of these two methods. The results showed that sonication produced emulsions with wider and bimodal size than emulsions produced by microfluidization (Jafari, He, and Bhandari 2006). However, energy-induced cavitation, shear, and turbulence can occur in a short time (~10^{-4} s) simultaneously for microfluidizers, and "overprocessing" can happen especially when biopolymers, such as proteins and long-chain polysaccharides, are involved in the emulsion system (Paquin 1999). For example, Jafari, He, and Bhandari (2006) also discovered that there were optimum parameters (moderate pressure and number of passes through the equipment) for using microfluidizers to produce emulsions with small particle size and good storage stability. Schulz and Daniels (2000) used hydroxypropyl methylcellulose as the emulsifier to stabilize emulsions containing 10% medium chain triglycerides; for the homogenization process at 900 bars for 5 passes, 99% of the particles were smaller than 150 nm, whereas when the pressure increased to 1600 bars, particle size increased to larger than 300 nm. This phenomenon was believed to be caused by the degradation of long-chain molecules and formation of polymers with significantly smaller molecular weights as a result of extremely high pressure or shear, the formation of cavitations, and turbulence at the same time in a short period (Schulz and Daniels 2000).

16.2.1.1.3 Ultrasonic Generator

Another widely used instrument is the ultrasonic generator. Ultrasonic frequency is greater than the upper limit of human hearing (>18 kHz) (Sanguansri and Augustin 2006). During the emulsification process, the sonicator probe generates mechanical vibrations with imploding cavitation bubbles. Collapse of these cavitations will lead to intensive shock waves in the surrounding liquid and high-velocity liquid jets to break the dispersed liquid (Maa and Hsu 1999). Wulff-Pérez et al. (2009) observed that an emulsion using Pluronic F68 (BASF, Florham Park, NJ) as a surfactant formed by ultrasonication had good physical stability even with a relatively small amount of surfactant and high amount of oil. Landfester, Eisenblätter, and Rothe (2004) studied the efficiency of the ultrasonic emulsification process by measuring the turbidity of polymerizable styrene emulsions formulated with different ultrasonication times at different amplitudes. The results showed that the emulsification efficiency is strongly dependent on those ultrasonication parameters (Landfester, Eisenblätter, and Rothe 2004). The advantage of ultrasonication is that it is a cheaper and cleaner method to prepare emulsions than high-pressure homogenization and microfluidization. However, its application is limited because it is only appropriate for small batches (Sanguansri and Augustin 2006).

16.2.1.2 Low-Energy Method

Studies have already proved that nanoemulsions can be formed with little or no energy input by using their physicochemical characteristics (Bouchemal et al. 2004).

16.2.1.2.1 Phase Inversion Method

The low-energy method requires study of the phase behavior of the water–oil–emulsifier system. Forgiarini et al. (2000) investigated the phase behavior of a model emulsion system formed by water, polyoxyethylene nonionic surfactant, and decane, and three low-energy emulsification methods: method A, addition of oil to an aqueous surfactant dispersion; method B, addition of water to a surfactant solution in oil; and method C, pre-equilibrate the samples first and then mix all the components together (Forgiarini et al. 2000). Results from Forgiarini's research showed that small droplet size with narrow distribution can be obtained by method B.

Another low-energy emulsification method that is already used in industry is the phase inversion temperature (PIT) method (Solans et al. 2005). The basic principle of the PIT method is that some surfactants (e.g., polyoxyethylene-type surfactant) change from hydrophilic to lipophilic with the increase of temperature as a result of the dehydration of the polyoxyethylene chains. When an oil-in-water (O/W) emulsion stabilized by this kind of surfactant is heated to the critical temperature (also known as *hydrophilic-lipophilic balance (HLB) temperature*), it will invert to water-in-oil (W/O) type emulsion. The coalescence rate is fast at this point; therefore, heating or cooling is required to get kinetically stable emulsions with small droplet size and narrow size distribution (Solans et al. 2005). Morales et al. (2003) studied the water, hexaethylene glycol monohexadecyl ether surfactant, and mineral oil nanoemulsion system formed by the PIT method, and emulsions with droplet size as small as 40 nm were obtained with narrow size distribution. They also found that the oil/surfactant ratios influenced the emulsion droplet size and size distribution. Izquierdo et al. (2005) found nanoemulsions formed by water, polyoxyethylene 4-lauryl ether and polyoxyethylene 6-lauryl ether, and isohexadacane by the PIT method with droplet radii of 60–70 nm and 25–30 nm when total surfactant concentrations were 4% and 8%, respectively (Izquierdo et al. 2005). Both droplet size and polydispersity decreased with the increase of total surfactant concentration in the system as a result of the increase of the interfacial area and the decrease of the interfacial tension.

Besides temperature, phase inversion can also be induced by adjusting the HLB value using surfactant mixtures at a constant temperature. Brooks and Richmond (1994) studied the transitional phase inversion phenomenon by changing the ratio of different nonionic surfactants in the emulsion system. Droplet size began to decrease and emulsification rates began to increase as the transitional inversion point was approached; changes were induced by altering the surfactant mixture's HLB values. Spontaneous emulsification occurred at the inversion point, which had a strong dependence on the types and concentrations of the surfactants. The study indicated that much a finer emulsion with much less energy input can be produced by transitional inversion processes compared with a direct emulsification method (Brooks and Richmond 1994).

16.2.1.2.2 Ouzo Emulsification Method

Spontaneous emulsification can also take place by using the Ouzo effect. The basic principle of the Ouzo effect is that when oil is dissolved in a water-miscible solvent (e.g., ethanol) and added to a water phase containing surfactant, the oil becomes supersaturated to nucleate along with the water–surfactant diffusion into the oil–solvent phase to form emulsion droplets. Ganachaud and Katz (2005) reviewed several aqueous dispersions prepared by this Ouzo emulsification method, from which nanospheres, nanocapsules containing an oil or a void, and even liposomes or vesicles can be produced in small sizes. Although Ouzo emulsification is a promising alternative method to conventional high-energy input, sometimes the emulsions formed by this method still need a certain amount of energy input to help stabilize the final product, and the use of organic solvent greatly limits its applications.

16.2.1.2.3 Membrane Emulsification

Membrane emulsification involves a specially designed microporous membrane as a necessary emulsifying element (Nakashima, Shimizu, and Kukizaki 2000). For the preparation of an oil–water emulsion, the continuous phase (surfactant aqueous solution) goes through the membrane module, whereas the dispersed phase (oil) is forced toward the membrane to make contact with the continuous phase, as shown in Figure 16.2. Usually high pressure is applied to facilitate the movement of dispersed oil through the pores of the membrane.

Membrane parameters that can affect the emulsion formation in this method include pore size and shape, porosity, and membrane type. Other parameters include the transmembrane pressure of the dispersed phase and the velocity of the continuous phase (Abrahamse et al. 2002). All these controllable parameters can offer the membrane emulsification method an advantage of producing emulsions with desirable droplet size; however, the low level of dispersed phase flux through the membrane limits its scaling-up feasibility (Sanguansri and Augustin 2006).

16.2.2 Applications

Many studies have been done to exploit the advantages of nanoemulsions, and most studies focused on using them as carriers for active molecules (e.g., drugs, proteins, and nutraceuticals). A major problem associated with nutrients and nutraceuticals is their limited solubility, especially in the water phase. Chen and Wagner (2004) used a high-pressure homogenization method to produce vitamin E nanoparticles stabilized by starch, and the product can be used for beverage fortification. For beverages with non-nanoparticle vitamin E emulsion formulations, the emulsion droplets gradually rise to the top of the bottle, which leads to the "ringing" phenomenon; the product will also become turbid during storage. For vitamin E nanoparticle–fortified water, the turbidity and physical stability problems were greatly alleviated (Chen and Wagner 2004). Another advantage to encapsulate the active compounds is the improved bioavailability or biological efficacy caused by the

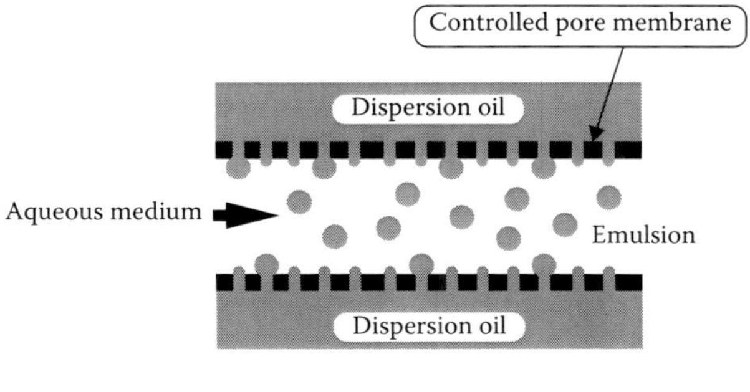

Figure 16.2 Schematic diagram of a membrane module. The pressurized oil to be dispersed is passed through controlled pores of membrane, forming countless oil droplets on the inner surface of the membrane. (From Nakashima, T., M. Shimizu, and M. Kukizaki, *Adv. Drug Deliv. Rev.*, 45, 47, 2000. With permission.)

increase of their contact area with intestinal membrane. Date and Nagarsenker (2007) used self-emulsifying nanoemulsions to deliver cefpodoxime proxetil (CFP), a high-dose antibiotic with poor bioavailability and pH-dependent solubility. They found an optimized formulation with a droplet size of 170 nm that can release CFP completely within 20 minutes irrespective of the pH of dissolution medium. A cholesterol-rich nanoemulsion was observed to concentrate in cancer tissues after injection into the bloodstream with low toxicity and increased antitumor activity (Dias et al. 2007). Wang, Wang et al. (2009) used nanoemulsions to encapsulate polyphenols, such as epigallocatechin gallate (EGCG) and curcumin, and tested their anti-inflammation and antitumor functions in mice. The results showed that after encapsulation both the physical and chemical stability and oral bioavailability of the polyphenols were greatly improved (Wang, Wang et al. 2009).

In conclusion, all the current studies of nanoemulsions for encapsulation and delivery of active compounds proved nanoemulsions' promise for use as carrier systems to enhance the bioavailability and stability of those compounds. However, disadvantages of nanoemulsions should also be considered, such as high cost of production, instability caused by Ostwald ripening, and lack of enough understanding and direct evidence to prove the benefits of nanoemulsions. Although many reviews of nanoemulsions stress the importance of their advantages and interest in nanoemulsions has been growing for several years, direct applications are still limited. The main reason for this is stability. Although an extremely small droplet size provides nanoemulsions kinetic stability because the brownian motion rate is sufficient to overcome the sedimentation and creaming rate induced by gravity force, Ostwald ripening, which is caused by polydispersity of emulsion droplets whereby larger droplets are energetically favored over smaller ones, is the major mechanism to destabilize an emulsion system. Studies have proved that Ostwald ripening can lead to phase separation of emulsion, which may not always be observed because the droplet growing rate gradually decreases with the increase of droplet size (Tadros et al. 2004).

Strategies have been proposed to reduce Ostwald ripening (Tadros et al. 2004): (1) addition of a second disperse phase component with low solubility in the continuous phase, and (2) utilization of surfactants that are strongly adsorbed at the interface and not easily desorbed during ripening. However, detailed studies are still needed to overcome the stability issue of nanoemulsions in real products.

Despite the aforementioned limitations, researchers are continuously making efforts in this area for the application of nanoemulsions, which requires optimization with respect to the formulation and preparation variables to produce nanoemulsion products with desired characteristics.

16.3 SOLID LIPID NANOPARTICLES

16.3.1 Introduction

The basic concept of solid lipid nanoparticles (SLNs) is to replace the liquid lipid (oil) component in emulsion with a solid lipid, which was introduced by the research groups of Gasco (1993) in Italy and Müller in Germany (Müller and Lucks 1996). The main reason to use solid lipid instead of liquid oil is that the mobility of encapsulated lipophilic compounds in a solid lipid matrix is considerably lower than in a liquid oil matrix. It has been claimed that nanoparticles based on solid lipids have advantages such as (Mehnert and Mäder 2001):

1. Increased protection for the encapsulated active compounds
2. Possible targeting and controlled release
3. Reduced toxicity
4. Feasibility of large-scale production

After several years of study on SLNs, nanostructured lipid carriers (NLCs) were introduced by Müller et al. to overcome disadvantages of SLNs (Müller, Radtke, and Wissing 2002). For example, the solid lipid matrix tends to form an ordered crystalline structure during storage to exclude the encapsulated active compounds. The new concept of NLCs is to use more than one kind of solid lipid or even liquid oil to create imperfections between the crystal lattices. This will bring a new generation of lipid nanoparticles with high loading capacity and low drug expulsion. A schematic representation of different structures of SLNs and NLCs is shown in Figure 16.3.

General ingredients of lipid nanoparticles are solid lipids, emulsifiers, and water. The commonly used solid lipids and emulsifiers are listed in Table 16.2.

16.3.2 Preparation

Because of the solid state of the lipid component, the production methods for SLNs are different from those of ordinary nanoemulsions. Basically they are still divided into two categories: the high-energy input method and the low-energy method. The most commonly used methods will be discussed in this session.

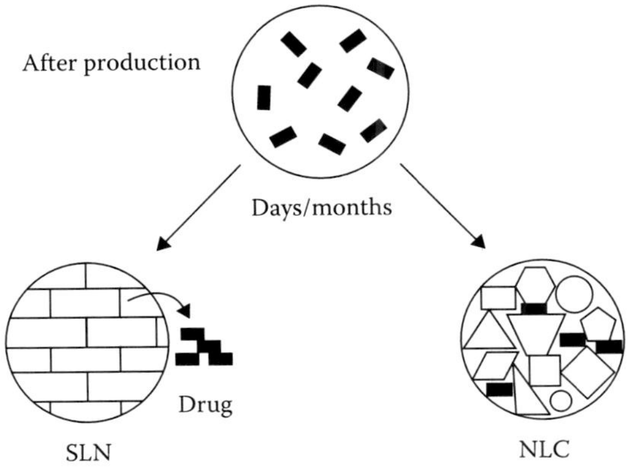

Figure 16.3 Schematic representation of different structures of solid lipid nanoparticles and nanostructured lipid carriers, respectively. (From Müller, R. H., M. Radtke, and S. A. Wissing, *Int. J. Pharm.*, 242, 121, 2002. With permission.)

Table 16.2 Lipids and Emulsifiers Commonly Used for Solid Lipid Nanoparticles

Lipids	References
Trimyristin	Bunjes et al. 1996; Westesen, Bunjes, and Koch 1997; Venkateswarlu and Manjunath 2004
Tristearin	Bunjes et al. 1996; Venkateswarlu and Manjunath 2004
Tripalmitin	Bunjes et al. 1996; Venkateswarlu and Manjunath 2004
Trilaurin	Westesen and Bunjes 1995; Bunjes et al. 1996
Cetyl palmitate	Sznitowska et al. 2001; Wissing and Müller 2003
Stearic acid	Cavalli et al. 1997; Yang, Lu et al. 1999
Palmitic acids	Jenning and Gohla 2001
Behenic acid	Cavalli et al. 1997
Dynasan 112 (glycerol trilaurate)	Schwarz and Mehnert 1997
Hard fat (e.g., Witepsol E85, W35, S58)	Almeida, Runge, and Muller 1997; Liedtke et al. 2000; Manjunath et al. 2005
Emulsifiers	**References**
Soy bean lecithin (Lipoid S100, S75)	Westesen and Bunjes 1995; Bunjes et al. 1996; Schwarz and Mehnert 1997; Westesen, Bunjes, and Koch 1997
Soy phosphatidylcholine 95% (Epikuron 200)	Cavalli et al. 1997; Venkateswarlu and Manjunath 2004
Egg lecithin	Morel et al. 1996; Sznitowska et al. 2001
Poloxamer 182	Almeida et al. 1997; Liedtke et al. 2000
Poloxamer 188	Cavalli et al. 1997; Liedtke et al. 2000; Jores et al. 2004; Venkateswarlu and Manjunath 2004
Poloxamer407	Müller, Maaben et al. 1996
Poloxamer908	Müller, Maaben et al. 1996; Goppert and Müller 2003
Tween 20	Charcosset, El-Harati, and Fessi 2005
Tween 60	Charcosset, El-Harati, and Fessi 2005
Tween 80	Almeida et al. 1997; Liedtke et al. 2000; Müller, Lippacher et al. 2000; Goppert and Müller 2003; Mei et al. 2003
Span 20	Mei and Wu 2005
Span 60	Patravale and Ambarkhane 2003
Span 80	Müller, Lippacher et al. 2000
Span 85	Asasutjarit et al. 2007
Sodium cholate	Müller, Rühl, and Runge 1996; Almeida et al. 1997; Liu et al. 2007
Sodium glycocholate	Westesen and Bunjes 1995

16.3.2.1 High-Pressure Homogenization

Basically, two techniques are commonly used when using high-pressure homogenizers to produce SLNs: the hot homogenization technique and the cold homogenization technique.

Hot homogenization is carried out by melting the lipid material at a temperature higher than its melting point (usually 5°C–10°C higher). The encapsulated active molecules (usually lipophilic) are dissolved in the melted lipid phase. At the same time, the surfactant(s) and stabilizer(s) are dissolved in the water phase at the identical temperature. In the next step, the lipid phase and water phase are mixed together by high-speed or shear homogenization and/or ultrasound technique to form a pre-emulsion. Then the high-pressure homogenizer is used to further break down the pre-emulsion droplets. The product is cooled to room temperature to form SLNs.

For the cold homogenization process, the active molecules are also dissolved into the lipid, and the mixed solution is solidified, for example, by pouring into liquid nitrogen (zur Müllen, Schwarz, and Mehnert 1998) and then grinding to microparticles with a powder mill sized to a micrometer scale. Then the powder is dispersed in cold surfactant(s) or stabilizer(s) water solution to form a cold pre-emulsion. Next, the cold pre-emulsion is passed through a high-pressure homogenizer to produce SLNs. In general, SLNs obtained from a cold homogenization technique often possess larger particle sizes and broader size distributions than those obtained from a hot homogenization technique (Mehnert and Mäder 2001). The advantage of cold homogenization is that it reduces the thermal treatment of active compounds compared with the hot homogenization process, except for the first step to dissolve the compounds in the lipid phase.

The homogenization parameters (such as pressure and cycle numbers) varied between different SLN formulations. The mean SLN sizes ranging from tens to several hundred nanometers can be obtained by changing homogenization parameters. For example, studies in the Müller group revealed that for SLNs encapsulated with budesonide, a 1500-bar pressure and 10 cycle numbers led to a mean particle diameter of 511 nm; an increase of the cycle numbers to 15 led to a mean particle diameter of 462 nm, whereas increasing the pressure to 2500 bar with 10 cycles led to a mean particle diameter of 363 nm (Müller, Jacobs, and Kayser 2001). Thus, the SLN sizes can be controlled simply by adjusting the homogenization parameters, which may need several trial batches for the researchers to obtain a set of optimal parameters. The influence of these parameters has already been studied extensively (Schwarz et al. 1994; Liedtke et al. 2000).

16.3.2.2 Solvent Emulsification–Evaporation and Solvent Diffusion

SLNs can also be produced by the solvent emulsification–evaporation method (Sjöström and Bergenståhl 1992). Basically both the lipid matrix and the active compounds are dissolved in water-immiscible organic solvent (e.g., chloroform and cyclohexane) to form a homogeneous solution. Then the solution is emulsified in the aqueous phase containing surfactant(s) by high-speed/high-pressure homogenization or ultrasound technique. The organic solvent is then evaporated (usually under reduced pressure), and the lipid phase will automatically precipitate to form nanoparticle dispersion. The advantage of the solvent evaporation method is the avoidance of thermal treatment in all steps, thus preventing thermal degradation of the sensitive active compounds, although sometimes heat is used to facilitate the dissolving process. Furthermore, SLNs produced by the solvent evaporation method usually have smaller particle sizes (usually <500 nm) and narrower size distributions (Sjöström and Bergenståhl 1992) than SLNs from other methods. For example, particles of cholesteryl acetate have sizes as small as 25 nm when obtained by emulsifying in phosphatidylcholine and sodium glycocholate aqueous solutions (Sjöström and Bergenståhl 1992). However, an obvious disadvantage of this method is the usage of organic solvent. Although the organic solvent is evaporated in the last step, the possibility of remaining residues creates potential toxicological problems.

For the solvent diffusion method, partially water-miscible solvents are used to substitute water-immiscible solvents in the emulsification–evaporation method. First, the solvent and water are mutually saturated to obtain a thermodynamic equilibrium state. Then, lipid is dissolved in the water-saturated solvent, and surfactant is dissolved in the solvent-saturated water. The two phases are emulsified together either by homogenization or ultrasonication to form pre-emulsion. Finally, excess water is added to the pre-emulsion to extract the organic solvent to the continuous phase, thus facilitating the formation of nanoparticles. Same as the solvent emulsification–evaporation method, small particles with narrow particle size distribution can be obtained from the solvent diffusion method (Wissing, Kayser, and Müller 2004). Trotta, Debernardi, and Caputo (2003) prepared glyceryl monostearate particles with particle sizes ranging from 205–695 nm with different kinds of organic solvents and different lipid concentrations. The typical ratio between the warm microemulsion and excess water is from 1:25–1:50 (Mehnert and Mäder 2001).

16.3.2.3 Dilution of Microemulsion

SLNs produced by dilution of microemulsion were developed by the group of Gasco (1993) in Italy. The term *microemulsion* is defined as "a system of water, oil and amphiphile which is a single optically isotropic and thermodynamically stable liquid solution" (Danielsson and Lindman 1981, p. 391). Compared with traditional emulsions, the preparation of microemulsions does not need high-energy input such as homogenization. Simply mixing and stirring all the components is sufficient to produce a microemulsion system with thermodynamic stability. It should be mentioned that there are various theories concerning the formation, stability, and other unique properties of microemulsions. An extensive review of the structure, dynamics, and transport behaviors was published by Moulik and Paul (1998) in which different models and theories were discussed.

To produce SLNs based on the microemulsion method, first a warm microemulsion is formed by stirring the mixture of melted lipid, surfactant (e.g., Tween 80), cosurfactant (e.g., butanol), and water together at a temperature slightly higher than the melting point of the solid lipid. This warm microemulsion is then dispersed in excess cold water under mild stirring to ensure the formation of particles results from precipitation and not mechanical stirring (Müller, Mäder et al. 2000). According to previous studies (Boltri et al. 1993; Mehnert and Mäder 2001), the droplet structure is already contained in the microemulsion; thus, no energy is required to achieve small particle sizes. The feasibility to produce microemulsion makes the large-scale production of SLNs based on microemulsion dilution technique possible in industry, and this is under development at Vectorpharma in Italy (Müller, Mäder et al. 2000). Extensive studies have been done on the influence of experimental parameters (e.g., microemulsion composition and dispersing methods) on SLN sizes and structures.

However, researchers should be aware that a high concentration of surfactants is usually needed for the microemulsion method, which might not be desirable especially in regard to cost and regulatory purposes.

16.3.3 Applications

The major reason that SLNs are considered the next generation of delivery systems after liposomes (Müller and Dingler 1998) is that the solid matrix provides better protection for labile compounds from degradation. For example, the incorporation of ascorbyl palmitate (an oil-soluble derivative of vitamin C) in SLNs greatly improves its stability under oxidative stress as a result of a decreased rate of oxygen diffusion in the solid matrix (Kristl et al. 2003).

Another characteristic that makes SLNs suitable for oral applications is related to their adhesive properties. Once in the gastrointestinal tract (GIT) of humans, they are able to release the encapsulated compounds when the solid core is degraded by various enzymes found in the GIT. The degradation of the solid matrix can be controlled by adjusting the lipid and surfactant compositions. Olbrich and Müller (1999) used an in vitro degradation assay based on the combination of pancreatic lipase and its colipase to study the SLN degradation behavior. The results proved that both the length of the fatty acid chains in the triglycerides and the surfactant type can exert influence on the degradation rate of SLNs: (1) the longer the chain length, the slower the degradation rate; and (2) surfactant can either accelerate (e.g., cholic acid sodium salt) or hinder (e.g., polyoxyethylene-polyoxypropylene block copolymers) the degradation of the SLNs (Olbrich and Müller 1999). This feature shows promise in helping design formulations with a desirable release profile.

The solid form of SLNs (e.g., tablets and pellets) is sometimes more favorable than the liquid form because the dry form of SLNs has better physical stability during storage and can also protect the encapsulated compounds, which are easily hydrolyzed (Suoto and Müller 2006). Liquid SLNs can be transferred to a powder form by spray- or freeze-drying, and the powder can be used to produce tablets or fill capsules. For cost reasons, spray-drying is more popular than freeze-drying, and the addition of protectants is required for both techniques (Schwarz and Mehnert 1997; Freitas and Müller 1998).

During the past 10 years, the number of publications about SLNs has increased, which indicates a strong academic interest in exploring the advantages of SLNs. Most current studies about SLNs are focused on delivery systems for drugs, such as camptothecin (Yang, Zhu et al. 1999), piribedil (Demirel et al. 2001), diclofenac sodium (Attama, Reichl, and Müller-Goymann 2008), and insulin (Zhang et al. 2006). SLN's patent first entered the pharmaceutical industry in 1999 and was acquired by SkyeParma AG (Muttenz, Switzerland). After that, NLC technology based on modification of the SLN concept began to be introduced into the cosmetic industry with large-scale production lines in Germany (Rimpler GmbH, Wedemark, Germany). Nutraceutical products as food supplements to promote human health share many of the same interests of pharmaceutical products, such as the capability to improve physical and chemical stability of sensitive compounds and the capability to provide a controlled release profile; thus, SLNs will be a perspective delivery system for the nutraceutical industry in the future.

16.4 COACERVATION

16.4.1 Introduction

Coacervation is a phenomenon in which a macromolecular aqueous solution separates into two immiscible liquid phases. The more concentrated phase is called *coacervate*, which is relatively dense and is in equilibrium with the relatively dilute liquid phase (Wang, Li et al. 2007).

It was Tiebackx (1911) who first reported on the coacervation phenomenon, but Jong and Kruyt were the first to classify coacervation into two systems: simple and complex coacervation (Jong and Kruyt 1929; Schmitt et al. 1998; Weinbreck, Minor et al. 2004). The coacervation phenomenon is defined as simple or complex, depending on whether the system contains one or more biopolymer(s) (Schmitt et al. 1998; Weinbreck, Tromp et al. 2004).

Complex coacervation is based on the formation of a complex (coacervate) between oppositely charged polymers, usually proteins and polysaccharides (Schmitt et al. 1998; Kruif, Weinbreck, and Vries 2004). Biopolymer complexes (coacervates) combine the functional properties of each component, and the resulting complexes' functional properties are generally improved. Consequently, the interesting hydration (solubility, viscosity), structuration (aggregation, gelation), and surface properties (foaming, emulsifying) of these complexes can be used in a number of domains, such as macromolecular purification, encapsulation, food formulation (fat replacers, texturing agents), and synthesis of biomaterials (edible films) (Schmitt et al. 1998).

Because of the huge industrial potential of protein–polysaccharide coacervates, an increasing number of research teams have focused on the study of these complex coacervations. The commonly used biopolymers for forming coacervates are listed in Table 16.3.

16.4.2 Preparation

In the mixture system of a protein and a polysaccharide solution, one may observe either one of the following possibilities, as depicted in Figure 16.4.

Coacervates can be formed mainly because of electrostatic interaction between proteins and polysaccharides. Protein–polysaccharide complex coacervation through electrostatic interactions gives either soluble complexes in a stable solution or insoluble complexes, leading to phase separation (Ye 2008).

It is easy to produce coacervation (Weinbreck, Vries et al. 2003; Weinbreck, Tromp et al. 2004; Wang, Lee et al. 2007; Wang, Li et al. 2007). The stock solutions of coacervation are prepared by dissolving proteins and polysaccharides in deionized water. The resultant mixture is stirred for a determined time at a predetermined

Table 16.3 (Nonexhaustive) Overview of the Biopolymers Commonly Used for Coacervates

Polysaccharide/Protein	References
Gum Arabic/albumin	Burgess et al. 1991; Burgess and Singh 1993
Carbopol/gelatin	Elgindy and Elegakey 1981
Pectin/gelatin	McMullen, Newton, and Becker 1984
Gum arabic/gelatin	Burgess and Carless 1984
Acacia gum/gelatin	Peters et al. 1992
	Takenaka, Kawashimam, and Lin 1980; Suzuki and Kondo 1982; Dong and Rogers 1993; Jizomoto et al. 1993; Rabiskova et al. 1994; Tirkkonen, Turakka, and Paronen 1994
Gellan gum/gelatin	Chilvers and Morris 1987
Xanthan gum/gelatin	Lii et al. 2002
Sodium carboxymethyl guar gum/gelatin	Thimma and Tammishetti 2003
Acacia gum/β-lactoglobulin	Sanchez, Despond et al. 2001; Sanchez et al. 2002; Sanchez and Renard 2002
	Schmitt et al. 1999, 2000, 2001b; Schmitt, Sanchez et al. 2001a
Maltodextrins, starch, corn syrup solids/whey proteins	Young, Sarda, and Rosenberg 1993
Gum Arabic/whey proteins	Weinbreck, Nieuwenhuijse et al. 2003, 2004; Weinbreck, Vries et al. 2003; Weinbreck, Rollema et al. 2004; Weinbreck, Tromp et al. 2004; Weinbreck, Wientjes et al. 2004
Chitosan/alginate	Yan, Khor, and Lim 2000, 2001
ι-Carrageenan/poly(L-lysine)	Girod et al. 2004
κ-Carrageenan/fish gelatin	Haug et al. 2003

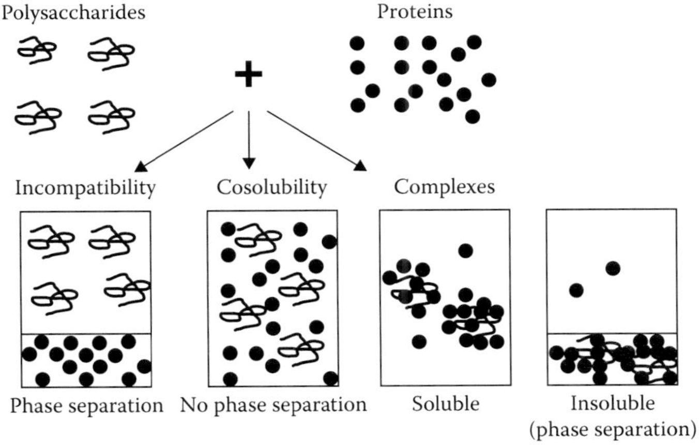

Figure 16.4 Main trends in the behavior of protein–polysaccharide mixture. (From Ye, A., *Int. J. Food Sci. Technol.*, 43, 406, 2008. With permission.)

temperature. Various concentrations of the mixtures can be obtained by diluting the stock solutions in deionized water. The mixtures can be further acidified or alkalinized to adjust their pH values; by adding microions, desired ionic strengths of the mixtures can also be obtained. The coacervates can be collected after removal of the supernatant through centrifugation.

The formation and stability of protein–polysaccharide complex coacervation are influenced by pH, ionic strength, charge densities of protein and polysaccharide, ratio of protein to polysaccharide, biopolymer concentration, and biopolymer molecular weight (Schmitt et al. 1998; Ye 2008). Besides the chemical parameters described previously, processing conditions (external physical and mechanical parameters), such as temperature, pressure, shearing rate, and time, can affect the formation and stability of these complexes.

16.4.2.1 Effect of pH

The value of pH is an important factor when forming complexes because protons can exert an effect on the ionization degree of the functional groups carried by the biopolymers, such as the amino group and the phosphate group. Xia and Dubin (1994) concluded that the maximum coacervation yield (containing a polysaccharide and a protein) can be obtained below the isoelectric point of the protein, more precisely at the electrical equivalence pH, at which the two biopolymers carry exactly opposite net charges resulting in a maximum electrostatic attraction.

The influence of pH on the formation of protein–polysaccharide complexes has been studied for more than 100 years. Kaibara et al. (2000) used turbidity and light-scattering methods, along with phase contrast microscopy, to study pH-induced coacervation in complexes of bovine serum albumin (BSA) and cationic poly-(dimethyldiallylammonium chloride) (PDADMAC) from pH 4–10. It was found that the state of macromolecular assembly of complexes between BSA and PDADMAC before and during the pH-induced coacervation can be characterized by specific pH values. It was also concluded that soluble primary BSA-PDADMAC complexes formed at the pH near the isoelectric pH of BSA based on the titration profiles. The influence of the pH has also been reported (Laos, Brownsey, and Ring 2007) in the study of the interactions between the algal polysaccharide furcellaran and the proteins bovine serum albumin and β-lactoglobulin at 0.03 M NaCl. The result showed that a soluble complex first formed between furcellaran and proteins with decreasing pH. As pH was further decreased, precipitation and complex coacervation occurred. Furthermore, the pH value at which the onset of the complex coacervation happened was in the vicinity of the isoelectric point of the proteins.

16.4.2.2 Effect of Ionic Strength

Ionic strength of biopolymer solution plays an important role in affecting complex coacervation. Electrostatic attractions between the biopolymers can mediate the formation of protein–polysaccharide complexes so the charge neutralization from ionic strength can hinder it (Schmitt et al. 1998). Coacervation is slightly affected by the salt concentration: when ionic strength is low, the number of charges carried by biopolymers is sufficient to allow electrostatic interactions; however, at high ionic strength, the charges carried by biopolymers are neutralized by the salt ions, which results in a decrease of electrostatic attraction between the biopolymers. Moreover, the counterions associated with the biopolymer before dissolution cannot be released in the medium at high salt concentrations, so the complexation is suppressed (Xia and Dubin 1994). Generally speaking, salt has a dissociating effect on all polyelectrolyte complexes (Kruif, Weinbreck, and Vries 2004).

An example of the effect of ionic strength on the coacervates of β-lactoglobulin with pectin is demonstrated by the experiments carried out by Wang, Lee et al. (2007). The initial protein–polysaccharide ratio was 5:1, and the result showed that the coacervate yield decreased with the increase of salt (NaCl) concentrations from 0.21–0.41 M. An interesting result has been found by Weinbreck, Vries et al. (2003) by studying the effect of NaCl concentrations on the complex coacervation of whey proteins and gum Arabic. It was found that the static light-scattering intensity corresponding to the soluble complex phase decreased with increasing ionic strength when salt concentration was above a certain value. However, the addition of a small amount of microions even slightly enhanced coacervation. Such a phenomenon was also found in previous studies carried out by Wang, Wang et al. (2007), Burgess (1990), and Weinbreck, Vries et al. (2003). It was explained that the addition of salt promoted the solubility of polymers (coiling of the molecule) at low salt concentrations; the access to the charge was then favored, and so is the electrostatic attraction.

16.4.2.3 Effect of Ratio of Protein to Polysaccharide and Biopolymer Concentration

Obviously, the ratio of protein to polysaccharide affects the behavior of complexes because it influences the charge balance of the complexes. For each complex coacervation system studied, the maximum coacervation yield is obtained at a specific protein/polysaccharide ratio at a given condition (pH and ionic strength). When one of the components (protein or polysaccharide) is in excess, soluble complexes are obtained because of the presence of non-neutralized charges. However, at high biopolymer concentrations, no coacervation occurs even when the polysaccharide or the protein is in excess (Ye, Flanagan, and Singh 2006). Weinbreck, Vries et al. (2003) gave the reason for this phenomenon that the increase of the biopolymer concentration is conducive to the release of more counterions to the solution, which screens the charges of the biopolymers, suppresses complexation, and increases the solubility of the complexes. Moreover, at high biopolymer concentrations, the system will show phase separation through thermodynamic incompatibility because of the competition between the macromolecules for the solvent (Tolstoguzov 1986, 1997). An example based on whey protein (WP) and xanthan gum (XG) complexes gave information about the effect of the protein/polysaccharide ratio on the complex characteristics (Laneuville, Paquin, and Turgeon 2000). It was found that the WP/XG ratio strongly influenced the size and the composition of the complexes.

A low WP/XG ratio produced principally small complexes, whereas in higher ratios the percentage of complexes with large particle sizes increased.

16.4.2.4 Effect of Charge Density of Protein and Polysaccharide

The number of charges present on the biopolymer per unit length can define the charge densities of the biopolymers. The strength of attractive coulombic interactions between protein and polysaccharide depends on the macromolecular charge densities to a large extent. Coacervation cannot occur under a critical charge density value.

16.4.2.5 Effect of Biopolymer Molecular Weight

Biopolymer molecular weight is also an influencing factor in the case of complex coacervation. Theoretically, biopolymer compatibility in solution decreases with the increase of the biopolymer molecular weight as a result of the decrease of the combinatorial entropy of mixing. For the complex coacervation phenomenon, Overbeek and Voorn (1957) predicted an increase in coacervation with the increase of the molecular weight of the biopolymers.

16.4.2.6 Effect of Processing Conditions

Processing conditions, including temperature, pressure, and shearing rate and time, can affect the formation and stability of protein–polysaccharide complexes.

The influence of temperature on the formation of complexes is a well-known phenomenon. The complex formation also results from hydrogen bonds, hydrophobic interactions of biopolymers, and covalent bonds. Hydrogen bond formation is enhanced by decreasing temperature. However, high temperature is conducive to hydrophobic interaction and covalent bonding because increasing temperature can result in the exposure of reactive sites via thermal denaturation of proteins and conformational changes of polysaccharide structure (Ye 2008).

High pressure also affects the formation of protein–polysaccharide complexes, which includes high hydrostatic pressure treatment (into vacuum chamber) and microfluidization treatment (into a microfluidization chamber) (Schmitt et al. 1998). The latter procedure combines high pressure with turbulence, cavitation, and shear phenomena. Fernandes and Raemy (1996) used high hydrostatic pressure (400, 600, and 800 MPa) treatment to change the structures of WP–polysaccharide (κ-carrageenan, XG, and high methyl ester pectin) mixed systems. It was demonstrated that a higher pressure does not always lead to a stronger gel.

Despite the technological importance of shear, few studies have been performed on this topic. It was difficult to obtain convergent results on the effect of the shear rate (Tirkkonen, Turakka, and Paronen 1994). An example about the effect of shearing time on the complex formation of β-lactoglobulin with acacia gum was given by Schmitt et al. (1998). The result showed that the number and size of the obtained coacervates were directly dependent on the shearing time, and a number of small-sized coacervates were obtained after short shearing times; in contrast, less numerous but larger coacervates were obtained after relative long shearing time.

16.4.3 Utilization of Coacervation in Encapsulation of Nutraceuticals

Coacervation is one of the typically used encapsulation methods. The research groups of Deasy (1984) in Ireland, Arshady (1990) in England, and Ijichi et al. (1997) in Japan used coacervation as a physicochemical procedure for the preparation of polymeric capsules. Protein–polysaccharide complex coacervation can also be used in food protection, packaging, and edible film formation. The main reasons for the use of protein–biopolymer coacervates in the food industry are the biodegradability and safety of these macromolecules.

Coacervation usually consists of three processes: emulsification, homogenization, and drying. Briefly, the sensitive nutraceutical (usually an oil) is emulsified in a protein solution. Then, a polysaccharide solution is poured into the emulsion and stirred for a predetermined time. The resultant mixture is homogenized using a high-speed homogenizer and/or high-pressure homogenizer to obtain a homogeneous suspension (this step is not necessary), followed by drying (if powder form is favored). The coacervate wall material coats or entraps the nutraceutical within it, providing a barrier to oxygen and water vapor.

Basically, three techniques are used in the final drying step: spray-drying, freeze-drying, and drum-drying. Each has different time and temperature regimens and can lead to different stability of products as a result of different wall thicknesses and densities (Desobry, Netto, and Labuza 1997). Spray-drying is the most common method used in drying encapsulation because the cost of spray-drying is 30–50 times less than freeze-drying, which is usually used in the flavor industry. Generally when using spray-drying, at least 80% of the total weight must be the wall material. If the wall material is not sufficiently thick or dense, the large surface area provided from spray-drying will enhance oxidation for the encapsulated compounds (Moreau and Rosenberg 1996). Desobry, Netto, and Labuza (1997) were the first to use the drum-drying method for encapsulation. They compared the effect of these three drying processes on pure β-carotene encapsulated with maltodextrin. Although drum-drying caused more initial loss in drying, the lower surface carotenoids and larger particle size resulted in greater stability compared with the other methods. Encapsulation of anhydrous milk fat by spray-drying in systems consisting of combinations of WP with carbohydrates (maltodextrin) has been studied by Young, Sarda, and Rosenberg (1993), and high encapsulation yield and efficiency were obtained. In all cases, spherical capsules were obtained in which the milk fat was physically isolated from the environment. The study proved that the combinations of WPs and carbohydrates were effective encapsulating agents. Weinbreck, Minor et al. (2004) encapsulated oils (sunflower oil, lemon and orange oil flavor) using WP–gum Arabic (GA) coacervates. The result showed successful formation of a smooth shell of WP-GA coacervate around the oil droplets at a specific pH (close to 4.0), and the payload of oil (i.e., amount of oil in the capsule) was higher than 80%.

Furthermore, at this pH (~4.0), the strength of the electrostatic interaction and the viscosity of the coacervates reached maximum values. Chen and Subirade (2005) used chitosan/β-lactoglobulin nanoparticles as nutraceutical carriers to encapsulate brilliant blue (BB). An in-vitro release study of BB into simulated gastric and intestinal fluids showed that less than 5% of the BB was released into the simulated gastric fluid in the first 30 minutes, implying that BB was protected from the stomach circumstances. Most of the BB was released in the simulated intestinal juice, implying that controlled release of BB can be achieved. The result showed that chitosan/β-lactoglobulin nanoparticles were promising biocompatible carriers for oral administration of sensitive nutraceuticals.

All the studies proved that coacervates can be successfully used for encapsulation purposes, and more work is needed to check the barrier capabilities of these coacervates against water and oxygen diffusion.

REFERENCES

Abrahamse, A. J., R. Van Lierop, R. G. M. van der Sman, A. van der Padt, and R. M. Boom. 2002. Analysis of Droplet Formation and Interactions during Cross-Flow Membrane Emulsification. *Journal of Membrane Science* 204: 125–37.

Almeida, A. J., S. Runge, and R. H. Muller. 1997. Peptide-Loaded Solid Lipid Nanoparticles (SLN): Influence of Production Parameters. *International Journal of Pharmaceutics* 149: 255–65.

Arshady, R. 1990. Microspheres and Microcapsules, a Survey of Manufacturing Techniques. Part II: Coacervation. *Polymer Engineering and Science* 31: 905–14.

Asasutjarit, R., S. I. Lorenzen, S. Sirivichayakul, K. Ruxrungtham, U. Ruktanonchai, and G. C. Ritthidej. 2007. Effect of Solid Lipid Nanoparticles Formulation Compositions on Their Size, Zeta Potential and Potential for in Vitro pHIS-HIV-Hugag Transfection. *Pharmaceutical Research* 24: 1098–107.

Attama, A. A., S. Reichl, and C. C. Müller-Goymann. 2008. Diclofenac Sodium Delivery to the Eye: In Vitro Evaluation of Novel Solid Lipid Nanoparticle Formulation Using Human Cornea Construct. *International Journal of Pharmaceutics* 355: 307–13.

Boltri, L., T. Canal, P. A. Esposito, and F. Carli. 1993. Lipid Nanoparticles: Evaluation of Some Critical Formulation Parameters. *Proceedings of the International Symposium on Controlled Release of Bioactive Materials* 20: 346–47.

Bouchemal, K., S. Briançon, E. Perrier, and H. Fessi. 2004. Nano-emulsion Formulation Using Spontaneous Emulsification: Solvent, Oil and Surfactant Optimisation. *International Journal of Pharmaceutics* 280: 241–51.

Brooks, B. W., and H. N. Richmond. 1994. Phase Inversion in Non-ionic Surfactant-Oil-Water Systems. I: The Effect of Transitional Inversion on Emulsion Drop Sizes. *Chemical Engineering Science* 49: 1053–64.

Brüsewitz, C., A. Schendler, A. Funke, T. Wagner, and R. Lipp. 2007. Novel Poloxamer-Based Nanoemulsions to Enhance the Intestinal Absorption of Active Compounds. *International Journal of Pharmaceutics* 329: 173–81.

Bunjes, H., K. Westesen, and M. H. J. Koch. 1996. Crystallization Tendency and Polymorphic Transitions in Triglyceride Nanoparticles. *International Journal of Pharmaceutics* 129: 159–73.

Burgess, D. J. 1990. Practical Analysis of Complex Coacervate Systems. *Journal of Colloid and Interface Science* 140: 227–38.

Burgess, D. J., and J. E. Carless. 1984. Microelectrophoretic Studies of Gelatin and Acacia for the Prediction of Complex Coacervation. *Journal of Colloid and Interface Science* 98: 1–8.

Burgess, D. J., K. K. Kwok, and P. T. Megremis. 1991. Characterization of Albumin-Acacia Complex Coacervation. *Journal of Pharmacy and Pharmacology* 43: 232–36.

Burgess, D. J., and O. N. Singh. 1993. Spontaneous Formation of Small Sized Albumin/Acacia Coacervate Particles. *Journal of Pharmacy and Pharmacology* 45: 586–91.

Cavalli, R., O. Caputo, M. E. Carlotti, M. Trotta, C. Scarnecchia, and M. R. Gasco. 1997. Sterilization and Freeze-Drying of Drug-Free and Drug-Loaded Solid Lipid Nanoparticles. *International Journal of Pharmaceutics* 148: 47–54.

Charcosset, C., A. El-Harati, and H. Fessi. 2005. Preparation of Solid Lipid Nanoparticles Using a Membrane Contactor. *Journal of Controlled Release* 108: 112–20.

Chen, C. C., and G. Wagner. 2004. Vitamin E Nanoparticle for Beverage Applications. *Chemical Engineering Research and Design* 82: 1432–37.

Chen, L., and M. Subirade. 2005. Chitosan/β-Lactoglobulin Core-Shell Nanoparticles as Nutraceutical Carriers. *Biomaterials* 26: 6041–53.

Chilvers, G. R., and V. J. Morris. 1987. Coacervation of Gelatin-Gellan Gum Mixtures and Their Use in Microencapsulation. *Carbohydrate Polymers* 7: 111–20.

Danielsson, I., and B. Lindman. 1981. The Definition of Microemulsion. *Colloids and Surfaces* 3: 391–92.

Date, A. A., and M. S. Nagarsenker. 2007. Design and Evaluation of Self-Nanoemulsifying Drug Delivery Systems (SNEDDS) for Cefpodoxime Proxetil. *International Journal of Pharmaceutics* 329: 166–72.

Deasy, P. B. 1984. *Microencapsulation and Related Drug Process*. New York: Marcel Dekker, 61–65.

DeFelice, S. L. 1995. The Nutraceutical Revolution: Its Impact on Food Industry R&D. *Trends in Food Science & Technology* 6: 59–61.

Demirel, M., Y. Yazan, R. H. Müller, F. Kiliç, and B. Bozan. 2001. Formulation and in Vitro-in Vivo Evaluation of Piribedil Solid Lipid Micro- and Nanoparticles. *Journal of Microencapsulation* 18: 359–71.

Desobry, S. A., F. Netto, and T. P. Labuza. 1997. Comparison of Spray-Drying, Drum-Drying and Freeze-Drying for β-Carotene Encapsulation and Preservation. *Journal of Food Science* 62: 1158–62.

Dias, M., J. Carvalho, D. Rodrigues, S. Graziani, and R. Maranhão. 2007. Pharmacokinetics and Tumor Uptake of a Derivatized Form of Paclitaxel Associated to a Cholesterol-Rich Nanoemulsion (LDE) in Patients with Gynecologic Cancers. *Cancer Chemotherapy and Pharmacology* 59: 105–11.

Dong, C., and J. A. Rogers. 1993. Acacia-Gelatin Microencapsulated Liposomes: Preparation, Stability, and Release of Acetylsalicylic Acid. *Pharmaceutical Research* 10: 141–46.

Elgindy, M. A., and M. A. Elegakey. 1981. Carbopol-Gelatin Coacervation: Influence of Some Variables. *Drug Development and Industrial Pharmacy* 7: 587–603.

Fernandes, P. B., and A. Raemy. 1996. High Pressure Treatment of Whey Protein/Polysaccharide Systems. *Progress in Biotechnology* 13: 337–42.

Forgiarini, A., J. Esquena, C. Gonzàlez, and C. Solans. 2000. Studies of the Relation between Phase Behavior and Emulsification Methods with Nanoemulsion Formation. *Progress in Colloid & Polymer Science* 115: 36–39.

Freitas, C., and R. H. Müller. 1998. Spray-Drying of Solid Lipid Nanoparticles (SLNTM). *European Journal of Pharmaceutics and Biopharmaceutics* 46: 145–51.

Ganachaud, F., and J. L. Katz. 2005. Nanoparticles and Nanocapsules Created Using the Ouzo Effect: Spontaneous Emulsification as an Alternative to Ultrasonic and High-Shear Devices. *ChemPhysChem* 6: 209–16.

Gasco, M. R. 1993. Method for Producing Solid Lipid Microspheres Having a Narrow Size Distribution. US Patent 5,250,236, filed August 2, 1991, issued October 5, 1993.

Girod, S., M. Boissière, K. Longchambon, S. Begu, C. Tourne-Petheil, and J. M. Devoisselle. 2004. Polyelectrolyte Complex Formation between Iota-Carrageenan and Poly(L-lysine) in Dilute Aqueous Solutions: A Spectroscopic and Conformational Study. *Carbohydrate Polymers* 55: 37–45.

Goldberg, I. ed. 1994. *Functional Foods, Designer Foods, Pharmafoods, Nutraceuticals.* London: Chapman & Hall.

Goppert, T. M., and R. H. Müller. 2003. Plasma Protein Adsorption of Tween 80-and Poloxamer 188-Stabilized Solid Lipid Nanoparticles. *Journal of Drug Targeting* 11: 225–32.

Hardy, G. 2000. Nutraceuticals and Functional Foods: Introduction and Meaning. *Nutrition* 16: 688–89.

Haug, I., M. A. K. Williams, L. Lundin, O. Smidsrød, and K. I. Draget. 2003. Molecular Interactions in, and Rheological Properties of, a Mixed Biopolymer System Undergoing Order/Disorder Transitions. *Food Hydrocolloids* 17: 439–44.

Ijichi, K., H. Yoshizawa, Y. Uemura, Y. Hatate, and Y. Kawano. 1997. Multi-layered Gelatin/Acacia Microcapsules by Complex Coacervation. *Journal of Chemical Engineering of Japan* 30: 793–98.

Izquierdo, P., J. Feng, J. Esquena, T. F. Tadros, J. C. Dederen, M. J. Garcia-Celma, N. Azemar, and C. Solans. 2005. The Influence of Surfactant Mixing Ratio on Nano-emulsion Formation by the Pit Method. *Journal of Colloid and Interface Science* 285: 388–94.

Jafari, S. M., Y. He, and B. Bhandari. 2006. Nano-emulsion Production by Sonication and Microfluidization—A Comparison. *International Journal of Food Properties* 9: 475–85.

Jafari, S. M., Y. He, and B. Bhandari. 2007. Effectiveness of Encapsulating Biopolymers to Produce Sub-micron Emulsions by High Energy Emulsification Techniques. *Food Research International* 40: 862–73.

Jenning, V., and S. H. Gohla. 2001. Encapsulation of Retinoids in Solid Lipid Nanoparticles (SLN). *Journal of Microencapsulation* 18: 149–58.

Jizomoto, H., E. Kanaoka, K. Sugita, and K. Hirano. 1993. Gelatin-Acacia Microcapsules for Trapping Micro Oil Droplets Containing Lipophilic Drugs and Ready Disintegration in the Gastrointestinal Tract. *Pharmaceutical Research* 10: 1115–22.

Jong, H. G. B. d., and H. R. Kruyt. 1929. Coacervation (Partial Miscibility in Colloid Systems). *Proceedings of the Koninklijke Nederlandse Akademie Van Wetenschappen* 32: 849–56.

Jores, K., W. Mehnert, M. Drechsler, H. Bunjes, C. Johann, and K. Mäder 2004. Investigations on the Structure of Solid Lipid Nanoparticles (SLN) and Oil-Loaded Solid Lipid Nanoparticles by Photon Correlation Spectroscopy, Field-Flow Fractionation and Transmission Electron Microscopy. *Journal of Controlled Release* 95: 217–27.

Kaibara, K., T. Okazaki, H. B. Bohidar, and P. L. Dubin. 2000. pH-induced Coacervation in Complexes of Bovine Serum Albumin and Cationic Polyelectrolytes. *Biomacromolecules* 1: 100–07.

Kristl, J., B. Volk, M. Gasperlin, M. Sentjurc, and P. Jurkovic. 2003. Effect of Colloidal Carriers on Ascorbyl Palmitate Stability. *European Journal of Pharmaceutical Sciences* 19: 181–89.

Kruif, C. G. d., F. Weinbreck, and R. Vries. 2004. Complex Coacervation of Proteins and Anionic Polysaccharides. *Current Opinion in Colloid & Interface Science* 9: 340–49.

Landfester, K., J. Eisenblätter, and R. Rothe. 2004. Preparation of Polymerizable Miniemulsions by Ultrasonication. *Journal of Coatings Technology and Research* 1: 65–68.

Laneuville, S. I., P. Paquin, and S. L. Turgeon. 2000. Effect of Preparation Conditions on the Characteristics of Whey Protein-Xanthan Gum Complexes. *Food Hydrocolloids* 14: 305–14.

Laos, K., G. J. Brownsey, and S. Ring. 2007. Interactions between Furcellaran and the Globular Proteins Bovine Serum Albumin and β-Lactoglobulin. *Carbohydrate Polymers* 67: 116–23.

Lawrence, M. J., and G. D. Rees. 2000. Microemulsion-Based Media as Novel Drug Delivery Systems. *Advanced Drug Delivery Reviews* 45: 89–121.

Liedtke, S., S. Wissing, R. H. Muller, and K. Mader. 2000. Influence of High Pressure Homogenisation Equipment on Nanodispersions Characteristics. *International Journal of Pharmaceutics* 196: 183–85.

Lii, C.-Y., S. C. Liaw, V. M. F. Lai, and P. Tomasik. 2002. Xanthan Gum-Gelatin Complexes. *European Polymer Journal* 38: 1377–81.

Liu, J., T. Gong, C. Wang, Z. Zhong, and Z. Zhang. 2007. Solid Lipid Nanoparticles Loaded with Insulin by Sodium Cholate-Phosphatidylcholine-Based Mixed Micelles: Preparation and Characterization. *International Journal of Pharmaceutics* 340: 153–62.

Maa, Y. F., and C. C. Hsu. 1999. Performance of Sonication and Microfluidization for Liquid-Liquid Emulsification. *Pharmaceutical Development and Technology* 4: 233–40.

Manjunath, K., J. S. Reddy, and V. Venkateswarlu. 2005. Solid Lipid Nanoparticles as Drug Delivery Systems. *Methods and Findings in Experimental and Clinical Pharmacology* 27: 127–44.

McMullen, J. N., D. W. Newton, and C. H. Becker. 1984. Pectin-Gelatin Complex Coacervates. II. Effect of Microencapsulated Sulfamerazine on Size, Morphology, Recovery, and Extraction of Water-Dispersible Microglobules. *Journal of Pharmaceutical Sciences* 73: 1799–803.

Mehnert, W., and K. Mäder. 2001. Solid Lipid Nanoparticles: Production, Characterization and Applications. *Advanced Drug Delivery Reviews* 47: 165–96.

Mei, Z., H. Chen, T. Weng, Y. Yang, and X. Yang. 2003. Solid Lipid Nanoparticle and Microemulsion for Topical Delivery of Triptolide. *European Journal of Pharmaceutics and Biopharmaceutics* 56: 189–96.

Mei, Z., and Q. Wu. 2005. Triptolide Loaded Solid Lipid Nanoparticle Hydrogel for Topical Application. *Journal of Drug Targeting* 31: 161–68.

Morales, D., J. M. Gutierrez, M. J. García-Celma, and Y. C. Solans. 2003. A Study of the Relation between Bicontinuous Microemulsions and Oil/Water Nano-emulsion Formation. *Langmuir* 19: 7196–200.

Moreau, D. L., and M. Rosenberg. 1996. Oxidative Stability of Anhydrous Microencapsulated in Whey Proteins. *Journal of Food Science* 61: 39–43.

Morel, S., E. Ugazio, R. Cavalli, and M. R. Gasco. 1996. Thymopentin in solid lipid nanoparticles. *International Journal of Pharmaceutics* 132: 259–61.

Moulik, S. P., and B. K. Paul. 1998. Structure, Dynamics and Transport Properties of Microemulsions. *Advances in Colloid and Interface Science* 78: 99–195.

Müller, R. H., and A. Dingler. 1998. The Next Generation after the Liposomes: Solid Lipid Nanoparticles (SLN, Lipopearls) as Dermal Carrier in Cosmetics. *Eurocosmetics* 7: 19–26.

Müller, R. H., C. Jacobs, and O. Kayser. 2001. Nanosuspensions as Particulate Drug Formulations in Therapy: Rationale for Development and What We Can Expect for the Future. *Advanced Drug Delivery Reviews* 47: 3–19.

Müller, R. H., A. Lippacher, and S. Gohla. 2000. Solid Lipid Nanoparticles (SLN) as a Carrier System for the Controlled Release of Drugs. *Handbook of Pharmaceutical Controlled Release Technology*. Edited D. L. Wise. Boca Raton, FL: CRC, 377.

Müller, R. H., and J. S. Lucks. 1996. Arzneistoffträger aus festen Lipidteilchen, Feste Lipidnanosphären (SLN). E. Patent 0605497.

Müller, R. H., S. Maassen, H. Weyhers, and W. S. Mehnert. 1996. Phagocytic Uptake and Cytotoxicity of Solid Lipid Nanoparticles (SLN) Sterically Stabilized with Poloxamine 908 and Poloxamer 407. *Journal of Drug Targeting* 4: 161–70.

Müller, R. H., K. Mäder, and S. Gohla. 2000. Solid Lipid Nanoparticles (SLN) for Controlled Drug Delivery—A Review of the State of the Art. *European Journal of Pharmaceutics and Biopharmaceutics* 50: 161–77.

Müller, R. H., M. Radtke, and S. A. Wissing. 2002. Nanostructured Lipid Matrices for Improved Microencapsulation of Drugs. *International Journal of Pharmaceutics* 242: 121–28.

Müller, R. H., D. Rühl, and S. Runge. 1996. Biodegradation of Solid Lipid Nanoparticles as a Function of Lipase Incubation Time. *International Journal of Pharmaceutics* 144: 115–21.

Nakashima, T., M. Shimizu, and M. Kukizaki. 2000. Particle Control of Emulsion by Membrane Emulsification and Its Applications. *Advanced Drug Delivery Reviews* 45: 47–56.

Olbrich, C., and R. H. Müller. 1999. Enzymatic Degradation of SLN—Effect of Surfactant and Surfactant Mixtures. *International Journal of Pharmaceutics* 180: 31–39.

Overbeek, J. T. J., and M. J. Voorn. 1957. Phase Separation in Polyelectrolyte Solution. Theory of Complex Coacervation. *Journal Cellular Comparative Physiology* 49: 7.

Paquin, P. 1999. Technological Properties of High Pressure Homogenizers: The Effect of Fat Globules, Milk Proteins, and Polysaccharides. *International Dairy Journal* 9: 329–35.

Patravale, V. B., and A. V. Ambarkhane. 2003. Study of Solid Lipid Nanoparticles with Respect to Particle Size Distribution and Drug Loading. *Pharmazie* 58: 392–95.

Peters, H. J. W., E. M. G. v. Bommel, and J. G. Fokkens. 1992. Effect of Gelatin Properties in Complex Coacervation Processes. *Drug Development and Industrial Pharmacy* 18: 123–34.

Pinnamaneni, S., N. G. Das, and S. K. Das. 2003. Comparison of Oil-in-Water Emulsions Manufactured by Microfluidization and Homogenization. *Pharmazie* 58: 554–58.

Rabiskova, M., J. Song, F. O. Opawale, and D. J. Burgess. 1994. The Influence of Surface Properties on Uptake of Oil into Complex Coacervates Microcapsules. *Journal of Pharmacy and Pharmacology* 46: 631–35.

Robin, O., V. Blanchot, J. C. Vuillemard, and P. Paquin. 1992. Microfluidization of Dairy Model Emulsions. I. Preparation of Emulsions and Influence of Processing and Formulation on the Size Distribution of Milk Fat Globules. *Le Lait* 72: 511–31.

Sanchez, C., S. Despond, C. Schmitt, and J. Hardy. 2001. Effect of Heat and Shear on β-Lactoglobulin-Acacia Gum Complex Coacervation. *Food Colloids, Fundamentals of Formulation.* Edited by E. Dickinson and R. Miller. Cambridge, UK: Royal Society of Chemistry, 332–43.

Sanchez, C., G. Mekhloufi, C. Schmitt, D. Renard, P. Robert, C. M. Lehr, A. Lamprecht, and J. Hardy. 2002. Self-Assembly of β-Lactoglobulin and Acacia Gum in Aqueous Solvent: Structure and Phase-Ordering Kinetics. *Langmuir* 18: 10323–33.

Sanchez, C., and D. Renard. 2002. Stability and Structure of Protein-Polysaccharide Coacervates in the Presence of Protein Aggregates. *International Journal of Pharmaceutics* 242: 319–24.

Sanguansri, P., and M. A. Augustin. 2006. Nanoscale Materials Development—A Food Industry Perspective. *Trends in Food Science & Technology* 17: 547–56.

Schmitt, C., C. Sanchez, S. Desobry-Banon, and J. Hardy. 1998. Structure and Technofunctional Properties of Protein-Polysaccharide Complexes: A Review. *Critical Reviews in Food Science and Nutrition* 38: 689–753.

Schmitt, C., C. Sanchez, D. Despond, D. Renard, P. Robert, and J. Hardy. 2001a. Structural Modification of β-Lactoglobulin as Induced by Complex Coacervation with Acacia Gum. *Food Colloids, Fundamentals of Formulation.* Edited by E. Dickinson and R. Miller. Cambridge, UK: Royal Society of Chemistry, 323–31.

Schmitt, C., C. Sanchez, S. Despond, D. Renard, F. Thomas, and J. Hardy. 2000. Effect of Protein Aggregates on the Complex Coacervation between β-Lactoglobulin and Acacia Gum at pH 4.2. *Food Hydrocolloids* 14: 403–13.

Schmitt, C., C. Sanchez, A. Lamprecht, D. Renard, C.-M. Lehr, C. G. de Kruif, and J. Hardy. 2001b. Study of β-Lactoglobulin/Acacia Gum Complex Coacervation by Diffusing-Wave Spectroscopy and Confocal Scanning Laser Microscopy. *Colloids and Surfaces B: Biointerfaces* 20: 267–80.

Schmitt, C., C. Sanchez, F. Thomas, and J. Hardy. 1999. Complex Coacervation between β-Lactoglobulin and Acacia Gum in Aqueous Medium. *Food Hydrocolloids* 13: 483–96.

Schulz, M. B., and R. Daniels. 2000. Hydroxypropylmethylcellulose (HPMC) as Emulsifier for Submicron Emulsions: Influence of Molecular Weight and Substitution Type on the Droplet Size after High-Pressure Homogenization. *European Journal of Pharmaceutics and Biopharmaceutics* 49: 231–36.

Schwarz, C., and W. Mehnert. 1997. Freeze-Drying of Drug-Free and Drug-Loaded Solid Lipid Nanoparticles (SLN). *International Journal of Pharmaceutics* 157: 171–79.

Schwarz, C., W. Mehnert, J. S. Lucks, and R. H. Müller. 1994. Solid Lipid Nanoparticles (SLN) for Controlled Drug Delivery. I: Production, Characterization and Sterilization. *Journal of Controlled Release* 30: 83–96.

Sjöström, B., and B. Bergenståhl. 1992. Preparation of Submicron Drug Particles in Lecithin-Stabilized o/w Emulsions I. Model Studies of the Precipitation of Cholesteryl Acetate. *International Journal of Pharmaceutics* 88: 53–62.

Solans, C., P. Izquierdo, J. Nolla, N. Azemar, and M. J. Garcia-Celma. 2005. Nano-emulsions. *Current Opinion in Colloid & Interface Science* 10: 102–10.

Souto E. B. and R. H. Müller. 2006. Applications of Lipid Nanoparticles in Food Industry. *Journal of Food Technology* 4: 90–95.

Spernath, A., and A. Aserin. 2006. Microemulsions as Carriers for Drugs and Nutraceuticals. *Advances in Colloid and Interface Science* 128–130: 47–64.

Suzuki, S., and T. Kondo. 1982. Interaction of Gelatin-Acacia Microcapsules with Surfactants. *Colloids and Surfaces* 4: 163–71.

Sznitowska, M., M. Gajewska, S. Janicki, A. Radwanska, and G. Lukowski. 2001. Bioavailability of Diazepam from Aqueous-Organic Solution, Submicron Emulsion and Solid Lipid Nanoparticles after Rectal Administration in Rabbits. *European Journal of Pharmaceutics and Biopharmaceutics* 52: 159–63.

Tadros, T., P. Izquierdo, J. Esquena, and C. Solans. 2004. Formation and Stability of Nano-emulsions. *Advances in Colloid and Interface Science* 108–109: 303–18.

Takenaka, H., Y. Kawashima, and S. Lin. 1980. Micromeritic Properties of Sulfamethoxazole Microcapsules Prepared by Gelatin-Acacia Coacervation. *Journal of Pharmaceutical Sciences* 69: 513–16.

Thimma, R. T., and S. Tammishetti. 2003. Study of Complex Coacervation of Gelatin with Sodium Carboxymethyl Guar Gum: Microencapsulation of Clove Oil and Sulphamethoxazole. *Journal of Microencapsulation* 20: 203–10.

Tiebackx, F. W. Z. 1911. Simultaneous Coagulation of Two Colloids. *Zeitschrift fuer Chemie und Industrie der Kolloide* 8: 198–201.

Tirkkonen, S., L. Turakka, and P. Paronen. 1994. Microencapsulation of Indomethacin by Gelatin-Acacia Complex Coacervation in Presence of Surfactants. *Journal of Microencapsulation* 11: 615–26.

Tolstoguzov, V. B. 1986. Functional Properties of Protein-Polysaccharide Mixtures. *Functional Properties of Macromolecles*. Edited J. R. Mitchell and D. A. Ledward. London: Elsevier Applied Science Publishers, 385–415.

Tolstoguzov, V. B. 1997. Protein-Polysaccharide Interactions. *Food Proteins and Their Applications*. Edited by S. Damodaran and A. Paraf. New York: Marcel Dekker, 171–98.

Trotta, M., F. Debernardi, and O. Caputo. 2003. Preparation of Solid Lipid Nanoparticles by a Solvent Emulsification-Diffusion Technique. *International Journal of Pharmaceutics* 257: 153–60.

Velikov, K. P., and E. Pelan. 2008. Colloidal Delivery Systems for Micronutrients and Nutraceuticals. *Soft Matter* 4: 1964–80.

Venkateswarlu, V., and K. Manjunath. 2004. Preparation, Characterization and in Vitro Release Kinetics of Clozapine Solid Lipid Nanoparticles. *Journal of Controlled Release* 95: 627–38.

Wang, X., Y. Jiang, Y.-W. Wang, M.-T. Huang, C.-T. Ho, and Q. Huang. 2007. Enhancing Anti-inflammation Activity of Curcumin through O/W Nanoemulsions. *Food Chemistry* 108: 419–24.

Wang, X., J. Lee, Y. W. Wang, and Q. R. Huang. 2007. Composition and Rheological Properties of β-Lactoglobulin/Pectin Coacervates: Effects of Salt Concentration and Initial Protein/Polysaccharide Ratio. *Biomacromolecules* 8: 992–97.

Wang, X., Y. Li, J. Lal, and Q. Huang. 2007. Microstructure of β-Lactoglobulin/Pectin Coacervates Studied by Small Angle Neutron. *Journal of Physical Chemistry B* 111: 515–20.

Wang, X., Y.-W. Wang, and H. Qingrong. 2009. Enhancing Stability and Oral Bioavailability of Polyphenols Using Nanoemulsions. *Micro/Nanoencapsulation of Active Food Ingredients*. Washington, DC: American Chemical Society, 198–212.

Wang, X., Y.-W. Wang, C. Ruengruglikit, and Q. Huang. 2007. Effects of Salt Concentration on Formation and Dissociation of β-Lactoglobulin/Pectin Complexes. *Journal of Agricultural and Food Chemistry* 55: 10432–36.

Weinbreck, F., M. Minor, and C. G. de Kruif. 2004. Microencapsulation of Oils Using Whey Protein/Gum Arabic Coacervates. *Journal of Microencapsulation* 21: 667–79.

Weinbreck, F., H. Nieuwenhuijse, G. W. Robijn, and C. G. de Kruif. 2003. Complex Formation of Whey Proteins: Exocellular Polysaccharide EPS B40. *Langmuir* 19: 9404–10.

Weinbreck, F., H. Nieuwenhuijse, G. W. Robijn, and C. G. de Kruif. 2004. Complexation of Whey Proteins with Carrageenan. *Journal of Agricultural and Food Chemistry* 52: 3550–55.

Weinbreck, F., H. S. Rollema, R. H. Tromp, and C. G. de Kruif. 2004. Diffusivity of Whey Protein and Gum Arabic in Their Coacervates. *Langmuir* 20: 6389–95.

Weinbreck, F., R. H. Tromp, and C. G. de Kruif. 2004. Composition and Structure of Whey Protein/Gum Arabic Coacervates. *Biomacromolecules* 5: 1437–45.

Weinbreck, F., R. de Vries, P. Schrooyen, and C. G. de Kruif CG. 2003. Complex Coacervation of Whey Proteins and Gum Arabic. *Biomacromolecules* 4: 293–303.

Weinbreck, F., R. H. W. Wientjes, H. Nieuwenhuijse, G. W. Robijn, and C. G. de Kruif. 2004. Rheological Properties of Whey Protein/Gum Arabic Coacervates. *Journal of Rheology* 48: 1215–28.

Weiss, J., E. A. Decker, J. McClements, K. Kristbergsson, T. Helgason, and T. Awad. 2008. Solid Lipid Nanoparticles as Delivery Systems for Bioactive Food Components. *Food Biophysics* 3: 146–54.

Westesen, K., and H. Bunjes. 1995. Do Nanoparticles Prepared from Lipids Solid at Room Temperature Always Possess a Solid Lipid Matrix? *International Journal of Pharmaceutics* 115: 129–31.

Westesen, K., H. Bunjes, and M. H. J. Koch. 1997. Physicochemical Characterization of Lipid Nanoparticles and Evaluation of Their Drug Loading Capacity and Sustained Release Potential. *Journal of Controlled Release* 48: 223–36.

Wildman, R. E. C. 2007. *Handbook of Nutraceuticals and Functional Foods*. Boca Raton, FL: CRC Press.

Wissing, S. A., O. Kayser, and R. H. Müller. 2004. Solid Lipid Nanoparticles for Parenteral Drug Delivery. *Advanced Drug Delivery Reviews* 56: 1257–72.

Wissing, S. A., and R. H. Müller. 2003. The Influence of Solid Lipid Nanoparticles on Skin Hydration and Viscoelasticity—In Vivo Study. *European Journal of Pharmaceutics and Biopharmaceutics* 56: 67–72.

Wulff-Pérez, M., Torcello-Gómez., M. J. Gálvez-Ruíz, and A. Martín-Rodríguez. 2009. Stability of Emulsions for Parenteral Feeding: Preparation and Characterization of O/W Nanoemulsions with Natural Oils and Pluronic f68 as Surfactant. *Food Hydrocolloids* 23: 1096–102.

Xia, J., and P. L. Dubin. 1994. Protein-Polyelectrolyte Complexes. *Macromolecular Complexes in Chemistry and Biology.* Edited by P. L. Dubin, J. Bock, R. Davis, D. N. Schulz, and C. Thies. Berlin: Springer-Verlag, 247–71.

Yan, X., E. Khor, and L. Y. Lim. 2000. PEC Film Prepared from Chitosan-Alginate Coacervates. *Chemical & Pharmaceutical Bulletin* 48: 941–46.

Yan, X., E. Khor, and L. Y. Lim. 2001. Chitosan-Alginate Films Prepared with Chitosans of Different Molecular Weights. *Journal of Biomedical Materials Research* 58: 358–65.

Yang, S., J. Zhu, Y. Lu, B. Liang, and C. Yang. 1999. Body Distribution of Camptothecin Solid Lipid Nanoparticles after Oral Administration. *Pharmaceutical Research* 16: 751–57.

Yang, S. C., L. F. Lu, Y. Cai, J. B. Zhu, B. W. Liang, and C. Z. Yang. 1999. Body Distribution in Mice of Intravenously Injected Camptothecin Solid Lipid Nanoparticles and Targeting Effect on Brain. *Journal of Controlled Release* 59: 299–307.

Ye, A. 2008. Complexation between Milk Proteins and Polysaccharides via Electrostatic Interaction: Principles and Applications—A Review. *International Journal of Food Science and Technology* 43: 406–15.

Ye, A., J. Flanagan, and H. Singh. 2006. Formation of Stable Nanoparticles via Electrostatic Complexation between Sodium Caseinate and Gum Arabic. *Biopolymers* 82: 121–33.

Young, S. L., X. Sarda, and M. Rosenberg. 1993. Microencapsulating Properties of Whey Proteins. II. Combination of Whey Proteins with Carbohydrates. *Journal of Dairy Science* 76: 2878–85.

Zhang, N., Q. Ping, G. Huang, W. Xu, Y. Cheng, and X. Han. 2006. Lectin-Modified Solid Lipid Nanoparticles as Carriers for Oral Administration of Insulin. *International Journal of Pharmaceutics* 327: 153–59.

zur Müllen, A., C. Schwarz, and W. Mehnert. 1998. Solid Lipid Nanoparticles (SLN) for Controlled Drug Delivery—Drug Release and Release Mechanism. *European Journal of Pharmaceutics and Biopharmaceutics* 45: 149–55.

CHAPTER 17

Apple Nutraceuticals and the Effects of Processing, and Their Therapeutic Applications

John Eatough and Brian Lockwood

CONTENTS

17.1 INTRODUCTION

Apples are one of the most commonly consumed fruits and are the prime source of flavonoids in the Western diet (Biedrzycka et al. 2008). Furthermore, when compared with other commonly consumed fruits, apples are ranked highest for their free phenolic content and second highest for both their total phenolic content and their level of antioxidant activity (Boyer et al. 2004). These dietary polyphenolic nutraceuticals have received significant attention because of their possible roles in the prevention of a variety of degenerative conditions such as cardiovascular disease and cancers (Scalbert et al. 2002).

17.2 APPLE POLYPHENOLS

Polyphenols are the major nutraceuticals in apples. They have diverse chemical structures, which can be characterized by the extent of oxidation of the oxygen heterocycle and the positions of hydroxylation on the phenolic moieties (Scalbert et al. 2000). The polyphenols that are present in apples have chemical structures that are ideally suited to act as free-radical scavengers. For example, in vitro they can act against superoxide, hydroxyl radicals, peroxyl radicals, hypochlorous acid, and peroxynitrous acid (Halliwell et al. 2005). Examples of apple polyphenols include the flavonoids such as flavanols ([−]-epicatechin and [+]-catechin), flavonols (quercetin and its glycoside rutin), as well as dihydrochalcones (phloretin and its glycoside phloridzin) and nonflavonoids such as hydroxycinnamates (chlorogenic acid) (Lotito et al. 2004b). Figure 17.1 shows the structures of major apple polyphenols.

Polymeric flavanols called *procyanidins* (proanthocyanidins) are also present in apples (Scalbert et al. 2000) and comprise heterogeneous groups of catechins (Lotito et al. 2004b), and polymerization ranges from 4–11 units (Scalbert et al. 2000). In addition, apples also contain carotenoids such as neoxanthin, violaxanthin, luteoxanthin, antheraxanthin, and lutein, which are found both esterified with fatty acids as well as unesterified, but limited biological activity has been attributed to these (Molnar et al. 2005). Apple polyphenols are more water soluble in their glycosidic form compared with their free form (Rice-Evans et al. 1997). However, the glycosidic bond can be cleaved, releasing the aglycone. Even though the majority of polyphenols that are found in apples appear as glycosides, this is not the case for catechins (Lotito et al. 2004b).

17.3 FACTORS AFFECTING POLYPHENOL LEVELS

The total polyphenol content varies between apple varieties and will also depend on a number of factors. Maturity, season, cultivars, yield, and region can all determine the polyphenol content (Alberto et al. 2006). More specifically, polyphenol levels may be influenced by agronomic practices such as irrigation, fertilization, and pest management. Furthermore, environmental conditions such as temperature, light, moisture, and ultraviolet radiation will also influence the polyphenol content (Erdman et al. 2007). Maturation of the fruit is affected by both temperature and light, and phenolic levels are highest 3 weeks after flowering. As apples become riper, the amounts of polyphenols will decrease (Biedrzycka et al. 2008). In addition

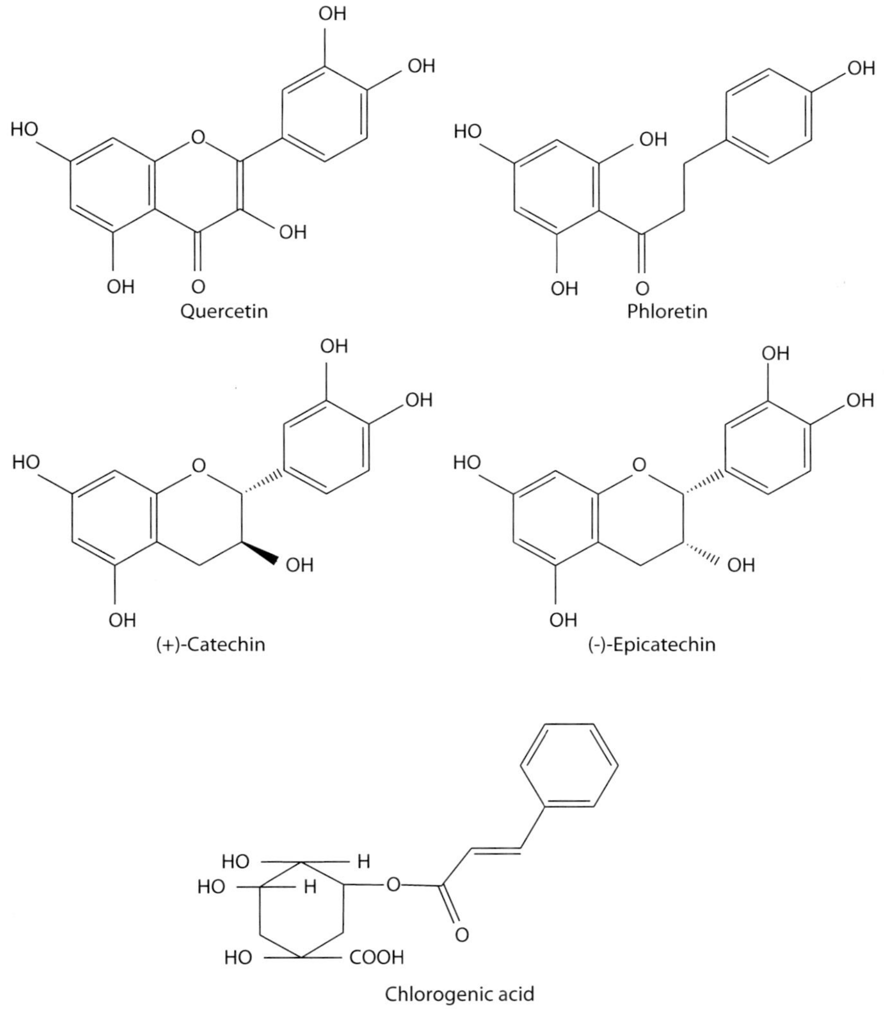

Figure 17.1 Common apple polyphenols.

to these factors and postharvest handling and storage conditions, apples will also show differences in levels of polyphenols (Erdman et al. 2007). Different varieties of apples will have different total phenolic contents, and overall apple nutraceutical composition varies markedly, depending on the cultivar; there is a 1.6-fold difference in total phenolics when measured over 10 cultivars, but a 10-fold range in chlorogenic acid levels (Kalt 2005). Further systematic investigations into the effects of cultivar, year of harvest, and storage conditions have been carried out (van der Sluis et al. 2001). Apples from Jonagold, Golden Delicious, Cox's Orange, and Elstar were analyzed for composition of phenolics and antioxidant activity. Total quercetin

Table 17.1 Concentration of Phenolics in Four Apple Cultivars in Milligrams/Kilograms Fresh Weight

Components	Jonagold	Golden Delicious	Cox's Orange	Elstar
Total quercetin glycosides	98 ± 16	67 ± 16	54 ± 15	60 ± 4
Total catechins	197 ± 17	173 ± 26	143 ± 59	162 ± 2
Phloridzin	28 ± 13	35 ± 16	14 ± 9	26 ± 15
Chlorogenic acid	201 ± 15	171 ± 18	69 ± 25	70 ± 4

Source: van der Sluis, A. A., M. Dekker, A. de Jager, and W. M. Jongen, *J. Agric. Food Chem.*, 49, 3606, 2001.

glycosides, total catechins, phloridzin, and chlorogenic acid levels were determined for the four cultivars. Levels are outlined in Table 17.1.

The data show up to threefold differences in levels of individual components. Overall, the Jonagold cultivar had the highest levels of the four major nutraceutical components, as well as the highest estimated antioxidant levels. Apples of three of these cultivars studied over 3 harvest years revealed the same level of phenolics, but there were significant seasonal variations in levels of total catechin and chlorogenic acid.

It has been estimated that the average polyphenol content of an apple is in the region of 200 mg (Molnar 2005). The major apple nutraceuticals are polyphenols, which are present in aqueous solution in the vacuole and have structural localization based on their chemical types. Investigations into the polyphenol distribution within the fruit found that the total quantity of phenolic compounds is higher in the peel compared with the flesh of apples (Wolfe et al. 2003). It was also found that the total phenolic concentration in the apple peel was 117% higher compared with the whole fruit and ranged from 3.05–9.54 mg/g (dry weight) (Lata 2007). The flavanols and procyanidins in the peel provided the greatest antioxidant activity at approximately 90% of the total calculated activity (Biedrzycka et al. 2008), and the apple peel contained anthocyanins and flavonols at significantly higher concentrations compared with the concentration of total phenolics. In particular it was found that anthocyanins were mostly found in red and dark red apple peel; therefore, none were detected in apple varieties such as Golden Delicious and Fuji (Lata 2007). Apple products showed a lower antioxidant activity than fresh fruit, and dried fruits had higher activity than fruit puree, probably because of lower levels of heat used in processing. An increase in activity during the first month of storage was correlated with levels of polyphenols, and packaging conditions also affected activity by altering oxygen availability (Sacchetti et al. 2008).

17.4 HARVESTING AND STORAGE

As with other fruits, harvesting, transport, and storage are designed to cause minimum physiological stress, which may cause degradation of constituents (Kalt

2005). Phenolic constituents are more susceptible to environmental conditions than certain other constituents before and after harvest, but levels may increase during storage.

Storage either in controlled atmospheric conditions or at 4°C produced no significant variation in either phenolic levels or antioxidant activity.

17.5 PROCESSING

Worldwide apple production is of the order of 200 million tonnes, and 21.5% is used for extraction of juice. Clear, cloudy, and alcoholic (cider) derivatives are produced from extraction using pressure, leaving 25% of apple weight as pomace. Figure 17.2 is a diagrammatic representation of the processes involved in production of apple derivatives. The most common form of processing for apples is extraction of the juice under pressure. The majority of quercetin is present in the peel; therefore, low levels are found in juice. Oxidation of apple phenolics occurs in the first stages of apple processing, during crushing and pressing of the fruit. This is commonly referred to as *browning* and is caused by cellular disruption resulting in contact between the polyphenol oxidase enzyme and its substrate, such as chlorogenic acid. This is not wholly detrimental because the oxidation products

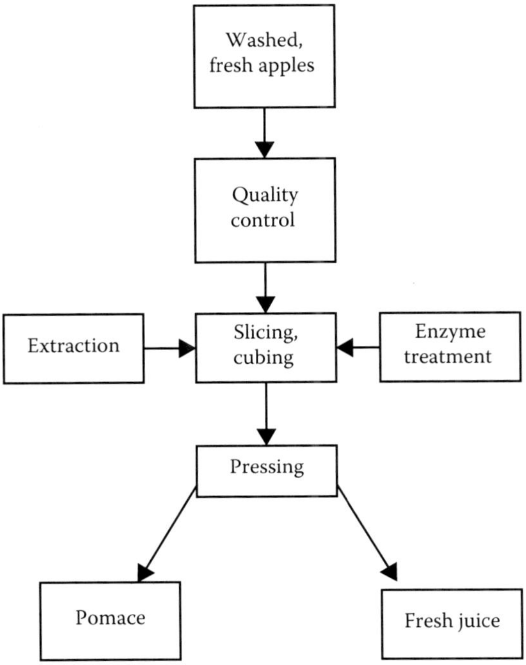

Figure 17.2 Initial stages in apple processing. (From Soler, C., J. M. Soriano, and J. Manes, *Nat. Prod. Comm.*, 4, 659, 2009. With permission.)

also contribute to total antioxidant activity (Guyot et al. 2008). Pulping and pressing to collect the juice and pulp-enzyming of Jonagold revealed that only 10% and 3% of the antioxidant activity had been extracted. The levels of chlorogenic acid and catechins were reduced to between 50% and 3%, respectively, whereas most of the antioxidant activity remained in the pomace (van der Sluis 2002). Similar results were shown for three other cultivars. Low levels of active components in these extracts were improved by using alcoholic extraction of both pulp and pomace. Levels of the nutraceuticals chlorogenic acid and quercetin glycosides were increased by 1.4 and 9 times, respectively, when compared with conventional apple juice (van der Sluis 2004).

A problem with this traditional process is that large levels of active polyphenols remain in the pomace, hence the use of alternative extraction methods to release them. Industrial-scale juicing takes place either with or without addition of pectolytic enzymes. A large amount of polyphenolics have been shown to bind to apple cell wall material, and "total liquefaction" using cellulases and pectinases has been tested but is costly and produces less organoleptically attractive products. A combination approach is sometimes used: first removal of fresh juice by pressure, followed by liquefaction as above. Alcoholic extraction is now also carried out by the addition of an alcohol to the pulp before pressing, which results in tenfold higher yields of nutraceuticals. A further publication on extraction found that in conventional juicing, 80% of quercetin remained in the pomace, whereas only 10% was in the juice, but methanol extraction of the pulp yielded 4 times the levels in pulp-enzymed apple juice (van der Sluis 1997).

Traditional clear juices are to an extent now superseded by cloudy juices because the turbidity that was earlier disliked by manufacturers is now known to be the result of the presence of bioactive components (Soler et al. 2009).

Increasing interest and use of nutraceuticals by the lay public has increased the demand for good quality raw material, with higher nutraceutical content. The processing of food-grade apple products such as juice, sauce, pies, canned products, and fresh-cut material creates millions of tonnes of waste annually, in particular from the skin, core, and seeds. This material is increasingly being treated in innovative ways to produce apple nutraceuticals (Rupasinghe 2008). Conventional processing is invaluable for enabling lengthy transport and extension to shelf life, but also affects the bioavailability of the nutraceuticals (Soler et al. 2009).

A wide range of procedures widely used in the processing of fruit products may have deleterious actions on nutraceutical components. These have been summarized (Pokorny et al. 2001) and include the effects of pasteurization, sterilization, blanching, evaporation, drying, and fermentation, which may all occur during apple processing. Pasteurization is responsible for changes caused by thermal destruction, and also oxidation. Thermal sterilization uses higher temperatures than pasteurization and may cause thermal degradation, in addition to killing microorganisms. The blanching stage is widely used and usually involves saturated steam or hot water tanks, and this rapid heating causes enzyme deactivation, which inhibits oxidation of phenolics into quinones. A drying stage may be used for reduction in moisture content, usually down to 6%–12%, but usually has no negative effects on polyphenol composition; however, vacuum evaporation is more effective but results in poorer organoleptic quality.

When clarification of juices is required, a number of adsorbents have been used, including bentonite, gelatin, silica, and more recently chitosan. The latter is most promising because it has limited effects on levels of bioactives (Soler et al. 2009). Lastly, fermentation, which may be part of processing, results in exchange of oxygen with carbon dioxide, and hydrolytic processes may cause degradation of phenolic esters or glycosides into acids or aglycones, which may be more active than the starting material, as is the case with quercetin.

17.6 QUALITY CONTROL

For therapeutic applications, juices and formulated products need to use extracts containing reliably high levels of specific components.

A number of analytical procedures have been devised for assessment of quality. An assay for antioxidant activity is widely used to establish the bioactivity of the total components. The Trolox (Hoffman-LaRoche, Basel, Switzerland) equivalent antioxidant capacity procedure is typically used. For identification and quantification of individual polyphenols, reversed-phase high-performance liquid chromatography (HPLC) is widely used. One such application was a major survey of Tunisian apple extracts using HPLC with electrospray ionization mass spectrometry of separated components (Fattouch et al. 2008). Another large survey identified 17 polyphenolic constituents from a large number of cultivars using HPLC with diode array detection (Kahle 2005). Other techniques that have been used include countercurrent chromatography (Cao et al. 2009), subcritical (CO_2, ethanol) extraction (Adil et al. 2007), and capillary electrophoresis of juice polyphenols with electrochemical detection (Peng et al. 2005).

One aspect of considerable importance in apple processing is the possible contamination of apples with the fungal metabolite patulin; approximately 50% of all juice samples tested have been found to be affected (Prieta et al. 1994). Juice samples from specific geographical regions have more recently been reported to contain levels of patulin above the recommended 50 μg/L (Soler et al. 2009).

17.7 PHARMACOKINETICS

The approximate bioavailability of the polyphenolic flavonoids contained in apples is between 1% and 5% (Lotito et al. 2004b). Furthermore, the recovery of polyphenols in the urine is low (1%–25%) after oral administration, and this is probably because of incomplete absorption, biliary excretion, or metabolism via both human and colonic bacterial enzymes (Scalbert et al. 2000). It has been found that chlorogenic acid is absorbed up to a maximum of 33% from the human small intestine, but only 0.3% was found in the urine as a result of extensive metabolism in the liver after absorption (Olthof et al. 2001). It is thought that polyphenols primarily undergo metabolic conjugation, which will lower the free polyphenol concentration measured in the blood (Pohl et al. 2006). Usually polyphenols will be conjugated to form *O*-glucuronides, sulfate esters, and the *O*-methyl ether (Scalbert et al. 2002). The glucuronidation and methylation of catechin are extensive in the small intestine, so much so that no free catechin was found in the plasma taken from the mesenteric vein after oral administration (Donovan et al. 2001). However, in studies where larger doses of (+)-catechin were given (83 mg/kg body weight), 7.5% of the dose was found as unchanged catechin in urine over a 24-hour period (Das 1971).

The plasma concentration of the majority of flavonoids quickly declines after absorption from the small intestine, with an elimination half-life of 2–28 hours (Manach et al. 2004). It was found that polymeric proanthocyanidins reach the colon and are degraded to low molecular weight compounds such as aromatic acids (Deprez et al. 2000). Procyanidin oligomers can also be degraded in the gut to form epicatechin monomers that can be absorbed (Holt et al. 2002). After having been given cloudy apple juice, recovery of polyphenols in ileostomy effluent varied from 0%–33% in volunteers (Kahle et al. 2006). However, after oral administration, some polyphenols such as catechin may reach relatively high concentrations in the colon (Das 1971). Glycosylated polyphenols may also show poor absorption because of their reduced hydrophobicity compared with their free aglycone counterparts. Bacterial hydrolysis of glycosides occurs in the lower part of the ileum and the cecum, and one study showed that obligate anaerobes such as *Bacillus uniformis* and *Bacillus ovatus* can hydrolyze rutin (quercetin glycoside) to quercetin through the production of β-glucosidase (Bokkenheuser et al. 1987). Cleavage of the polar moieties that are glycosylated to produce free polyphenols will enable them to undergo passive diffusion across the small intestine brush border (Scalbert et al. 2000). However, not all glycosidic forms of polyphenols show poor absorption properties. While studying the bioavailability of quercetin glycosides, it was found that the mean peak plasma concentration was 20 times higher and was reached 10 times faster for quercetin linked with glucose compared with quercetin linked with rutinose (Hollman et al. 1999).

17.8 ANTIOXIDANT ACTIVITIES OF APPLE POLYPHENOLS

It is known that apples contain polyphenols that can act as antioxidants. One study looked at the antioxidant capacity of plasma from six healthy volunteers after they had consumed five Red Delicious apples. It was found that there was a statistically significant increase in the antioxidant capacity of the plasma after 1 hour (measured as ferric-reducing antioxidant potential). However, it was found that the concentration of endogenous plasma urate and the plasma antioxidant capacity were correlated to a significant extent even though apples do not contain urate or its precursors (Hollman et al. 1999). In fact, urate has been found to be a powerful antioxidant in vivo (Becker 1993). One report concluded that it was not the consumption of polyphenols that increases the antioxidant capacity of human plasma, but instead the increase in plasma urate after apple consumption. It was postulated that the urate concentration increases after apple consumption as a result of the metabolic effect of fructose on urate (Lotito et al. 2004a).

The antioxidant activity of 100 g of fresh apples was equated to the activity observed with 1700 mg of vitamin C (Liu et al. 2005). Previous research found that the polyphenols with the greatest antioxidant capacities were found to be catechin, quercetin, and its glucoside rutin. In addition, the apple polyphenols had greater antioxidant capacities than Trolox (Hoffman-LaRoche), a water-soluble vitamin E analog (Lotito et al. 2004b). After regression analysis of experimental in vitro data, no significant relationship was found between the concentration of total phenols or the concentration of individual phenolic classes and the antioxidant activity observed (Pearson et al. 1999). It has been suggested that the antioxidant activity observed is probably not caused by individual polyphenols, but rather is the summation of the actions of all the polyphenols and their possible synergistic and antagonistic effects.

17.9 ROLES OF APPLE POLYPHENOLS IN CARDIOVASCULAR DISEASE

Numerous researchers have demonstrated that flavonoids display protective effects against the initiation and further development of atherosclerosis (Erdman et al. 2007). It has been postulated that because flavonoids can act as efficient antioxidants, foods rich in flavonoids may exert cardio- and cerebroprotective effects via reducing the oxidative damage to vascular cells and low-density lipoprotein (LDL) cholesterol (Lotito et al. 2004b), and animal studies have established that the progression of atherosclerosis is inhibited by antioxidants (Das 1971). It has been found that both apple juices and extracts of the apple peel, flesh, and whole apple effectively inhibit the copper-catalyzed oxidation of LDL cholesterol in vitro (Pearson et al. 1999). Furthermore, when healthy volunteers consumed 1500 mg of apple polyphenols three times per day over 4 weeks, high-density lipoprotein (HDL) cholesterol levels increased, whereas both total cholesterol and LDL cholesterol decreased significantly. The study showed that the average serum LDL cholesterol decreased from 148.6 $\pm$ 5.96 mg/dL to 137.0 $\pm$ 7.74 mg/dL ($p < 0.01$) (Nagasako-Akazome et al. 2005). A more recent study involved the administration of four capsules each containing 150 mg of extracted apple polyphenols to subjects who had a body mass index between 23 and 30. The study found that both LDL cholesterol and total cholesterol levels and visceral fat area had been significantly reduced ($p < 0.05$) in those who had the apple polyphenols compared with the control group (Nagasako-Akazome et al. 2007).

Epidemiological data showed an inverse relationship between the consumption of apples and deaths caused by coronary heart disease in a study of 5133 subjects aged 30–69 years (Knekt et al. 1996). It was thought that this was because of the flavonoid content of apples. The processes that result in a decrease in serum cholesterol may be caused by an enhanced excretion of cholesterol in the feces as a result of the apple polyphenols binding to cholesterol and/or bile acid, thereby inhibiting cholesterol enterohepatic circulation. Furthermore, it has been suggested that the number of serum cholesterol-activating LDL cholesterol receptors found in the liver could be reduced as a result of enhanced liver cholesterol isomerization to bile acid, which might increase LDL cholesterol particle uptake into the blood; however, there could also be other mechanisms (Nagasako-Akazome et al. 2005). One study involving obese Zucker rats showed that both lipoprotein lipase and pancreatic lipase were inhibited by procyanidins in a dose-dependent manner, thereby decreasing the weight of the rats (Yoshikawa et al. 2002). In another study it was found that apple polyphenols can markedly inhibit the activities of a starch-digesting enzyme found in the small intestine of a rat and pancreatic lipase from a pig (Nagasako-Akazome et al. 2007). Overall, apple polyphenols may be useful for preventing lifestyle-related diseases such as cardiovascular disease in communities where obesity is an important problem.

17.10 ROLES OF APPLE POLYPHENOLS IN CANCER

Cancer rates increase throughout the world as a result of the continual increase in life expectancy, ever-increasing urbanization, and the subsequent alterations in environmental conditions, including lifestyle (Surh 2003). Several studies have specifically linked apple consumption with a reduced risk of cancer (Boyer et al. 2004). The multivariate odds ratios for subjects who consumed one or more apples a day compared with those who consumed less than one apple per day were 0.75 for cancers of the pharynx and oral cavity. Furthermore, in the same study, multivariate odd ratios of 0.75, 0.80, 0.58, 0.82, 0.85, and 0.91 were found for cancers of the esophagus, colorectum, larynx, breast, ovary, and prostate, respectively (Gallus et al. 2005). This study showed that it is possible that there may be some link between the incidence of cancer and the consumption of apples.

17.10.1 Colon Cancer

One of the leading types of cancer in the Western world is colon cancer, and it is thought to be one of the most preventable forms of visceral cancer (Gosse et al. 2005). In the rat model it was found that the administration of cloudy apple juice effectively inhibited carcinogenesis of the colon caused by 1,2-dimethlhydrazine (Barth et al. 2007). The same study also found that the cloudy apple juice lowered the extent of hyperproliferation and DNA damage, as well as reduced the number of large aberrant crypt foci within the distal colon. However, this model may not be applicable to humans; therefore, experiments in a human cell line would be more appropriate. The growth of human colon adenoma cells (LT97) and human colon carcinoma cells (HT29) was significantly reduced after being incubated with apple extracts. The human colon adenoma cells were from an early colon adenoma in the premalignant stage of tumor development, whereas the carcinoma cell line consisted of fully transformed colon cancer cells (Veeriah et al. 2007). The experiments showed that the level of growth inhibition was much greater in the human colon adenoma cell line compared with the carcinoma cell line. Therefore, it can be inferred that apple polyphenols may have a greater chemoprotective effect in the preneoplastic lesion compared with fully transformed carcinoma cells. Interestingly, this study found that if the apple extracts were fermented, then a tenfold decrease in activity was observed, inferring that the breakdown products of apple polyphenols during fermentation may not have a protective effect (Veeriah et al. 2007).

A study was carried out into which fractions of apple extracts were responsible for the effects seen in colon cancer. One prepared extract contained 72.6% polyphenols and no procyanidins (30% flavonoids and 42.5% hydroxycinnamic acids), whereas the other contained 78.4% procyanidins. This study established that there was no effect on the growth of metastatic colon cancer SW620 cells when they were incubated with the extract containing only monomeric polyphenols at concentrations between 10 and 100 μg/mL. However, a reduction in cell growth was seen when the SW620 cells were incubated with the procyanidin containing extract with a resulting IC_{50} of 45 μg/mL. Moreover, it showed that in rats the procyanidin extracts alone could significantly reduce preneoplastic lesions ($p < 0.01$) and the aberrant crypt numbers by 50% (Gosse et al. 2005). The authors suggested that procyanidins can work inside the cell membrane, resulting in an alteration in signal transduction pathways such as a down-regulation of protein kinase C (PKC). Therefore, it is thought that procyanidins are responsible for the effects observed. The PKC family consists of at least 12 subgroups, which all have serine–threonine kinase activity (Koivunen et al. 2006). PKCβII is significantly elevated in colon tumors compared with normal colonic epithelium.

In fact, it is thought that PKCβII is involved in the numerous stages of colon cancer, but in addition it is possible that other PKC isozymes could play a role (Gokmen-Polar et al. 2001). It has been found that PKCα can suppress apoptosis and induce cell proliferation, whereas it is thought that PKCδ plays an active role in inducing apoptosis (Koivunen et al. 2006). One group observed a statistically significant decrease in protein levels of PKCα, PKCβII, and PKCγ and a significant increase in PKCδ when polyphenol-rich apple juice extracts (500 μg/mL) were incubated with the human colon carcinoma cell line (HT29) for 24 hours. This resulted in a significant reduction in intracellular PKC activity after 24 hours. They also found that when the human colon carcinoma HT29 cells were exposed to apple polyphenols there was a significant increase in the activity of caspase-3, which is a cysteine-dependent aspartase involved in apoptosis (Kern et al. 2007). Furthermore, polyphenol-rich apple juice extracts at concentrations greater than 650 μg/mL were found to induce DNA fragmentation in human colon carcinoma HT29 cells after 24 hours. In addition, another study found that a procyanidin extract activated caspase-3 activity and down-regulated PKC (Gosse et al. 2005). One study found that quercetin, phloretin, and polyphenol-rich apple extracts can all act to antagonize CYP1A1 induction in the human colon carcinoma cell line Caco-2 (Pohl et al. 2006). Therefore, this shows that apple polyphenols can inhibit CYP1A1 activity by preventing the activation of carcinogens. It has been suggested that compounds such as apple polyphenols, which can effectively inhibit CYP1A1, may be ideal candidates as chemoprotective agents (Schwarz et al. 2003). We know that the natural nontoxic dietary apple polyphenols can reach the colon, where they can exert their effects (Gosse et al. 2005); therefore, even though the mechanisms of action of dietary apple polyphenols in colon cancer are not completely defined, the evidence to date shows that apple constituents may well act as agents against colon cancer.

17.10.2 Prostate Cancer

It is also thought that the constituents of apples may also alter the progression of prostate cancer. In an in vitro study it was found that quercetin can lower the levels of androgen receptors, which are responsible for prostate cancer development in LNCaP prostate cancer cells (Xing et al. 2001). Androgen action and the functional status of the androgen receptor are significant factors in prostate cancer and its progression (Heinlein et al. 2004). Moreover, prostate cancer is treated with endocrine therapy with the intent of lowering the concentrations of circulating androgen and inhibiting androgen effects in the prostate itself (Culig et al. 1998). In fact, it appears that when the levels of androgens are reduced, the tumor cells can adapt by increasing the expression of the androgen receptor gene to overcome this (Culig et al. 1998). It was reported that the down-regulation of the androgen receptor was caused by quercetin acting at the transcription level, therefore reducing the levels of mRNA encoding for the androgen receptor. This in vitro study found that the androgen-dependent secretion of human kallikrein 2 and prostate-specific antigen had been significantly reduced in LNCaP prostate cancer cells as a result of quercetin lowering the levels of androgen receptors in a dose-dependent manner (Xing et al. 2001). Therefore, quercetin can possibly inhibit the development and progression of prostate cancer and has the potential to become a chemopreventive and/or chemotherapeutic agent for prostate cancer (Xing et al. 2001). However, the in vitro study does not take into account the pharmacokinetics of quercetin as these properties may make quercetin ineffective in vivo. Therefore, studies in animals should be conducted to determine whether we can achieve the same results in a whole biological system.

17.10.3 Breast Cancer

Many studies have been conducted concerning the relationship between breast cancer and the consumption of apples. The incidence of mammary tumors after carcinogen administration was lowered in a dose-dependent manner after rats were given doses of whole apple extracts (Liu et al. 2005). The doses of whole apple extracts were equivalent to 3.3, 10, and 20 g of apples/kg of body weight corresponding to a 17%, 39%, and 44% reduction in the number of mammary tumors, respectively, after 24 weeks. Moreover, apple extracts were found to inhibit the proliferation of human breast cancer MCF-7 cells in a dose-dependent manner with polyphenol concentrations of 40 mg/mL and higher. Proliferation was inhibited by 81.7% when the MCF-7 cells were exposed to concentrations of 100 mg/mL, with the median effective concentration (EC_{50}) at 65.1 ± 1.5 mg/mL. Furthermore, these researchers explored the mechanisms that led to the inhibition of cancer cell growth.

They found that apple extracts could significantly inhibit the activation of nuclear factor (NF)-κB by tumor necrosis factor–α in human breast cancer MCF-7 cells (Yoon et al. 2007). Moreover, NF-κB is a transcription factor that regulates the extent of cell proliferation as well as the process of apoptosis and is known to be implicated in the development of cancer in humans (Bours et al. 2000). Therefore, the study using MCF-7 cells shows us a mechanism for the results seen in breast cancer (Yoon et al. 2007). Furthermore, the earlier in vivo study suggests that these effects may be seen in humans (Liu et al. 2005). In fact, we may be able to use apple extracts as a support in cancer therapy to reduce cancer cell resistance to chemotherapy, thus increasing its efficacy by activating apoptosis and halting cell proliferation that is promoted by NF-κB (Yoon et al. 2007).

17.10.4 Lung Cancer

It is known that the consumption of fruits and vegetables is linked to a reduced risk of lung cancer and some other epithelial cancers (Knekt et al. 1997), and it has been reported that the incidence of lung cancer showed an inverse association with the intake of flavonoids. It was inferred that quercetin is the main flavonoid responsible because it contributed to more than 95% of the total flavonoid intake in one study population (Knekt et al. 1997). It was shown in this study that there was a statistically significant relationship between apple consumption and the incidence of lung cancer. In addition, the relationship was closer in nonsmokers compared with current smokers. In a further study involving 521,457 people mostly aged 25–70 years, it was found that the consumption of apples was inversely correlated with the risk of developing lung cancer in all participants, as well as those who smoked ($p < 0.05$) (Linseisen et al. 2007).

Finally, an in vitro study found that the growth of cancerous $HEpG_2$ liver cells can be inhibited in a dose-dependent manner after being exposed to apple extracts for 96 hours (Wolfe et al. 2003). The study found that inhibition was dose dependent and that apple peel was found to possess the greatest inhibitory activity, followed by the flesh combined with peel, whereas the flesh alone had the lowest inhibitory activity.

The P-glycoprotein is a 170-kDa ATP-dependent plasma membrane transporter that can extrude chemotherapeutic agents out of cells (Kitagawa 2006). In multidrug-resistant cancers, the multidrug resistance MDR1 gene product P-glycoprotein is overexpressed, leading to the reduced accumulation of cytotoxic drugs as a result of increased drug efflux (Cowan 1999). Extracts of apples effectively inhibited P-glycoprotein function in a mouse transfected with the human MDR1 gene. This study showed that the level of inhibition of the P-glycoprotein with apple extracts was greater than that seen with (±)-verapamil (Molnar et al. 2005). An analysis of the structure–activity relationship between the interaction of flavonoids and the P-glycoprotein has been conducted. It has been reported that a planar and hydrophobic structure is important for relatively small molecular weight flavonoids, which do not have large substituents, because this makes an interaction with the hydrophobic region of the P-glycoprotein more favorable. In comparison, a large hydrophobic region and the presence of neighboring hydrophilic hydroxyl groups appear to be important for nonplanar flavonoids, which have large substituents. Polyphenols are thought to be ideal candidates for modulators of multidrug resistance because they are able to inhibit the function of P-glycoproteins, and in addition they show good physiological safety (Kitagawa 2006).

17.11 ANTIBACTERIAL ACTIVITY OF APPLE POLYPHENOLS

It was found in one study that extracted carotenoids from the peel of Golden Delicious apples have potent anti-*Helicobacter pylori* activity with a minimum inhibitory concentration (MIC_{50}) of 36 μg/mL, which is effective, compared with an MIC_{50} for metronidazole of 45 μg/mL (Molnar et al. 2005). A direct correlation between the total phenolic content of apples and the antimicrobial effect has been shown, with the extracts from the Granny Smith variety being much more potent than Royal Gala (Alberto et al. 2006). Moreover, this study showed that extracts from the apple skin had the highest polyphenol concentrations, and these extracts showed the greatest antibacterial activities. The most susceptible organisms were found to be *Pseudomonas aeruginosa* ATCC 27853, *Escherichia coli* ATTCC 25922, and *Staphylococcus aureus* ATTCC 29213, and the least susceptible organisms were found to be *E. coli* ATTCC 35218 and *E. coli* (Alberto et al. 2006). Apple polyphenols are consequently thought to act as a mechanism of defense against microorganisms (Cowan 1999). Their mode of action may include binding to extracellular and soluble proteins, the bacterial cell wall, as well as cell wall polypeptides, surface-exposed adhesins, and membrane-bound enzymes (Biedrzycka et al. 2008).

17.12 ROLES OF APPLE POLYPHENOLS IN ALLERGIC DISEASES

It is thought that polyphenol-enriched apple extracts could prove useful for both preventing and treating allergic disease (Tokura et al. 2005). Apple-condensed tannins (oligomeric catechins) and procyanidins have been found to strongly inhibit the release of histamine from rat basophilic leukemia (RBL-2H3) cells caused by antigen stimulation (Kanda et al. 2005). In one human study, it was found that the symptoms of atopic dermatitis were improved after patients were given apple-condensed tannins at a daily dosage of 10 mg/kg. It was proposed that apple-condensed tannins can be absorbed through the intestine, and if this is the case, then they could well have pharmacological activities in vivo (Kojima et al. 2000). These authors also found that the number of blood eosinophils were significantly decreased ($p < 0.05$) in those patients who were given apple-condensed tannins. In atopic dermatitis, both the numbers of eosinophils and eosinophil granule protein levels increase as the severity of the disease increases (Simon et al. 2004). The antiallergic properties of apples were also demonstrated in a double-blind study, in which it was found that the sneezing score was significantly reduced in those subjects with cedar pollinosis (a subtype of allergic rhinitis) who were administered 500 mg of apple polyphenols once daily (Kishi et al. 2005). It is thought that the mechanism of action of apple polyphenols is as follows: FcεRI is a receptor that has high affinity for the IgE antibody and plays a substantial role in allergic reactions (Takahashi et al. 2005), and the cross-linking of IgE and FcεRI on effector cells (mast cells and basophils) induces the release of chemical mediators and the expression of cytokine genes (Takahashi et al. 2005). The inhibition of mast cell degranulation was shown to be caused by apple polymeric proanthocyanidins preventing the binding of the IgE antibody to FcεRI on the mast cells (Tokura et al. 2005). Furthermore, experiments in mice have found that condensed apple tannins inhibit antibody production (IgG1 and IgE) stimulated by oral sensitization, as well as anaphylaxis caused by IgG1, IgE, and mast cells. It has been hypothesized that apple-condensed tannins could prevent the development of food allergies if taken during early childhood (Kojima et al. 2000). Other workers claim that apple-condensed tannins should be a recommended adjunct to the therapeutic protocol in atopic dermatitis. It has also been claimed that apple polyphenols may be especially useful against sneezing in allergic rhinitis. In addition, it is suggested that the dosing of conventional antiallergic drugs could possibly be reduced if apple polyphenols are taken before pollen dispersion (Kishi et al. 2005). It is, however, necessary to study whether the antiallergic activities of apple-condensed tannins increase if the number of catechin units in the polymeric forms increases (Kanda et al. 2005).

Extrinsic asthma is thought to be an allergic process; therefore, it is possible that apples may alleviate the symptoms of asthma. However, in one study involving 2640 children aged 5–10 years, no link was found between an improvement in asthma symptoms and an increase in consumption of fresh apples in the diet (Okoko et al. 2007).

17.13 CONCLUSIONS

Apples are the major source of flavonoids in the Western diet and have a high level of antioxidant activity. They contain common flavonoids and flavonols, as well as dihydrochalcones, along with nonflavonoids such as hydroxycinnamates. Polymeric flavanols such as proanthocyanidins (procyanidins) also occur in apples, along with carotenoids, which are found both free and esterified.

There are wide-ranging factors that affect levels of apple nutraceuticals, including cultivar, agronomic practice, environmental conditions, storage conditions, and processing technology, including extraction technique, which has a major effect on levels. Development of quality control techniques has progressed in line with processing, and a wide range of sophisticated procedures are now available to determine quality, which can also be used to direct agronomic processes.

The bioavailability of apple flavonoids is between 1% and 5%, and their metabolism involves conjugation. Plasma concentration of the majority of flavonoids quickly declines after absorption from the small intestine, with an elimination half-life of 2–28 hours. Apples contain a number of polyphenols with known antioxidant activity that has been demonstrated in vivo. It has also been suggested that the fructose contained in apples will increase the endogenous plasma urate levels, which will therefore increase the antioxidant capacity of plasma.

Flavonoids have been shown to have protective effects against the initiation and further development of atherosclerosis, and there is evidence that the regular consumption of apples may protect against cardiovascular disease by improving the LDL/HDL balance, probably as a result of antioxidant activity.

Epidemiological evidence suggests that apple consumption lowers the risk of a number of cancers, probably because of the presence of procyanidins. Apple polyphenols are also known to have antibacterial activity, and polymeric flavonols show activity in allergic diseases.

From the studies done thus far, there is evidence that the regular consumption of apples may protect against cardiovascular disease, cancer, and allergic diseases. However, many studies have concentrated on demonstrating the effectiveness of apple polyphenols in vitro rather than in vivo, and this should be addressed with more studies in animal models. The pharmacokinetic parameters for most polyphenols are still ill-defined. It must be noted that even changes in the colonic microflora of individuals may affect the extent of polyphenol absorption, especially polymeric polyphenols. To date there is also limited work on the action of polyphenol metabolites. Consumption of apples appears to be good for health, does not cause adverse effects, and may help to prevent the onset of chronic diseases.

REFERENCES

Adil, I. H., H. I. Cetin, M. E. Yener, and A. Bayindirli. 2007. Subcritical (Carbon Dioxide + Ethanol) Extraction of Polyphenols from Apple and Peach Pomaces, and Determination of the Antioxidant Activities of the Extracts. *Journal of Supercritical Fluids* 43: 55–63.

Alberto, M. R., M. Canavosio, R. Andres, and M. C. Manca de Nadra. 2006. Antimicrobial Effect of Polyphenols from Apple Skins on Human Bacterial Pathogens. *Electronic Journal of Biotechnology* 9: 205–09.

Barth, S. W., C. Faehndrich, A. Bub, B. Watzl, F. Will, H. Dietrich, G. Rechkemmer, and K. Briviba. 2007. Cloudy Apple Juice Is More Effective than Apple Polyphenols and an Apple Juice Derived Cloud Fraction in a Rat Model of Colon Carcinogenesis. *Journal of Agricultural and Food Chemistry* 55: 1181–87.

Becker, B. F. 1993. Towards the Physiological Function of Uric Acid. *Free Radical Biology & Medicine* 14: 615–31.

Biedrzycka, E., and R. Amarowicz. 2008. Diet and Health: Apple Polyphenols as Antioxidants. *Food Reviews International* 24: 235–51.

Bokkenheuser, V. D., C. H. L. Shackleton, and J. Winter. 1987. Hydrolysis of Dietary Flavonoid Glycosides by Strains of Intestinal Bacteroides from Humans. *Biochemical Journal* 248: 953–56.

Bours, V., M. Bentires-Alj, A. C. Hellin, P. Viatour, P. Robe, S. Delhalle, V. Benoit, and M.-P. Merville. 2000. Nuclear Factor-κB, Cancer, and Apoptosis. *Biochemical Pharmacology* 60: 1085–89.

Boyer, J., and R. H. Liu. 2004. Apple Phytochemicals and Their Health Benefits. *Nutrition Journal* 3: 5.

Cao, X., C. Wang, H. Pei, and B. Sun. 2009. Separation and Identification of Polyphenols in Apple Pomace by High-Speed Counter-Current Chromatography and High-Performance Liquid Chromatography Coupled with Mass Spectrometry. *Journal of Chromatography A* 1216: 4268–74.

Cowan, M. M. 1999. Plant Products as Antimicrobial Agents. *Clinical Microbiology Review* 12: 564–82.

Culig, Z., A. Hobisch, A. Hittmair, H. Peterziel, A. C. Cato, G. Bartsch, and H. Klocker. 1998. Expression, Structure, and Function of Androgen Receptor in Advanced Prostatic Carcinoma. *Prostate* 35: 63–70.

Das, N. P., 1971. Studies on flavonoid metabolism. Absorption and metabolism of (+)-catechin in man. *Biochemical Pharmacology* 20: 3435–45.

Deprez, S., C. Brezillon, S. Rabot, C. Philippe, I. Mila, C. Lapierre, and A. Scalbert. 2000. Polymeric Proanthocyanidins Are Catabolized by Human Colonic Microflora into Low-Molecular-Weight Phenolic Acids. *Journal of Nutrition* 130: 2733–38.

Donovan, J. L., V. Crespy, C. Manach, C. Morand, C. Besson, A. Scalbert, and C. Remesy. 2001. Catechin Is Metabolized by Both the Small Intestine and Liver of Rats. *Journal of Nutrition* 131: 1753–57.

Erdman, J. W. Jr., D. Balentine, L. Arab, G. Beecher, J. T. Dwyer, J. Folts, J. Harnly, P. Hollman, C. L. Keen, G. Mazza, M. Messina, A. Scalbert, J. Vita, G. Williamson, and J. Burrowes. 2007. Flavonoids and Heart Health: Proceedings of the ILSI North America Flavonoids Workshop, May 31–June 1, 2005, Washington, DC. *Journal of Nutrition* 137: 718S–37S.

Fattouch, S., P. Caboni, V. Coroneo, C. Tuberoso, A. Angioni, S. Dessi, N. Marzouki, and P. Cabras. 2008. Comparative Analysis of Polyphenolic Profiles and Antioxidant and Antimicrobial Activities of Tunisian Pome Fruit Pulp and Peel Aqueous Acetone Extracts. *Journal of Agricultural and Food Chemistry* 56: 1084–90.

Gallus, S., R. Talamini, A. Giacosa, M. Montella, V. Ramazzotti, S. Franceschi, E. Negri, and C. La Vecchia. 2005. Does an Apple a Day Keep the Oncologist Away? *Annals of Oncology* 16: 1841–44.

Gokmen-Polar, Y., N. R. Murray, M. A. Velasco, Z. Gatalica, and A. P. Fields. 2001. Elevated Protein Kinase C β II Is an Early Promotive Event in Colon Carcinogenesis. *Cancer Research* 61: 1375–81.

Gosse, F., S. Guyot, S. Roussi, A. Lobstein, B. Fischer, N. Seiler, and F. Raul. 2005. Chemopreventive Properties of Apple Procyanidins on Human Colon Cancer-Derived Metastatic SW620 Cells and in a Rat Model of Colon Carcinogenesis. *Carcinogenesis* 26: 1291–95.

Guyot, S., S. Bernillon, P. Poupard, and C. M. G. C. Renard. 2008. Multiplicity of Phenolic Oxidation Products in Apple Juices and Ciders, from Synthetic Medium to Commercial Products. *Recent Advances in Polyphenol Research* 1: 278–92.

Halliwell, B., J. Rafter, and A. Jenner. 2005. Health Promotion by Flavonoids, Tocopherols, Tocotrienols, and Other Phenols: Direct or Indirect Effects? Antioxidant or Not? *American Journal of Clinical Nutrition* 81: 268S–76S.

Heinlein, C. A., and C. Chang. 2004. Androgen Receptor in Prostate Cancer. *Endocrine Reviews* 25: 276–308.

Hollman, P. C. H., M. N. C. P. Bijsman, Y. Van Gameren, E. P. Cnossen, J. H. de Vries, and M. B. Katan. 1999. The Sugar Moiety Is a Major Determinant of the Absorption of Dietary Flavonoid Glycosides in Man. *Free Radical Research* 31:569–73.

Holt, R. R., S. A. Lazarus, M. C. Sullards, Q. Y. Zhu, D. D. Schramm, J. Hammerstone, C. G. Fraga, H. H. Schmitz, and C. L. Keen. 2002. Procyanidin Dimer B2 [Epicatechin-(4.beta.- 8)-Epicatechin] in Human Plasma after the Consumption of a Flavanol-Rich Cocoa. *American Journal of Clinical Nutrition* 76: 798–804.

Kahle, K., M. Kraus, and E. Richling. 2005. Polyphenol Profiles of Apple Juices. *Molecular Nutrition & Food Research* 49: 797–806.

Kahle, K., M. Kraus, W. Scheppach, and E. Richling. 2006. Colonic Availability of Apple Polyphenols—A Study in Ileostomy Subjects. *Molecular Nutrition & Food Research* 49: 1143–50.

Kalt, W. 2005. Effects of Production and Processing Factors on Major Fruit and Vegetable Antioxidants. *Journal of Food Science* 70: R11–19.

Kanda, T., H. Akiyama, A. Yanagida, M. Tanabe, Y. Goda, M. Toyoda, R. Teshima, and Y. Saito. 1998. Inhibitory Effects of Apple Polyphenol on Induced Histamine Release from RBL-2H3 Cells and Rat Mast Cells. *Bioscience, Biotechnology, and Biochemistry* 62: 1284–89.

Kern, M., G. Pahlke, K. K. Balavenkatraman, F. D. Bohmer, and D. Marko. 2007. Apple Polyphenols Affect Protein Kinase C Activity and the Onset of Apoptosis in Human Colon Carcinoma Cells. *Journal of Agricultural and Food Chemistry* 55: 4999–5006.

Kishi, K., M. Saito, T. Saito, M. Kumemura, H. Okamatsu, M. Okita, and K. Takazawa. 2005. Clinical Efficacy of Apple Polyphenol for Treating Cedar Pollinosis. *Bioscience, Biotechnology, and Biochemistry* 69: 829–32.

Kitagawa, S. 2006. Inhibitory Effects of Polyphenols on P-Glycoprotein-Mediated Transport. *Biological & Pharmaceutical Bulletin* 29: 1–6.

Knekt, P., R. Jarvinen, A. Reunanen, and J. Maatela. 1996. Flavonoid Intake and Coronary Mortality in Finland: A Cohort Study. *British Medical Journal* 312: 478–81.

Knekt, P., R. Jarvinen, R. Seppanen, M. Hellovaara, L. Teppo, E. Pukkala, and A. Aromaa. 1997. Dietary Flavonoids and the Risk of Lung Cancer and Other Malignant Neoplasms. *American Journal of Epidemiology* 146: 223–30.

Koivunen, J., V. Aaltonen, and J. Peltonen. 2006. Protein Kinase C (PKC) Family in Cancer Progression. *Cancer Letter* 235: 1–10.

Kojima, T., H. Akiyama, M. Sasai, S. Taniuchi, Y. Goda, M. Toyoda, and Y. Kobayashi. 2000. Anti-allergic Effect of Apple Polyphenol on Patients with Atopic Dermatitis: A Pilot Study. *Allergology International* 49: 69–73.

Lata, B. 2007. Relationship between Apple Peel and the Whole Fruit Antioxidant Content: Year and Cultivar Variation. *Journal of Agricultural and Food Chemistry* 55: 663–71.

Linseisen, J., S. Rohrmann, A. B. Miller, H. B. Bueno-de-Mesquita, F. L. Buechner, P. Vineis, A. Agudo, I. T. Gram, L. Janson, V. Krogh, K. Overvad, T. Rasmuson, M. Schulz, T. Pischon, R. Kaaks, A. Nieters, N. E. Allen, T. J. Key, S. Bingham, K.-T. Khaw, P. Amiano, A. Barricarte, C. Martinez, C. Navarro, R. Quiros, F. Clavel-Chapelon, M.-C. Boutron-Ruault, M. Touvier, P. H. M. Peeters, G. Berglund, G. Hallmans, E. Lund, D. Palli, S. Panico, R. Tumino, A. Tjonneland, A. Olsen, A. Trichopoulou, D. Trichopoulos, P. Autier, P. Boffetta, N. Slimani, and E. Riboli. 2007. Fruit and Vegetable Consumption and Lung Cancer Risk: Updated Information from the European Prospective Investigation into Cancer and Nutrition (EPIC). *International Journal of Cancer* 121: 1103–14.

Liu, R. H., J. Liu, and B. Chen. 2005. Apples Prevent Mammary Tumors in Rats. *Journal of Agricultural and Food Chemistry* 53: 2341–43.

Lotito, S. B., and B. Frei. 2004a. Relevance of Apple Polyphenols as Antioxidants in Human Plasma: Contrasting in Vitro and in Vivo Effects. *Free Radical Biology & Medicine* 36: 201–11.

Lotito, S. B., and B. Frei. 2004b. The Increase in Human Plasma Antioxidant Capacity after Apple Consumption Is Due to the Metabolic Effect of Fructose on Urate, not Apple-Derived Antioxidant Flavonoids. *Free Radical Biology & Medicine* 37: 251–58.

Manach, C., and J. L. Donovan. 2004. Pharmacokinetics and Metabolism of Dietary Flavonoids in Humans. *Free Radical Research* 38: 771–85.

Molnar, P., M. Kawase, K. Satoh, Y. Sohara, T. Tanaka, S. Tani, H. Sakagami, H. Nakashima, N. Motohashi, N. Gyemant, and J. Molnar. 2005. Biological Activity of Carotenoids in Red Paprika, Valencia Orange and Golden Delicious Apple. *Phytochemistry Reviews* 19: 700–07.

Nagasako-Akazome, Y., T. Kanda, M. Ikeda, and H. Shimasaki. 2005. Serum Cholesterol-Lowering Effect of Apple Polyphenols in Healthy Subjects. *Journal of Oleo Science* 54: 143–51.

Nagasako-Akazome, Y., T. Kanda, Y. Ohtake, H. Shimasaki, and T. Kobayashi. 2007. Apple Polyphenols Influence Cholesterol Metabolism in Healthy Subjects with Relatively High Body Mass Index. *Journal of Oleo Science* 56: 417–28.

Okoko, B. J., P. G. Burney, R. B. Newson, J. F. Potts, and S. O. Shaheen. 2007. Childhood Asthma and Fruit Consumption. *European Respiratory Journal* 29: 1161–68.

Olthof, M. R., P. C. H. Hollman, and M. B. Katan. 2001. Chlorogenic Acid and Caffeic Acid Are Absorbed in Humans. *Journal of Nutrition* 131: 66–71.

Pearson, D. A., C. H. Tan, J. B. German, P. A. Davis, and M. E. Gershwin. 1999. Apple Juice Inhibits Human Low Density Lipoprotein Oxidation. *Life Sciences* 64: 1913–20.

Peng, Y., F. Liu, Y. Peng, and J. Ye. 2005. Determination of Polyphenols in Apple Juice and Cider by Capillary Electrophoresis with Electrochemical Detection. *Food Chemistry* 92: 169–75.

Pohl, C., F. Will, H. Dietrich, and D. Schrenk. 2006. Cytochrome P450 1A1 Expression and Activity in Caco-2 Cells: Modulation by Apple Juice Extract and Certain Apple Polyphenols. *Journal of Agricultural and Food Chemistry* 54: 10262–68.

Pokorny, J., and S. Schmidt. 2001. Natural Antioxidant Functionality during Food Processing. *Antioxidants in Food.* Sawston, UK: Woodhead Publishing, 331–54.

Prieta, J., M. A. Moreno, S. Diaz, G. Suarez, and L. Dominguez. 1994. Survey of Patulin in Apple Juice and Children's Apple Food by the Diphasic Dialysis Membrane Procedure. *Journal of Agricultural and Food Chemistry* 42: 1701–03.

Rice-Evans, C., J. N. Miller, and G. Panganga. 1997. Antioxidant Properties of Phenolic Compounds. *Trends in Plant Science* 2: 152–59.

Rupasinghe, H. P. V., and C. Kean. 2008. Polyphenol Concentrations in Apple Processing By-Products Determined Using Electrospray Ionization Mass Spectrometry. *Canadian Journal of Plant Science* 88: 759–62.

Sacchetti, G., E. P. G. Cocci, D. Mastrocola, and M. Dalla Rosa. 2008. Influence of Processing and Storage on the Antioxidant Activity of Apple Derivatives. *International Journal of Food Science Technology* 43: 797–804.

Scalbert, A., C. Morand, C. Manach, and C. Remesy. 2002. Absorption and Metabolism of Polyphenols in the Gut and Impact on Health. *Biomedicine & Pharmacotherapy* 56: 276–82.

Scalbert, A., and G. Williamson. 2000. Dietary Intake and Bioavailability of Polyphenols. *Journal of Nutrition* 130: 2073S–85S.

Schwarz, D., and I. Roots. 2003. In Vitro Assessment of Inhibition by Natural Polyphenols of Metabolic Activation of Procarcinogens by Human CYP1A1. *Biochemical Biophysical Research Communications* 303: 902–07.

Simon, D., L. R. Braathen, and H.-U. Simon. 2004. Eosinophils and Atopic Dermatitis. *Allergy* 59: 561–70.

Soler, C., J. M. Soriano, and J. Manes. 2009. Apple-Products Phytochemicals and Processing: A Review. *Natural Product Communications* 4: 659–70.

Surh, Y.-J. 2003. Cancer Chemoprevention with Dietary Phytochemicals. *Nature Reviews Cancer* 3: 768–80.

Takahashi, K., and C. Ra. 2005. The High Affinity IgE Receptor (Fc. epsilon .RI) as a Target for Anti-allergic Agents. *Allergology International* 54: 1–5.

Tokura T., N. Nakano, T. T. Ito, H. Matsuda, Y. Nagasako-Akazome, T. Kanda, M. Ikeda, K. Okumura, H. Ogawa, and C. Nishiyama. 2005. Inhibitory Effect of Polyphenol-Enriched Apple Extracts on Mast Cell Degranulation in Vitro Targeting the Binding between IgE and FcεRI. *Bioscience, Biotechnology, and Biochemistry* 69: 1974–77.

van der Sluis, A. A., M. Dekker, A. de Jager, and W. M. Jongen. 2001. Activity and Concentration of Polyphenolic Antioxidants in Apple: Effect of Cultivar, Harvest Year, and Storage Conditions. *Journal of Agricultural and Food Chemistry* 49: 3606–13.

van der Sluis, A. A., M. Dekker, and W. M. Jongen. 1997. Flavonoids as Bioactive Components in Apple Products. *Cancer Research* 114: 107–08.

van der Sluis, A. A., M. Dekker, G. Skrede, and W. M. F. Jongen. 2002. Activity and Concentration of Polyphenolic Antioxidants in Apple Juice. 1. Effect of Existing Production Methods. *Journal of Agricultural and Food Chemistry* 50: 7211–19.

van der Sluis, A. A., M. Dekker, G. Skrede, and W. M. F. Jongen. 2004. Activity and Concentration of Polyphenolic Antioxidants in Apple Juice. 2. Effect of Novel Production Methods. *Journal of Agricultural and Food Chemistry* 52: 2840–48.

Veeriah, S., T. Hofmann, M. Glei, H. Dietrich, F. Will, P. Schreier, B. Knaup, and B. L. Pool-Zobel. 2007. Apple Polyphenols and Products Formed in the Gut Differently Inhibit Survival of Human Cell Lines Derived from Colon Adenoma (LT97) and Carcinoma (HT29). *Journal of Agricultural and Food Chemistry* 55: 2892–900.

Wolfe, K., X. Wu, and R. H. Liu. 2003. Antioxidant Activity of Apple Peels. *Journal of Agricultural and Food Chemistry* 51: 609–14.

Xing, N., Y. Chen, S. H. Mitchell, and C. Y. F. Young. 2001. Quercetin Inhibits the Expression and Function of the Androgen Receptor in LNCaP Prostate Cancer Cells. *Carcinogenesis* 22: 409–14.

Yoon, H., and R. H. Liu. 2007. Effect of Selected Phytochemicals and Apple Extracts on NF-kappa B Activation in Human Breast Cancer MCF-7 Cells. *Journal of Agricultural and Food Chemistry* 55: 3167–73.

Yoshikawa, M., H. Shimoda, N. Nishida, M. Takada, and H. Matsuda. 2002. Salacia Reticulata and Its Polyphenolic Constituents with Lipase Inhibitory and Lipolytic Activities Have Mild Antiobesity Effects in Rats. *Journal of Nutrition* 132: 1819–24.

CHAPTER 18

Green Concepts in the Food Industry

Anwesha Sarkar, Shantanu Das, Dilip Ghosh, and Harjinder Singh

CONTENTS

18.1 INTRODUCTION

The world population has grown inexorably during the past two centuries. It is estimated that 7 billion humans will inhabit Earth by late 2011 and 9 billion in 2050.[1] Obviously the need to produce, process, and store more food to satisfy the demand of this ever-growing population is increasing day by day. Data indicate that the overall area of the world's agriculture-rich land has remained almost constant since 1980.[2] Moreover, rapid industrialization is gradually competing with arable land for its allocation to the industrial sector.[3] Thus, the gap between primary food production and consumption has increased exponentially.

This industrial development and the consequent change in employment structure from agriculture and cottage industries to more service-oriented sectors have caused the gradual migration of an increased proportion of rural populations to towns and cities. Urbanization has triggered environmental footprints, not only by transforming agricultural lands to areas for construction of urban dwellings,[4] but also by contributing to the emission of greenhouse gases from increased use of private automobiles for transportation and use of fossil fuels for generation of electricity.[5] Furthermore, urbanization has significantly contributed to the changes in the types of food demanded. The fast-moving and convenience-driven urban population and its changing lifestyles have diametrically shifted the dietary pattern from consumption of basic staples such as sorghum and millet to wheat (which can be processed into bread and pasta more easily).[6] Interestingly, increasing income and rapid urbanization have caused a marked shift in the food consumption pattern in the Asian subcontinent from cereals to meat and dairy products, which consequently has resulted in increased consumption of cereals for animal feed.[7] From the 1970s to the 1990s, consumption of meat and milk in developing countries increased by threefold compared with the twofold increase seen in developed countries.[8] For example, in India, there has been a sharp decrease in the consumption of the cereals (27.2% to 21.8%), particularly barley, maize, and tapioca, during the trade liberalization period (1987–1988 to 1999–2000), with a subsequent transition in food preference toward dairy products (from 13.23% to 15.9%) and processed foods (from 13.59% to 16.7%) in urban areas.[9] In general, we can say that the current population explosion and associated urbanization involving increased usage of resources pose direct threats to the environment, such as utilization of resources (food, water, and energy), environmental pollution, increasing levels of carbon footprints, waste generation, and global warming.[10]

Being food science professionals, we are mainly concerned about the environmental implications of the food miles covered within the cycle of a food product from procurement of raw materials until it reaches consumers,[11] which involves degradation of natural resources and triggers ecological footprints. For example, production of pasteurized milk involves a great deal of clean water, grain, and other foodstuffs cycled through cows to produce a small amount of quality milk during the milking stage; contributes to greenhouse emission during transport to dairy factories in refrigerated containers; involves energy and water in the processing step; and finally, contributes to noncomposting waste through plastic packaging. To address these food industry–linked environmental implications, our common goal should be to collectively act as a team (from food researchers to agricultural producers to food processers to distributers) to develop sustainable solutions. According to the United Nations, sustainable development implies development that "meets the needs of the present without compromising the ability of future generations to meet their own needs."[12] To ensure long-term environmental sustainability, one of the critical approaches is to improve the environmental performance of food products and processes continuously from farm to fork and beyond in such a way that the impact on the environment is minimal at each step, while satisfying consumers' demands for food safety, nutrition, health, convenience, lifestyle, and product preferences.

Such concerns for environmental issues and food security have culminated in the development of opportunities for "green foods." Green food can be defined in many different ways. For example, according to *Mosby's Dictionary of Complementary and Alternative Medicine*, "green food is a food mixture that includes avocado, sprouts, wheatgrass, and leafy greens."* However, in this chapter, green foods has been used to focus on the umbrella of foods that are generally fresh, chemical free, nutritious, natural, or produced in an environmentally sustainable manner.[13] Basically with green food, we limit ourselves to foods whose production has minimal environmental implications.

The aim of this chapter is to essentially focus on aspects of processed foods in terms of environmental issues, and consequently the corporate social responsibilities for dealing with environmental footprints, which fits within the context of this book. The chapter is organized as follows. The first part provides a review of the endemic scientific and policy research about the green trends adopted in recent times, followed by some major achievements from the food corporate sector. For many years, food processing industries have shown prominent focus in promoting environmental sustainability. This largely includes reducing water consumption, nonrenewable energy usage, and waste generation; effectively managing resources; packaging and transportation reduction initiatives; lowering of carbon footprints; and engaging in a range of initiatives with food chain partners so that the resources not only meet the needs of the present but also satisfy the demands of future generations. Some of the recent practical approaches that have been taken to work toward environmental sustainability in terms of effective resource utilization by food industries have been discussed using relevant case studies from leading food companies.

The second part includes a short section on responsibilities of food retailers involving sustainability throughout the food supply chain from producer to consumer. This involves case studies to focus on a more holistic approach that highlights the green initiatives taken by food retailers at each step of the production cycle to reduce their carbon footprint and energy usage.

The third part discusses a special area within green food: organic foods. Organic production refers to naturally grown food that limits or excludes the use of synthetic chemical fertilizers and pesticides, as well as rearing animals in more natural environments without the routine use of drugs, growth hormones, and antibiotics.[13] Here we attempt to discuss the issues in terms of commercialization and scale-up of organic foods, as well as gaining consumer interest.

Finally, this book chapter identifies future directions and priority areas for further action by food chain members, from farmers to retailers.

* Jonas, W. B. 2005. *Mosby's Dictionary of Complementary and Alternative Medicine*. St. Louis, MO: Elsevier Mosby.

18.2 WATER MANAGEMENT

As many as 3 billion people, or nearly one third of the world's population, will face a major water crisis by 2025.[14] An ever-growing population, diminishing water resources, and increasing industrialization are among the main causes of this forecasted water shortage. A number of initiatives have been taken to promote sustainable water management involving reduced water usage, recycling of water, proper treatment of effluent water with lowest contaminant levels for discharge, and water conservation in different sectors of economy, such as agricultural irrigation, urban water supply, and industries. In this section, our discussion is limited to the recent efforts and positive contributions of the food processing sector to maintain sustainable water with an attempt to address the serious predicted issue of global water scarcity using a few selected examples from the diversified food and beverage industries.

To address efficient water management, we will discuss the contribution of the food industries to continuously reduce water consumption in processing using proper recycling techniques without compromising food hygiene requirements and by conserving water throughout the food cycle, followed by a short section on water harvesting initiatives to contribute collectively to a positive water balance. In this chapter, no coverage is given to effluent treatment. This is not to indicate that treatment of wastewater leaving the processing plants, thereby reducing their chemical oxygen demand (COD), biochemical oxygen demand (BOD), and suspended solids to the acceptable limits before the treated water is discharged to natural ecosystems, is less relevant to water management. Effluent treatment technologies based on coagulation, adsorption treatments, sedimentation, filtration, aerobic and anaerobic digestion methods, and membrane technologies have been well documented[15,16] and hence are not considered in this chapter. However, progress in treatment of wastewater for reuse within the plant is briefly discussed.

18.2.1 Water Reuse and Recycling

The treatment of low-contaminated process water, as well as its reuse and recycling within the food processing plant, is one of the most important approaches for substantial reduction of freshwater consumption and wastewater generation. According to *Codex Alimentarius* guidelines, the term *reuse* has been defined as "the recovery of water from a processing step, including from the food component itself; its reconditioning treatment if applicable; and its subsequent use in a food manufacturing operation."[17] Despite the huge consumption of water per ton of processed food and the potential for reuse, only a few food processing operations use reconditioned process water back to their processing lines because of hygiene concerns and the strict regulations in this area.[18,19] However, recycling of water has become technically more feasible as a result of the development of better purification processes in recent times to meet quality standards.

Wastewater treatments basically include primary (removal of suspended particles by filtering, centrifugation, flocculation, or sedimentation), secondary (reduction of biochemical and chemical oxygen demand, removal of total suspended solids and organic carbon, removal of most of the microorganisms by aeration, clarification, nanofiltration, reverse osmosis, membrane filtration, aerobic digestion), and tertiary (removal of organic and inorganic material, dissolved solids by disinfection, ion exchange resins) clean-up procedures.[18,19] For example, a study conducted at the Max Planck Institute (Munich, Germany) suggested that the freshwater requirement for washing bottles in a beverage industry can be reduced by nearly 55% using recycled water (by careful design of a reverse osmosis membrane system connected to the washing equipment, followed by ultraviolet disinfection).[20] Besides use of a single-membrane processing system, a combination of membranes has also been used for water treatment. Such a combination of membrane processes like microfiltration or ultrafiltration with a conventional bioreactor is known as a *membrane bioreactor*.[21] Bioreactors essentially use microorganisms to break down the organic component in a reaction tank where the oxygen supply is controlled based on the type of reaction (aerobic or anaerobic), thus reducing the BOD of the water and making the contaminated water fit for reuse. Generally, bioreactors are followed by a clarifier or settling methods to separate the treated water from the biomass. In membrane bioreactors, the separation of the treated water from the microorganisms is done by membrane module. The coupling of such a bioreactor with an external membrane unit or internal unit (submerged membrane directly within the biological reactor) enables not only the recycling of the activated sludge back to the bioreactor but also critically allows the retention of the suspended solids, germs, and viruses that would have escaped in the treated water if settling was used as the final separation step.[22–24] Thus, it provides an attractive option for the treatment and reuse of processed water. It has been shown that using an integrated membrane filtration process, comprising a membrane-supported bioreactor and a combined nanofiltration/ultraviolet disinfection stage,[25] COD removal rates of greater than 95% were achieved. Moreover, after the disinfection, the biochemical parameters of the treated water even met the limits of the German Drinking Water Act. The membrane reactor coupled with reverse osmosis, designed by Alfa Laval (Lund, Sweden), provides an effective means to treat waste process water for its reuse as utility water.[26]

At present, water recycling and reuse practices in the food processing sector take place primarily to supply water for agricultural irrigation and within the overall food industry to supply water to cooling towers, wash water for equipment rinsing, or even for process water, especially after reconditioning.[27] In a span of 12 years (1995–2007), Unilever (London, UK) has reduced a major portion of its freshwater consumption (per ton of production) by maximizing water recycling on site.[28] For example, at the Caivano foods factory in Italy, treated wastewater is used in cooling towers instead of being discharged. Cadbury at its Valladolid factory in Spain has installed systems to recycle the water used in the vacuum systems on site.

Like the food sectors, critical efforts have been made by the beverage sector to recapture and purify process wastewater. The PepsiCo (Purchase, NY) bottling group in North America has equipped a reverse osmosis water purification system to recycle and reuse approximately 280 million gallons of process water used in production annually, thus minimizing freshwater usage. SABMiller (London, UK) also uses reverse osmosis systems to allow wastewater to be reused as service water. The company recently built a new water-recovery plant at its Tanzanian brewery that recycles 65% of the plant's wastewater from the brewing stages for use in cleaning processes that do not require high quality water. Kraft Foods, at its Jacksonville, Florida coffee plant, installed a closed-loop system to reuse water to cool coffee grinding equipment, thus reducing freshwater usage by more than 35%.

Not only the water recycled from the processing steps but also the water derived from raw materials can be intelligently used to support sustainable water management. A few selected case studies provide examples of some excellent innovations made by different food companies to use the inherent water of the raw material. For example, it is well known that potatoes contain a lot of water (~45%–60%). PepsiCo Walkers' business has reduced water usage at its potato chip facility using an innovative solution that captures 85% of a potato's natural moisture content from the potato chip fryer and reuses it as utility water. A significant proportion of freshwater was used to make Pomarola pasta sauces and Extrato Elefante tomato paste by Unilever at its Goiânia, Brazil factory in the past. Water was used to wash the tomatoes, clean the equipment, and also as a cooling agent. Innovative strategies by Unilever have enabled the use of water evaporated during cooking of tomatoes to wash equipment and for cooling. Nestlé China reduced total water usage in its China factories by 23% in 2005 by extracting water from fresh milk during the production of milk powder and reusing it to supplement the water used in boilers and cooling towers, as well as for general cleaning purposes.

Not only multinational organizations but also local industries have promoted promising initiatives that are worth mentioning here. For example, Nordzucker, a German sugar company, has reduced its water usage through reuse of water contained in sugar beet (~75% moisture content) for beet transport, washing, and evaporation purposes.[28] Approximately 90% of the water used in the company's sugar production originates from the inherent moisture content of the beet itself, which is recovered in different operations within the process, such as evaporation, crystallization, and refining. The water is reused approximately 20 times, and once it has been used to the full, it is treated and discharged into the natural water systems.

Moving to Southeast Asia, tapioca starch production is one of the major food industries in Thailand. The tapioca starch extraction process from cassava roots consumes a vast quantity of water and generates wastewater of nearly 12–20 m^3/ton starch produced.[29,30] The Department of Industrial Works in Thailand has launched a program to reduce environmental footprints from tapioca starch plants. To minimize freshwater consumption, the recycling of wastewater generated from the starch separator and starch dewatering centrifuge in both the fiber-separating stage and root washing within the production line has been suggested.[30] This proposed water reuse would help plants, which have production capacity of 180 tons/day and average water use of 33 m^3/ton starch, to reduce water consumption by approximately 5 m^3 water/ton starch.

18.2.2 Water Usage Reduction

Water is a prerequisite processing element in the food processing industry. Not only the actual processing but also the commercial cleaning, disinfection, washing, and sanitizing of fresh produce result in a generally water-intensive process. Implementation of dry clean-up procedures—vacuuming rather than hosing, using mixed wet and dry cooling systems—can be an effective means to reduce some of the utility water usage. For example, Cadbury at its factory in Beirut, Lebanon has been shifting toward dryer cleaning methods. A promising initiative by PepsiCo to reduce water usage gave rise to a waterless rinsing technology at Gatorade's US plants in 2008. Using air rinsing in their Gatorade and Propel bottling lines saved more than 5 million liters of water per year from going through the sewage system. The Cargill Malt (Minneapolis, MN) plant in the United States is saving 264 million gallons of water a year through tank automation technology with proper controls to prevent overflows, and improved awareness and education.

Cleaning in place (CIP) is commonly used in the food industry for ensuring hygienic safety of foods and for recovering plants' performance by cleaning tanks, piping, and even workspaces by circulation of detergents, alkalis, acid solutions, and rinsing water. Traditionally, a 7-step CIP involves a water rinse, alkaline wash, hot water treatment, acid rinse, water rinse, disinfectant cycle, and potable water rinse or hot water sanitization when a chemical disinfectant is used, thus consuming a large volume of water. With the latest advancements, CIP has been achieved in only three steps, namely, water rinse, detergent and disinfectant cycle, and potable rinse to ensure minimal water usage.[31] With proper monitoring of the plant processes, it might be possible to maintain the hygiene standard by carrying out a 3-step CIP more frequently, followed by a 7-step CIP rarely (only when required), thus contributing to reduction in water usage. CIP with ozone treatment is also an emerging technique in which disinfection by ozone can remove water-rinsing steps and can contribute to gallons of water saving.[32] Because ozone (O_3) breaks down into oxygen (O_2), ozonated water from the final rinse can be reozonated and used for the first rinse, reducing the intermittent water usage.

Apart from manufacturing plants, food industries have also tried to work on reducing water consumption by engaging in a range of initiatives with food chain partners. For example, Nestlé, Unilever, and PepsiCo have collaborated with farmers to reduce water usage by up to 50% in agriculture using sustainable drip irrigation technology rather than traditional flood irrigation systems for coffee, tomato, and potato farming, respectively.

18.2.3 Water Harvesting

Broadly, reusing the water within the plant or cutting down the water usage as discussed in the previous section seems to be effective in reducing freshwater consumption. However, maintaining sustainable water supplies is also equally important. Rainwater harvesting is an efficient way of saving our valuable water supplies. It basically implies collecting and storing rainwater that runs off a catchment area in a reservoir during the rainy season so that it can be used later.[33,34] For example, PepsiCo has constructed rain and roof water harvesting structures across its network of manufacturing facilities to conserve water. At its Gatorade plant in Mexico, a rainwater harvest system was installed, which collected 640 m^3 of water in 2009, thus helping to recharge the underground water. Over the years, Coca-Cola (Atlanta, GA) and its bottling partners in India have installed more than 500 rainwater harvesting structures across the country to replenish water sources. With the goal of replenishing the amount of water used in their processing plants, Coca-Cola partnered with the World Wildlife Fund to measurably conserve seven of the world's most critical freshwater river basins (e.g., Yangtze in China, Rio Bravo in Mexico). Local watersheds harvested in various countries also are worth mentioning. For example, in 2010 Coca-Cola India partnered a project with the state government to revive the 1000-acre Nemam Lake, near Chennai, Tamil Nadu, to create a water-recharge capacity in excess of 150,000 kL in the first phase. At its company headquarters in Northfield, Illinois, Kraft Foods has three lakes on the corporate campus that capture rainwater to supply approximately 50% of the property's irrigation needs.

Besides the direct efforts in water conservation, many food and drink companies have contributed to water sustainability indirectly through partnerships with government and nongovernmental welfare organizations. Project WET (who uses the slogan "Preparing Future Generations on Water Issues"), supported by Nestlé Waters, has helped to train millions of children and teachers to increase awareness and promote an understanding of water conservation.

Finally, the food industries are striving toward positive water balance by saving and recharging more water than their usage in the processing plants. Furthermore, they also return only cleaned water back into the environment by efficiently removing most of the organic load of the water, allowing the factories to reduce their environmental footprint.

18.3 WASTE MANAGEMENT

According to the definition by the United Nations Statistics Division:

> Wastes are materials that are not prime products (that is products produced for the market) for which the generator has no further use in terms of his/her own purposes of production, transformation or consumption, and of which he/she wants to dispose. Wastes may be generated during the extraction of raw materials, the processing of raw materials into intermediate and final products, the consumption of final products, and other human activities. Residuals recycled or reused at the place of generation are excluded.[35]

Although this broad definition applies to most sectors of industries and agriculture, it is confusing with reference to the status of so-called by-products that are generated during food processing in addition to the primary product (e.g., buttermilk from butter production). For most food industries, these by-products are not considered waste and are used. However, in many smaller food industries, these by-products are discharged to the environment; thus, it seems difficult to apply the definition precisely. In this section, our discussion will be mainly focused on the actions to reduce waste generated in the food industries mainly by improving by-product utilization.

Food industries use the majority of raw materials from animal and plant sources to produce a broad range of finished processed products. From an environmental point of view, processing of such raw materials of agricultural origin produces substantial amounts of by-products and solid waste. The main treatment method of solid wastes is generally composting.[36] However, utilization of resources in an efficient manner involving recovery and reuse of by-products and wastes as raw materials is a promising approach to reduce the amount of waste generated. Not only utilization of such coproducts and by-products in the plant itself but also the contribution of animal feed (e.g., spent grains, distillery waste), fertilizers (e.g., spent Kieselguhr, carbonation sludge),[37] biodegradable plastics (e.g., brewery malt waste–based polyhydroxybutyrate),[38] and biofuels (e.g., biodiesels)[39] can lead to significant reductions of biowaste generation and related greenhouse gas emissions. In this section, we will discuss primarily by-product utilization within the food processing plants by reviewing research from academic institutes and case studies from food industries. We will highlight the initiatives implemented by food industries to promote responsible packaging solutions, including reuse, recycling, and recovery together with use of biodegradable plastics. Detailed information on by-product utilization and waste management in food industries can be found in a number of recently published books and reviews.[37,40–44]

18.3.1 By-Product Utilization

Among the green approaches, waste reduction is one of the most critical concerns of the food and beverage industry. To reduce waste generation, full by-product utilization seems to be the most promising area where continuous innovations can significantly contribute not only to waste management but also can generate value-added products. Utilization of dairy by-products, such as skim milk, has been known for centuries.[45]

Another important dairy by-product is whey. Whey is the yellow-green watery liquid that separates from the curd during the manufacture of cheese and casein. Whey mainly contains lactose and whey proteins; thus, its high BOD is a major threat for disposal.[46] However, this particular dairy waste should be considered as an expensive raw material from which value-added products can be generated. Whey protein comprises only 20% of milk proteins, but its nutritional quality is much higher than casein.[47,48] It is rich in all the essential amino acids and has a protein efficiency ratio of 3.2 compared with 2.6 of casein, which is generally considered a reference protein. The digestibility is very high, approximately 97%–98%, so it is near to an ideal protein. It possesses bioactive peptides, which have potent roles in dealing with severe medical conditions, such as cardiovascular diseases, hypertension, and obesity. The industrial manufacture and utilization of whey components, particularly whey protein, has been known since the nineteenth century,[45] but has progressed considerably as a result of pioneering research in whey protein carried out at the Fonterra Research Centre, New Zealand (formerly the New Zealand Dairy Research Institute). Advanced approaches have enabled the utilization of whey on a commercial scale for the manufacture of whey protein isolates using membrane filtration processes, allowing further use of whey as a functional ingredient in other processed food products, such as yogurts, beverages, dairy desserts, nutritional products, and infant food. It is worth including here the latest research in whey utilization by the Riddet Institute (Palmerston North, New Zealand). One of our authors, with his team in collaboration with the Fonterra Research Centre, is investigating the development of highly functional nanofibrils from β-lactoglobulin (a protein extracted from whey).[49] Dairy protein nanofibrils have potential applications in the food industry because of their ability to enhance viscosity and form gels at lower protein concentrations than with random aggregates.[50] Hence continuous innovations in the field of dairy by-products may lead to the development of many high-valued products with improved functionalities, together with preventing the problem of disposing huge amounts of waste. Although whey protein ingredients derived from whey from large-scale food industries have a prominent share in the market, the major production of whey from small-scale industries still remains unused. It is a great challenge to target such dairy industries and collect whey, a high-quality, low-cost protein source that could shield large sections of the global population from protein deficiency and contribute to a reduction of environmental pollution.

In the meat industry, except for dressed meat, everything including organs, fat or lard, skin, feet, abdominal and intestinal contents, bone, and blood of animals are by-products.[51] Significant quantities of ovine blood (e.g., nearly 40 million liters annually in New Zealand alone) are commercially available as a by-product[52] and are currently used as low-valued products (e.g., blood meal, fertilizer) or discarded as effluent. It would be more beneficial if this blood were further processed into high-valued products, which also would contribute to effective waste management. Recent studies in our laboratory at the Riddet Institute investigated the utilization of ovine blood[53,54] for the development of immunoglobulin concentrate. Immunoglobulin concentrate containing approximately 84% of total protein (including ~73% IgG) was efficiently isolated from ovine blood through a purification procedure. Authors have demonstrated the in vitro antipathogenic activity of ovine serum immunoglobulins and have showed that ovine immunoglobulin fraction selectively improved the growth performance, organ weight, and gut morphology in growing rats. Besides utilization of sheep blood, other parts such as bone residues, mechanically separated meat, and bone cakes from mechanical separation can be used to manufacture meat protein hydrolysates.[55] The utilization of such low-valued mechanically separated lamb meat for the development of high-valued meat protein hydrolysate with proven nutritional characteristics is an interesting ongoing research activity of the Riddet Institute (S. Das, personal communication, June 2010). Thus, these studies indicate a promising way of processing a by-product that may result in the development of a high-valued nutraceutical, together with significant contribution to waste reduction in meat industries.

In the fruit and vegetable processing industries, by-products, particularly seeds such as mango kernels and tomato seeds, can serve as almost no-cost proteinaceous raw materials for high-quality protein preparation. For example, the mango kernel, which is a by-product of mango juice and the pulp industry, is used for preparation of fats for confectionaries because of its characteristic resemblance to expensive cocoa butter fat.[56] Interesting research carried out by one of our authors at the Central Food Technological Research Institute (Mysore, India) showed efficient use of vegetable processing waste, in particular tomato seeds and skins, for the development of a nutraceutical protein hydrolysate containing tomato.[57,58] Tomatoes are an integral part of diets worldwide and are processed industrially to produce juices, ketchup, sauces, paste, puree, and powder.[59] During commercial processing, almost 40% of the raw material is removed as waste after the juice or pulp extraction process and mainly consists of peel, seeds, fibrous matter, trimmings, core, and cull tomato.[60] This waste is used to some extent as fertilizers but generally contributes to significant environmental implications. Of the waste generated, the tomato skin is well known as an excellent source of lycopene, an antioxidant carotenoid. Interestingly, tomato skins can contain up to 5 times more lycopene than the pulp.[61] Moreover, tomato seeds have been indicated as a potential source of good quality plant protein with high lysine content and relatively higher protein efficiency ratio compared with other plant proteins such as soy, wheat, and oats.[57,58,62,63] Interestingly, blending of wheat flour with tomato seed meal (which is of no cost because it is a by-product) can result in lysine supplementation of bread.[64,65] Thus, the utilization of these tomato processing wastes (seeds and skin), which are available at zero cost, can contribute to the development of a value-added nutraceutical, together with its implication in reducing solid waste. A potential way forward in this area has been taken recently by a newly formed Italian company, BioLyco (Rome), which has patented a process for industrial manufacturing of lycopene from tomato processing waste in collaboration with the University of Rome.[66]

Not only by-product utilization within the plants but also use of wastes for development of animal feed has been practiced for years. For example, once the best-before date is reached, some dairy products need to be disposed of in landfills. Fonterra (Auckland, New Zealand) is recycling and reusing their expired or end-of-run product into the animal feed market, thus significantly reducing the amount of total waste to landfills. Besides dairy, vegetable, and meat processing, there are huge opportunities for utilization of by-products in other sectors of food processing as well (such as production of biopolymer chitosan from seafood waste[67]). Efficient utilization of these by-products will contribute enormously to cutting down the amount of waste going to the final disposal.

18.3.2 Green Packaging Initiatives

Food packaging plays a vital role in protecting, containing, and preserving the food product during transportation and storage, and provides commercial and consumer information. Most importantly, packaging is essential to ensure safety, quality, and hygiene of the food product. A wide range of packaging materials are used for food packaging, including paper or pulp-based materials, metal, glass, wood, plastics, or a combination of more than one material as composites. Readers are referred to the references[68,69] for in-depth information about progress in food packaging. The food and beverage industries are fully aware not only of the necessity and benefits of packaging use but also recognize the environmental implications of packaging, and thereby their responsibilities in reducing packaging waste for landfills. Significant improvement has been achieved in reducing packaging waste through reduction in packaging use, recycling or reuse, and finally use of more environmentally friendly packaging solutions, such as biodegradable packaging.

In effort to minimize packaging impact, Frito-Lay (the snacks division of PepsiCo) is reusing its paperboard cartons five or six times each, and then recycling, thus requiring usage of 120,000 fewer tons a year. Continuing PepsiCo's initiatives in gaining sustainability, Propel bottling line uses 33% less plastic than the previous 500-mL bottle and 30% less labeling material using new packaging technology. Nestlé Waters in North America has engineered a 12.5-g Eco-Shape half-liter bottle in 2008, contoured especially to keep its shape with less amount of plastic usage, thus contributing to reduction of packaging footprints. Similarly, over the years Coca-Cola has reduced its global polyethylene terephthalate (PET) usage by more than 10,000 tonnes using innovative processing and packaging redesign efforts.[28] Unilever's progress in packaging reduction is also worth mentioning. A new design for the Knorr Recipe Secrets soup pouch by Unilever has eliminated the need for an outer carton, thus generating 50% reduction in overall packaging materials. Replacement of Knorr Pomarola cans weighing 53.1 g with pouches weighing only 7.5 g has saved nearly 4000 tonnes of packaging in 2009 at Unilever, Brazil. One of the most revolutionary innovations in the dairy industry in terms of reducing packaging footprints seems to be the use of handle-free milk bottles. Dairy Crest, an Esher, UK-based chilled dairy products company, has developed a prototype handle-free plastic bottle to reduce the usage of high-density polyethylene plastics by nearly 5000 tons annually. The 1- and 2-pint bottles require 10% less plastic to make than standard bottles[70]; thus, the large-scale manufacturing of such handle-free bottles can potentially reduce packaging consumption and waste. In the beverage industries, hot filling of products is an age-old technique, which generally requires plastic bottles with higher thickness to sustain the temperature of nearly 85°C. Intuitively, a properly monitored aseptic packaging environment with cold filling (using food-grade preservatives) in plastic bottles of reduced thickness may be used to cut down the quantities of plastic consumption enormously without compromising on safety, hygiene, and shelf life of the product.

Confectionary industries have shown significant reduction in their packaging usage. For example, Nestlé UK showed promising efforts in replacing nonrecyclable plastic with recyclable cardboard packaging in most of its 20 million Easter eggs. Cadbury is also working continuously on packaging reduction. For example, in 2008 Cadbury Easter eggs had no cardboard box; rather, they were wrapped in foil, which reduced usage of plastic and cardboard by 78% and 65%, respectively, than a standard egg. The company has also reduced the amount of plastic casing in small- and medium-sized Easter eggs, thereby saving 202 tons of plastic. Finally, Cadbury has reduced cardboard usage in transporting the eggs from warehouses to stores, thus saving 117 tons of cardboard, or metaphorically 2000 trees.

Besides the "3 Rs" of reduce, recycle, and reuse, food companies have identified that using or making environmentally friendly biodegradable packaging materials, that is, polymers obtained from renewable resources and that can be recycled and composted,[71] is one of the most sensible options to reduce environmental implications of packaging. In the past few years, biodegradable plastics, such as polyhydroxyalkanoates, polylactic acid, and starch-based polymers, have received considerable attention. In 2005, Nestlé pioneered a new technology in Europe by launching new packaging trays for Nestlé Dairy Box and Black Magic in the United Kingdom. The trays are bioplastics, that is, made from maize starch, a renewable resource, and hence are compostable. In the same line of working toward a sustainable environment using greener packaging, Frito-Lay started manufacturing decomposable (within 14 weeks) innovative bags for packaging its potato chips. This biodegradable bag is made by using polylactic acid in place of traditional packaging, which was generally constructed from multiple layers of polyolefin materials. Regarding sustainable packaging innovations, Coca-Cola initiated PlantBottle packaging, which is partially made from plant materials, that is, PET plastic that is manufactured using up to one-third of its materials derived from sugar cane or molasses, a by-product of sugar cane processing, and thus resulting in a reduction of carbon emissions by approximately 25%. These examples of technological developments (although not exhaustively reviewed) may serve as motivation for other food and beverage companies to address sustainable packaging challenges.

Although most efforts made by the food and beverage industries in waste management are dedicated toward reduced waste generation by efficient by-product utilization and reducing packaging footprints, the inevitable organic wastes are generally directed toward composting in the presence of oxygen or by anaerobic digestion in the absence of oxygen. Wastes from diversified sectors of the food industry, such as meat processing, grain processing, and fruit and vegetable processing, can contribute to composting, creating a suitable source of natural fertilizer for agricultural use.[36] Case studies involving on-site composting in food industries have not been well reported. However, examples from Fonterra and Nestlé are appropriate here. Fonterra has on-site worm composting operations throughout its New Zealand factories, whereby food wastes are treated by the worms, and the resultant manure is used to grow plants and trees around the factories. Similarly, Nestlé built its own composting operation in New Milford, Connecticut in 1990 to recover waste, including spent coffee grounds, tea leaves, and residues from the production of confectionery and seasonings. Nestlé's factories in many regions also are using spent coffee grounds as biofuel, thus reducing 800,000 tons of spent coffee grounds in landfills annually. Hence these kinds of promising green initiatives by food industries are not only contributing to waste management but also considerably to energy management, which is discussed in our next section.

18.4 ENERGY MANAGEMENT

Recently, increased accumulation of CO_2 and other greenhouse gases has led to the most pressing and complex global environmental issues, that is, global warming and climate change. Approaches to ensure reduction in greenhouse gas emissions involve working toward decreased dependence on nonrenewable energy sources and identifying energy-saving opportunities. The generation of power using windmills in New Zealand and Australia[72] is an excellent example demonstrating that switching to greener power can enormously reduce carbon footprints. Together with other economic sectors, food and beverage companies are also showing indisputable attempts to ensure efficient energy performance and reduced generation of carbon footprints. In this section, we will discuss the steps taken by the food industries to ensure reduction of energy usage, which broadly include voluntarily cutting of energy use, reducing impacts of transportation, and using sustainable energy, including switching to renewable sources of fuel.

18.4.1 Processing Approaches

In general, food industries are considered to be less energy intensive compared with many other industrial sectors.[73,74] However, to maintain the quality and hygiene of food products and to make them safe for human consumption, the food industries incorporate significant amounts of energy and electricity primarily in two kinds of processing: heat processing (e.g., pasteurization, sterilization, evaporation, drying, blanching) and cold processing (e.g., refrigeration, freezing).

The dairy industry, for example, has shown genuine leadership in energy savings since the nineteenth century, particularly in reference to heat treatment. In this section, we will focus on some of the age-old techniques used in the milk processing sector (well documented in the literature and textbooks), which have contributed enormously to energy management. High-temperature, short-time (HTST) pasteurization is a well-established procedure for destroying the pathogenic microorganisms in fluid milk. For production of pasteurized fluid milk, HTST pasteurizers (basically plate heat exchangers) are generally used, which typically have a regeneration section to optimize the heat recovery between a cold raw product and the same hot pasteurized product. Basically, heating and cooling energy can be enormously saved (~90% or more) by preheating incoming raw milk with outgoing hot pasteurized milk and vice versa. Even ultrahigh temperature (UHT) treatment of milk is a process of high bactericidal effect, in which the milk is heated at approximately 135°C–150°C for a few seconds followed by integration with aseptic packaging in sterile containers, allowing milk to be stored at ambient temperature for more than 6 months without any deterioration in its quality.[75,76] These UHT plants also use some heat regeneration, that is, hot milk heats the incoming cold milk. In this way, indirect heating systems can achieve 80%–95% energy recovery. Moreover, because of the shelf stability of the processed UHT products, this process eliminates the huge amount of energy usage in refrigeration during transportation, distribution, and storage.

In cases of drying milk to milk powders, if spray-drying were done directly to the fluid milk, it would consume a huge quantity of heat energy. This is the reason why milk is first evaporated to 40%–60% solids before proceeding further in drying operations, which potentially contributes to energy savings. Compared with separation by evaporation, the energy consumed in spray-drying is 10–20 times higher per kilogram of water removed. Because milk is susceptible to heat-induced deterioration, evaporation is carried out under vacuum to reduce the boiling point of milk. This not only hinders heat damage of milk but also contributes to energy savings, much more than if the evaporation was carried out under atmospheric pressure. Moreover, the use of multiple-effect evaporators compared with single-effect ones further adds to the energy efficiency by using the latent heat contained in the discharged vapor from the boiling milk in successive effects. Thus, each successive chamber uses the vapors from the preceding effect as the heat source, thus basically reusing the energy rather than using fresh steam, contributing to potentially huge energy savings. Mathematically, adding one effect to the original evaporator decreases the energy consumption by 50%. Adding another effect reduces it to 33% and so on, up to the addition of seven effects. The vapor from the preceding effect can be further recompressed for higher steam economy by use of thermocompression or mechanical vapor recompression and then used to heat the milk in the next effect of the evaporator, which may be operated at a lower pressure and temperature than the preceding effect. Compared with evaporation, concentration by membrane filtration involves significantly lesser amounts of energy and can be a suitable option when milk concentrations of up to 12%–20% solid are required.[77] Thus, evaporation of milk in the dairy industry is a sophisticated, state-of-the-art example of energy management.

Another important aspect to reduce energy usage in the dairy industry is cold pasteurization, that is, the use of inherent lactoperoxidase-H_2O_2-thiocyanate (LP) system in milk.[78,79] Lactoperoxidase, an enzyme naturally present in milk, acts in the defense against microbial activity in raw milk in the presence of thiocyanate and hydrogen peroxide. Owing to its bacteriostatic effect, the LP system can maintain the initial quality of milk for 4–7 hours at 30°C–35°C and for 24–26 hours at 15°C.[80,81] Although it cannot replace pasteurization of milk in the processing plants, it can definitely help to preserve raw milk for short periods, for example, preserving the quality and hygiene of milk in collection centers in developing countries, and thus contributing potentially to energy savings. For detailed information on energy-saving aspects, readers are referred to the references on unit operations in the food and dairy industries.[82–87]

Food industries use cold processing such as refrigeration in a number of ways, for example, to chill or freeze food for preservation, to make frozen desserts such as ice cream, and many more. The synthetic refrigerants, that is, fluorocarbons, are used as refrigerants in most refrigeration systems since the 1930s.[88–90] Certainly, fluorocarbons were considered to be more effective than ammonia, one of the natural and main refrigerants that had been used until then, because of its toxicity, pungent odor, and moderate flammability. However, lately scientists have realized that these fluorocarbons have significant environmental impacts with respect to the depletion of the ozone layer and global warming experienced during the past decades. Hence returning to the use of natural substances such as air, water, ammonia, and CO_2 for refrigeration purposes appears to be more promising. The food industries are collaborating with different research institutes for the development of viable alternative refrigerants that have fewer environmental implications. In 2003, Nestlé pioneered a remarkable initiative in reviving the use of CO_2 in industrial refrigeration at its Jonesboro facility in Arkansas for the company's Stouffer's frozen meal line. Using the patented technology of Praxiar Technology Inc.,[91] Nestlé installed a cascade system, which is basically a hybrid of ammonia refrigeration with a secondary loop using CO_2. This system allows operation of freezing systems at temperatures below traditional ammonia-based refrigeration systems while eliminating the presence of ammonia in their processing rooms. In 2005, Nestlé extended the same technology to its Bangchan ice cream factory in Thailand, thus promoting the commercial use of natural refrigeration. These kinds of innovations in science and technology, together with their commercial applications in the food industries, can motivate the stakeholders in diversified food sectors for generating energy savings, and thus contributing holistically to a greener climate.

Recently, food industries have also shown promising initiatives in dealing with energy management by switching to renewable sources of energy. Data show that over 11 years (1995–2006), Unilever reduced CO_2 emissions per tonne of production by 33.5%.[28] Of the total energy used by Unilever sites, 14.8% comes from renewable sources. With continuous energy-saving efforts and by switching to solar energy sources, Nestlé Purina PetCare's Denver plant has reduced energy usage by 12.4% per tonne of product. Fonterra's Awaroa dairy farm in Edgecumbe, New Zealand is also using solar power to heat the water that is used for cleaning the milking plant, which typically accounts for 30% of the dairy farm's electricity use. Frito-Lay's popular line of multigrain snack called Sunchips is made using solar power, thereby eliminating more than 1.7 million pounds of CO_2 emissions every year. Similar green initiatives by Cadbury Schweppes include the use of hydroelectric energy for its facility in Claremont, Tasmania; solar energy for its plant in Karachi, Pakistan; and wind energy in Ireland, reducing the carbon emissions enormously.

These case studies are only a few examples of technological advancements that are expected to deliver significant greenhouse gas savings. After processing in the factories, the food products need to be distributed to the point of consumption. The next section considers the fuel usage and carbon footprint generated during this transportation.

18.4.2 Green Transportation Initiatives

The term *food miles* as introduced previously in this chapter is defined as the distance covered by the food from the location where it is produced to the destination where it is consumed; that is, the distance covered by the food during its passage through the food cycle from farm to fork.[92] Because the population around the world is concentrated in urban areas and because most food industries use agricultural products as raw materials, which are mainly generated in rural sectors, food miles have increased at an escalating rate during the past few decades. Consequently, it has resulted in environmental implications in terms of increased fuel use and greenhouse gas emissions. In view of contributing to the wider context of greener transportation, food industries are pursuing a range of initiatives to curb the energy intensiveness of the food transportation system by optimizing transport efficiency.

One of the easiest ways to reduce transportation-induced carbon footprints is maximum utilization of pallets on loading trucks. Using proper calculations to consolidate more products to be loaded on each pallet and loading more pallets on each truck can lead to fewer journeys, less fuel use, and therefore less climate impact. For example, in 2003 Nestlé Waters in Canada increased the payload (load in truck) from 23 to 27 tons as a result of the approval by the Canadian government for the use of three-axle trucks. Furthermore, the addition of an extra layer of bottles to each pallet, together with the use of lightweight trucks, increased the payload to 30 tons and thereby reduced Nestlé's CO_2 emissions by 24% in Canada.

The environmental impact of transport is largely dependent on the mode of transportation. In general, road transport is the main source of greenhouse emissions compared with other means of transport, such as rail transport and shipping. An intelligent approach to reduce road transport–induced carbon footprints is to use a combination of transport modes, that is, road and rail transport. This approach involves transporting food products in containers with the longest part of the journey accomplished by rail, followed by local delivery via road transport. A case study from Taiwan compared the CO_2 emissions of road transportation with intermodal coastal shipping and truck transportation and inferred that replacing long-haul trucks with intermodal transportation can significantly reduce CO_2 emission.[93] Using a similar approach, Nestlé Waters in Switzerland has combined road and rail transportation for product deliveries, which resulted in a 62% reduction in CO_2 emissions along that route. Nestlé Waters has also attempted to bring the production sites of multi-spring brands closer to areas of consumption, thereby reducing food miles and road-induced carbon footprints. By changing the mode of transportation of cocoa beans from Ghana to the United Kingdom, Cadbury Schweppes has significantly contributed to its greener responsibilities. Previously, these beans were shipped to southeast England and then transported via road to the cocoa-processing factory at Chirk in northern Wales, involving 2150 loads traveling more than 500,000 miles. Now the beans are directly transported via shipping to Liverpool in northwest England (40 miles away from the processing plant), which has resulted in 1600 loads traveling only 135,000 miles, thereby reducing environmental impact by 75%. For many years, river transportation has been a significant part of the Danone (Paris, France) supply chain, keeping 5000 trucks off the road annually.

In summary, water, waste, and energy management are not generic across the diversified areas of the food processing industries. The approaches for efficient utilization of resources and reducing environmental impact from the food and beverage industries are largely dependent on the entire manufacturing and distribution process of that particular product. Raw materials, intermediate product, final product, processing steps, type of packaging, distribution, and many other factors need to be carefully studied, and potential methods to reuse, recycle, and replenish need to be meticulously identified for innovation and execution of greener processes. Finally, proper integrated approaches need to be adopted to reduce environmental footprints to an absolute minimum.

18.5 GREEN INITIATIVES BY RETAILERS

Food industries are connected with the retail industry, which can be broadly classified into three areas: food service (e.g., quick service store), modern trade (e.g., supermarket, hypermarket, quick service store), and traditional trade (e.g., corner store, "mom and pop" store). Without understanding the sustainability drive by the retail sector, the discussion on green initiatives in the food sector will remain incomplete. Sustainability initiatives of retail sectors in terms of minimizing environmental impact broadly involve energy and water conservation, reduction of paper usage, introduction of biodegradable plastics, and reduction of wastes throughout the entire life cycle of a product. In this section, we attempt to delineate the green initiatives by retailers using some specific case studies.

18.5.1 Food Services

Food services chains, such as Starbucks (Seattle, WA), McDonald's (Oak Brook, IL), and Yum! Brands Inc. (KFC, Pizza Hut, and Taco Bell; Louisville, KY), have expanded at an escalating rate, and thus have huge environmental implications. Similar to efforts of "going green" in the food manufacturing sectors, fast food services have also launched several initiatives to reduce their environmental footprint, allowing people to consume sustainable green products. For example, over the years, Starbucks, an international coffee house chain, has promoted environmental sustainability by incessant innovations in their disposable coffee cups. Interesting breakthroughs have been the development in 1997 of the recycled-content cup sleeve as a way to protect customers from hot beverages and avoid the waste of double cupping. During the next 9 years, Starbucks developed the first hot beverage paper cup with 10% postconsumer recycled fiber. In addition to the packaging initiatives, the contribution for ethical sourcing of raw material, that is, coffee, is phenomenal. The company promotes purchasing of certified organic coffee, which is produced with minimum use of toxic pesticides and fertilizers, thus contributing to a healthy supply chain (detailed information on organic production is provided in the next section).

McDonald's, one of the world's largest chains of hamburger restaurants, is continuously striving for sustainability toward a greener environment. More than 50% of the McDonald's restaurants in the United States participate in an innovative oil recycling program, resulting in the recycling of nearly 13,000 pounds of used cooking oil per restaurant each year. The McCafe business division of McDonald's sources its coffee beans ethically from Rainforest Alliance-certified farms of Brazil, Colombia, and Costa Rica, which meet specific and holistic standards balancing all aspects of green production.

Key approaches adopted by Yum! Brands Inc. to promote a sustainable environment include use of solar energy, rainwater harvesting, composting, recycling of waste including reclamation of oil, and effectively managing the supply chain from raw material procurement, including livestock and agricultural produce, to the restaurant food preparation and delivery.

18.5.2 Modern Trade

In view of a greener approach, the Bracknell, UK supermarket chain Waitrose developed an innovative strategy to sell milk in environmentally friendly plastic sacs called Eco Paks in a couple of its stores in 2007. The bags contributed to a great reduction in plastic usage (almost 75% less plastic than bottles). However, because of consumer unacceptability and low sales, this product was withdrawn, and Waitrose shifted back to using plastic bottles in 2010. Interestingly, its competitor Sainsbury's (London, UK) started selling milk in plastic bags at all its stores after testing the packaging for 18 months in 50 stores. Other packaging-reduction techniques that companies are using include hybrid packaging that puts a plastic bag within a paper shell, like the GreenBottle used for milk at Asda (Leeds, UK) stores. Walmart (Bentonville, AR) and Whole Foods Market (Austin, TX) have shown leadership in largely switching to renewable sources of energy, such as solar and wind power, thus reducing their environmental footprints. They are also consolidating their trucks in an efficient manner to have reduced transportation impact and fewer greenhouse gas emissions.

Similar to the greener efforts of Walmart in the United States, another UK supermarket, Tesco (Cheshunt, UK), has adopted sustainable approaches. While cutting emissions, one of the remarkable green achievements of Tesco is its first eco store, that is, a "zero-carbon store," which opened in Ramsey, Cambridgeshire in 2009. The building is made from sustainably sourced timber, uses solar power and roof lighting, harvests rainwater for flushing the toilets and car washing, and uses broadly energy-efficient equipment.

Another important sector of retail is traditional trade, in which information about green initiatives is relatively less documented. This is largely because the nature of the business is smaller, localized, and largely unorganized. However, through personal insights and experiences, it can be stated that these retailers are also showing prominent green approaches in terms of shifting from plastic to biodegradable packages in keeping up with the global context of a sustainable environment.

Although green approaches in terms of efficient management of water, resources, waste, energy, packaging, transportation, and distribution are most relevant to the overall food, industrial, and retail sectors, sustainable agricultural production to source raw materials cannot be neglected because farming is the foremost point of the food chain. Although the food and retail industries are not directly engaged in agricultural activities, they strongly promote a holistic approach to sustainable agriculture and organic farming to secure safe and ethical food supplies. Hence our next section will discuss the recent perfectives on organic and ethical food production.

18.6 ORGANIC FOOD PRODUCTION

Consumer attention toward the purchase and consumption of organic food has increased markedly, particularly in relation to health and environment issues. There is a widespread belief that organically grown products are safer, greener, tastier, and, above all, healthier than conventionally grown products.[94] This trend has driven an increase in organic food research, especially as marketers seek to understand motivations such as environmental concern, organic purchase intentions and behavior, availability, quality, price consciousness, subjective norms, and risk aversion. Because of the lack of convincing scientific data, these discussions sometimes seem to be rather ideological than scientific in nature.

18.6.1 Does Green Look Beautiful in Nature?

This is a widely used phrase: "It's NOT EASY being green." In reality, being green is tough, so tough that the color itself fails dismally. The truth is that the color green, the most powerful symbol of sustainable design, is not ecologically responsible and can be damaging to the environment. From a chemistry perspective, green-colored plastic and paper cannot be recycled or composted safely; green also is a difficult color to manufacture, and toxic substances are often used to stabilize it.

As reported in the *New York Times*[95]:

> Take Pigment Green 7, the most common shade of green used in plastics and paper. It is an organic pigment but contains chlorine, some forms of which can cause cancer and birth defects. Another popular shade, Pigment Green 36, includes potentially hazardous bromide atoms as well as chlorine; while inorganic Pigment Green 50 is a noxious cocktail of cobalt, titanium, nickel and zinc oxide.

18.6.2 Ecological Sustainability and Citizenship

In recent years there has been a growing interest in the resurgence of "alternative agro-food networks" and locally sourced organically produced food, which has also been suggested as a model of sustainable consumption for a range of economic, social, and environmental reasons. The rationale for organic food is that it is a production method more in harmony with the environment and local ecosystems.

> By working with nature rather than against it, and replenishing the soil with organic material, rather than denuding it and relying upon artificial fertilizers, proponents claim that soil quality and hence food quality will be improved, biodiversity will be enhanced, and farmers can produce crops that have not resulted in large-scale industrial chemical inputs, with attendant pollution of waterways and land degradation.[96]

Today the most commonly cited reasons for consuming organic food are food safety, the environment, animal welfare, and taste (Figure 18.1).

Ecological citizenship is another hot topic for consumers using a sustainable consumption strategy proposed by the UK government to "do their bit" by buying

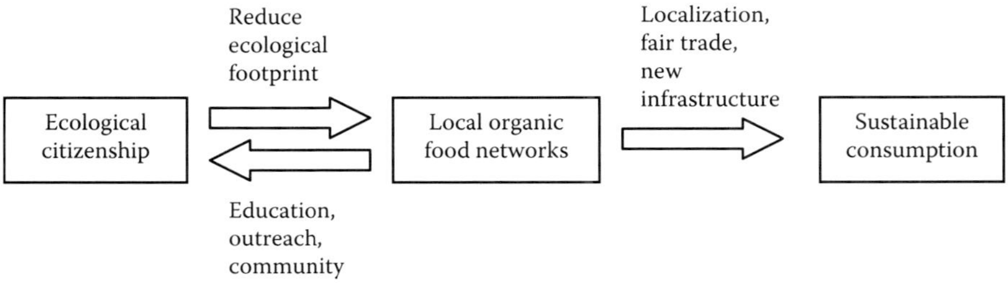

Figure 18.1 Relationships between ecological citizenship, local organic food networks, and sustainable consumption. (Adapted from Seyfang, G., *J. Rural Stud.*, 22, 383, 2006.)

"green" or "ethical" goods. However, this mainstream approach, along with the food miles theory, has serious limitations, which are mostly political. "Ecological Citizenship is non-territorial and non-contractual and is concerned with responsibilities and the implications of our actions on the environment and on other, distant people,"[96] which is supported by another similar model, called *planetary citizenship*. "Planetary Citizenship is about identifying with the Earth as a whole and the whole of humanity, about working towards a collaborative instead of a competitive world, with a re-shaped economy driven by social and environmental need rather than financial pressures."[96]

18.6.3 Ethical and Consumer Perspectives of Organic Foods

The organic nature of food is publicized as an alternative to the perceived health risks of chemically induced foodstuffs and to suggest natural, sustainable, and wholesome eating.[97] DuPuis and Goodman[98] emphasized the ethical norms of organic food systems, which are embedded in the ethics of care, stewardship, and the agrarian vision that leads to rural development.

In this way, organic food can be linked with shortened commodity chains, localized production and consumption, environmental sustainability, and the management of food-associated risk.[99] Consumers increasingly criticize globalization of agricultural production and question economic, environmental, and social consequences of global trade.[100] The most significant example is "fair trade" products (e.g., coffee, tea, or bananas), which have achieved double-figure market shares in some countries in recent years. Market observations indicate that consumers appreciate ethical concerns with domestic products, and a large share of consumers are willing to pay additional prices for these ethical products.[101]

18.7 CONCLUSIONS AND FUTURE DIRECTIONS

Food science and technology has traversed a long way to satisfy consumer demands. In past centuries, food science and technology was mainly concerned about food security. Deforestation created arable lands, and rapid industrialization generated processed food products. Then there was the necessity to develop the technologies to make the food safe for human consumption by eliminating toxicological impacts, destroying pathogenic microorganisms, and preserving the quality of foods for long-term storage. Once the food was safe, the taste, flavor, texture, appearance, and other sensorial attributes of the food to satisfy consumer acceptance became the priority of the food industry. This was followed by the aspirational foods, such as chocolates, cola, and coffees, which have limited contribution to nutrition but an enormous emotional impact. Advertisements and communication allowed nurturing the romanticism aspects associated with the foods. Finally, in the current scenario, the food industry is largely directed toward food with added health benefits. Phytochemicals today play a cardinal role in the marriage of nutrition and pharmaceuticals, which has led to the development of a new category of nutrients and ingredients called *nutraceuticals* (nutrition + pharmaceutical, or pharmaceutical nutrients). Thus, foods have now been accepted to be capable of playing not only a "health-promoting" but also a "disease-preventing" role over and above their routine function of supporting growth and maintenance, and the terms *functional foods* and *nutraceuticals* have come into the limelight.

In this era of development of food technology, we assume that the next trend will be "go green" and "ethical" foods. We have already reviewed case studies across major food companies and retail sectors understanding their leadership in not only producing safe, nutritious, tasty, pleasurable, and therapeutic food, but also in creating less environmental impact during processing. As these food industries spread consumer awareness about environmental issues through media communications, increasingly more people are becoming concerned about the greener aspects of food; thus, we assume that the green approaches will drive the momentum of future food developments.

Using a fundamental understanding of the nutrient balance in the entire food chain, we realize that foods extracted from plant sources require less energy to be produced and result in fewer CO_2 emissions and thus are "greener." Besides the literal meaning, even from an energy requirement and CO_2 emissions viewpoint, we can say that green foods may largely contain foods from plant sources, such as grains, vegetables, and fruits, compared with foods from animal sources (e.g., dairy products, fish, and meat) because food from higher up the food chain requires more energy to be produced.[13] For example, production of meat involves the production of plants, which are then fed to the animal. This shifting of cereals from the direct consumption to the indirect consumption by feeding them to animals involves potential metabolic losses in terms of calorie-equivalent grain/meat conversion ratios, which vary from 2:1 for poultry, 3:1 for pork, to 7:1 for beef.[102,103] Unfortunately, in most developing parts of the world, such as South Asia, sub–Saharan Africa, and parts of Latin America, where large sections of people live below the poverty line, people cannot afford to get sufficient protein in their diet. Intuitively, if the cereal proteins are consumed directly rather than losing proteins in the method of conversion to meat proteins, the loss of nutrients can be significantly reduced, together with saving millions of people from protein malnutrition. However, it is worth noting that not only the quantity but also the quality of protein intake is critical for the human diet. It is a well-known fact that animal proteins such as milk, meat, and eggs are generally of high quality, containing sufficient amounts of all essential amino acids, along with higher digestibility and bioavailability. In contrast, most of the plant proteins are deficient in one or more essential amino acids, such as wheat in lysine, maize in tryptophan and lysine, rice in lysine and threonine, and legumes in methionine.[104] Hence to meet the protein quality requirements, one of the intelligent approaches is to complement plant protein sources in a suitable way to counterbalance the amino acids or to supplement with lesser amounts of animal proteins to create novel protein blends, which definitely would be more affordable than net animal protein. This will not only help to address widespread global malnutrition but also aid in environmental sustainability because of less use of animal protein, and thus less loss of resources on conversion of plant protein to animal protein.

Finally, extensive research in science and technology is required to reduce carbon footprints and develop strategies for sustainable management of water, energy, and other natural resources to design greener, that is, low environmental impact, food.

ACKNOWLEDGMENTS

The authors thank the major food and beverage companies Cadbury, Coca-Cola, Danone, Fonterra, Kraft, Nestlé, PepsiCo, and Unilever for sharing their case studies of promising green initiatives toward environmental sustainability through their corporate websites, which have significantly contributed to this chapter. The authors would also like to thank the major retailers Asda, McDonald's, Starbucks, Sainsbury's, Tesco, Waitrose, Walmart, Whole Foods Market, and Yum! Brands Inc. for sharing their efforts to address various sustainability challenges at various steps of the food cycle.

REFERENCES

1. Population Division of the Department of Economic and Social Affairs. 2009. World Population Prospects. The 2008 revision. *Population Newsletter*. New York: United Nations.
2. Kawashima, H., M. J. Bazin, and J. M. Lynch. 1997. A Modelling Study of World Protein Supply and Nitrogen Fertilizer Demand in the 21st Century. *Environmental Conservation* 24: 50–57.
3. Hengzhou, X., Q. Futian, and G. Zhongxing. 2007. Dynamic Changes and Optimal Allocation of Arable Land Conversion in Transition of Jiangsu province. *China Population, Resources and Environment* 17: 54–58.
4. Hara, Y., K. Takeuchi, and S. Okubo. 2005. Urbanization Linked with Past Agricultural Landuse Patterns in the Urban Fringe of a Deltaic Asian Mega-City: A Case Study in Bangkok. *Landscape and Urban Planning* 73: 16–28.
5. Satterthwaite, D. 2009. The Implications of Population Growth and Urbanization for Climate Change. *Environment and Urbanization* 21: 545–67.

6. Cohen, M., and J. L. Garrett. 2009. *The Food Price Crisis and Urban Food (In)Security*. Urbanization and Emerging Population Issues. Working Paper Series No. 2. London: IIED/UNFPA.
7. Rosegrant, M. W., N. Leach, and R. V. Gerpacio. 1999. Alternative Futures for World Cereal and Meat Consumption. *Proceedings of the Nutrition Society* 58: 219–34.
8. Delgado, C. L. 2003. Rising Consumption of Meat and Milk in Developing Countries Has Created a New Food Revolution. *Journal of Nutrition* 133: 3907S–10.
9. Chatterjee, S., A. Rae, and R. Ray. 2007. Food Consumption and Calorie Intake in Contemporary India. *Discussion Papers*. Palmerston North, New Zealand: Department of Applied and International Economics, Massey University.
10. Preston, S. H. 1996. The Effect of Population Growth on Environmental Quality. *Population Research and Policy Review* 15: 95–108.
11. Engelhaupt, E. 2008. Do Food Miles Matter? *Environmental Science & Technology* 42: 3482.
12. Brundtland, G. H. 1987. Report of the World Commission on Environment and Development. *General Assembly Resolution*. New York: United Nations, p. 8.
13. Ghosh, D. 2010. Green Food & Environmental Sustainability: Two Sides of a Coin. *Nutraceuticals World* January: 40–42.
14. Seckler D., U. Amarasinghe, D. Molden, R. de Silva, and R. Barker. 1998. World Water Demand and Supply, 1990–2025: Scenarios and Issues, Research Report no. 19. 1998. Colombo, Sri Lanka: International Water Management Institute.
15. Waste-Water Treatment Technologies: A General Review. 2003. *Economic and Social Commission for Western Asia*. New York: United Nations.
16. Wang, L. K., Y.-T. Hung, H. H. Lo, and C. Yapijakis. 2006. *Waste Water Treatment in the Food Processing Industry*. Boca Raton, FL: CRC Press, Taylor and Francis Group.
17. Discussion Paper on Proposed Draft Guidelines for the Hygienic Reuse of Processing Water in Food Plants. 1999. *Codex Alimentarius Commission: Codex Committee on Food Hygiene*. Washington, DC: Joint FAO/WHO Food Standards Programme, p. 5.
18. Mavrov, V., A. Fähnrich, and H. Chmiel. 1997. Treatment of Low-Contaminated Waste Water from the Food Industry to Produce Water of Drinking Quality for Reuse. *Desalination* 113: 197–203.
19. Palumbo, S. A., K. T. Rajkowski, and A. J. Miller. 1997. Current Approaches for Reconditioning Process Water and Its Use in Food Manufacturing Operations. *Trends in Food Science & Technology* 8: 69–74.
20. Scharnagl, N., U. Bunse, and K.-V. Peinemann. 2000. Recycling of Washing Waters from Bottle Cleaning Machines Using Membranes. *Desalination* 131: 55–63.
21. Judd, S. 2006. *The MBR Book: Principles and Applications of Membrane Bioreactors in Water and Wastewater Treatment*. Oxford, UK: Elsevier Publications.
22. Sutton, P. M. 2006. Membrane Bioreactors for Industrial Wastewater Treatment: Applicability and Selection of Optimal System Configuration. *Proceedings of the Water Quality Technology Conference and Exposition* 41–50: 3233–48.
23. Wisniewski, C. 2007. Membrane Bioreactor for Water Reuse. *Desalination* 203: 15–19.
24. Marrot, B., A. Barrios-Martinez, P. Moulin, and N. Roche. 2004. Industrial Wastewater Treatment in a Membrane Bioreactor: A Review. *Environmental Progress* 23: 59–68.

25. Blöcher, C., M. Noronha, L. Fünfrocken, J. Dorda, V. Mavrov, H. D. Janke, and H. Chmiel. 2002. Recycling of Spent Process Water in the Food Industry by an Integrated Process of Biological Treatment and Membrane Separation. *Desalination* 144: 143–50.
26. Ahrens, D. 2009. Water Processing: Recovering Water in Beverage and Food Production. *Filtration & Separation* 46: 15–17.
27. Casani, S., M. Rouhany, and S. Knøchel. 2005. A Discussion Paper on Challenges and Limitations to Water Reuse and Hygiene in the Food Industry. *Water Research* 39: 1134–46.
28. *Managaging Environmental Sustainibility in the European Food and Drink Industries: Issues, Industry Action and Future Strategies*, 2nd ed. 2008. Brussels, Belgium: Confederation of the Food and Drink Industries in the EU (CIAA), 34–39.
29. Vigneswaran, S., V. Jegatheesan, and C. Visvanathan. 1999. Industrial Waste Minimization Initiatives in Thailand: Concepts, Examples and Pilot Scale Trials. *Journal of Cleaner Production* 7: 43–47.
30. Chavalparit, O., and M. Ongwandee. 2009. Clean Technology for the Tapioca Starch Industry in Thailand. *Journal of Cleaner Production* 17: 105–10.
31. Watkinson, W. J. 2004. Advances in Cleaning and Sanitation Technology and Methodology. Presented at the IDF/FAO International Symposium on Dairy Safety and Hygiene, Cape Town, South Africa, March 2004.
32. Pascual, A., I. Llorca, and A. Canut. 2007. Use of Ozone in Food Industries for Reducing the Environmental Impact of Cleaning and Disinfection Activities. *Trends in Food Science & Technology* 18: S29–S35.
33. Boers, T. M., and J. Ben-Asher. 1982. A Review of Rainwater Harvesting. *Agricultural Water Management* 5: 145–58.
34. Qadir, M., B. R. Sharma, A. Bruggeman, R. Choukr-Allah, and F. Karajeh. 2007. Non-conventional Water Resources and Opportunities for Water Augmentation to Achieve Food Security in Water Scarce Countries. *Agricultural Water Management* 87: 2–22.
35. United Nations. 1997. *Glossary of Environment Statistics, Studies in Methods, Series F, No. 67*. New York: United Nations.
36. Schaub, S. M., and J. J. Leonard. 1996. Composting: An Alternative Waste Management Option for Food Processing Industries. *Trends in Food Science & Technology* 7: 263–68.
37. Russ, W., and R. Meyer-Pittroff. 2004. Utilizing Waste Products from the Food Production and Processing Industries. *Critical Reviews in Food Science and Nutrition* 44: 57–62.
38. Yu, P., H. Chua, A. Huang, W. Lo, and G. Chen. 1998. Conversion of Food Industrial Wastes into Bioplastics. *Applied Biochemistry and Biotechnology* 70–72: 603–14.
39. Winfried, R., M.-P. Roland, D. Alexander, and L.-K. Jürgen. 2008. Usability of Food Industry Waste Oils as Fuel for Diesel Engines. *Journal of Environmental Management* 86: 427–34.
40. Hang, Y. D. 2004. Management and Utilization of Food Processing Wastes. *Journal of Food Science* 69: CRH104–07.
41. Kosseva, M. R. 2009. Processing of Food Wastes. *Advances in Food and Nutrition Research*. Edited by Steve L. Taylor. San Diego: Academic Press, 57–136.
42. Oreopoulou, V., and W. Russ. 2007. Integrating Safety and Environmental Knowledge into Food Studies towards European Sustainable Development. *Utilization of By-products and Treatment of Waste in the Food Industry*. Edited by Kristberg Kristbergsson. New York: Springer.

43. Arvanitoyannis, I. S. 2008. *Waste Management for the Food Industries*. Amsterdam: Academic Press.
44. Waldron, K. W. 2007. *Handbook of Waste Management and Co-product Recovery in Food Processing*, vol. 1. Cambridge, UK: Woodhead Publishing Ltd.
45. Kelly, E. 1919. The Utilization of Dairy By-products. *Journal of Dairy Science* 2: 46–49.
46. Smithers, G. W., F. J. Ballard, A. D. Copeland, K. J. De Silva, D. A. Dionysius, G. L. Francis, C. Goddard, P. A. Grieve, G. H. McIntosh, I. R. Mitchell, R. J. Pearce, and G. O. Regester. 1996. New Opportunities from the Isolation and Utilization of Whey Proteins. *Journal of Dairy Science* 79: 1454–59.
47. Smithers, G. W. 2008. Whey and Whey Proteins—From "Gutter-to-Gold." *International Dairy Journal* 18: 695–704.
48. Ha, E., and M. B. Zemel. 2003. Functional Properties of Whey, Whey Components, and Essential Amino Acids: Mechanisms Underlying Health Benefits for Active People (Review). *Journal of Nutritional Biochemistry* 14: 251–58.
49. Loveday, S. M., X. L. Wang, M. A. Rao, S. G. Anema, L. K. Creamer, and H. Singh. 2010. Tuning the Properties of β-Lactoglobulin Nanofibrils with pH, NaCl and $CaCl_2$. *International Dairy Journal* 20: 571–79.
50. Graveland-Bikker, J. F., and C. G. de Kruif. 2006. Unique Milk Protein Based Nanotubes: Food and Nanotechnology Meet. *Trends in Food Science & Technology* 17: 196–203.
51. Liu, D.-C., and H. W. Ockerman. 2001. Meat Co-products. *Meat Science and Application*. Edited by Y. H. Hui, W.-K. Nip, R. W. Rogers, and O. A. Yang. New York: Marcel Dekker, 581–604.
52. Anderson, R. C., and P.-L. Yu. 2003. Isolation and Characterisation of Proline/Arginine-Rich Cathelicidin Peptides from Ovine Neutrophils. *Biochemical and Biophysical Research Communications* 312: 1139–46.
53. Balan, P., K.-S. Han, S. M. Rutherfurd, H. Singh, and P. J. Moughan. 2009. Orally Administered Ovine Serum Immunoglobulins Influence Growth Performance, Organ Weights, and Gut Morphology in Growing Rats. *Journal of Nutrition* 139: 244–49.
54. Han, K.-S., M. Boland, H. Singh, and P. Moughan. 2009. The in Vitro Anti-Pathogenic Activity of Immunoglobulin Concentrates Extracted from Ovine Blood. *Applied Biochemistry and Biotechnology* 157: 442–52.
55. Tarté, R. 2009. Meat-Derived Protein Ingredients. *Ingredients in Meat Products: Properties, Functionality, and Applications*. Edited by Rodrigo Tarté. New York: Springer, 145–71.
56. Moharram, Y. G., and A. M. Moustafa. 1982. Utilisation of Mango Seed Kernel (Mangifera indica) as a Source of Oil. *Food Chemistry* 8: 269–76.
57. Sarkar, A. 2005. *Utilization of Industrial By-product for the Development of a Nutraceutical Protein Hydrolysate*. Master's. Department of Protein Chemistry and Technology, Central Food Technological Research Institute (CFTRI), Mysore, India.
58. Sarkar, A., P. Kaul, and V. Prakash. 2007. Utilization of Tomato By-products for the Development of Lycopene and Protein Hydrolysate Rich Nutraceutical Additive. Presented at the 77th Annual Session and Symposium on Novel Approaches for Food and Nutritional Security, Mysore, India.
59. Rao, A. V., Z. Waseem, and S. Agarwal. 1998. Lycopene Content of Tomatoes and Tomato Products and Their Contribution to Dietary Lycopene. *Food Research International* 31: 737–41.

60. Topal, U., M. Sasaki, M. Goto, and K. Hayakawa. 2006. Extraction of Lycopene from Tomato Skin with Supercritical Carbon Dioxide: Effect of Operating Conditions and Solubility Analysis. *Journal of Agricultural and Food Chemistry* 54: 5604–10.
61. Sharma, S. K., and M. Le Maguer. 1996. Lycopene in Tomatoes and Tomato Pulp Fractions. *Italian Journal of Food Science* 8: 107–13.
62. Brodowski, D., and J. R. Geisman. 1980. Protein Content and Amino Acid Composition of Protein of Seeds from Tomatoes at Various Stages of Ripeness. *Journal of Food Science* 45: 228–29.
63. Sogi, D. S., R. Bhatia, S. K. Garg, and A. S. Bawa. 2005. Biological Evaluation of Tomato Waste Seed Meals and Protein Concentrate. *Food Chemistry* 89: 53–56.
64. Sogi, D. S., J. S. Sidhu, M. S. Arora, S. K. Garg, and A. S. Bawa. 2002. Effect of Tomato Seed Meal Supplementation on the Dough and Bread Characteristics of Wheat (PBW 343) Flour. *International Journal of Food Properties* 5: 563–71.
65. Carlson, B. L., D. Knorr, and T. R. Watkins. 1981. Influence of Tomato Seed Addition on the Quality of Wheat Flour Breads. *Journal of Food Science* 46: 1029–31.
66. Lavecchia, R., and A. Zuorro. 2010. Process for Extraction of Lycopene. US Patent Application 20100055261, filed November 6, 2007, issued May 15, 2008.
67. No, H. K., and S. P. Meyers. 1989. Crawfish Chitosan as a Coagulant in Recovery of Organic Compounds from Seafood Processing Streams. *Journal of Agricultural and Food Chemistry* 37: 580–83.
68. Yam, K. L. 2009. *The Wiley Encyclopedia of Packaging Technology*, 3rd ed. New York: John Wiley & Sons.
69. Robertson, G. L. 2006. *Food Packaging: Principles and Practices*, 2nd ed. Boca Raton, FL: CRC Press, Taylor and Francis Group.
70. Fortescue, S. 2008. Dairy Industry in Drive to Boost Green Credentials: Proud of Dairy's Environmental Work. *News Release*. London: Dairy UK.
71. Siracusa, V., P. Rocculi, S. Romani, and M. D. Rosa. 2008. Biodegradable Polymers for Food Packaging: A Review. *Trends in Food Science & Technology* 19: 634–43.
72. Ackermann, T., and L. Söder. 2000. Wind Energy Technology and Current Status: A Review. *Renewable and Sustainable Energy Reviews* 4: 315–74.
73. de Groot, H. L. F., E. T. Verhoef, and P. Nijkamp. 2001. Energy Saving by Firms: Decision-Making, Barriers and Policies. *Energy Economics* 23: 717–40.
74. Sandberg, P., and M. Söderström. 2003. Industrial Energy Efficiency: The Need for Investment Decision Support from a Manager Perspective. *Energy Policy* 31: 1623–34.
75. Burton, H. 1977. The Review of UHT Treatment and Aseptic Packaging in Dairy Industry. *International Journal of Dairy Technology* 30: 135–42.
76. Deeth, H. C., and N. Datta. 2002. Ultra-High Temperature Treatment (UHT) | Heating Systems. *Encyclopedia of Dairy Sciences*. Edited by H. Roginski. Oxford, UK: Elsevier, 2642–52.
77. Ramírez, C. A., M. Patel, and K. Blok. 2006. From Fluid Milk to Milk Powder: Energy Use and Energy Efficiency in the European Dairy Industry. *Energy* 31: 1984–2004.
78. Fox, P. F., and P. L. H. McSweeney. 1998. *Dairy Chemistry and Biochemistry*. Dordrecht, The Netherlands: Kluwer Academic Publishers Group.
79. Reiter, B., and G. Härnulv. 1984. Lactoperoxidase Antibacterial System: Natural Occurrence, Biological Functions and Practical Applications. *Journal of Food Protection* 47: 724–32.
80. Benefits and Potential Risks of the Lactoperoxidase System of Raw Milk Preservation. 2005. *Report of an FAO/WHO Technical Meeting*. Rome: The Food and Agriculture Organization (FAO) of the United Nations and World Health Organization (WHO).

81. Bjorck, L., O. Claesson, and W. Schulthes. 1979. The Lactoperoxidase/Thiocyanate/Hydrogen Peroxide System as a Temporary Preservative for Raw Milk in Developing Countries. *Milchwissenschaft* 34: 726–29.
82. Robinson, R. K. 1994. *Modern Dairy Technology: Advances in Milk Products*, 2nd ed. vol. 1. New York: Elsevier.
83. Robinson, R. K. 1994. *Modern Dairy Technology: Advances in Milk Processing*, vol. 2. New York: Elsevier.
84. *Alfa Laval/Tetra Pak, Dairy Processing Handbook*. 1995. Lund, Sweden: Tetra Pak Processing Systems.
85. Kessler, H. G. 1981. *Food Engineering and Dairy Technology*. Freising, Germany: Verlag A. Kessler.
86. Singh, R. P., and D. R. Heldman. 2009. *Introduction to Food Engineering*, 4th ed. London: Academic Press.
87. Earle, R. L., and M. D. Earle. 2004. *Unit Operations in Food Processing*. Palmerston North, New Zealand: The New Zealand Institute of Food Science & Technology Inc..
88. Lorentzen, G. 1995. The Use of Natural Refrigerants: A Complete Solution to the CFC/HCFC Predicament. *International Journal of Refrigeration* 18: 190–97.
89. Riffat, S. B., C. F. Afonso, A. C. Oliveira, and D. A. Reay. 1997. Natural Refrigerants for Refrigeration and Air-Conditioning Systems. *Applied Thermal Engineering* 17: 33–42.
90. Lee, T.-S., C.-H. Liu, and T.-W. Chen. 2006. Thermodynamic Analysis of Optimal Condensing Temperature of Cascade-Condenser in CO2/NH3 Cascade Refrigeration Systems. *International Journal of Refrigeration* 29: 1100–08.
91. Howard, H. E. 2003. Method for Operating a Cascade Refrigeration System. US Patent 6,557,361, filed March 26, 2002, issued May 6, 2003.
92. Hill, H. 2008. Food Miles: Background and Marketing. *A Publication of ATTRA—National Sustainable Agriculture Information Service* P312: 1–12.
93. Liao, C.-H., P.-H. Tseng, and C.-S. Lu. 2009. Comparing Carbon Dioxide Emissions of Trucking and Intermodal Container Transport in Taiwan. *Transportation Research Part D: Transport and Environment* 14: 493–96.
94. Niewold, T. A. 2010. Organic More Healthy? Green Shoots in a Scientific Semi-Desert. *British Journal of Nutrition* 103: 627–28.
95. Rawsthorn, A. 2010. The Toxic Side of Being, Literally, Green. *New York Times* http://www.nytimes.com/2010/04/05/arts/05iht-design5.html.
96. Seyfang, G. 2006. Ecological Citizenship and Sustainable Consumption: Examining Local Organic Food Networks. *Journal of Rural Studies* 22: 383–95; see pp. 385, 388.
97. Zander, K., and U. Hamm. 2010. Consumer Preferences for Additional Ethical Attributes of Organic Food. *Food Quality and Preference* 21: 495–503.
98. DuPuis, E. M., and Goodman, D. 2005. Should We Go "Home" to Eat?: Toward a Reflexive Politics of Localism. *Journal of Rural Studies* 21: 359–71.
99. Clarke, N., P. Cloke, C. Barnett, and A. Malpass 2008. The Spaces and Ethics of Organic Food. *Journal of Rural Studies* 24: 219–30.
100. Aldanondo-Ochoa, A. M., and C. Almansa-Sáez. 2009. The Private Provision of Public Environment: Consumer Preferences for Organic Production Systems. *Land Use Policy* 26: 669–82.
101. Global Organic Market. 2009. Time for Organic Plus Strategies. *Organic Monitor*. http://www.organicmonitor.com/r2905.htm.

102. Yotopoulos, P. A. 1985. The "New" Food-Feed Competition. *Proceedings of the FAO Expert Consultation on the Substitution of Imported Concentrate Feeds in Animal Production Systems in Developing Countries, FAO Animal Production and Health Paper, No. 63*. Edited by R. Sansoucy, T. R. Preston, and R. A. Leng. Bangkok, Thailand: Food and Agricultural Organization of the United Nations.
103. de Haan, C., H. Steinfeld, and H. Blackburn. 1997. *Livestock & the Environment: Finding a Balance. Report of Study by the Commission of the European Communities, the World Bank and the Governments of Denmark, France, Germany, The Netherlands, United Kingdom and The United States of America*. New York: Food and Agriculture Organization of the United Nations, the United States Agency for International Development, and the World Bank.
104. Passmore, R., B. M. Nicol, M. N. Rao, G. H. Beaton, and E. M. Demayer. 1974. *Handbook on Human Nutritional Requirements*. Rome: Food and Agricultural Organization of United Nations and World Health Organization.

CHAPTER 19

Flavoring of Nutraceuticals

Aparna Keskar and William Igou, Jr.

CONTENTS

19.1 INTRODUCTION

The acceptability of almost everything that we consume, whether it is food, drink, confectionery, medicine, or nutraceuticals, is dependent to some extent on its flavor. Flavor is undoubtedly one of the most important attributes of the food we eat. The term *flavor* is widely used and has several meanings depending on its context. Flavor can be characterized as a sensory experience induced by chemical compounds present in what we consume. Some flavor components arise from normal biosynthetic processes caused by the presence of certain ingredients. Other components exist only as precursors and develop characteristic flavors during processing as a result of chemical reactions induced by heat or fermentation. Some are intentionally added as flavorings at an appropriate stage in the processing of the product.

19.1.1 Flavors in Foods

The flavor of foods may be classified as:

1. Natural flavor: Pre-existing in the food product
2. Process flavor: Developing in the finished product as a result of processing
3. Compounded flavor: Intentionally added flavor formulated to produce a desired sensory effect

4. Taste modifiers: Additives that affect the basic taste sensations
5. Abnormal flavors and off-notes: Off-odors and off-flavors resulting from the presence of special ingredients (nutraceuticals)

19.1.2 Achieving Flavor Balance

The role of flavor in an end product is one of profile alteration, which includes:

1. The selection and balancing of existing or potential flavor factors working within the constraints of nutritional necessity, the nature and sources of raw materials and supplementary ingredients, and the total concept of the end product
2. The adjustment of the flavor profile, resulting from the method of processing used to suit particular palates or consumer anticipations
3. Corrections to overcome any pre-existing or developed flavor defects
4. Imparting an entirely new flavor in products, which are bland or absolutely nasty to taste
5. Extending the range and flexibility of the end products

Each of these actions calls for an individual judgment on the part of the product development team and involves the knowledge of available raw materials, minimum and optimum processing conditions, legal constraints, and likely consumer response to the final product.

19.2 CRITERIA FOR APPLICATION OF FLAVORS TO NUTRACEUTICALS

Flavors can be added to nutraceutical products for several reasons:

1. To impart flavor to an otherwise bland product created to give some desirable flavor experience. This is a classic function of flavors, in which the end product is attributed a certain overall flavor.
2. To modify or complement an existing flavor profile. This is a function created to satisfy particular requirements of a nutraceutical base.
3. To disguise or cover undesirable flavor attributes or off-notes. In this product-specific demanded function of flavors, the taste buds are tricked into not perceiving the off-note.
4. A combination of numbers 1, 2, and/or 3.

19.3 SIGNIFICANCE OF FLAVORING NUTRACEUTICALS

Nutraceutical and functional food products are increasingly making their way onto mainstream store shelves. Whether it is a beverage with added lutein for eye health or bread with omega-3–rich fish oil, functional foods offer consumers the opportunity to get added nutrients while fulfilling their caloric needs. Instead of buying products that just "taste good," consumers are looking for something that has functionality to it and are often willing to pay more for those premium products. Unfortunately, although manufacturers and consumers are ready to escape the pill-popping days of the past, they are not happy about the tastes that often come with the functional ingredients. It is generally an observation that what is good for your health does not always taste good.

Formulating products with unique ingredients that have added value often involves the use of ingredients that have tastes and aromas not typically found in everyday use. Luckily for consumers, the science to cover up these off-notes is developing rapidly in flavor houses. Masking and modifying agents are natural and/or artificial additions that counteract or mask the harsh flavors from such ingredients. Use of masking agents makes products more palatable to the end user. Among the most common attributes flavor houses are asked to mask are bitterness, metallic aftertaste, and "beany" notes. Bitterness is contributed by many botanical and herbal ingredients, whereas soy is the most common "beany" contributor.

19.3.1 Masking Effect

A masking agent is a specialty flavor or an ingredient developed for covering up the undesirable taste attributes of a specific nutraceutical product.

The diversity of functional foods presents a significant challenge to the flavor industry in developing acceptable flavorings that will allow these products to be successful in the marketplace. Nutraceutical companies often have a nutritional base developed, or a list of ingredients desired to form the nutraceutical component of a food item. Flavor companies work with such a nutraceutical or functional base to mask off-notes. Working with a base to neutralize it of undesirable odor and taste allows the flavorist to select the most complementary flavors for the product. Food product developers want a flavor to magically transform a poor-tasting base into a palate-pleasing product. Replacing sugar and covering up the taste of milk, soy protein, vitamins, and fish oil are some examples that flavor houses are asked to work on. Understanding the nature of the ingredients in the base and their interactions, as well as knowing the source and type of undesirable notes present, goes a long way toward the end result of creating pleasant-tasting foods and beverages. Flavors are expected to provide a characteristic flavor, as well as cover undesirable base notes. Flavor companies take a multifunctional approach to solving these flavor challenges by using masking agents in combination with flavoring ingredients. Maltodextrin is multifunctional. In recovery beverages, maltodextrin helps smooth out high-protein solutions, enhancing mouthfeel and masking some of their off-flavors.

Although masking agents can cover up a wide range of undesirable notes in a variety of applications, flavor companies spend the most time covering up flavors and attributes such as bitterness, metallic notes, vitamin and mineral flavors, and soy notes. These might be a result of the addition of vitamins, minerals, soy and other proteins, herbal and botanical supplements, or amino acids. Masking bitterness is perhaps the most frequently desired attribute. Many ingredients cause bitter notes to develop, including botanicals and caffeine, or bitterness can result from chemical reactions during storage or processing. Products with multiple nutraceutical ingredients pose a tremendous masking challenge because of the high probability of chemical interactions leading to complex taste challenges, far more than the individual nutraceutical ingredients themselves. For example, a fortified soy beverage begins with the "beany" taste of soy, with additional metallic and chalky notes from minerals (like potassium and calcium, respectively) or sour notes from vitamins. In the process of developing a masking agent for a specific product, food chemists use their knowledge and experience of masking and formulations. Then the product with the masking agent added may be subjected to sensory panel testing to gather an objective analysis (as to the success of the masking agent). Sensory panels serve as an important benchmark during the process of product development.

Regardless of what ingredients go into a functional or nutraceutical product, here is a list of some misconceptions and/or shortcomings about the abilities of masking agents:

1. There is no silver bullet or cure-all. Only in rare cases can a single ingredient do it all.
2. Adding more masking agent is not always the answer. It may cover up the undesirable off-notes, imparting an unwanted note or taint of its own.
3. Flavors are dynamic systems by themselves and are subject to change over time. A well-masked nutraceutical product might taste good initially but become imbalanced over time as a result of inherent changes in the masking agent, the nutraceutical base, or both, for example, protein breakdown and oxidation.
4. Even a minor change in the nutraceutical base formulation can alter the balance of the system and require the flavorist to go back to the drawing board.

19.3.2 Palatability

Flavor perception is a complex interaction between taste, aroma, and chemosensory input. What we perceive as flavor is a combination of what we taste on our tongues and mostly what we smell.

As consumers are turning more to value-added functional foods, manufacturers of such nutraceuticals are approaching flavor houses not just for masking agents to cover off-notes but also for enhancers and modifiers to improve the texture, taste, and aroma of the product itself.

19.4 UNDERSTANDING FLAVOR PERCEPTION

Consumers most frequently associate flavor with "taste." The sense of taste involves basic sensations of sweet, sour, salty, bitter, and umami. By identifying taste bud receptors, independent of olfaction, researchers recently identified a sweet receptor and receptors for umami and bitterness. Molecules bind to taste buds that transmit taste information to the brain. There it is combined with information about the aroma. Chemoreceptors in the olfactory nerves pick up aromatic compounds and relay the information directly to the brain. The combination of sensations imparted by a food results in "flavor" as represented in Figure 19.1.

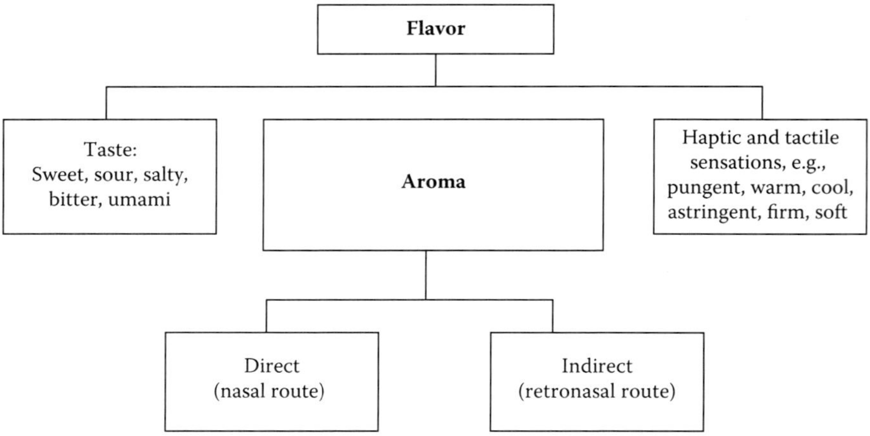

Figure 19.1 Flavor of a food.

The difficult part in the development of many functional food and beverage products is the addition of nutraceutical ingredients that have unacceptable odors or are bitter and astringent. As a result, manufacturers have adopted the use of masking agents to neutralize inherent flavors in the functional base. From that point, formulators then develop an acceptable flavor profile using flavors as enhancers and modifiers to reach a palatable product.

19.4.1 Flavors as Enhancers

A flavor enhancer is a substance added to a food to supplement, improve, or intensify its original flavor without imparting any flavor of its own. The International Organization of the Flavor Industry (IOFI) defines *flavor enhancer* as a substance with little or no odor at the level used, the primary purpose of which is to increase the flavor effect of certain food components well beyond any flavor contributed by the substance itself (IOFI 1990).

Enhancers work on the aromatic and taste attributes of a product. From a functional standpoint, however, enhancers and flavors may overlap. Enhancers can accentuate aromatic qualities without affecting taste. Depending on the need of the nutraceutical product, enhancers can also provide the aromatic sensation needed to complement the existing system. Understanding the functional portions of nutraceuticals is important to develop a full flavor (enhancer) profile. Artificial or added sweeteners, such as aspartame or glycyrrhizin (a licorice root extract), can be used at concentrations not normally perceived as sweet to enhance other, more costly flavor ingredients or the natural sweetness inherent in foods. Other ingredients contribute to mouthfeel, such as the creamy or fatty taste perception.

Based on the classification in Table 19.1, monosodium glutamate and 5′-ribonucleotides, for example, are simultaneous taste enhancers (category 1 in

Table 19.1 Categories of Flavor Enhancers and Flavor Modifiers

SR	Category	Examples	Remarks
1	Flavor enhancers that have no flavor	Monosodium glutamate, 5′-ribonucleotides	Enhance taste impressions
2	Flavor modifiers that have flavor at very subthreshold low or atypically low usage levels	Vanillin, ethyl vanillin, maltol, ethyl maltol	Enhance aroma impressions, e.g., fruity, chocolate
		4-hydroxy-2,5-dimethyl-3-(2*H*)-furanone (strawberry furanone), 4-hydroxy-5-methyl-3-(2*H*)-furanone	Enhance aroma impressions, e.g., fruity, creamy
		3-methyl-2-cyclo-pentene-2ol-l-one (Cyclotene, Dow Chemical Co., Midland, MI)	Enhance odor impressions, e.g., nutty, maple, brown
3	Flavor modifiers that impart flavor at typical usage level	All the above mentioned in category 2	Flavor ingredients that impart specific flavor impressions like sweet, brown, fruity, creamy
4	Flavor modifiers that impart little or no flavor at or below typical conventional functional usage levels	Sucrose, glycyrrhizin, non-nutritive sweeteners (aspartame, sucralose, stevia)	Balance and modify the concerned off-taste or off-note

Table 19.1) and odor suppressors (category 3 in Table 19.1). The term *flavor modifier* includes both taste-enhancing and odor-suppressing effects.

Conversely, flavor modifiers can block flavors. In savory dishes in which a sweet taste is typically not desired, sugar can enhance other relevant characteristics. In this case, a chemical such as lactisole will temporarily block sweetness receptors on the tongue, allowing the savory flavor to dominate. Adenosine monophosphate (AMP), a naturally occurring nucleotide substance, can block bitter food flavors, including those found in coffee. AMP works much like lactisole. It will not directly alter the bitter flavor, but instead alters human perception of bitter by blocking the associated receptor. It has been found that AMP is useful in a couple of contexts, such as blocking or suppressing the bitter or metallic off-taste of potassium chloride, which is typically used as a salt substitute alternative to sodium chloride. AMP can also suppress the vitamin and mineral off-notes picked up in some of the leading meal replacements and diet beverages.

19.4.1.1 Monosodium Glutamate

Monosodium glutamate (MSG) is the monosodium salt of L-(+)-glutamic acid. MSG is used almost exclusively to improve the flavor of foods. It is permitted worldwide, although maximum concentration limits apply in some countries and in some types of foods. Glutamic acid is ubiquitous in nature and occurs as the building block of all proteins. It is also the most abundant amino acid in almost all proteins. MSG works best between pH 5.0 and 8.0. In more acidic conditions, like pH 2.2–4.4 and/or with high temperatures, the flavor-improving effect of MSG is lost because it is partly dehydrated. MSG is mostly used in savory applications. It has no flavor-improving effect for a wide range of foods, such as confectionary and dairy products, beverages, and desserts.

19.4.1.2 5′-Ribonucleotides

5′-Ribonucleotides are the building blocks of ribonucleic acid and have flavor-modifying properties. Ribonucleotides consist of a purine base, ribose, and phosphoric acid linked to the 5′-position of ribose. To give the flavor-improving effect, the purine base within the 5′-ribonucleotide must have a hydroxyl group in the 6-position. Another prerequisite is the phosphate in the 5′-position of the ribose moiety. Ribonucleotides are also mainly popular as flavor modifiers in savory applications.

19.5 FLAVORS AS MODIFIERS

Some food components that have little or no contribution to taste or aroma at typical usage levels are capable of increasing, decreasing, or modifying the taste and aroma of foods. These components may be termed *flavor modifiers*.

The flavor of a food consists of aroma, taste, and tactile sensations in the mouth (mouthfeel) and the nose. This means that a flavor modifier may affect aroma, taste, and/or mouthfeel of a product. Flavor chemists use flavors as modifiers to change the mouthfeel or oral perception of a nutraceutical product. These offer noncharacterizing effects on the palate, such as creaminess and fullness. Modifying agents do not conceal an off-taste but instead change the product to make it palatable—modifying the undesirable notes. Modifiers affect the chemosensory perception. Because of a lack of a clear definition, flavor enhancers and flavor modifiers can be classified into four categories as shown in Table 19.1.

The following chemicals can be used as flavorings (if used in higher amounts) or as modifiers (if used in lower amounts).

19.5.1 Maltol and Ethyl Maltol

In 1861, maltol was first isolated from the bark of the larch tree by the British chemist Stenhouse. Today, industrial production is carried out by fermentation combined with chemical synthesis. Maltol naturally occurs in many foods, for

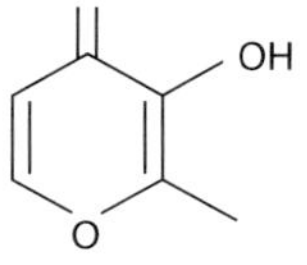

Figure 19.2 Chemical structure of maltol.

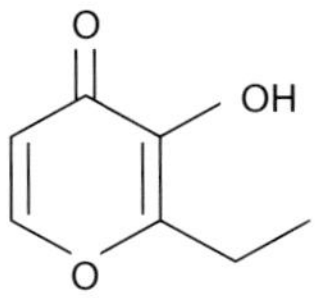

Figure 19.3 Chemical structure of ethyl maltol.

example, baked goods, cocoa, chocolate, coffee, caramel, malt, condensed milk, and cereals. It is formed when carbohydrates are heated. It is a white crystalline powder with a caramel-like odor (Figure 19.2).

Ethyl maltol has not yet been found in nature (Figure 19.3). It is 2–5 times stronger than maltol, and depending on where and how much is used, it can replace maltol in sweet foods. Ethyl maltol and maltol can be used in conjunction as modifiers.

Both maltol and ethyl maltol have also been reported to be highly effective in improving the perception of low-fat food systems. Low-fat yogurt and ice cream taste richer, fuller, and creamier with the addition of maltol or ethyl maltol. In other words, their mouthfeel is improved. Maltol enhances the flavor of sweet foods. The addition of 5–75 ppm of maltol may permit a 15% sugar reduction in many sweet foods.

19.5.2 Furanones and Cyclopentenolones

Both 4-hydroxy-2,5-dimethyl-3(2*H*)-furanone and 4-hydroxy-5-methyl-3(2*H*)-furanone are enhancers of fruity and creamy odor impressions (see Table 19.1). Both furanones have caramel-like odors, with 4-hydroxy-2,5-dimethyl-3(2*H*)-furanone possessing an additional burnt pineapple odor. Both furanones are formed on heating of sugars present in foods and are applied as flavor modifiers in foods where maltol and ethyl maltol are used (Figure 19.4).

3-Methyl-2-cyclopentene-2-ol-1-one (MCP; trade name: Cyclotene, Dow Chemical Co., Midland, MI) is an enhancer of nutty, maple flavor, and chocolate odor impressions (Figure 19.5). MCP has been isolated from beechwood tar and identified in various foods (e.g., maple syrup).

It is formed when sugars are heated at pH 8–10, and it has a caramel-like odor. MCP has also been used to mask the salty taste impression.

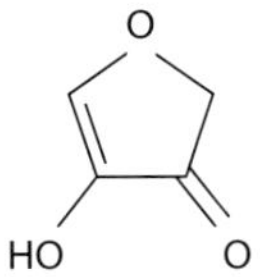

Figure 19.4 Chemical structure of strawberry furanone.

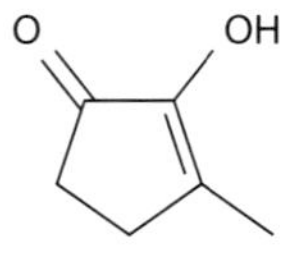

Figure 19.5 Chemical structure of Cyclotene (Dow Chemical Co., Midland, MI).

19.5.3 Vanillin and Ethyl Vanillin

Vanillin has been known as a flavoring substance since approximately 1816, and the pure chemical had been obtained from ethanolic extracts of vanilla beans by 1858. In 1874, Tiemann and Haarmann reported it as 3-methoxy-4-hydroxy-benzaldehyde. Finally, Reimer synthesized vanillin from guaiacol and thus proved its chemical structure (Berlitz and Grosch 1982; Figure 19.6). For many years, the most important source of vanillin was eugenol, from which it was obtained by oxidation. Today the major portion of commercial vanillin is obtained by processing waste sulfite liquors, and the rest is obtained by synthetic processes starting from guaiacol.

Vanillin has a vanilla-like odor. Ethyl vanillin, which is 2–4 times stronger than vanillin, has not yet been found in nature (Figure 19.7). It is synthesized from eugenol, isoeugenol, or safrole. Both vanillin and ethyl vanillin are important flavoring substances. In addition, they enhance fruity and chocolate odor impressions (see Table 19.1).

Last but not least, it must be mentioned that sodium chloride has a flavor-enhancing effect at usage levels below and above its taste threshold (370–5000 ppm). Without salt, many foods (both sweet and savory) have a flat taste. Salt may enhance sweetness and mouthfeel and decrease bitter, sour, and metallic sensations.

At the end, flavor formulation remains a partnership between flavorists and marketers to hide the negatives and accentuate the positives. Working with flavors is like a culinary skill. It involves achieving the correct balance of ingredients in the given system, and that is something that is mastered over time.

19.6 PHYSICOCHEMICAL FACTORS AFFECTING THE SELECTION PROCESS OF FLAVORS FOR NUTRACEUTICALS

Unit operations encountered in the processing of foods, beverages, and other consumable products are mainly associated with raw material preparation, mixing

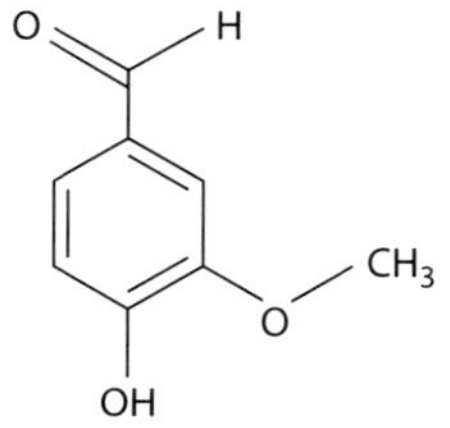

Figure 19.6 Chemical structure of vanillin.

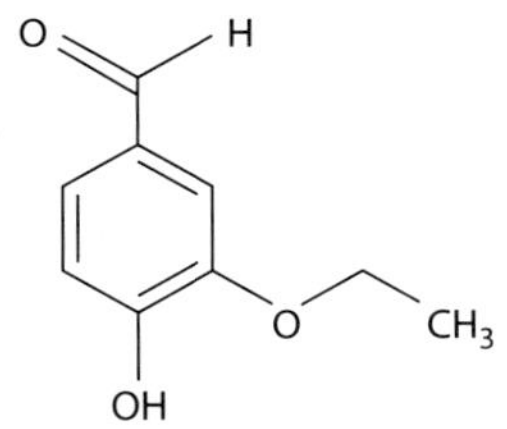

Figure 19.7 Chemical structure of ethyl vanillin.

and blending, thermal processing, and packaging. Included in these processes, the following conditions are the most likely to affect the incorporation of flavors and the flavor profile of the end product.

19.6.1 Temperature and Time

Because of their aromatic nature, most flavors are to some extent heat sensitive. At elevated temperatures, particularly in the presence of water or other volatile solvents and ingredients, they may flash off through evaporation or steam distillation. Less stable compounds may change as a result of chemical interactions with other ingredients. The extent of change is usually a function of temperature and time. The effect of ultrahigh temperatures for short time (HTST) intervals usually result in significantly less flavor loss or degradation than compared with much lower temperatures over a longer time. This also depends on other factors inherent to the product (e.g., pH, the presence of proteins). Knowledge of the temperatures to which the total nutraceutical system will be exposed, as well as the hold-times, is important in deciding the nature of the flavor to be used, especially with respect to any flavor solvents or flavor ingredients subjected to temperatures in excess of their boiling points.

Other changes that occur with heat may be considered more or less acceptable depending on consumer taste preferences. In direct-steam ultrahigh temperature (UHT) heat processing systems, for example, the exposure to heat is so rapid that the interactions of sugars and proteins do not have a chance to develop as they would in traditional pasteurization, retort, or even HTST pasteurization. This reaction, known as Maillard browning, produces cooked, caramel-like notes. Direct UHT with steam injection does not readily create these notes, which enables the development of a product that has a fresher, lighter flavor. The higher the temperatures used, the greater the chances of flavor loss or degradation.

Those containing high levels of lipids are prone to rancidity caused by oxidation. These changes generally result in unacceptable off-flavor notes.

19.6.2 Open or Closed System

The addition of flavors in an open system (i.e., blending in an open vat) is likely to result in greater volatile losses than in a closed system (i.e., in-line processing or retorting in sealed containers). When open handling cannot be avoided, precautions should be taken to minimize exposure by using covered containers.

19.6.3 pH

Most fruits contain natural organic acids (e.g., citric, malic, and tartaric acids) that contribute significantly to the flavor profile. The use of acids may be necessary when imitating a fruit flavor that has an intrinsically neutral taste; otherwise, the correct flavor impression is not achieved in the end product. Some flavors and certain spices (e.g., turmeric) contain ingredients that are sensitive to changes in pH, and it is essential that this particular condition be carefully reproduced during the product development stage and shelf-life testing to ensure that undesirable effects do not occur. pH, whether low or high, is case specific, depending on what is in the nutraceutical product.

19.6.4 Exposure to Air

This is of particular concern in products that are aerated. High-speed mixing operations can result in considerable volatile losses, but, more importantly, any occluded air produces conditions conducive to oxidation of any unsaturated lipids present. The pneumatic conveyance of powdered flavor ingredients may also result in significant volatile losses unless encapsulated flavorings or seasonings are used.

The guiding principle is to expose any added flavor to the minimum of treatment. Obviously, for some products addition into the primary mix cannot be avoided, but wherever possible, flavorings should be added at as late a stage as consistent with uniformity of dispersion in the end product.

19.7 BIOCHEMICAL FACTORS AFFECTING THE SELECTION PROCESS OF FLAVORS FOR NUTRACEUTICALS

Biological structures in any food and nutraceutical product interact with flavor ingredients that can affect the selection process of that particular flavor for a specific product system. Examples of more commonly used flavor ingredients include diacetyl, isoamyl acetate, acetaldehyde, and ethyl acetate. Such flavor ingredients might interact with nonactive ingredients from the nutraceutical base.

19.7.1 Interactions with Carbohydrates

Simple sugars and/or starches often serve as carriers for flavor ingredients. Model tests with dry (or virtually dry) sugars in the crystalline state resulted in a number of flavor ingredients (e.g., ethyl acetate, butyl amine) only binding weakly to glucose, saccharose, and lactose. This bond, which presumably involves adsorption on the relatively small surface area of the crystalline sugar, is completely reversible under vacuum at 23°C. If the sugars are in the amorphous state (with a larger surface area), the adsorption is considerably greater. Consequently, there is a possibility of a change in the ratio of volatile flavor components, resulting in alteration of the flavor profile and also possible alteration in the aroma.

Many flavor ingredients (e.g., acetaldehyde, diacetyl, ethyl acetate, and 2-hexanone) bind with varying degrees of strength to the polysaccharides *pectin*, *guar gum*, *alginate*, *agar-agar*, *cellulose*, and *methyl cellulose*. A lot more is known about the binding of flavor ingredients to *starch*. Most starches consist of two fractions: amylose (unbranched glucose chains) and amylopectin (branched glucose chains). Starches change their native structure during and after boiling in water. Hydrogen bridge bonds are weakened in this gelatinization process; the starch grains swell, and part of the amylose and amylopectin is dissolved. At the same time, amylopectin binds large quantities of water, and amylose forms helical structures under water binding. Both amylose and amylopectin are involved in the binding of flavor. Flavor ingredients can be entrapped in the helical structures of amylose. Different starches show varying flavor binding capacity. Starches with low amylose content (e.g., tapioca at 17%) and waxy starches consisting of only amylopectin have a weak binding capability; those with high amylose content (like potato or corn) have a greater one. Again, the net effect of this interaction is similar as mentioned earlier with the sugars.

All the aforementioned interactions of flavor ingredients with gelatinized starches are in aqueous media. If such systems are dried, they are much more stable. The moist starch binds more flavor than the dry one. When water is added to reconstitute a liquid system, interactions similar to those mentioned earlier in sugars can occur. Partially hydrolyzed starch products such as *dextrin* and *maltodextrin* can react in a different way. After partial hydrolysis, the starches lose a major part of their flavor-binding properties. As a result, these two starch products are widely used in manufacturing of spray-dried or powdered flavors.

Cyclodextrins are used for the encapsulation of flavorings and colorings, and also for masking off-flavors. β-Cyclodextrin is the best suited for encapsulation of flavors. In an aqueous medium it forms inclusion complexes with flavor ingredients. The more lipophilic the flavor ingredient, the more easily it is entrapped. The interstitial spaces of cyclodextrin are nonpolar, whereas the surface is polar. Among enzyme-modified starch derivatives, cyclodextrins behave as empty molecular capsules with the ability to entrap guest molecules of appropriate geometry and polarity. The included molecules are protected from surroundings: light, heat, oxidation, and so on. The flavor cyclodextrin complexes show the aforementioned advantageous properties while they are in the dry, solid state. On contact with water, cyclodextrin complexes release their flavor content. Unstable flavor ingredients may become extraordinarily stable in such inclusion complexes.

Table 19.2 summarizes the types of binding of flavor ingredients to various carbohydrates. Essentially this is a matter of reversible physical and physicochemical binding (i.e., adsorption, inclusion complexes, and hydrogen bridges).

Table 19.2 Types of Interactions between Flavor Ingredients and Carbohydrates

Carbohydrates	System	Type of Bond
Simple sugars	Aqueous	Not known
(Mono- and disaccharides)	Dry	Adsorption
Pectin and alginate	Dry and moist	Electrovalent bond? Covalent bond?
Cellulose	Dry and moist	Hydrogen bridges?
Guar gum, agar-agar, and methylcellulose	Dry and moist	Not known (with methylcellulose as with cellulose?)
Starches and cyclodextrins	Aqueous	Inclusion complexes
Dextrins and maltodextrins	Aqueous	Adsorption

19.7.2 Interactions with Proteins

Both native and denatured proteins can bind and chemically react with flavor ingredients. Aldehydes (such as nonanal), ketones (such as 2-heptanone, 2-nonanone), and alcohols (such as butanol and hexanol) form exceptionally strong bonds to *native proteins* (e.g., soy protein, whey protein) in aqueous systems. During the binding process, the flavor ingredients cause conformational changes in the protein molecules. Native proteins bind aldehydes and ketones through hydrophobic interactions, whereas alcohols (like butanol and hexanol) are bound hydrophobically and through hydrogen bridges.

In the nutraceutical industry, the *heat-denatured proteins* are more important than the native ones. Heat treatment is characteristic of many nutraceutical products containing protein. The binding of flavor ingredients to heated soy proteins in an aqueous medium results in increased binding of aldehydes (e.g., hexanal), ketones (e.g., 2-nonanone), and alcohol (e.g., hexanol).

As anticipated, the degree of binding of flavor ingredients to native proteins and heat-denatured proteins depends on temperature and pH.

With the possibility of flavor ingredients binding to native and denatured proteins in aqueous systems, it is important to see whether the binding is reversible. In general, it would be true to say that hydrocarbons, alcohols, and ketones are reversibly bound (through hydrophobic interactions and/or hydrogen bridge formation). Aldehydes are known to react chemically with free amino groups from amino acids to form *Schiff's bases.* There is a possibility of this chemical interaction occurring in the nutraceutical system. This binding is generally irreversible as a result of subsequent reactions, and the part of the aldehyde that reacts chemically is lost or altered as far as sensory properties are concerned. This can result in alteration of the flavor profile.

In general, interactions between flavor ingredients and proteins are more complex than those between flavor ingredients and carbohydrates. Table 19.3 summarizes some of the interactions between various flavor ingredients and proteins.

19.7.3 Interactions with Free Amino Acids

Free amino acids can bind with many flavor ingredients in aqueous media. Ketones and alcohols are reversibly bound by hydrogen bridges to the amino or carboxyl groups of the amino acids, whereas with proteins some aldehydes react chemically with the amino groups to form Schiff's bases. The amino acid cysteine reacts in an aqueous medium with aldehydes and ketones to thiazolidine-4-carboxylic acid. This reaction is reversible under heating, particularly under acidic pH. Dry amino acids adsorb volatile aldehydes (e.g., hexanal), ketones (e.g., diacetyl), acids, and amines.

19.7.4 Interactions with Lipids

The most important lipids in foods are *fats* and *oils*. These consist mainly of triglycerides. Triglycerides can bind large quantities of lipophilic (i.e., nonpolar) and partly lipophilic flavor ingredients. The binding capacity of fats (triglycerides that are solids at ambient temperature) is less than that of oils (triglycerides that are liquid at ambient temperature). The quantity of flavor ingredient bound also depends on the chain length of the fatty acids in the triglyceride, as well as on the presence of saturated or unsaturated fatty acids in it. Triglycerides with long-chain fatty acids bind less ethanol and ethyl acetate than those with short-chain fatty acids. In lipid–water mixtures, in oil–water flavor emulsions, and in water–oil flavor emulsions, the flavor ingredients are distributed between the lipid and water phases as a function of the structure of the flavoring substances (more lipophilic or more hydrophilic), the type of the lipid, and the temperature.

Many volatile flavor ingredients have a lower vapor pressure in lipids and thus a higher odor threshold than they do in aqueous systems. What this means is that the addition of even small amounts of lipid can significantly reduce the concentration of a flavor ingredient above an aqueous system.

19.7.5 Interactions with Inorganic Salts, Fruit Acids, Purine Alkaloids, Phenolic Compounds, and Common Solvents

A well-known phenomenon in *inorganic salts* is the salting-out effect. Adding sodium sulfate, ammonium sulfate, or sodium chloride, for example, in portions to

Table 19.3 Type of Interactions between Flavor Ingredients and Protein

Class of Substances	Type of Interactions
Hydrocarbons	Hydrophobic interaction
Alcohols	Hydrophobic interaction Hydrogen bridges
Aldehydes	Hydrophobic interaction Chemical reaction (Schiff's base)
Ketones	Hydrophobic interactions Hydrogen bridges
Esters	Unexplained

aqueous systems has the effect of driving out some of the volatile compounds into the gas phase or into a solvent that is immiscible with water. Additions of 5%–15% to aqueous systems result in increases of headspace concentration of ethyl acetate, isoamyl acetate, and menthone up to 25%. This NaCl concentration, however, is not normally present in any consumable products. In products with a normal salt content, the salt has virtually no effect on the vapor pressure of volatile compounds. The same is true for calcium chloride.

Aqueous systems containing *purine alkaloids* (e.g., caffeine or theobromine) or phenolic compounds (e.g., naringin) lower the headspace concentration of a number of flavor ingredients (e.g., benzaldehyde, benzyl acetate, furfural, various pyrazines, and terpenes). In the case of phenolic compounds, there is possible hydrogen bridge bond formation.

Ethanol, propylene glycol, and glycerin, which are the most commonly used solvents, can form acetals with aldehydes. The formation of acetal is generally reversible. With an acidic pH value, ethanol, propylene glycol, glycerin, and aldehyde are released again. In products containing ethanol with a pH of 7, part of the aldehyde is bound as acetal.

19.7.6 Interactions with Fat Replacers

Many products on the market are available as reduced-fat or fat-free versions of the original. In relation to flavor, fat has three functions in any product:

- Mouthfeel
- Carrier of the flavor
- Precursor to the flavor

Fatty or oily mouthfeel is a combination of several parameters. These include viscosity (thickness, body, fullness), texture (creaminess, smoothness), and adsorption or absorption (physiological effect on taste buds). Reduced-fat or fat-free products exhibit a high flavor impact initially that dissipates quickly, whereas full-fat products gradually build up intensity and dissipate more slowly. Current available ingredients that have, or are claimed to have, fat-mimicking properties may be divided into five categories:

1. Traditionally used emulsifiers (lecithin, polysorbate 60, span 60)
2. Lipid-based fat replacers (e.g., Caprenin, Olestra [both Proctor & Gamble, Cincinnati, OH])
3. Carbohydrate-based replacers (e.g., modified starch, maltodextrin, cellulose, Avicel [FMC BioPolymer, Philadelphia, PA])
4. Protein-based replacers (e.g., milk-derived solids, egg-derived solids)
5. Mixed-blend replacers

Of these, emulsifiers and lipid-based ingredients are true fat replacers. Protein-based fat replacers exhibit more fatlike flavor interactions than carbohydrate-based and mixed-blend replacers.

19.8 COMMONLY FACED PROBLEMS IN FLAVORING OF NUTRACEUTICAL PRODUCTS AND THEIR SOLUTIONS

The following is a list of solutions to common flavoring problems that the flavor industry formulators encounter when trying to flavor products with undesirable flavors:

1. Unavailability of the plain nutraceutical base: Provide the flavor supplier with an unflavored nutraceutical or beverage base whenever possible.
2. Failure to cover unwanted notes first: Flavor suppliers recommend covering up odd base notes first. It is beneficial to cover the undesirable notes first, thus optimizing the base, before adding the characterizing flavor. Many customers only consider addition of flavors late at the end of the product development cycle, which can be a mistake. Flavorings should be considered from the beginning.
3. Overflavoring of products: "If a little is good, more is better." Many food formulators are still using this incorrect approach to flavor their products. Flavors are designed to be used at a certain strength that when exceeded becomes imbalanced, resulting in chemical burn and introducing unwanted off-notes.
4. Not using dual-functionality ingredients: Try to use dual-functionality ingredients. Although many functional ingredients have a negative impact on flavor, some can improve flavor or texture while imparting their nutritional boost. Agave nectar, for example, is a potent sweetener and a good source of inulin. Also, maltodextrin has dual functionality. It is used in recovery beverages to help smooth out high-protein solutions, thus enhancing mouthfeel and masking some of their off-flavors. Such dual functionalities can help match application to ingredient and reduce formulation challenges.
5. Flavor function restrictions: Flavors are tricky. Also, they are often application specific. Just because one flavor suits one application does not mean it will work well in another.
6. Vitamin and mineral flavor issues: The consensus among nutritional ingredient suppliers, flavor experts, and food manufacturers is that B vitamins, especially thiamin, and minerals like iron and copper can present technical challenges to formulators. In some cases, nutrients are water insoluble (e.g., some minerals), making these more difficult to provide in a beverage. Vitamins like thiamin (vitamin B1) can be challenging in some products because they impart a sulfur note similar to a rotten egg. This is a result of the thiamin molecule's sulfur-containing portion. Minerals like phosphorus, potassium, magnesium, zinc, and iron, as well as the sulfur-containing amino acids like methionine, cystine, and cysteine, also can bring about lingering metallic flavor issues. Calcium does not have a metallic flavor but may precipitate out and give a chalky texture or flavor when used at higher levels. Minerals like iron can chemically react with flavor ingredients to form "inclusion complexes."
7. Challenges involved in flavoring protein products: High levels of protein can invite flavor issues; soy often is problematic when it comes to off-notes. However, all proteins have amino acids and an amine backbone that can allow them to participate in Maillard reactions along with sugars. Harsher processing conditions such as retorting, HTST processing, or UHT processing often create off-notes such as burnt, caramelized, nutty, "beany," sulfuric or bitter. Heat processing can also result in protein denaturation, thus releasing free amino acids that can react with

flavor ingredients. The increased use of proteins in nutraceutical products has created the need for a vast array of masking agents, modulating flavors (modifiers), and characterizing flavors that ultimately create a palatable product.

8. Herbal and botanical ingredient challenges: The most common unwanted taste attribute on addition of herbal or botanical ingredients to a nutraceutical product is bitterness and astringency. Bitter and astringent notes in the mouth caused by herbals are masked when flavorists modify the taste perception by increasing the sweetness perception or by reducing the bitter perception.
9. Non-nutritive sweetener challenges: With the introduction of stevia to the beverage market, we are seeing a natural product that behaves differently than sugar. It takes longer for the sweetness to peak, has a different mouthfeel, and has a distinct aftertaste perceived as bitter and licorice-like. The solution is to mask the bitter aftertaste and enable food technologists to use much less stevia than sugar to create the desired sweetness level, thus creating an excellent low-calorie product. Also, other non-nutritive sweeteners like sucralose and aspartame pose their own challenges, although their use is fairly common compared with that of stevia, which is fairly new.

These are some of the general problems encountered while trying to flavor nutraceutical products. It is not by any means an all-inclusive list. Working with flavor systems to maximize the outcome can be complex. As presented earlier, numerous factors can affect overall flavor delivery. The presence and levels of protein, fat, sugar, and non-nutritive sweeteners, salt, and acid can affect flavor notes. Functional ingredients, such as starch and carrageenan, can affect flavor delivery. Because so many ingredients from a nutraceutical product can interfere with flavor delivery, flavors developed for such applications are usually customized and application specific. Systems containing high fat or lipid ingredients often need more concentrated flavors to overcome the effect of fat. Certain metals can react with the flavor and accelerate oxidation. Citrus flavors are especially vulnerable to oxidation. If the processing steps are known while the flavor is being developed, the flavorist can choose the best solvent system. The solvent is especially important if heat is applied to the nutraceutical product. The flavorist may also recommend a higher dosage to compensate for flavor loss during processing. It might also be wiser to add the flavor at a different point in the product development process than originally planned.

Flavors can be affected by hot filling or aseptic packaging, as well as chemical preservatives. The longer the hold time involved during heat processing of a product, the more damage to the flavor occurs. Powdered flavors are typically encapsulated to help prevent activity between other ingredients that can produce off-flavors or loss of flavor. The encapsulation process involves creating a barrier between the core flavor components and the final or finished application. This eventually results in extending the shelf life of the flavor. Generally, encapsulated flavors are application specific and are typically water soluble. When feasible, it is recommended to use an encapsulated flavor for a product that will be subject to high temperature processing. Also, using encapsulated flavor is the best option for powdered products expected to have a long shelf life. In any system, the flavor must be applied and delivered in sensible quantities. This reinforces the need for flavors that are properly concentrated and suited for the nutraceutical product in concern.

The author recently solved a flavor problem in a hemp milk base by reducing the amount of vanilla and adding a flavor modifier that greatly smoothed the base. The customer was expecting the flavor to both flavor the system and cover up the bitter end notes of the hemp milk.

19.9 STABILITY OF FLAVORS IN NUTRACEUTICAL PRODUCTS

The solutions to flavor stability issues range from the use of edible films and encapsulation to nanotechnology. Flavor companies have been creative with masking systems, and many of the technologies are patented and/or proprietary. Solutions are product dependent. For example, encapsulation or even a basic fat coating over B vitamins and minerals can improve stability and flavor in a bar, although that would not work in a beverage.

19.9.1 Encapsulation

Encapsulation is basically the art of encasing an ingredient in a starch or other material, such as natural gums, proteins, or hydrocolloids, and then spray-drying the mixture to form a capsule around the ingredient. It also may be useful to encapsulate the flavors themselves. Normally the encapsulation process limits exposure to oxygen, extending the shelf life of the dried material. Recent research on various starches as encapsulation agents may offer a less-expensive approach for encapsulating both flavor compounds and nutritional ingredients. Starches may react uniquely with specific ingredients, holding in flavor compounds and protecting fats from oxidation. Among enzyme-modified starch derivatives, cyclodextrins behave as empty molecular capsules with the ability to entrap guest molecules of appropriate geometry and polarity. The included molecules are protected from surroundings: light, heat, oxidation, etc. The encapsulated system is stable and exhibits its advantageous properties while in a dry system, but once reconstituted to a liquid, different reactions as mentioned earlier might occur. Encapsulation is a good way to stabilize nutrients in a bar, to which it is relatively easy to add higher amounts of nutrients.

19.9.2 Moisture

Because moisture in general can accelerate degradation, there are some advantages to working with dry systems—when it comes to some nutrients. Protecting B vitamins and certain minerals with a fat coating can reduce interactions of these nutrients with other ingredients (such as in nutritional bars). However, when it comes to concerns with proteins and peptides participating in the Maillard reaction, it is easier to stabilize these nutritional ingredients in a liquid form (compared with a dry form). Many bars are intermediate-moisture foods; this is the most optimal moisture level for Maillard reactions to occur, says Reineccius. In contrast, the extremely high moisture level of a beverage offers some protection.

Depending on the finished product or on the specific off-note created from a particular nutrient, vitamin, or other flavor, a different approach would be taken for stabilizing the system. Increasing the acidity in a beverage can help "clean up" the aftertaste of non-nutritive sweetener blends. However, in some cases, increasing the acidity of the beverage can also intensify a certain off-note. High acidity can also promote the occurrence of other chemical reactions.

19.9.3 Nanotechnology

Nanotechnology is a powerful new masking technology that uses fine particles (with hydrophobic and hydrophilic properties) to protect nutrients from interactions with other components in a food. In the range of approximately 1–100 nm, particles or systems may have novel properties because of their small size. These properties have the ability to be controlled or manipulated on the atomic or molecular level. Mono- and diglycerides, whey proteins, and other molecules with hydrophobic and hydrophilic properties can potentially be used as part of these systems. It is still expensive but becoming more mainstream; it should become less expensive and more applicable with time.

19.9.4 Solubility

Another important issue to address in flavoring of nutraceutical beverages is solubility. In beverages, solubility issues can result in precipitation. There are clear products such as vitamin-fortified waters and smoothie-type products that can be viscous (pastelike). Essentially the flavors added to the waters usually must be clear, nonopaque, and/or colorless. Smoothie-type products, on the other hand, may allow use of flavors in an emulsion form or with juice or some pulp.

19.10 METHODS USED TO INCORPORATE FLAVORS IN NUTRACEUTICAL PRODUCTS

Efforts have been made to incorporate various nutraceuticals into foods or beverages. However, the amounts of these beneficial nutraceutical ingredients are typically low relative to what one must consume.

What is preferred is a system for delivering nutraceuticals in a way that is a more enjoyable and flavorful alternative than taking a pill, tablet, capsule, or other medicinal form, but still has the appropriate concentration of the nutraceutical. Listed below are a few ways in which a flavor can be incorporated into a nutraceutical product. These are examples of products currently on the market.

1. Using a chocolate base: A chocolate formulation for delivering nutraceuticals comprising a chocolate base, either blended with the chocolate itself or added as a liquid

or cream filling, that has flavor and a low-calorie sweetener like sucralose. The chocolate is high in antioxidants, adenosine monophosphate, and fat-soluble nutrients. Using this inventive system, delivery of nutraceuticals in unit dosage form is facilitated because the selected dose is carried within individual chocolate product pieces that taste substantially the same as conventional chocolate, although with few calories from carbohydrates or effects on insulin response encountered with typical chocolate formulations. In such a case, the flavor, either a water- or oil-soluble liquid, can be incorporated in the chocolate base or in the cream filling. A good example of this is VIACTIV (McNeil Nutritionals, LLC, Ft. Washington, PA).

2. Nutraceutical beverages: The decline of the carbonated soft drink market has motivated a large number of manufacturers to enter the nutraceutical arena, making it the most dynamic sector in soft drinks. Consequently, leading soft drink manufacturers have diversified their portfolios by acquiring nutraceutical brands, extending current lines with nutraceutical variants, or developing their own nutraceutical brands to capitalize on this market opportunity. During the past couple years, the demand for functional beverages has increased, particularly for liquid energy shots, drinks with vitamin and mineral fortification, antioxidant or high polyphenol beverages containing ingredients like teas or berries, and beverages enhanced with herbal ingredients. Whatever the health-enhancement goal of the liquid delivery product, it must taste good and thus flavors are added to it. Pomegranate and açaí are examples of the more popular flavors, whereas the long-standing citrus flavors like orange, lemon, and lemon–lime are the most common. In nutraceutical beverages, only water-soluble and clear flavors are preferred. Flavors can be liquid or powdered.
3. Detoxifying products: Products that help consumers to detoxify their digestive system exemplify how the early health-related functions of gut health products are supplemented by lifestyle products. In this case, water-soluble flavors are preferred, whether a clear liquid or an emulsion.
4. Chewable tablets and gummy bears: This method is especially popular in products like vitamin C or omega-3 dietary supplements for kids. Citrus flavors like orange are common in vitamin C chewable applications. Of course, in this case the flavor is in the form of a powder and spray dry. Spray-drying is an efficient method of converting liquid flavors into powders. By spray-drying it is possible to produce a powder with controlled physical properties such as flowability, residual moisture content, and bulk density. In the case of omega-3 fatty acids, which are oil soluble, the gummy bears are a perfect delivery method. In gummy bears, strong water-soluble flavors are needed, preferably with propylene glycol used as the solvent.
5. Aloe vera beverages: Aloe vera has a distinct bitter and astringent taste. It is a medicinal plant that exerts a range of health effects both topically and internally. There are many aloe beverages on the market. Some are like a liquid multivitamin with aloe; others are just plain aloe vera juice with some flavors. Mostly liquid water-soluble flavors are preferred for this kind of product.
6. Energy bars: Market shelves are stocked with food bars designed to fit nearly any consumer need, whether meal replacement, nutritional boost, or snack. No matter the desired end attribute, developing bars follows a basic blueprint. Formulators combine proteins, grains, carbohydrates, and other specific value-adding nutrients. Other texture-adding items are present like nuts, dried fruit pieces, and sometimes a coating for the bar that has added flavor. Both liquid and powdered flavors can be incorporated in the bars.

19.11 SIGNIFICANCE OF THE FLAVOR INDUSTRY

Flavor houses are working harder than ever to deliver nutraceutical products that taste great, but it takes more than just a client–customer relationship to get the job done right. The process of flavoring nutraceutical products requires as much art as science. Working as a flavor house in the context of the nutraceuticals market is time-consuming because there is no such thing as a one-size-fits-all when it comes to flavoring these types of products. Nutraceutical applications require a lot more attention and effort, and customization frequently becomes inevitable. For a flavor house, working with nutraceutical products is like mastering a complex balancing act. We do not just offer masking solutions or add flavors to products—many times it is doing a lot of both. It is not as simple as covering up the off-notes; there is a whole balancing act that needs to take place between the functional ingredients and the flavor. Practically speaking, despite the basic knowledge of possible interactions as outlined previously, when flavoring an industrially produced nutraceutical product, almost every product must be considered individually. However, we have observed some general trends. For example, some citrus and berry flavors work well with bitter ingredients, and the chocolate, vanilla, caramel, and butterscotch flavors work better with proteins.

Products with unique ingredients and health value impart tastes and aroma not typically found in traditional foods or beverages. Because of these demands, the flavorists, applications technologists, and sensory scientists must work closely with the end user to ensure that the commercial product has high consumer acceptance of both flavor and functionality. Availability of the customer's base to the flavor development team is the only way to ensure that possible interactions between flavor ingredients and the base ingredients can be recognized and considered. The use of new descriptive analysis allows the flavor group to understand the attributes of these defects and to create flavors that allow functional products to have enhanced palatability. In fact, sensory terminology has expanded in recent years to accommodate the taste attributes imparted by the nutraceutical ingredients. The solutions that flavor companies offer can range from the use of the defect as a flavor note in the final flavor to the suppression or removal of the defect. This is a sure way to end up with a product that is flavored optimally and marketable.

Figure 19.8 shows a typical workflow at a flavor house once a customer approaches the company with a specific project.

19.12 CLASSIFICATION OF FLAVORS

There are specific labeling requirements for flavors, which are found in the Food and Drug Administration's (2010) regulations, specifically Code of Federal Regulations Title 21, section 101.22.

1. Artificial flavor or artificial flavoring: The term *artificial flavor* or *flavoring* means any substance, the function of which is to impart flavor, which is not derived from

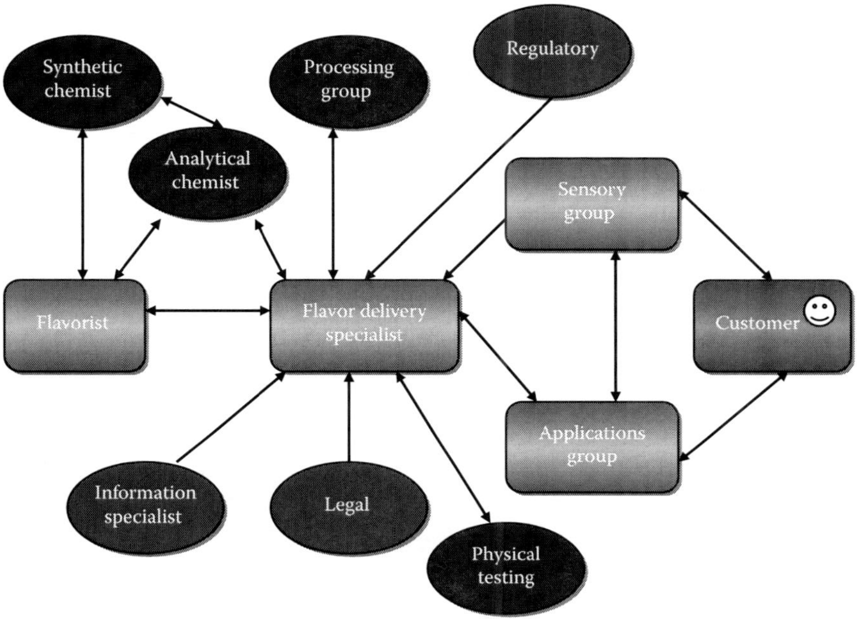

Figure 19.8 Collaboration between the nutraceutical industry and flavor industry.

spice, fruit or fruit juice, vegetable or vegetable juice, edible yeast, herb, bark, bud, root, leaf or similar plant material, meat, fish, poultry, eggs, dairy products, or fermentation products thereof. Artificial flavor includes the substances listed in sections 172.515(b) and 182.60, except where these are derived from natural sources. In cases where the flavor contains only artificial flavor(s), the flavor should be so labeled, for example, "artificial strawberry flavor."

2. Natural flavor: The term *natural flavor* or *natural flavoring* means the essential oil, oleoresin, essence or extractive, protein hydrolysate, distillate, or any product of roasting, heating, or enzymolysis, which contains the flavoring constituents derived from a spice, fruit or fruit juice, vegetable or vegetable juice, edible yeast, herb, bark, bud, root, leaf or similar plant material, meat, seafood, poultry, eggs, dairy products, or fermentation products thereof, whose significant function in food is flavoring rather than nutritional. Natural flavors include the natural essence or extractives obtained from plants listed in sections 182.10, 182.20, 182.40, 182.50, and part 184, and the substances listed in section 172.510.

The following definitions were accepted by IOFI:

1. Natural flavor "with other natural flavor": If the food contains both a natural characterizing flavor from the product whose flavor is simulated and other natural flavor that simulates, resembles, or reinforces the characterizing flavor, the food shall be labeled as the name of the food and be immediately followed by the words "with

other natural flavor" in letters not less than one-half the height of the letters used in the name of the characterizing flavor.

They are preparations and single substances, respectively, acceptable for human consumption, obtained exclusively by physical processes from vegetable, sometimes animal, raw materials either in their natural state or processed for human consumption.

2. Natural and artificial flavor/nature-identical* flavor: In cases where the flavor contains both a natural flavor and an artificial flavor, the flavor shall be so labeled, for example, "natural and artificial strawberry flavor." They are substances chemically isolated from aromatic raw materials or obtained synthetically; they are chemically identical to substances present in natural products intended for human consumption, either processed or not.
3. Artificial flavor: These are those substances that have not yet been identified in natural products intended for human consumption, either processed or not.

19.13 ANALYTICAL METHODS USED FOR DETECTION OF FLAVORS USED IN NUTRACEUTICALS

Flavors as they are currently prepared in the flavor industry are complex mixtures of chemically derived flavor ingredients and extracts of varying compositions, as well as carriers and other components (e.g., additives). Hence it is important for the quality control of individual flavor components, as well as whole flavors, to have analytical methods that allow testing for identity and specific screening for components limited by law.

Generally quality control for most flavorings requires quantitative details of the individual ingredients or of the entire system. In such cases gas chromatography (GC) is chosen for volatile compounds, and high-performance liquid chromatography (HPLC) is used for nonvolatile compounds. The current methods require a preceding cleanup and concentration of the sample.

19.13.1 Sample Preparation Methods

1. Solid–liquid extraction: The commonly used method is Soxhlet extraction using diethyl ether and/or pentane as the solvent. The sample is extracted continuously using a Soxhlet extractor over many hours. The extract is then dried and concentrated by evaporating the solvent. The residual extract is then applied directly to GC analysis. This method is used for solid or pasty flavors and also for whole fruits or plants and their parts where the composition of the volatile compounds is required.
2. Liquid–liquid extraction: In this case, the sample is diluted with deionized water and extracted with an organic solvent like pentane or dichloromethane using the funnel separator. The extract is then dried and concentrated by solvent evaporation. The concentrated extract is applied to GC analysis. This method is used for liquid extracts and flavors.

* "Nature-identical" is not an applicable category per US regulations. It is only applicable for Europe.

3. Simultaneous distillation and extraction: The sample is diluted with tenfold the quantity of distilled water and is then extracted for several hours with dichloromethane or diethyl ether using a simultaneous distillation extraction apparatus. The organic phase is dried and concentrated by solvent evaporation. The concentrated extract is then subjected to GC analysis.
4. Liquid–solid extraction: The sample is diluted with appropriate solvent. The solution is applied onto a cartridge containing an appropriate solid-phase material (e.g., silica gels, ion exchange columns). The required compounds are dissolved from the cartridge using different solvents. The eluent is used directly for chromatographic analysis. The advantages of this clean-up method are a small amount of solvent yields a high concentration of the sample, speed, and no interference from carbohydrates.
5. Solid-phase microextraction (SPME): SPME is a simple and efficient, solventless sample preparation method invented by Pawliszyn in 1989 (Vas and Vekey 2004). SPME has been widely used in different fields of analytical chemistry since its first applications to environmental and food analysis and is ideally suited for coupling with mass spectrometry (MS). All steps of conventional liquid–liquid extraction, such as extraction, concentration, and transfer to the chromatograph, are integrated into one step and one device, considerably simplifying the sample preparation procedure. It uses a fused-silica fiber that is coated on the outside with an appropriate stationary phase. The analytes in the sample are directly extracted to the fiber coating. The SPME technique can be routinely used in combination with GC, HPLC, and capillary electrophoresis, and places no restriction on MS. SPME reduces the time necessary for sample preparation, decreases purchase and disposal costs of solvents, and can improve detection limits. The SPME technique is ideally suited for MS applications, combining a simple and efficient sample preparation with versatile and sensitive detection.

19.13.1.1 High-Performance Liquid Chromatography

HPLC is an integral part of routine analysis of low- or nonvolatile flavor compounds, as well as of thermally instable or decomposing flavoring substances. According to the column material mainly used for separation, HPLC procedures can be divided into the following categories:

1. Adsorption chromatography
   a. Normal phase chromatography: Uses a polar material for the stationary phase, and an apolar solvent like hexane for the mobile phase.
   b. Reversed-phase chromatography: Uses an apolar material for the stationary phase, and a polar solvent like water for the mobile phase.
2. Ion exchange chromatography: The stationary phase contains anionic or cationic groups that interact with the ionic groups of the sample ingredients to result in separation.
3. Affinity chromatography: The stationary phase consists of a special chemically modified carrier that reversibly reacts with sample molecules. Water is generally used for the mobile phase. This method is used mainly for isolation and analysis of bioactive components (e.g., lipids, proteins).

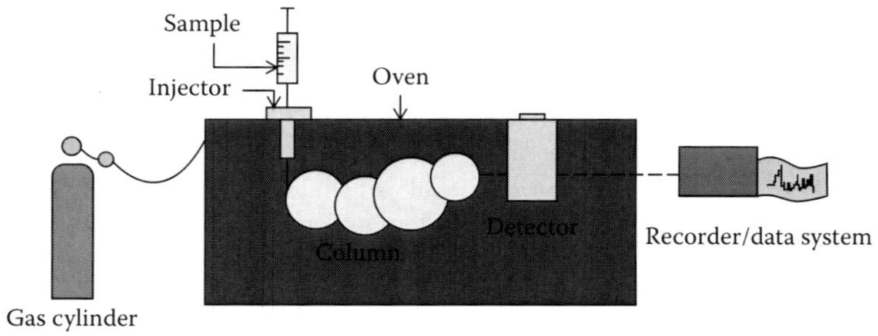

Figure 19.9 Schematic representation of gas chromatography.

19.13.1.2 Gas Chromatography

This technique is the most commonly used chromatographic method for analysis of volatile flavor compounds (Figure 19.9).

The stationary phase consists of capillary columns, which can be polar, midpolar, and nonpolar. Usually helium is used as carrier gas. The detection system may be flame ionization detector, electron capturing detector, or nitrogen–phosphorus detector.

19.13.1.3 Gas Chromatography/Mass Spectrometry Coupling

GC/MS is a method that combines the features of GC and MS to identify different substances in a test sample (Figure 19.10). This is the most common coupling technique used in the quality control of flavor compounds and other flavor ingredients. The mass selective detection (MSD) system used in routine analysis of flavors enables qualitative and quantitative analysis of the structure of single components.

After GC separation, the single components are transferred to the MSD using a specific interface. In the MSD ion source, the molecules are ionized by electron impact conditions at conventional 70 eV. This leads to the fragmentation of

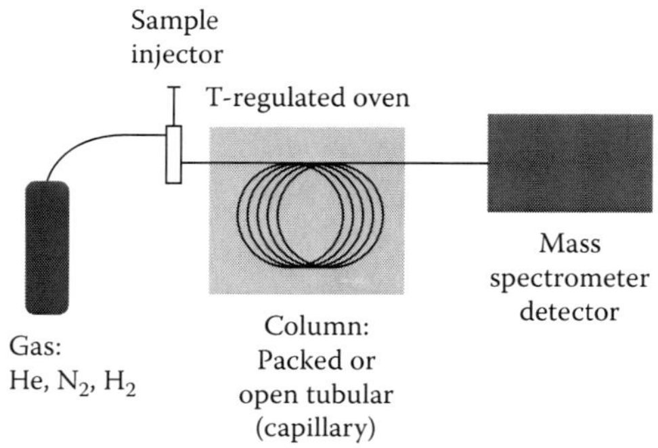

Figure 19.10 Schematic representation of gas chromatography/mass spectrometry.

these molecules by formation of different ions. The separation of ions is achieved with magnetic or electric fields or both combined. The separated ions impinge on the electron multiplier detector system that collects and amplifies these signals.

19.14 EXAMPLES OF FLAVORED NUTRACEUTICAL PRODUCTS

Listed below are a few examples of nutraceutical products on the market.

1. Natural and artificial vanilla- or chocolate-flavored whey powder isolate: Whey protein isolate is the most popular protein used for its biological value and protein efficiency for building muscle. Vanilla and chocolate flavor powders are blended into the product. When reconstituted, they help to cover the protein notes and also impart the characteristic flavor to the product.
2. Orange-, root beer-, cherry-, and mixed fruit-flavored effervescent drink tablets: This product has no sugar, no carbohydrates, and 100% of the daily requirement of vitamin C. Powdered flavors are used in this application.
3. Orange-flavored juice drink fortified with fiber: This drink provides as much fiber as three apples with fewer calories. It is flavored with a liquid water-soluble orange flavor.
4. Lemon–lime-flavored aloe juice drink: This product contains whole-leaf aloe vera gel. The aloe gel contains more than 75 nutrients and 200 active compounds, including 20 minerals, 18 amino acids, and 12 vitamins. In a product like this, liquid water-soluble emulsions can also be used.
5. Lemon–ginger-, chai spice-, and cranberry–almond-flavored nutrition bars: These nutrition bars contain heart-healthy phytosterols. Both liquid and powdered flavors can be incorporated into the energy bars.
6. Strawberry-, vanilla-, mixed berry-, and strawberry–banana-flavored cultured yogurt: A low-fat yogurt that contains *Bifidus regularis*, a natural probiotic culture that helps to regulate the digestive system. Yogurt is also available with fiber and whey proteins added.
7. Vitamin water: Dragonfruit-, raspberry–apple-, peach–mango-, jackfruit–guava-, tamarind–pineapple, and green tea-flavored waters are available with different value-added nutrients in each.
8. Value-added peanut butter: This is a product that is marketed as being healthier than regular peanut butter by adding a daily dose of omega-3 fatty acids. An oil-soluble flavor can also be added to such a system.
9. Zero IMPACT bars (VPX Sports, Weston, FL) with Clarinol conjugated linoleic acid are available in strawberry–yogurt and chocolate flavors. This product helps reduce body fat without compromising on the taste of a nice snack.

REFERENCES

Belitz, H. D., and W. Grosch. 1982. *Lehrbuch der Lebensmittelchemie* [Textbook of Food Chemistry]. Berlin: Springer Verlag.

Food and Drug Administration. 2010. Code of Federal Regulations. Title 21, subpart B. http://www.accessdata.fda.gov/scripts/cdrh/cfdocs/cfcfr/CFRSearch.cfm.

International Organisation of the Flavour Industry. 1990. *Code of Practice for the Flavour Industry*. Geneva: Author.

Vas, G., and K. Vekey. 2004. Solid-Phase Microextraction: A Powerful Tool Prior to Mass Spectrometric Analysis. *Journal of Mass Spectrometry* 39: 233–253.

FURTHER READING

Backas, N. 2009. Brimming Opportunities for Nutraceutical Beverages. *Food Product Design* January: 50–68.

Berry, D. 2009. Powerful Performers. *Food Product Design* April: 32–50.

Decker, K. J. 2009. Protein—A Functional Powerhouse. *Food Product Design* June: 34–48.

Eckert, M. A. 2009. Putting the Function in Beverages. *Natural Products Insider* 14: 50.

Granato, H. 2002. Manipulating Flavor Perception in Functional Products. *Natural Products Insider* http://www.naturalproductsinsider.com/articles/2002/04/manipulating-flavor-perception-in-functional-produ.aspx.

Granato, H. 2002. Masking Agents Maximize Functional Foods' Potential. *Natural Products Insider* http://www.naturalproductsinsider.com/articles/2002/01/masking-agents-maximize-functional-foods-potentia.aspx.

Hazen, C. 2009. Maximizing Flavor Delivery. *Food Product Design* June: 86–98.

Heath, H. B. 1981. *Source Book of Flavors.* New York: AVI Publishing Company, Inc.

Heath, H. B., and G. Reineccius. 1986. *Flavor Chemistry and Technology.* New York: AVI Publishing Company, Inc.

Ohr, L. M. 2008. Functional Ingredients and Healthy Developments. *Food Technology* August: 97–103.

Piggott, D. 2009. Walking the Tightrope—Balancing Flavor and Functionality in Nutritional Beverages. *Natural Products Insider* 14: 48.

Porzio, M. A. 2007. Flavor Delivery and Product Development. *Food Technology* January: 22–29.

Szente, L., and J. Szejtli. 1988. Flavor Encapsulation. *Stabilization of Flavors by Cyclodextrins,* ACS Symposium Series, vol. 370. New York: Oxford University Press.

Ziegler, E., and H. Ziegler. 1998. *Flavourings.* Weinheim, Germany: Wiley-VCH.

CHAPTER 20

Nutraceutical Clinical Batch Manufacturing

Weiyuan Chang

CONTENTS

20.1 INTRODUCTION

The word *nutraceutical* was first introduced in 1989 by Stephen DeFelice, founder and chair of the Foundation for Innovation in Medicine (Brower 1998). *Nutraceutical* is a combination of "nutrition" and "pharmaceutical" and is defined as "a food (or part of a food) that provides health benefits and fitness, including the prevention of some approved structure/function claims and/or the common, non-serious 'life stages' symptoms" (Zeisel 1999, p. 1853). Today, nutraceuticals are widely manufactured because of the significant increase in the public demand. Several clinical studies have been done for nutraceuticals. Good and qualified clinical manufacturing is critical. In this chapter, I will focus on the clinical trials, good manufacturing practice, and future development of nutraceuticals.

Carefully conducted clinical trials in human volunteers are the fastest and safest way to find treatments that improve health. Since the growth in the nutraceuticals market after 1990, the National Institutes of Health (NIH) in collaboration with the Food and Drug Administration (FDA) developed standards to regulate nutraceutical manufacturing and clinical studies to ensure public health and safety. There are two major types of clinical trials. *Interventional trials* determine whether experimental treatments or new ways of using known therapies are safe and effective under controlled environments. *Observational trials* address health issues in large groups of people in natural settings. Use *S*-adenosyl methionine (SAMe) as an example. SAMe is a nutritional supplement available as an over-the-counter formula and recommended for depression and arthritic pain. A PubMed search in May 2010 showed 6,760 published research articles and 206 clinical studies in progress. Studies in

Europe suggest that SAMe may help to decrease the fatigue and itching common in people with liver problems, as well as to decrease levels of liver enzymes in the blood, suggesting that it may decrease the amount of liver injury. Therefore, the majority of current SAMe clinical studies focus on liver-related issues, including a randomized controlled trial of an adjunctive therapy for alcoholic liver disease in patients with intrahepatic cholestasis and primary biliary cirrhosis as a therapeutic intervention of relieving pruritus and fatigue.

20.1.1 Standards for Clinical Research

There are standards applied to seven areas of clinical research, including clinical informatics, data management and protocol tracking, biostatistics support, quality assurance and quality control, protocol review, and human resources and physical plant, as well as training and education. However, SAMe is just one amino acid compound under the category of dietary supplement trials among 47,621 additional clinical trials in all fifty of the United States and 90,857 cases worldwide (US National Library 2010a).

The framework for successful manufacturing of nutraceutical clinical material is important. It is a serious manufacturing process starting from chemistry development of the bench scale through the engineering run, to the clinical batch manufacturing (from bench to batch). In process development, current good manufacturing practice (cGMP) plays an important role in controlling the quality. In correlation to the international standard International Conference on Harmonisation (ICH) Guideline (International Conference on Harmonisation 2007), the pharmaceutical industry followed the guideline to guarantee an effective quality management system. The ICH Guideline describes the key elements of a robust quality systems model and shows how implementation of such a model is one way to comply with the FDA's cGMP regulations. The ICH guidelines need to be implemented throughout the different stages of a product lifecycle. These guidelines for process development in the pharmaceutical industry are implemented in preparing the Chemistry, Manufacturing, and Controls (CMC) section and Common Technical Document (CTD) for new drug application (NDA) filing or for drug master file. There are still many pitfalls in nutraceutical manufacturing. Again, I use SAMe as an example. The dosage in the over-the-counter formula is not sufficient to reach therapeutic levels unless patients take a good quantity of SAMe per day. A similar situation is found in clinical studies of cysteine, lipid replacement therapy, and *Ginkgo biloba* supplements, as well as in many other clinical trial reports (Nicolson, Ellithorpe, and Ayson-Mitchell 2010). In addition, the bioavailability, toxicity, first-pass metabolism, pH stability, high doses, and, most of all, patient compliance for those taking more than five SAMe pills daily are considered. To solve this problem, many pharmaceutical industries develop new formulations, including nanoparticulate carrier systems (Thassu, Deleers, and Pathak 2007), high quality polymer or plasticizer coating, delayed-release capsules containing enteric-coated pellets, and implant systems, for example. These new approaches have to follow cGMP and manufacturing regulations as discussed in Section 20.3. However, as new biopharmaceutical technology is developed, it is expected to achieve process scale-up, process optimization, and process validation in nutraceuticals before clinical batch manufacturing. Clearly, it is in the best interest of US consumers that nutraceuticals are appropriately approved for manufacture and marketed following regulations and well-planned clinical studies.

20.2 FRAMEWORK FOR SUCCESSFUL MANUFACTURING OF NUTRACEUTICAL CLINICAL MATERIAL

20.2.1 Final Formulation

A traditional dosage form of a drug is composed of the active ingredients and excipients as the final formulated products (Figure 20.1). Nutraceutical excipients include gums, colorants, glidants, lubricants, disintegration agents, suspending agents, solvents, sweeteners, coatings, binders, disintegrants, antiadherents, antistatic agents, surfactants, plasticizers, capsules, emulsifying agents, and flavors (Shargel et al. 2001). Preformulation involves the characterization of an active ingredient's physical, chemical, and mechanical properties to choose what other excipients should be used in the preparation during the drug development and preclinical development stages. It is generally best to expect to move ahead with the formulation used for the Phase I clinical trial material at least through Phase II clinical development. These typically consist of hand-filled capsules containing a small amount of the active ingredients and diluents. The final formulation may vary as a result of variations in the blending or in the process itself. By the time Phase III clinical trials are reached, the formulation of the drug should have been developed to be close to the preparation that will ultimately be used in the market. The ideal formulation development work is best completed before moving into process development and scale-up before clinical batch manufacturing, unless additional bridging studies are conducted to accommodate changes. Formulation problems might also be uncovered at the pharmacodynamic stages, which will require changes to the formulation, but these should be minor and occur rarely.

20.2.2 Robust Manufacturing Process

The preclinical formulation may change as a result of variations, multiple validations, and quality controls (Food and Drug Administration 2010c). To avoid large variation, the statistical prediction intervals become wide. These wide prediction intervals suggest further experimentation focusing on a robust process that stabilizes process performance. The objective of a robust process is to reduce variability in formulation and processing by applying propagation of error calculations to reduce variation transmitted from controlled factors. From a technical standpoint, adequate process development must be conducted at scale before manufacturing of the Phase I clinical material. During the development process, the manufacturing run in terms of cGMP compliance is required. These issues are addressed within the Quality System (International Conference on Harmonisation 2007) of the organization to ensure that all criteria for product safety and quality are met before product release.

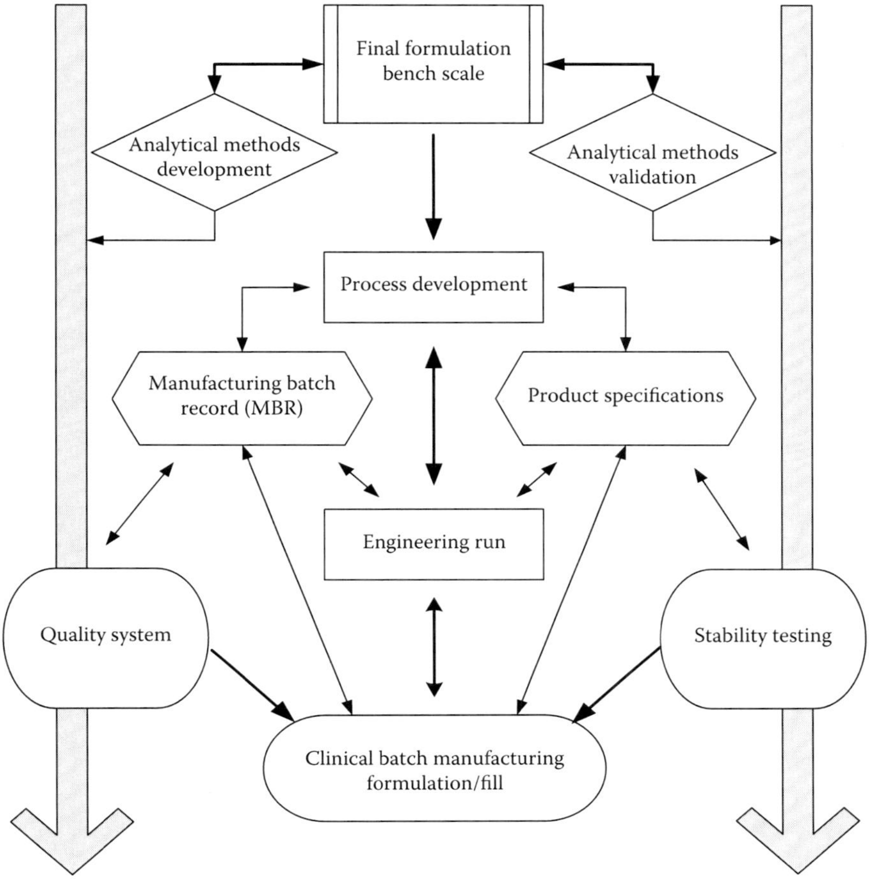

Figure 20.1 Diagrammatic representation of the relationship between key aspects of the nutraceutical product development process leading up to a successful formulation and fill phase for manufacturing of Phase I clinical material. (Modified from Senior, J. H., *Adv Nanotechnol Appl*, 1, 355, 2009.)

This is especially important if any pharmacology and toxicology or safety and efficacy studies have been conducted using bench-scale material.

20.2.3 Reproducible Manufacturing Process

As shown in Figure 20.1, the reproducible manufacturing process is controlled by two types of specifications: (1) manufacturing specifications, as defined in the master batch record (MBR), and (2) product specifications (Senior 2009). Specifications are acceptance criteria to which active ingredients and drug product must conform before they can be released as products for human use. Specifications should be established for critical materials used in the manufacture of nutraceuticals, for both intermediate and final products. Specifications for product release are chosen to confirm the product quality, rather than to characterize the product. Product specifications are expected to evolve during the drug development process. It is likely and desirable that specifications for active ingredients and final products will be tightened during the drug development cycle. Thus, as the product manufacturing acquires a batch history, it should be possible to narrow specifications for the parameters most crucial for a quality product. The MBR describes all component materials in the final product and materials used during manufacture that are intended to be removed during purification. The MBR also contains all test results for in-process and final testing, or provides a direct, traceable link to records of testing. It notes individual pieces of equipment used in manufacture and allows traceability to records of use, maintenance, and cleaning. As a result, the MBR is a carefully controlled document that also serves as the template from which the steps in the execution can be documented and laid out in such a way that all the steps in the process can be thoroughly checked during the quality assurance and manufacturing review process.

Both MBR and product specifications are considered scientific evaluations; there is also a regulatory support portion in process development, the CTD. In brief, the CTD controls three major areas before the clinical Phase II trial: (1) comprehensive dossier on drug quality, safety, and efficacy; (2) pharmacological, technological, and clinical development; and (3) meetings with the authorities for medicinal and botanical drug registration. I will discuss this subject more in section 20.3.5 Common Technical Document. These two types of specifications work together to define the process and to test the success and reproducibility of the end product, thus ensuring that product integrity and quality are ultimately maintained.

20.2.4 Choice of Clinical Manufacturer

Any substance intentionally added to food is a food additive and subject to premarket approval by the FDA unless the substance is "generally regarded as safe" (GRAS) under the conditions of intended use (Shargel et al. 2001; Food and Drug Administration Guidance for Industry 2008a). In addition, new dietary ingredients (NDIs) are defined by the Dietary Supplement Health and Educational Act (DSHEA; FDA 2010d); therefore, any manufacturer who wishes to market an NDI needs to submit a notice to the FDA that demonstrates this ingredient is reasonably expected to be safe when consumed. Many firms may chose to develop and manufacture the drug product entirely in-house. However, others prefer to use a contract manufacturer. As such, changing manufacturers during the lifecycle of drug development can demand extra expense and time. Therefore, it is desirable to choose a manufacturer who can support manufacture for all anticipated drug lot sizes and with an excellent record of cGMP compliance (Food and Drug Administration 2010a, 2010b). If the product is manufactured by the sponsoring firm, chances for securing approval of NDI and GRAS applications are high. It is helpful to have continuity in staff during the development processes. Another important aspect for both in-house and contract manufacturing is the readiness and experience of the organization for cGMP inspection by regulatory agencies.

20.2.5 Analytical Method Development for Nutraceuticals

The methods used in the laboratory bench studies may be used if they are qualified to be sufficiently robust, reproducible, accurate, and follow other key criteria for successful method validation (Food and Drug Administration 2010c). The extent of method validation evolves during the drug development process. Thus, a more

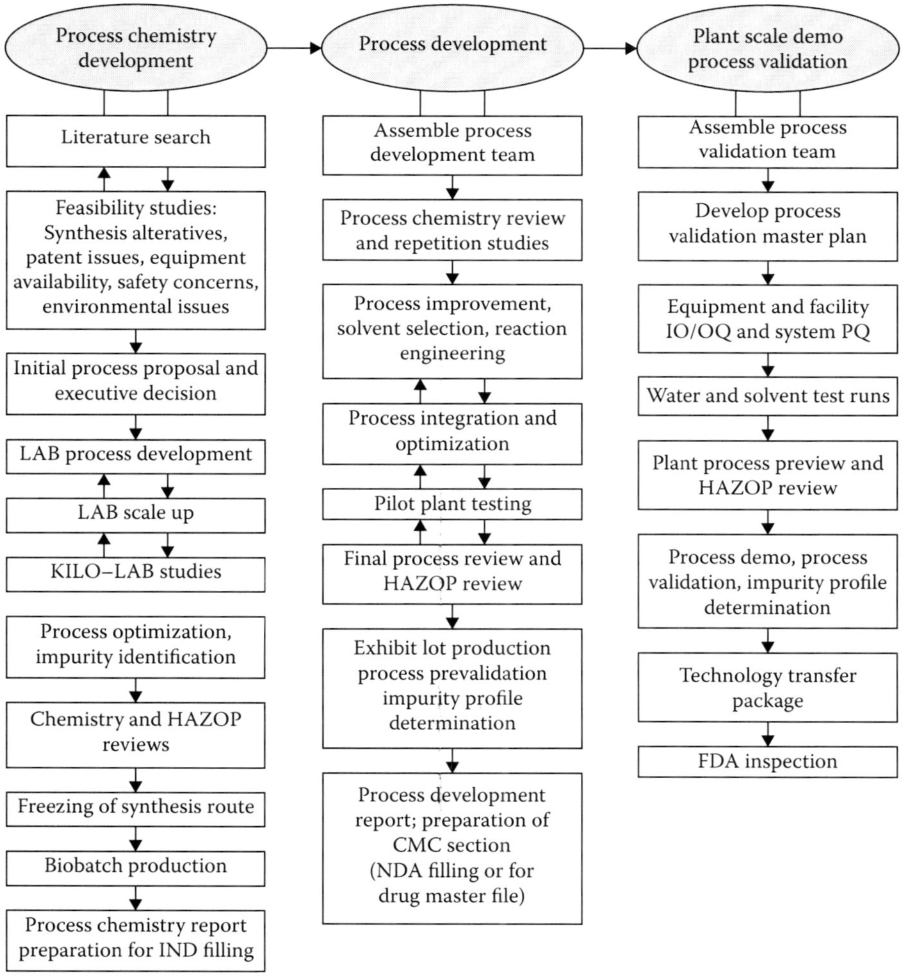

Figure 20.2 Flowchart of nutraceutical manufacturing and scale-up under current good manufacturing practice. It shows detailed diagrammatic representation of the overall process of clinical batch manufacturing. IQ, installation qualification; OQ, operation qualification; PQ, performance qualification.

modest approach to method validation is acceptable for preparation of Phase I clinical material than later for clinical study material.

Reference standards need to be made available (Senior 2009; Food and Drug Administration 2010a). These can consist of a representative bench-scale batch set aside as reference material under controlled storage and with, for example, an annual retest date (Figure 20.2). Any issues of stability of the nutraceutical products must be resolved so that the reference material can remain in specification for as long as possible.

In recent years, the DSHEA permitted the FDA to establish cGMPs for dietary supplements and properly validated "publicly available" methods (Betz et al. 2007). The law requires that any enforcement action taken against dietary supplement products use "publicly available" methods. The NIH has streamlined an ongoing method validation process through collaboration between AOAC International (Gaithersburg, MD) and various stakeholder groups (Betz et al. 2007; US National Library of Medicine 2010a). They are representatives of the dietary supplement industry, regulatory and other governmental bodies, consumer groups, nongovernmental organizations, and research scientists. The NIH was advised on basic quality issues such as identity and contamination and on the usage of existing frameworks for method development and validation (Betz et al. 2007).

20.2.6 Analytical Method Validation for Nutraceuticals

To date, nine supplement methods have been collaboratively studied, with seven approved as first-action official methods. Collaborative study reports have been published for seven of the approved methods: ephedrine alkaloids in botanicals or dietary supplements by high-performance liquid chromatography (HPLC)/ultraviolet (UV) (Roman 2004), ephedrine alkaloids in plasma or urine by liquid chromatography/mass spectrometry (LC/MS) (Trujillo and Sorenson 2003), glucosamine in raw materials and finished products (Zhou, Waszkuc, and Mohammed 2005), β-carotene in raw materials and finished products (Szpylka and DeVries 2005), flavonol glycosides in *G. biloba* raw materials and finished products (Gray 2007), phytosterols in saw palmetto raw materials and finished products (Sorenson and Sullivan 2007a), and aristolochic acid in raw materials and finished products (Sorenson and Sullivan 2007b). There are an additional 9 single-laboratory validation studies in progress and 38 additional ingredients in various stages of study. These include SAMe, St. John's wort, omega-3 fatty acids, soy isoflavones, L-carnitine, vitamin Bs, chondroitin sulfate, black cohosh, green tea catechins, lutein, turmeric, ginger, milk thistle, African plum, and flax seed. Adequate time must be allowed to build and enforce these validation studies (see Figure 20.2).

20.2.7 Quality System

A *quality system* is defined as a system of formalized business practices that define responsibilities for organizational structure, processes, procedures, and resources needed to implement product quality, quality by design, continuous improvement,

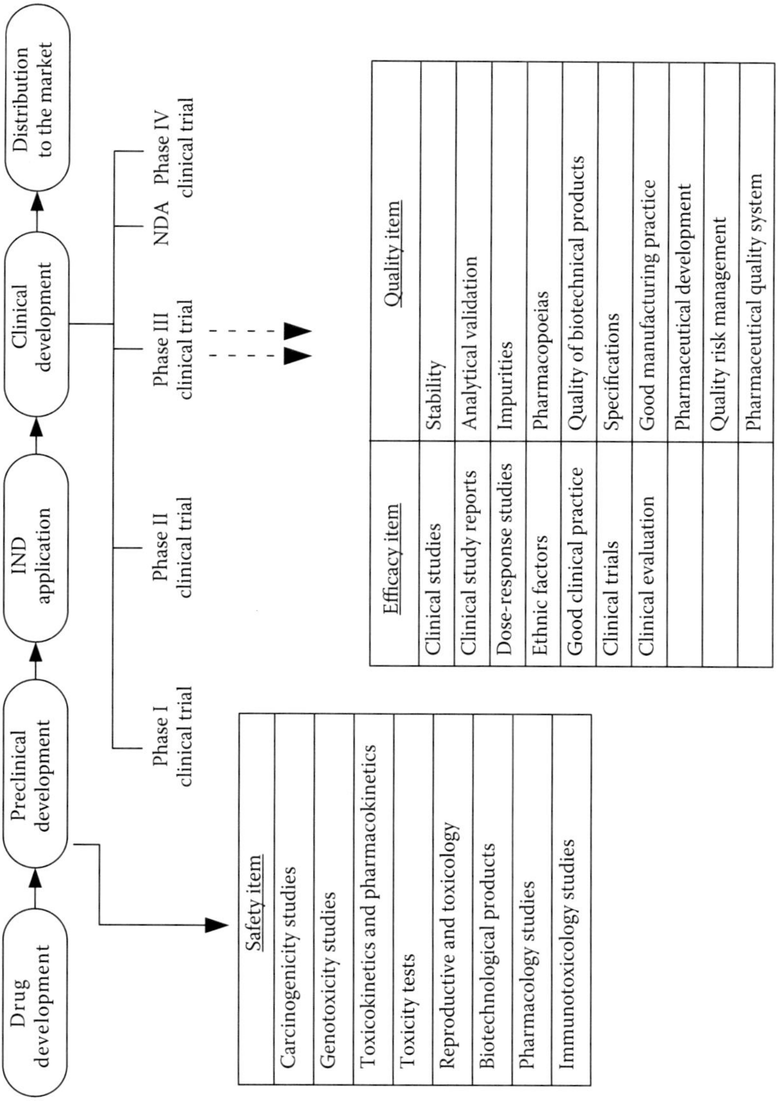

Figure 20.3 Drug products containing nutraceuticals during clinical trials: The following phases of product development proceed sequentially.

and risk management. Central to the quality system are applicable government regulations and guidance documents; written standard operating procedures and work instructions; and contracts, quality agreements, and templates (Food and Drug Administration 2010c).

By the time Phase III clinical trials are reached, the formulation of the drug or nutraceutical should be developed close to the trial product preparation that will be used in the market. In this case, safeguards need to be in place to ensure rigor in preparation of such clinical material. Figure 20.3 showed 10 quality items under ICH guidelines in the clinical trial (International Conference on Harmonisation 2007; Food and Drug Administration 2009, 2010c).

The guidelines describe a model for an effective quality management system for the pharmaceutical industry, referred to as the "Pharmaceutical Quality System." The guideline is intended to provide a comprehensive approach to an effective pharmaceutical quality system based on International Organization for Standardization concepts, includes applicable cGMP regulations, and complements the ICH "Q8 Pharmaceutical Development" and "Q9 Quality Risk Management." This guideline can also be implemented throughout the different stages of a product lifecycle. In addition, the ICH Q10 guidelines (International Conference on Harmonisation 2007) demonstrate industry and regulatory authorities' support of an effective pharmaceutical quality system to enhance the quality and availability of medicines around the world in the interest of public health (International Conference on Harmonisation 2007; Food and Drug Administration 2010c).

In 2009, the FDA published a guideline for industry entitled "Quality Systems Approach to Pharmaceutical Current Good Manufacturing Practice Regulations" (Food and Drug Administration 2009). The guideline describes the key elements of a robust quality systems model and shows how implementation of such a model is one way to comply with the FDA's cGMP regulations. It also shows the correlation of ICH Q10 quality system components with the FDA's cGMP regulations. In due course, these will also have to be followed by nutraceutical manufacturers.

20.3 STAGES OF CURRENT GOOD MANUFACTURING PRACTICE MANUFACTURING OF NUTRACEUTICAL CLINICAL MATERIAL

20.3.1 Setting Specifications for Nutraceutical Products

Before manufacturing the first clinical batch, detailed specifications need to be drafted not only for the drug product itself but also for the drug substance and all major components of the drug product. Thus, the chemical and physical description of the nutraceutical system will be clearly defined to include the source, grade, and certificate of analysis for each raw material. The nutraceutical product will be thoroughly identified with respect to all the parameters that govern the chemistry, physical properties, and biological activity of the nutraceutical system, for example, particle size range and distribution, response to temperature, light, pressure, humidity, pH, oxidation, stability in biological fluids, and any other parameters that the drug product might encounter (Food and Drug Administration Guidance for Industry 2008a).

20.3.2 Nutraceutical Tablets

The oral route is the most important method for administering nutraceuticals for systemic effect. Solids and liquids can be given via the oral route, and tablets and capsules are the most commonly used solid dosage forms. To seek high-quality dietary supplements, compliance with cGMP is imperative for nutraceutical manufacturers (Shargel 2001; Food and Drug Administration Guidance for Industry 2008a; Food and Drug Administration 2010b). Two common challenges to cGMP compliance during tablet manufacturing is struggling with batch-to-batch consistency and contamination.

Tablet consistency is an important factor in quality control and time to market for tablet-based supplements. The traditional tablet coaters can achieve merely the uniformity. From tablet pan design to spray technology, these machine robots have the ability to reduce processing times significantly. By decreasing production times, manufacturers can operate with fewer tablet coaters, making it easier to meet cGMP standards for this part of the manufacturing process (Food and Drug Administration 2010a). Typically, with conventional coaters, tablet weight must increase 4%–5% to achieve a uniform coat. In 10 minutes or less, manufacturers can attain tablet homogeneity in each batch to achieve better batch consistency (Food and Drug Administration Guidance for Industry 2008a).

20.3.3 Nutraceutical Capsules

Capsules are solid dosage forms in which one or more nutraceutical ingredients are enclosed within a small gelatin shell (Shargel et al. 2008). Gelatin capsules may be hard or soft. Empty hard gelatin capsule shells are manufactured from a mixture of gelatin and colorants. The US Pharmacopeia (USP) also permits the use of 0.15% sulfur dioxide to prevent decomposition of gelatin during manufacture. However, soft gelatin capsules are prepared from gelatin shell. Glycerin or polyhydric alcohol is added to these shells to make them elastic or plastic-like. Soft gelatin capsules are used to contain liquids, suspension, pastes, dry powders, or pellets. In the early stage of product development, hard gelatin capsule dosage forms are often developed for Phase I clinical trials. If the nutraceuticals show efficacy, the same formulation may be used in Phase II studies. The uniformity of dosage forms can be demonstrated by either weight variation or content uniformity methods. Disintegration tests are not usually required for capsules unless they have been treated to resist solution in gastric fluid. In this case, they must meet the requirements for disintegration of enteric-coated tablets.

20.3.4 Nutraceutical Components

Nutraceutical product = active nutraceutical substance + other excipients (e.g., enhancer, inactive ingredients).

The nutraceutical chemical components are composed with active substance and excipients. The excipients are present to create a formulation of the active drug substance that will be optimal for the desired therapeutic and stability outcomes. Therefore, such components will be subject to the regulations and constraints of any pharmaceutical ingredient.

20.3.5 Common Technical Document

The CTD is a set of specifications for the application profile for the registration of medicines and has been designed for use across Europe, Japan, and the United States. Thus, CTDs have global harmonization of submission content, and regional differences are handled in a modular fashion. Because the CTD is a "submission-ready document," it standardizes the submission format from clinical to marketing (from Notice of Claimed Investigational Exemption for a New Drug [IND] to NDA); therefore, it is highly recommended and accepted by the FDA (International Conference on Harmonisation 2007). Because each section or attachment is a unique technical report or a section within the report, it carries more efficient use of resources and expedites submissions, with less cost and stress to the organization.

The CTD is divided into five modules: (1) administrative and prescribing information; (2) overview and summary of modules 3–5 (Q, S, and E); (3) quality (pharmaceutical documentation); (4) safety (nonclinical study reports and toxicology studies); and (5) efficacy (clinical studies).

An advantage of CTD is that it merges document management and submission management to align marketing goals early in development. CTD also promotes clear communication, transparency, and teamwork. Lastly, CTD facilitates traceability in quality systems.

20.3.6 Notice of Claimed Investigational Exemption for a New Drug Application

In the preclinical stage, animal pharmacology and toxicology data are obtained to determine the safety and efficacy of the nutraceuticals. The innovator company for nutraceuticals must submit an IND for approval (Food and Drug Administration Guidance for Industry 2008a). After approval of the IND from the FDA, the manufacturer may then conduct clinical studies of its investigational nutraceuticals. The law requires the manufacturer to submit the following information: (1) name of the drug; (2) its composition; (3) methods of manufacturing and quality control; and (4) information from preclinical investigations regarding pharmacological, pharmacokinetics, and toxicological evaluations (Food and Drug Administration 2010c). Also, no attempt is made to develop a final formulation during this stage. The FDA may answer within 30 days from the date the IND is filed. If the FDA approves the IND, the innovator company may start human clinical testing of the nutraceuticals (Shargel et al. 2001; Food and Drug Administration Guidance for Industry 2008a).

20.4 CLINICAL DEVELOPMENT

20.4.1 Clinical Trials Management and Types

The innovator company's design of the clinical trial must be in accordance with internationally accepted guidelines (CPMP/ICH/135/95; International Conference on Harmonisation 2007). First, the innovator has to prepare the complete trial documentation, including study protocol; case report form; participant information and consent forms; and investigators' brochure. Additionally, the innovator must prepare any documentation required by any ethics committees and the relevant national authorities. Recruitment of trial site and physicians as well as screening and inclusion of participants are the second steps after document preparation. Next, the medical professionals record and monitor crucial data at the beginning, during, and at the end of the trial. Double data entry, data cleaning, and statistical analysis are required. There are at least five types of clinical trials: placebo-controlled double-blind randomized clinical trial, placebo-controlled single-blind randomized clinical trial, double-blind randomized noninferiority clinical trial, randomized cross-over clinical trial, and drug monitoring studies. The final report is based on statistical results and conduct of the trial.

20.4.2 Phase I Clinical Trial

The purpose of the Phase I clinical trial is to detect any adverse effects of the nutraceuticals. This phase involves a small number of health volunteers for study of the toxicity and tolerance of the nutraceuticals. Initially, a number of subjects receive a low dose of the nutraceuticals, which is gradually increased once safety of the nutraceuticals is assured (Food and Drug Administration Guidance for Industry 2008a).

20.4.3 Phase II Clinical Trial

The nutraceutical is now tested on a limited number of patients who suffer from the diseases for which the nutraceutical or drug is claimed to address. Dose–response studies are performed to determine the optimum dosage regimen for treating the diseases. In Phase II clinical trials, a final formulation is developed. This formulation is bioequivalent to the dosage form used in the initial clinical studies (Shargel et al. 2001).

20.4.4 Phase III Clinical Trial

This trial involves hundreds or thousands of patients. The study is often conducted at a physician's office or in hospitals that have contracted with the manufacturer to conduct studies. In this phase, a double-blind study is normally conducted. It is a type of study in which the nature of the drug is concealed from patients and attending physicians. In this type of study, one group of patients receive the testing nutraceuticals and another group of patients receive the placebo; the result of both groups is then compared to find out the true effectiveness of the nutraceuticals. If results of the Phase III studies are favorable, the drug sponsor's may submit an NDA to the FDA.

20.4.5 New Drug Application

An NDA contains a complete report including the drug's safety and efficacy, which has been noted on an IND. By law, the FDA has 180 days to review an NDA and to answer the sponsor's company (Food and Drug Administration Guidance for Industry 2008b).

20.4.6 Phase IV Clinical Trial

After the NDA is submitted and before approval to market the product is obtained from the FDA, manufacturing scale-up activities occur (Food and Drug Administration Guidance for Industry 2008a). Scale-up is the increase in the batch size from the clinical batch, the submission batch, or both, up to the full-scale production batch size using the finished, marketed product. This stage is also well known as *postmarketing surveillance*. Once the NDA has been approved, the innovator company may legally distribute the nutraceuticals in interstate commerce. Manufacturers must maintain and keep adequate postmarketing reports and records. In addition, they must submit any new information regarding a drug's safety and efficacy or any serious drug interactions to the FDA (Food and Drug Administration Guidance for Industry 2008b). I will discuss the details of postmarketing reports and adverse effects in section 20.6 Specific Lessons from Clinical Manufacturing Experiences with Marketed Nutraceuticals.

20.5 ROLE OF CURRENT GOOD MANUFACTURING PRACTICE IN PROCESS DEVELOPMENT

20.5.1 International Conference on Harmonisation Guideline

The ICH of Technical Requirements for Registration of Pharmaceuticals for Human Use is a unique project that brings together the regulatory authorities of Europe, Japan, and the United States, as well as experts from the pharmaceutical industry in the three regions to discuss scientific and technical aspects of product registration (International Conference on Harmonisation 2007).

The purpose is to make recommendations on ways to achieve greater harmonization in the interpretation and application of technical guidelines and requirements for product registration to reduce or obviate the need to duplicate the testing carried out during the research and development of new medicines. The objective of such harmonization is a more economical use of human, animal, and material resources, as well as the elimination of unnecessary delay in the global development and availability of new medicines, while maintaining safeguards on quality, safety, efficacy, and regulatory obligations to protect public health. The ICH topics are divided into four major categories: quality, safety, efficacy, and multidisciplinary (Q, S, E, and M). In this section, I focus on quality among all other categories (International Conference on Harmonisation 2007).

20.5.1.1 Quality Topics

Quality topics relate to chemical and pharmaceutical quality assurance and are divided into 10 specific items (detailed in Figure 20.3).

20.5.1.1.1 Q1 Stability

Q1 stability provides recommendations on stability testing protocols, including temperature, humidity, and trial duration. Furthermore, the revised document takes into account the requirements for stability testing in climatic zones III and IV to minimize the different storage conditions for submission of a global dossier (Food and Drug Administration 2009). It details the stability testing and photostability of new drug substances and products, testing of new drug substances and products, stability testing of new dosage forms, bracketing and matrixing designs for stability testing of new drug substances and products, evaluation of stability data, and stability data package for registration applications in climatic zones III and IV (International Conference on Harmonisation 2007).

20.5.1.1.2 Q2 Analytical Validation

Q2 analytical validation identifies the validation parameters needed for a variety of analytical methods. It also discusses the characteristics that must be considered during the validation of the analytical procedures included as part of registration applications (International Conference on Harmonisation 2007).

20.5.1.1.3 Q3 Impurities

The Q3 impurities section addresses the chemistry and safety aspects of impurities, including the listing of impurities in specifications, and defines the thresholds for reporting, identification, and qualification (Figure 20.4). It details the impurities in new drug substances and products and offers a guideline for residual solvents (International Conference on Harmonisation 2007; Food and Drug Administration 2010b).

20.5.1.1.4 Q4 Pharmacopoeias

Q4 provides the framework on how to set specifications for drug substances to address how regulators and manufacturers might avoid setting or agreeing to conflicting standards for the same product as part of the registration in different

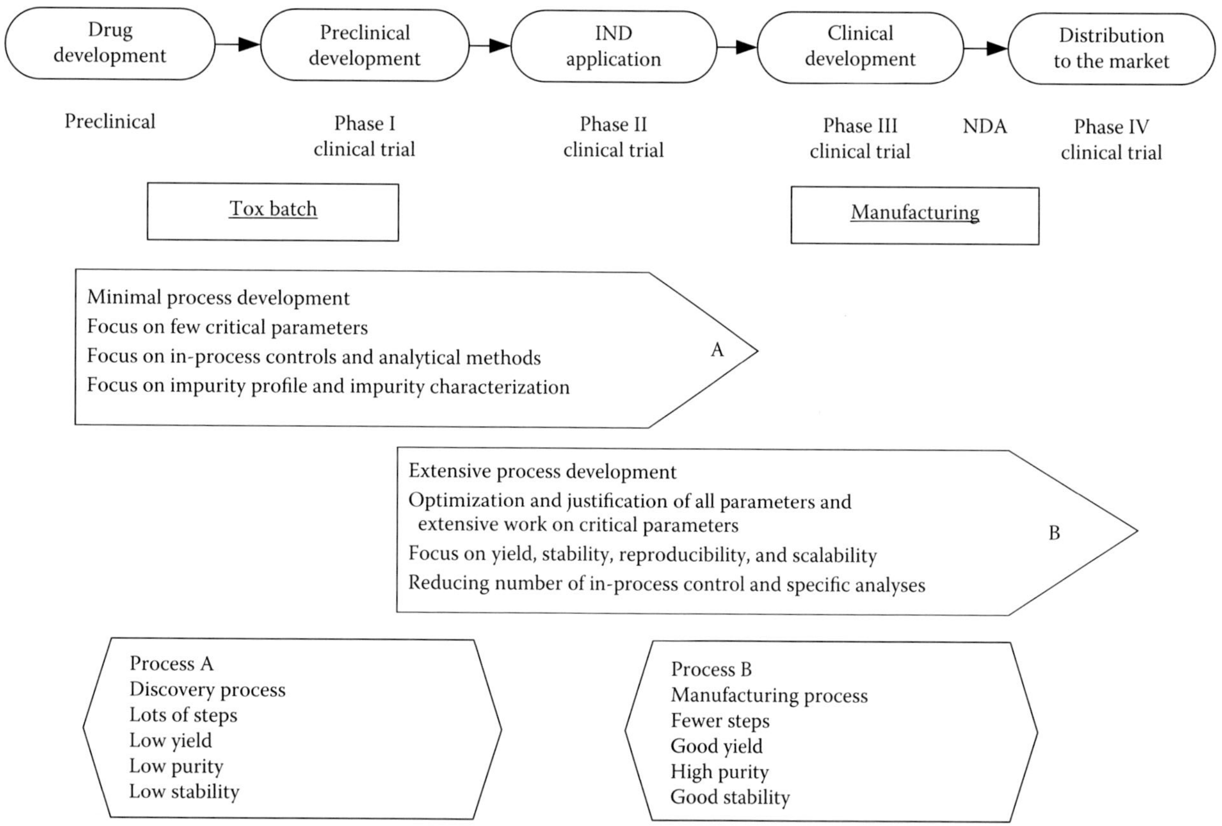

Figure 20.4 Process of clinical batch manufacturing. A diagrammatic representation of the timeline and advantages in various aspects of clinical trials.

regions. These chapters are at various stages of harmonization between Europe, Japan, and the United States. Q4 details the pharmacopoeial harmonization and provides evaluation and recommendation of pharmacopoeial texts for use in the ICH regions (International Conference on Harmonisation 2007, Annex 1–14).

20.5.1.1.5 Q5 Quality of Biotechnological Products

Q5 concerns the testing and evaluation of the viral safety of biotechnology products derived from characterized cell lines of human or animal origin (International Conference on Harmonisation 2007; Food and Drug Administration 2009). The purpose is to provide a general framework for virus testing experiments for the evaluation of virus clearance and the design of viral tests and clearance evaluation studies. It details the viral safety evaluation of biotechnology products derived from cell lines of human or animal origin, analysis of the expression construct in cells used for production of rDNA–derived protein products, stability testing of biotechnological and biological products, derivation and characterization of cell substrates used for production of biotechnological and biological products, and comparability of biotechnological and biological products subject to changes in their manufacturing process (Food and Drug Administration 2010b).

20.5.1.1.6 Q6 Specifications

Bulk drug substance and final product specifications are key parts of the core documentation for worldwide product license applications. This provides harmonized guidance and is divided into two parts: chemical substances, and biotechnological or biological substances (International Conference on Harmonisation 2007).

20.5.1.1.7 Q7 Good Manufacturing Practice

The ICH (International Conference on Harmonisation 2007) provides adequate international agreement on the technical aspects of cGMP for the components of pharmaceutical products, both active and inactive (Food and Drug Administration 2010b). It is detailed in the "GMP Guide for Active Pharmaceutical Ingredients" (Food and Drug Administration 2010a, 2010b).

20.5.1.1.8 Q8 Pharmaceutical Development

Q8 provides guidance on the contents of section 3.2.P.2 (pharmaceutical development) for drug products as defined in the scope of module 3 of the CTD at ICH M4 (International Conference on Harmonisation 2007). The guideline does not apply to contents of submissions for drug products during the clinical research stages of drug development. Also, to determine the applicability of this guideline for a particular type of product, applicants should consult with the appropriate regulatory authorities.

20.5.1.1.9 Q9 Quality Risk Management

Q9 provides principles and examples of tools of quality risk management that can be applied to all aspects of pharmaceutical quality, including development, manufacturing, distribution, and the inspection, submission, and review processes throughout the lifecycle of drug substances and drug (medicinal) products (Betz et al. 2007; Food and Drug Administration 2009) and biological and biotechnological products, including the use of raw materials, solvents, excipients, and packaging and labeling materials (International Conference on Harmonisation 2007).

20.5.1.1.10 Q10 Pharmaceutical Quality System

Q10 applies to pharmaceutical drug substances and drug products (Food and Drug Administration and the Department of Health and Human Services 2009; Food and Drug Administration 2010a), including biotechnology and biological products, throughout the product lifecycle. It should be applied in a manner that is appropriate and proportionate to each of the product lifecycle stages, recognizing the differences among, as well as the different goals of, each stage (International Conference on Harmonisation 2007).

20.5.1.2 Other Topics

The other three major ICH categories are outlined briefly below:

S – Safety topics: Those relating to in vitro and in vivo preclinical studies (e.g., carcinogenicity testing, genotoxicity testing). See the detailed items in Figure 20.3.
E – Efficacy topics: Those relating to clinical studies in human subjects (e.g., dose–response studies, good clinical practices). See the detailed items in Figure 20.3.
M – Multidisciplinary topics: These are cross-cutting topics that do not fit uniquely into one of the aforementioned categories (e.g., MedDRA, Electronic Standards for the Transfer of Regulatory Information [ESTRI], M3, CTD, and M5).

In summary, Quality Risk Management and Pharmaceutical Quality System (Q9 and Q10) of the ICH guidelines are appropriate for drug production but not for manufacture of nutraceuticals. However, in the future it is expected that these quality process developments will also be implemented for nutraceuticals.

20.5.2 Chemistry, Manufacturing, and Controls

CMC stands for Chemistry, Manufacturing, and Controls a section of the CTD. According to FDA Guidance: Format and Content of the Chemistry Manufacturing and Controls Section of an Application (FDA Guidance for Industry 1997), a complete CMC must contain a product summary; quality control of components and raw materials, reference standards, manufacturing, and testing facilities; manufacture and quality control of drug substance and drug product and container and closures; analytical controls and methods; stability and expiration dating of drug substance and product; vial labels (not applicable to nutraceuticals); and the scientific literature citations. In the quality and stability section of the CMC, the manufacturer prepares qualification runs, and the minimum qualification is three successive passing batches.

The purposes of the CMC focus on the maintenance of identity, strength, and purity, as well as other safety and effective related characteristics. The secondary focus is on the consistency of each batch to gain the reproducible products.

20.5.3 Preapproval Inspection

The manufacturing facility is inspected after an NDA is submitted and before the application is approved (see Figure 20.2). A preapproval inspection (PAI) may be initiated if a major change is reported in a supplemental application to an NDA. During the API, the investigators perform a general cGMP inspection relating specifically to the nutraceutical product intended for the market (Food and Drug Administration 2010a, 2010b). The investigators also review the development report to verify that the nutraceutical product has sufficient supporting documentation to ensure a validated product and a rationale for the manufacturing directions. Next, they consult the CMC section of the NDA and determine the capability of the manufacturer to produce the nutraceuticals as described and verify the traceability of the information submitted in the CMC section to the original laboratory notebooks, electronic information, and batch records. Finally, the investigators recommend whether to approve the manufacture of the nutraceutical product after the PAI.

20.5.4 Scale-Up and Postapproval Changes

The scale-up and postapproval change guidelines are intended to reduce the number of manufacturing changes that require preapproval by the FDA. The guidelines are published by the FDA on the Internet (Food and Drug Administration Guidance for Industry 2008b) and provide recommendations to sponsors of NDA during the following changes in the postapproval period.

1. Amount of the excipient to aid in the processing of the product during scale up
2. Site of manufacture
3. Scale up or scale down the batch size of the formulation
4. Manufacturing process or equipment

20.6 SPECIFIC LESSONS FROM CLINICAL MANUFACTURING EXPERIENCES WITH MARKETED NUTRACEUTICALS

20.6.1 Postmarketing Adverse Effects Reporting

A recent independent test by Consumerlab, LLC of 10 brands of the nutritional supplement St. John's wort demonstrates the current problem with nutritional supplements in the United States. Six of the 10 supplements failed Consumer Laboratory's standards because of high levels of cadmium and/or lead and/or extremely low levels of active ingredients (hypericin and hyperforin) (Harding 2007). Unfortunately, this is not an isolated case of inappropriate products by any one manufacturer but rather a part of a much larger problem that the nutraceutical market faces today. Often nutraceuticals are contaminated or do not contain the advertised amount of active ingredients. Some manufacturers seem to ignore the importance of ensuring safety against contaminants and ensuring quality of products by complying with food or pharmaceutical cGMP. Doing so protects not only the people who use the supplements but also the manufacturer who could face consumer lawsuits or seizures from the FDA.

The reporting of adverse effects is extremely important because clinical trials were done using small sample sizes, which may not identify rare adverse effects. The 1982 cyanide poisonings of acetaminophen is an example of something beyond the manufacturer's control (intentional murder by tampering with a product on store shelves) that simply "went wrong." A nurse and three firefighters in the Chicago area are credited with hypothesizing the link between the seven cyanide poisoning deaths and acetaminophen found in the victims' homes (Sotonoff 2009). With good communication of the problems between the manufacturer and the FDA, a quick response, such as notifying consumers, recalling products, and improving packaging to prevent tampering, likely saves lives (Wolnik et al. 1984). The following three examples are more typical of postmarketing adverse effect reporting, but the importance of having a method to track problems cannot be understated.

20.6.1.1 Example 1: Niaspan

Niaspan (Abbott Laboratories, Abbott Park, IL) (Niaspan Tablets Prescribing Information 2010) is an extended-release formulation of niacin (also known as vitamin B3) approved for use to treat certain types of hyperlipidemia. Many studies on niacin have been done during the past 35 years, with more recent clinical studies focusing on combining statin drugs such as Lipitor (Pfizer Inc., New York, NY) with Niaspan. Clinical trials showed that "flushing" of the skin, rash or itching, and diarrhea, nausea, and vomiting were the most common adverse reactions to Niaspan (Niaspan Tablets Prescribing Information 2010). Postmarketing reporting has resulted in new (March 2009) labeling changes to state that blurred vision, hepatitis, myopathy or rhabdomyolysis (muscle tissue breakdown), creatine kinase increase, and skin discoloration has been reported (FDA 2010f). Doctors can monitor certain biological enzymes using a basic blood test called a *comprehensive metabolic panel* to help detect injuries to the muscles and liver, and doing this simple test periodically may be appropriate for all patients receiving any drug for hyperlipidemia.

The newest labeling change for Niaspan (February 2010) cites the adverse reaction "burning sensation/skin burning sensation" as another postmarketing adverse reaction (FDA 2010f). The difference between this burning and the flushing described in trials was not indicated in the literature, which is an example of a general problem with effects reporting. Often the words used overlap, are redundant, or do not exactly describe what is happening. A good example is "skin discoloration." To a layperson, this may mean skin yellowing typical of liver problems, skin redness like with "flushing," an allergic rash, or something else entirely. Guidelines for effects reporting require that the words used by the people reviewing them must use *Medical Dictionary for Regulatory Activities* (US National Library of Medicine 2010b) terminology to make the terms used sufficiently similar in multiple languages to ensure problems are not lost in translations. The take-home message is that postmarketing adverse effect reporting can sometimes uncover or provide clues of problems with a drug, but it does not "prove" that a problem exists.

20.6.1.2 Example 2: Lovaza

Lovaza (GlaxoSmithKline Inc. [Brentford, UK] 2009) (formerly called Omacor, a fish oil derivative) is used for hyperlipidemia and contains two omega-3 fatty acid esters called eicosapentaenoic acid and docosahexaenoic acid. Some adverse events reported in clinical trials include infection or flu syndrome, burping, indigestion, back pain, rash, and taste perversion (GlaxoSmithKline Inc. 2009). The postmarketing adverse events of anaphylactic reaction and hemorrhagic diathesis have been reported (GlaxoSmithKline Inc. 2009). The prescribing information clues us in that these effects were fairly expected before they were reported, as the wording "is of concern" is used to communicate that these could happen but did not during the trial. This allowed the manufacturer and investigator to discuss the fact that seafood and shellfish are possible food allergens, as well as the possibility of this drug causing bleeding as a result of changes in platelet clotting.

To put these side effects in perspective, the postmarketing adverse events described in the prescribing information for Lipitor (Pfizer Inc.), a traditional pharmaceutical used to treat high cholesterol, listed the following reactions: anaphylaxis, angioneurotic edema, bullous rashes, rhabdomyolysis, fatigue, tendon rupture, hepatic failure, dizziness, memory impairment, depression, and peripheral neuropathy (Pfizer Inc. Lipitor Tablets 2009). Additionally, Lipitor causes a host of drug interaction problems because it is metabolized by the liver enzyme CYP3A4. This particular pathway often becomes overwhelmed when multiple drugs are taken that oxidize with this same enzyme. The metabolism of omega-3 fatty acids, on the other hand, uses multiple enzymes such as ones from the CYP2C family and CYP4F3B (Fer et al. 2001; Harmon et al. 2006). From a general drug interaction safety standpoint, Lovaza is much better (unless the patient is taking blood thinners) because there are so many pathways to metabolize it and few other drugs use the same pathways. Even from a side effect standpoint, the increase in infection seen with Lovaza may relate to the possibility that it has anti-inflammatory effects as well (Groeger et al. 2010). For some patients, it may be desirable to reduce inflammation as well; therefore, a slightly increased risk of infection may be acceptable, depending on the individual situation. However, the take-home message is that postmarketing adverse effects may be predicted and discussed in the prescribing information and that the whole of the drug's risks must be weighed (i.e., side effects and possible drug interactions) before deciding whether it is appropriate to use.

20.6.1.3 Example 3: S-Adenosenyl Methionine

Various formulations of SAMe are used as prescription drugs in Europe (e.g., Gumbaral, Samyr, and Heptral). A 2002 meta-analysis of clinical trials concluded that "SAMe appears to be as effective as NSAIDs in reducing pain and improving functional limitation in patients with osteoarthritis without the adverse effects often associated with NSAID therapies" (Soeken et al. 2002, p. 430). This study also showed a significantly lower dropout rate for patients in the trial taking SAMe than for NSAIDs. The first and major reason to drop out of a study is because of side effects. The lower dropout rate observed might be because SAMe has less troublesome side effects and more beneficial effects for patients.

An additional therapeutic use for SAMe is to treat depression (Kagan et al. 1990), and it has been shown to affect many neurotransmitters as traditional antidepressants. A recent meta-analysis of studies on the traditional drug therapies imipramine (a tricyclic antidepressant) and paroxetine (a selective serotonin reuptake inhibitor antidepressant) has shown that these drugs do not seem to help patients with mild to moderate depression any more than placebo (Fournier et al. 2010). This raises an alarming inquiry of whether the current drug therapies for depression may put patients at risk for serious side effects without any beneficial evaluations. In fact, many antidepressants now carry the FDA's strongest possible warning label (called a "black box" warning) that they may increase the risk of suicidal thinking and behavior in children, teens, and young adults. To our knowledge, it is unknown whether SAMe might also have side effects that could worsen a psychiatric condition. A 2002 meta-analysis of 90 different clinical trials for SAMe seems to explain the problem: 34% of these studies did not mention any side effects or adverse events at all (Hardy et al. 2002). Some postmarketing adverse effect reporting may point to SAMe increasing manic behavior (Pies 2000), but generally this effect cannot be established as caused by the drug in question. In this case, postmarketing reporting combined with reports from other drugs meant to treat the same condition show that more studies are needed. A well-designed clinical trial could possibly answer questions about the efficacy and safety of SAMe in young patients. This is an important field for professionals to examine further because depression can lead to death by suicide.

20.6.2 Raw Material Specifications

Nutraceuticals face a unique manufacturing problem because the raw materials (or active substances) often come from nature, which has absolutely no or rare quality controls beyond survival of the fittest. It is legally up to the manufacturer of a dietary supplement to take all reasonable measures and precautions to ensure that the identity, quality, and purity of the raw materials used in the product have been proven.

Often the manufacturers rely on outside testing (i.e., USP verification), and this is probably the best way because the laboratory testing the materials does not make additional money for "passing" any materials. It is important that all ingredients of the nutraceuticals were verified for quality control and assurance instead of merely the active ingredients.

For example, an herbal blend for promoting lung health received an alert from the FDA in 2010 because the New York City Department of Health and Mental Hygiene was checking supplements taken by a patient with lead poisoning and found it to contain 1100 ppm of lead (FDA 2010e). Subsequent FDA testing confirmed extremely high levels of lead at "more than 500 ppm." This problem may have happened because the manufacturer did not test certain raw materials for heavy metal contamination before packaging or marketing. Heavy metal contamination may be of particular concern for nutraceuticals because some plants and fish tend to bioaccumulate (concentrate) heavy metals from their environments. The best-described examples of this in the literature are mercury in certain fish and cadmium in tobacco. Testing for heavy metals is often done by a technique called *atomic absorption spectroscopy* (Levine et al. 2005).

20.6.3 Analytical Method Development and Validation

Most methods of analyzing samples can be simply explained as having two parts: separating the chemicals in the sample and detecting the chemicals in the sample. These techniques may have such names as high-performance liquid chromatography, thin-layer chromatography, and gas chromatography. Other methods to separate out chemicals are more suited for certain samples, such as organic extraction, centrifugation, and electrophoresis. Once the chemicals in the sample are separated, they must be detected under cGMP regulation. This can be done by spectroscopy methods such as infrared, ultraviolet-visible, and MS. Sometimes antibodies are used to detect things using techniques such as Western blot analysis and enzyme-linked immunosorbent assay.

To know which methods to use, there is a standard set of instructions for each dietary supplement called the *United States Pharmacopeia–National Formulary* (USP-NF). This document details exactly which tests are to be performed and exactly to what specifications. There are also USP reference standards, which are samples of chemicals that other laboratories have already analyzed. Reference standards should always be included in any testing to be sure that equipment and methods are working correctly. For some testing it may be necessary to use a certified reference material reference standard, which is a USP reference standard that comes with additional paperwork certifying it as having been properly tested and documented.

20.6.4 Stability Testing Parameter and Control

Stability means three important things: that the desired compound is still present, that it is not contaminated by toxins or undesirables formed by a spoilage process, and that it is still active in a biologically meaningful way. The USP-NF monographs describe requirements for stability purposes. Additional information on cGMP regulations applicable to stability can be found in the *Handbook of Stability Testing in Pharmaceutical Development* (Kim 2009). The testing of stability should start with the raw materials because many raw materials can be harvested seasonally only and must be stored for some time before processing. Commonly with plants, the drying method is used for preservation, but if the stored product gets too wet, problems such as mold growth can occur (e.g., ingredients susceptible to contamination with aflatoxin or other natural toxins). The temperature and humidity levels for storage of dried plants should be tightly controlled or at least well documented, and when the plants come out of storage they should be tested and inspected visually. Scent is part of what the FDA calls the "appearance" of a substance; thus, smelling the raw materials is part of this testing. For other raw materials, freezing may be another good option for storage. This usually works best when the active ingredient is a small organic molecule and not a large molecular-sized enzyme or protein (most of these raw nutraceutical materials must be frozen in special buffers with proteinase inhibitors to ensure the quality).

The next problem encountered is the stability of intermediates in the manufacturing process. If part of the process is to liquefy plant materials, then care must be taken to not let the liquefied materials (an intermediate) sit around too long or remain unprotected from such things as light, heat, and microbial growth. 110.80 Processes and Controls listed that the equipment or intermediates shall either not contain levels of microorganisms that may produce food poisoning or other disease in humans, or they shall be pasteurized or otherwise treated. Additionally, under the regulation of the Sanitary Operations, the single-service articles (e.g., utensils, paper cups, and paper towels) should be stored appropriately and be designated for one-time use. Sometimes special procedures such as using light-blocking plastics, running equipment in a cold room (maintaining at 45°F/7.2°C or below, per 110.80 Processes and Controls of cGMP), or adding preservatives to intermediates become necessary. It is necessary to sterilize equipment or intermediates under cGMP regulations (FDA Guidance for Industry 2008a).

20.7 CONCLUSION

Our mission is to develop nutraceutical production processes of appropriate scale, from preclinical exploration to commercialization. Under cGMP regulation, it is expected that process scale-up, process optimization, and process validation will be achieved during the manufacturing of the nutraceutical batches. The challenges lie in solving the inevitable problems in stages of clinical trials, including active ingredients, contents, systemic administration (avoid first-pass effect), patient compliance, delivery formulation, and proper dosing and dosage. Likewise, the manufacture of nutraceuticals has critical parameters for each batch of clinical materials to ensure safety, potency, purity, consistency, and efficacy. Once a successful batch of clinical materials has been manufactured, the ambitions of the investigator shift to the challenge of manufacturing new batches. The consistent characteristics of the clinical batches are extremely important, and the use of robust methodology and processes helps to sustain and standardize nutraceutical consistency.

ACKNOWLEDGMENT

Special thanks to Paula M. Logsdon, a current graduate student at University of Louisville studying biochemistry and molecular biology.

REFERENCES

Betz, J. M., K. D. Fisher, L. G. Saldanha, and P. M. Coates. 2007. The NIH Analytical Methods and Reference Materials Program for Dietary Supplements. *Analytical and Bioanalytical Chemistry* 389: 19–25.

Brower, V. 1998. Nutraceuticals: Poised for a Healthy Slice of the Healthcare Market? *Nature Biotechnology* 16: 728–31.

Fer, M., Y. Dréano, D. Lucas, L. Corcos, J. P. Salaün, F. Berthou, and Y. Amet. 2008. Metabolism of Eicosapentaenoic and Docosahexaenoic Acids by Recombinant Human Cytochromes P450. *Archives of Biochemistry and Biophysics* 471: 116–25.

Food and Drug Administration and the Department of Health and Human Services. 2009. International Conference on Harmonisation; Guidance on Q10 Pharmaceutical Quality System; Availability. Notice. *Federal Register* 74: 15990–1.

Food and Drug Administration. 2010a. *Current Good Manufacturing Practice in Manufacturing, Processing, Packaging or Holding of Drugs.* Code of Federal Regulations Title 21, section 210. http://www.accessdata.fda.gov/scripts/cdrh/cfdocs/cfcfr/cfrsearch.cfm?cfrpart=210&showfr=1.

Food and Drug Administration. 2010b. *Current Good Manufacturing Practice for Finished Pharmaceuticals.* Code of Federal Regulations Title 21, section 211. http://www.accessdata.fda.gov/scripts/cdrh/cfdocs/cfcfr/cfrsearch.cfm?cfrpart=211.

Food and Drug Administration. 2010c. *Good Laboratory Practice for Non-clinical Laboratory Studies.* Code of Federal Regulations Title 21, section 58. http://www.accessdata.fda.gov/scripts/cdrh/cfdocs/cfCFR/CFRSearch.cfm?CFRPart=58&showFR=1.

Food and Drug Administration. 2010d. *Dietary Supplement Health and Education Act of 1994.* http://www.fda.gov/RegulatoryInformation/Legislation/FederalFoodDrugandCosmeticActFDCAct/SignificantAmendmentstotheFDCAct/ucm148003.htm.

Food and Drug Administration. 2010e. FDA Warns Consumers to Avoid Vita Breath Dietary Supplement. http://www.fda.gov/newsevents/newsroom/pressannouncements/ucm210448.htm.

Food and Drug Administration. 2010f. MedWatch Reports for the Drug Niaspan. http://www.fda.gov/safety/medwatch/safetyinformation/ucm203614.htm.

Food and Drug Administration Guidance for Industry. 2007. Format and Content of the Chemistry Manufacturing and Controls Section of an Application. http://www.gmp-compliance.org/eca_guideline_511.html.

Food and Drug Administration Guidance for Industry. 2008a. INDs—Approaches to Complying with cGMP during Phase I. http://www.fda.gov/AnimalVeterinary/GuidanceComplianceEnforcement/GuidanceforIndustry/default.htm.

Food and Drug Administration Guidance for Industry. 2008b. 180-Day Generic Drug Exclusivity Under the Hatch-Waxman Amendments to the Federal Food, Drug, and Cosmetic Act. http://www.fda.gov/downloads/Drugs/GuidanceComplianceRegulatoryInformation/Guidances/ucm079342.pdf.

Fournier, J. C., R. J. DeRubeis, S. D. Hollon, S. Dimidjian, J. D. Amsterdam, R. C. Shelton, and J. Fawcett. 2010. Antidepressant Drug Effects and Depression Severity: A Patient-Level Meta-Analysis. *JAMA* 303: 47–53.

GlaxoSmithKline Inc. 2009. Lovaza Capsules Prescribing Information. http://us.gsk.com/products/assets/us_lovaza.pdf.

Gray, D., K. LeVanseler, M. Pan, and E. H. Waysek. 2007. Evaluation of a Method to Determine Flavonol Aglycones in *Ginkgo biloba* Dietary Supplement Crude Materials and Finished Products by High-Performance Liquid Chromatography: Collaborative Study. *Journal of AOAC International* 90: 43–54.

Groeger, A. L., C. Cipollina, M. P. Cole, S. R. Woodcock, G. Bonacci, T. K. Rudolph, V. Rudolph, B. A. Freeman, and F. J. Schopfer. 2010. Cyclooxygenase-2 Generates Anti-inflammatory Mediators from Omega-3 Fatty Acids. *Nature Chemical Biology* 6: 433–41.

Harding, A. 2007. Heavy Metals Found in Some St. John's Wort Products. Reuters Health. http://www.qualityhealth.com/news/heavy-metals-found-st-johns-wort-products-981.

Hardy, M., I. Coulter, S. C. Morton, J. Favreau, S. Venuturupalli, F. Chiappelli, F. Rossi, G. Orshansky, L. K. Jungvig, E. A. Roth, M. J. Suttorp, and P. Shekelle. 2002. *Evidence Report, S-adenosyl-L-methionine for Treatment of Depression, Osteoarthritis, and Liver Disease.* http://www.ncbi.nlm.nih.gov/bookshelf/br.fcgi?book=hserta&part=A1 01701.

Harmon, S. D., X. Fang, T. L. Kaduce, S. Hu, V. Raj Gopalb, J. R. Falckb, and A. A. Spectora. 2006. Oxygenation of Omega-3 Fatty Acids by Human Cytochrome P450 4F3B: Effect on 20-Hydroxyeicosatetraenoic Acid Production. *Prostaglandins, Leukotrienes and Essential Fatty Acids* 75: 169–77.

International Conference on Harmonisation. 2007. *Pharmaceutical Quality System Q10.* http://www.fda.gov/downloads/Drugs/GuidanceComplianceRegulatoryInformation/Guidances/ucm073517.pdf.

Kagan, B. L., D. L. Sultzer, N. Rosenlicht, and R. H. Gerner. 1990. Oral S-Adenosylmethionine in Depression: A Randomized, Double-Blind, Placebo-Controlled Trial. *American Journal of Psychiatry* 147: 591–95.

Kim, H. 2009. *Handbook of Stability Testing in Pharmaceutical Development.* New York: Springer.

Levine, K. E., M. A. Levine, F. X. Weber, Y. Hu, J. Perlmutter, and P. M. Grohse. 2005. Determination of Mercury in an Assortment of Dietary Supplements Using an Inexpensive Combustion Atomic Absorption Spectrometry Technique. *Journal of Automated Methods and Management in Chemistry* 2005: 211–16.

Niaspan Tablets Prescribing Information. 2010. Abbott Laboratories. http://www.rxabbott.com/pdf/niaspan.pdf.

Nicolson, G. L., R. R. Ellithorpe, and C. Ayson-Mitchell. 2010. Lipid Replacement Therapy with a Glycophospholipid-Antioxidant-Vitamin Formulation Significantly Reduces Fatigue within One Week. *Journal of the American Nutraceutical Association* 13: 10–14.

Pfizer Inc. Lipitor Tablets Prescribing Information. 2009. http://www.pfizer.com/files/products/uspi_lipitor.pdf.

Roman, M. C. 2004. Determination of Ephedrine Alkaloids in Botanicals and Dietary Supplements by HPLC-UV: Collaborative Study. *Journal of AOAC International* 87: 1–14.

Pies, R. 2000. Adverse Neuropsychiatric Reactions to Herbal and Over-the-Counter "Antidepressants." *Journal of Clinical Psychiatry* 61: 815–20.

Senior, J. H. 2009. Manufacture of Nanoparticulate Clinical Trial Material. *Advances in Nanotechnology and Applications*. Edited by Y. Pathak and H. Tran. Louisville, KY: Center for NanoTechnology: Education, Research, & Applications (CENTERA). Sullivan Univ. Press, vol. 1, 355–403.

Shargel, L., A. H. Mutnick, P. F. Souney, and L. N. Swanson. 2001. *Comprehensive Pharmacy Review*. Philadelphia: Lippincott William & Wilkins.

Soeken, K. L., W. L. Lee, R. B. Bausell, M. Agelli, and B. M. Berman. 2002. Safety and Efficacy of S-Adenosylmethionine (SAMe) for Osteoarthritis. *Journal of Family Practice* 51: 425–30.

Sorenson, W. R., and D. Sullivan. 2006. Determination of Campesterol, Stigmasterol, and beta-Sitosterol in Saw Palmetto Raw Materials and Dietary Supplements by Gas Chromatography: Single-Laboratory Validation. *Journal of AOAC International* 89: 22–34.

Sorenson, W. R., and D. Sullivan. 2007a. Determination of Campesterol, Stigmasterol, and Beta-sitosterol in Saw Palmetto Raw Materials and Dietary Supplements by Gas Chromatography: Collaborative Study. *Journal of AOAC International* 90: 670–78.

Sorenson, W. R., and D. Sullivan. 2007b. Determination of Aristolochic Acid I in Botanicals and Dietary Supplements Potentially Contaminated with Aristolochic Acid I Using LC-UV with Confirmation by LC/MS: Collaborative Study. *Journal of AOAC International* 90: 925–33.

Sotonoff, J. 2009. A Crime that Changed America. *Daily Herald*. http://www.dailyherald.com/story/?id=269677.

Szpylka, J., and J. W. DeVries. 2005. Determination of β-Carotene in Supplements and Raw Materials by Reversed-Phase High Pressure Liquid Chromatography: Collaborative Study. *Journal of AOAC International* 88: 1279–91.

Thassu, D., M. Deleers, and Y. Pathak, eds. 2007. *Nanoparticulate Drug Delivery Systems*. New York: Informa Healthcare.

Trujillo, W. A., and W. R. Sorenson. 2003. Determination of Ephedrine Alkaloids in Human Urine and Plasma by Liquid Chromatography/Tandem Mass Spectrometry: Collaborative Study. *Journal of AOAC International* 86: 643–56.

Trujillo, W. A., W. R. Sorenson, P. La Luzerne, J. W. Austad, and D. Sullivan. 2006. Determination of Aristolochic Acid in Botanicals and Dietary Supplements by Liquid Chromatography with Ultraviolet Detection and by Liquid Chromatography/Mass Spectrometry: Single Laboratory Validation Confirmation. *Journal of AOAC International* 89: 942–59.

US National Library of Medicine. 2010a. A Service of US National Institutes of Health. http://www.clinicaltrials.gov.

US National Library of Medicine. 2010b. *Medical Dictionary for Regulatory Activities* (MedDRA). A Service of US National Institutes of Health. nihlibrary.nih.gov/ResearchTools/Pages/MedDRA.aspx.

Wolnik, K. A., F. L. Fricke, E. Bonnin, C. M. Gaston, and R. D. Satzger. 1984. The Tylenol Tampering Incident—Tracing the Source. *Analytical Chemistry* 56: 466–74.

Zeisel, S. H. 1999. Regulation of "Nutraceuticals." *Science* 285: 183–85.

Zhou, J. Z., T. Waszkuc, and F. Mohammed. 2005. Determination of Glucosamine in Raw Materials and Dietary Supplements Containing Glucosamine Sulfate and/or Glucosamine Hydrochloride by High-Performance Liquid Chromatography with FMOC-Su Derivatization: Collaborative Study. *Journal of AOAC International* 88: 1048–58.

CHAPTER 21

New Technologies Protect Nutraceuticals against Counterfeiting

Roland Meylan

CONTENTS

21.1 INTRODUCTION

Anticounterfeiting and traceability are different problems, requiring different solutions. On the one hand, traceability needs standardization and interoperability among the various manufacturers and the intervening third parties within the supply chain up to the retail point; on the other hand, anticounterfeiting features, especially those invisible to the naked eye or covert ones, need secrecy and confidentiality. They should be constantly kept in pace with the technological advances of the counterfeiters. Hundreds of different anticounterfeiting solutions are available today, each one with its own area of application. For protection of a folding carton packaging; an aluminum blister pack; a flexible packaging; a tablet, a liquid, or powder; or a glass or molded vial, each calls for a different solution. Selection of the right one for the right product is the task of the brand product manufacturer, with consideration not only for the cost of the security feature but also taking into consideration the global cost of industrialization and deployment of the anticounterfeiting program worldwide.

21.2 VISIBLE AND INVISIBLE TO THE NAKED EYE SECURITY FEATURES

Many health care product manufacturers have added visible security features to their packaging to prevent counterfeiting. For example, these include holograms, kinegrams, embossing, microprinting, moiré, or special ink such as optical variable ink. However, these visible features provide not only minimal security but they also require training for effective authentication when faced with fraudulent reproductions of such visible security features.

The use of "covert" features (Covert and Overt 2006) invisible to the naked eye produces a higher level of protection because of the inability of counterfeiters to identify the presence of such features, and they consequently cannot attack them. Covert security should never be disclosed, and to prevent leaks these techniques should only be known to a limited number of trustworthy persons.

The best known covert security solution is invisible ink, such as ultraviolet ink (visible under ultraviolet light) or infrared ink (visible under infrared light). To authenticate these inks, a lamp that emits light in the required wavelength range will suffice. The drawback of these inks is that they can easily be bought by anyone on the market. There are other chemical tracers or ink additives that can provide security against counterfeiting, such as DNA or magnetic tracers. These provide higher security because they require verification devices that can only be obtained from the original supplier of the security feature (Figure 21.1).

The problem with such special inks, ink additives, or taggant resides in the related logistics and manufacturing procedures, such as press cleaning, temperature,

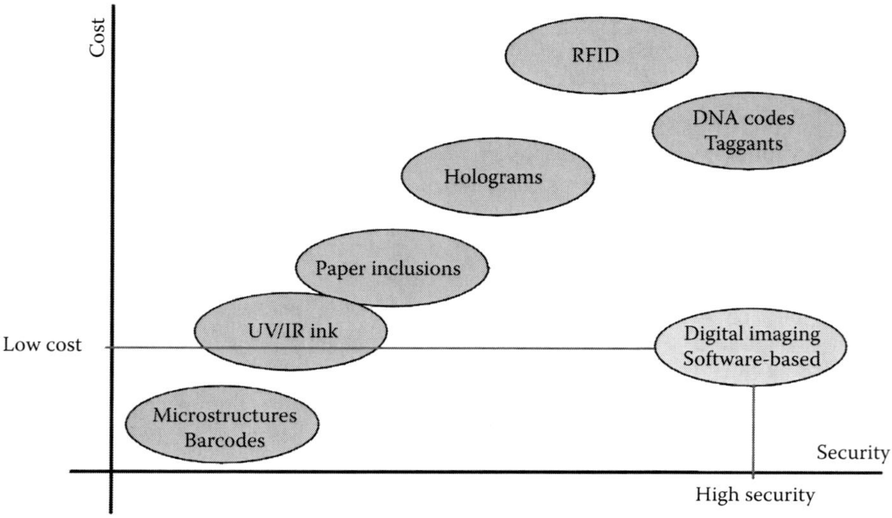

Figure 21.1 Various security features, visible or invisible to the naked eye.

and pressure sensitivity, as well as interaction with other chemicals. Although efficient and effective, their implementation and deployment are costly.

All these techniques based on a security additive can be qualified as *analog or hardware based* because they require additional security elements or special substances.

21.3 HARDWARE- AND SOFTWARE-BASED SECURITY FEATURES

As in other industries, the digital and software revolution opens exciting new possibilities. Digital technologies can now be used to combat counterfeiting of nutraceutical products as well as pharmaceutical ones at low cost while providing a high level of security (Kutter 2006). These digital technologies are breakthroughs compared with former methods. The chemical, micro-, or nanotechnology experts have been replaced by software engineers and digital imaging specialists.

Compared with the cost of dispatching security devices or materials to the various production centers, the deployment cost of software-based security solutions is much lower, especially when Internet technology and web applications are used.

21.4 SENSORY-BASED AND MACHINE-BASED "GENUINE/FAKE" VERIFICATION

On selection of a security feature, it is not sufficient to just evaluate the purchase cost, the robustness against fraudulent replication, the cost of implementation in the production process, the cost of global management, and any impact on the production process. An important part of the problem is how a "genuine/fake" verification is performed (Figure 21.2).

Figure 21.2 Example of genuine/fake verifications based on digital imaging software using standard consumer electronic equipment that can be offline or online.

In this case, the various anticounterfeiting features can be placed in two main categories:

- Features that use human sensory perception
- Features that are machine readable

If human sensory perception is used (e.g., visual, tactile, or oral), adequate training is required for a person to be able to distinguish a genuine security feature from a fake replication when both are to hand. In the case of a machine-readable feature, only a step-by-step process is required, which, if well documented, can be performed by anyone without any specific knowledge or training.

21.5 ONLINE AND OFFLINE MACHINE-READABLE VERIFICATION PROCESSES

For chemical or other ink additive security features, offline security processes are mainly carried out with specific scanners. In this case, achievement of verification programs at multiple sites requires the branded product manufacturer to purchase multiple scanners. The alternative would be that any suspected item be sent to a central location for verification. Such a procedure would be costly and would considerably delay the expected genuine/fake verdict.

Internet and mobile connections are today widely available around the world, including in developing countries. A security feature enabling genuine/fake verifications to be carried out online via a central secured server results in an almost instant verdict. This constitutes a major benefit, eliminating the need for sensitive security elements to be in the hands of an operator and thus avoiding the risk that retroactive engineering be carried out on the equipment with a view to counterfeiting. Another major benefit of an online verification is the consolidation of all the verifications performed worldwide, thus facilitating the detection of any correlation between various fraudulent sources within the supply chain. As for all criminal acts, the quicker you uncover them, the more you are well positioned to identify the criminal source and to stop it.

21.6 PROTECTING PACKAGING WITH INVISIBLE MARKING BUT NORMAL VISIBLE INK

The digital revolution mentioned previously has generated numerous anticounterfeiting solutions based on visible ciphered coding and invisible marking. One patented covert security solution uses normal visible ink to protect various layers of nutraceutical packaging. This solution already protects billions of items worldwide (Cryptoglyph Digital Security n.d.). It uses a pattern made by apparently random microdots invisible to the naked eye printed on the primary or secondary packaging or label (offset, flexography, rotogravure). These dots are invisible to the naked eye

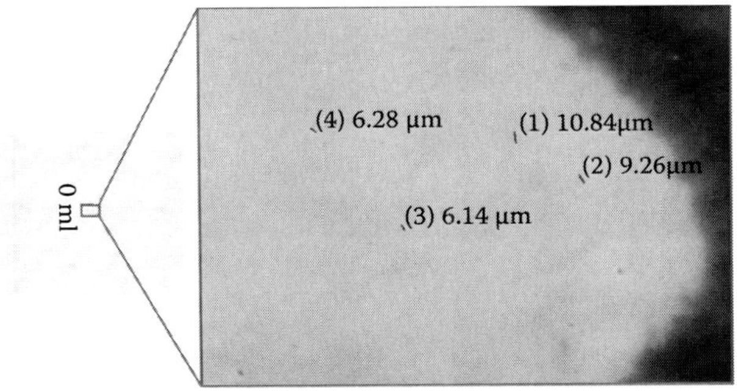

Figure 21.3 Example of microscopic detail of invisible microdots printed on the packaging, thus generating a unique pattern that identifies the product as genuine.

and spread over the whole surface of the packaging or label. The pattern is generated by a 128-bit software key big enough to offer billions of billions of different patterns, each one constituting a unique identity.

The microdots are difficult to distinguish, even with a magnifying glass, because the dots' color and size are chosen to be camouflaged within the imperfections found in all printed material structures (Figure 21.3).

For aluminum-based blister packs, the patented solutions are based on microvariations of the thickness of the lacquer layer; in other words, these microvariations are microholes instead of microdots. The varnish is applied by regular varnish printers (offset, flexography, rotogravure), thus incurring no additional production cost.

Accordingly, the invisible pattern of microdots or microholes can be easily integrated into any current packaging production line. The digital file of the security pattern is simply embedded in the prepress packaging artwork using standard graphic design software. It requires no modification of the packaging design, and it is incorporated as usual before creation of the printing cylinder.

21.7 PROTECTION OF MOLDED NUTRACEUTICAL CONTAINER AND VIAL CLOSURES

A new patented solution recently disclosed makes it possible to identify a molded part at a distance, without additional marking (Fingerprint n.d.), and thus without supplementary manufacturing costs. This element is decisive for mass-produced products, with items numbering in the tens or hundreds of millions.

The patented procedure is based on the surface irregularities naturally present in the die cavities of the injection molds, inherent to the manufacturing procedure of these die cavities. These surface irregularities constitute a unique and nonreproducible signature at a microscopic scale (Figure 21.4); it makes it possible to identify the imprint of a die cavity, just as fingerprints enable the identification of an individual. This unique signature is transmitted to all the molded pieces through this die cavity.

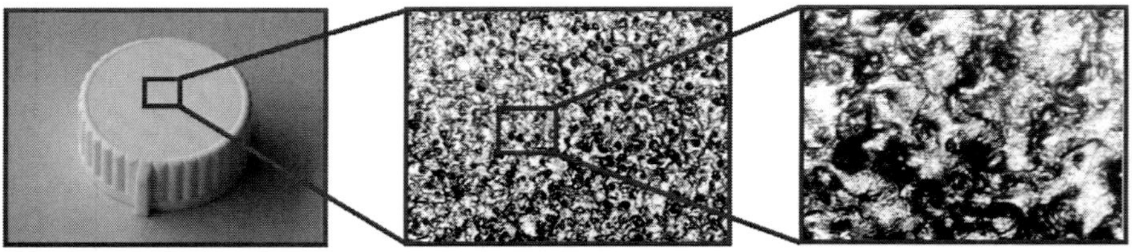

Figure 21.4 Details of a molded closure of a nutraceutical jar showing microscopic irregularities generated by the die cavity used to produce the part.

Reference images identifying the die cavities constituting the injection molds are stored in a secure database, accessible only to authorized users via the Internet. To verify whether a molded part is original or counterfeit, all that needs to be done is to scan or take a photo of the surface of the item and send the image to the secure server. The server returns the verdict in a few seconds: die cavity of the mold identified (original product) or not identified (counterfeit). The verification procedure is automatic and operates 24 hours a day.

21.8 CONCLUSION

Software-based digital security features protect packaging and labeling and jars against counterfeiting. They should be considered separately from serialization and track and trace features. Cost of deployment is much lower for software-based security features compared with hardware- or special substance-based security elements.

Machine-readable security features enable any authorized person to carry out instant online genuine/fake verifications worldwide, with almost no previous training or any specific security knowledge. If online verification is feasible, it allows instant consolidation of all genuine/fake verification results performed anytime or anywhere, thus maximizing the chances of uncovering fraudulent sources and putting a stop to them.

REFERENCES

Cryptoglyph Digital Security Solution. n.d. http://www.alpvision.com/cryptoglyph-covert-marking.html.

Fingerprint. n.d. web page http://www.alpvision.com/solid-parts-authentication.html.

Jordan, F. 2006. Covert and Overt Protection for Valuable Documents. *ISSA (Information System Security Association) Journal* November: 32–34.

Kutter, M. 2006. Protecting Pharmaceutical Products from Counterfeiting Using Digital Imaging Technologies. *Pharmaceutical Industry* 68: 1005–08. http://www.alpvision.com/pdf/2006_08_30_pharmind.pdf.

CHAPTER 22

Automated Manufacturing of Nutraceuticals

Naresh Rajanna

CONTENTS

22.1 OVERVIEW OF NUTRACEUTICAL MANUFACTURING

Nutraceutical manufacturing involves five major processes:

1. Ingredients (and raw materials) sourcing and warehousing
2. Water treatment (for liquid nutraceuticals)
3. Blending
4. Filling (encapsulation or bottling)
5. Packaging

The process begins with *sourcing* all the required ingredients, such as flavors, additives, vitamins, and coloring agents; and raw materials, such as bottles, corrugate material, shrink wrap, caps, and labels; and storing them in temperature-controlled or normal warehouses based on the needs of the raw materials.

The next phase in the process involves the *treatment of water*. Water is treated and cleansed to meet required quality control standards, usually exceeding the quality of the local water supply. This process is critical to achieving high quality products and a consistent taste profile.

During the *blending* stage, various ingredients are added and mixed with treated water. As ingredients are blended, water is piped into large, stainless steel tanks. Also during this phase, artificial, non-nutritive sweeteners such as aspartame or saccharin, liquid sugars like fructose or sucrose, or natural sugars such as stevia can be added. During this production process, food coloring may be added. The encapsulation/bottling/filling room is in a highly protected environment to prevent the product from exposure to possible contaminants. This operation is highly automated with a computerized process control system managed by a minimal number of personnel. The capacity of the blending tanks typically dictates the minimum batch of the product that can be produced with an optimal yield and minimum raw material wastage. The blending tank personnel monitor the equipment for efficiency, adding bulk lids or caps to the capping operation as necessary. Empty bottles and cans are transported automatically to the filling machine via bulk material-handling equipment.

Stringent quality control procedures are followed throughout the production process. To ensure that the finished products meet required quality standards, high-speed digital imaging technology is used to detect bottle defects, and technicians measure many variables, including sugar content, taste, and fluid level. The last stage in the process is packaging. This is also a highly automated process where bottles or cans enter the packaging machinery equipped with wrapped labels, shrink film, or cardboard to form cases, which are then placed into reusable plastic trays or shells. The packaged products then enter a palletizing machine, which automatically stacks them onto pallets. The loaded pallets are moved—typically via forklift—to a warehouse for storage.

Figure 22.1 illustrates a typical nutraceutical drink manufacturing process from raw material procurement to finished good stocking and delivery.

22.1.1 Water Treatment

Impurities such as suspended particles, organic matter, and bacteria may degrade taste and color. Impurities in the water are removed through a process of coagulation, filtration, and chlorination. Coagulation involves mixing gelatinous precipitate into the water to absorb suspended particles. During the clarification process, alkalinity must be adjusted with an addition of lime to reach the desired pH level. The clarified water is then poured through a sand filter and courser beds of gravel to remove particles. Sterilization is necessary to destroy bacteria and organic compounds that might spoil the water's taste or color. To sterilize the water, small amounts of chlorine are added to the water and then filter out. Next an activated carbon filter dechlorinates the water and removes residual organic matter, much like the sand filter. A vacuum pump deaerates the water before it passes into a dosing station (Figure 22.2).

22.1.2 Blending

Blending is a process where the purified water is mixed with sugar, flavors, and concentrates in the dosing station in a predetermined sequence according to

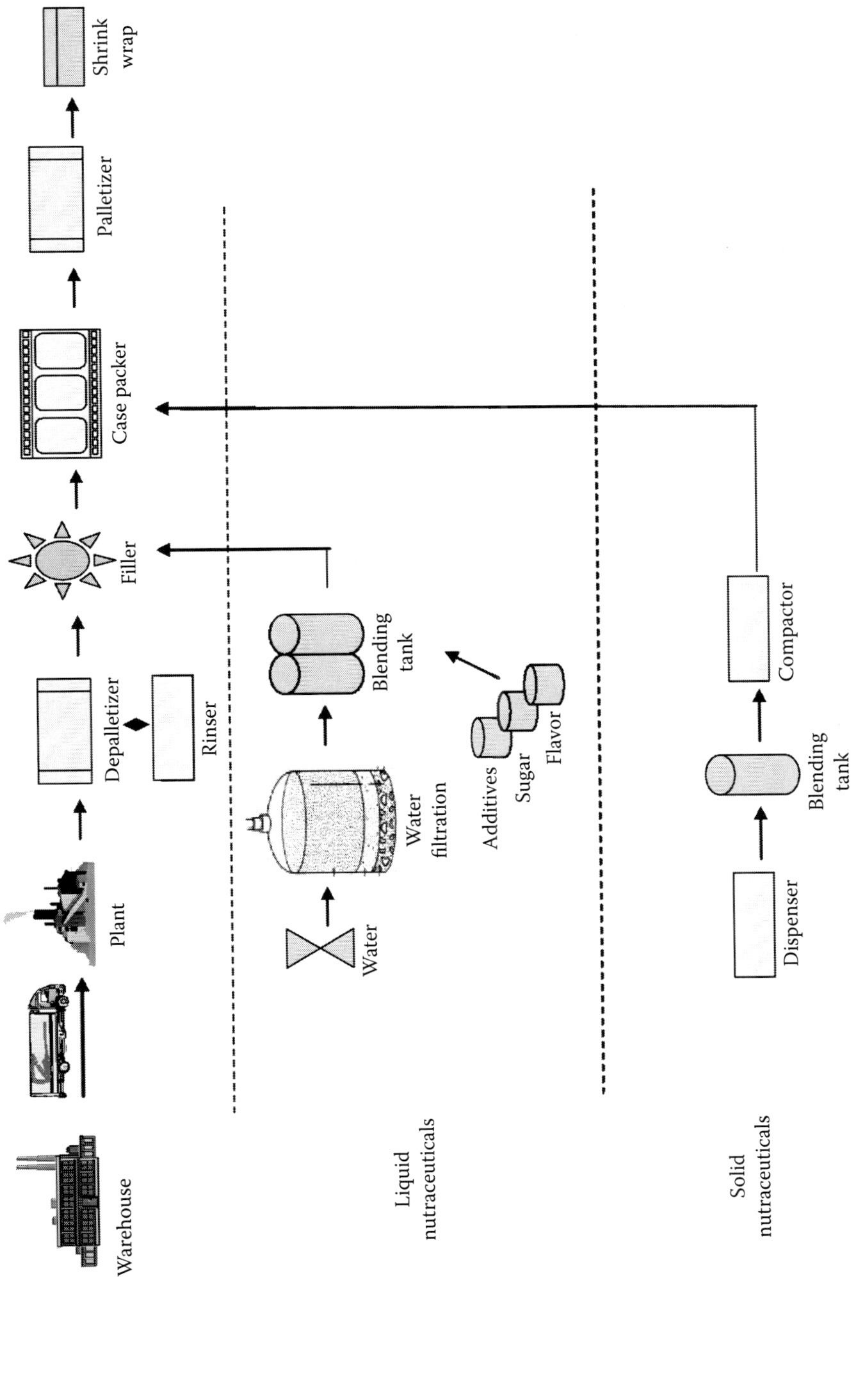

Figure 22.1 This schematic describes the continuous process of liquid and solid nutraceutical manufacturing.

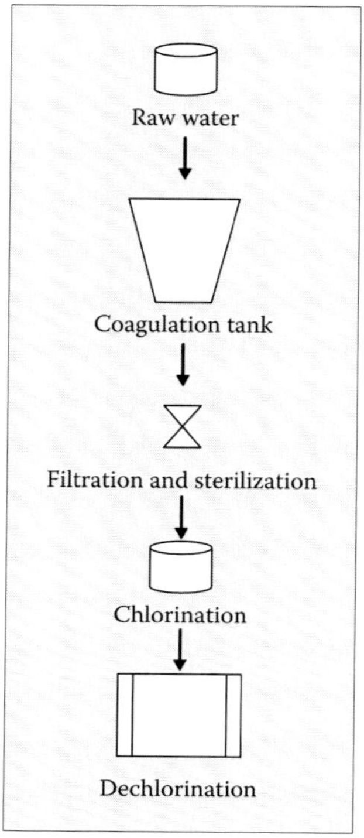

Figure 22.2 This schematic describes a water treatment flow for liquid nutraceutical manufacturing.

their compatibility. The ingredients are conveyed into the blending tanks where they are carefully mixed without causing unwanted aeration. If required, the syrups are pasteurized. The blended solution is then conveyed into the filler through a network of pipelines. Based on the flavors and additives, blending tanks have to be cleaned to eliminate any lingering effects of the previous flavor or color. The clean-in-place processes typically take anywhere from 5 minutes to 2–3 hours to complete.

22.1.3 Filling

22.1.3.1 Bottling: Liquid Nutraceuticals

Before the bottling operation, the glass, can, or polyethylene terephthalate bottles, which are brought at the start of the manufacturing lines over large pallets, are fed through a depalletizer. These automated machines depalletize (Figure 22.3) the large pallets of empty containers so that the containers flow through subsequent manufacturing processes in a single line. During this flow, the containers undergo rigorous automatic and manual inspection for any defects.

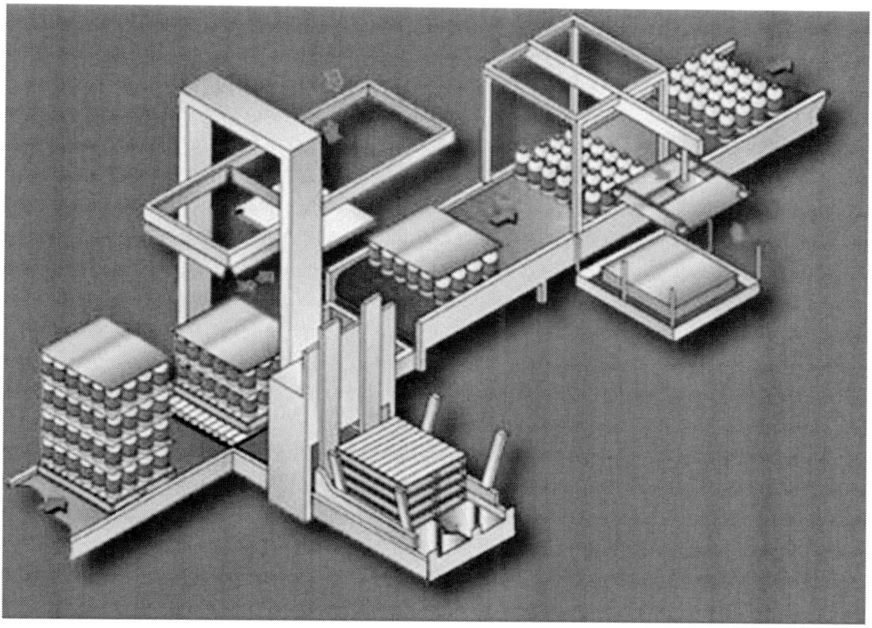

Figure 22.3 An Xodus Bulk Depalletizer machine. (From Priority One Packaging. With permission.)

In the next phase, the right amount of the liquid nutraceutical is poured into the container. Before the filling step, the containers are cleaned one more time. In a high-speed automated beverage-filling machine, the filling of a container is performed by positioning the container into a filling position relative to a filling valve. For high-speed operation, the filing stations are rotated at high speed, and containers are sequentially introduced to the filling machine from a conveyor system for filling and then removed for further processing steps. On positioning the container in a filling station and sealing in association with a filling valve, an amount of a beverage is introduced. The container is then removed from the filling valve, and subsequent capping or other processes are performed.

Bottling facilities differ in the types of bottling lines companies operate and the types of products they can run: cans versus bottles, hot fill versus cold fill, aseptic versus tetra pack, natural versus conventional, etc.

22.1.3.2 Tableting: Solid Nutraceuticals

Tablets offer an efficient means of reducing powders and granules into a compact nutraceutical product. The tablet is versatile, compact, and robust and can be produced in large quantities at high speeds. Tablets are produced using punches and dies in a tablet-pressing machine wherein the ingredients of the product, usually in powder or granule form, are bonded into a solid shape by compaction at high pressure.

Vitamin, mineral, and food supplement tablets tend to be large and bulky. This is to enable sufficient delivery of the beneficial ingredients. This necessitates usage of high compaction forces to bond the ingredients into a robust tablet.

As in the case of the nutraceutical beverage manufacturing process, solid nutraceuticals flow through a similar process. Ingredients in dry powder form are brought through dispensing stations into the blending area. A dry tumble blender is used to ensure that the product is fully blended. Blending within a single dedicated discrete vessel is the only way to ensure that there is no cross-contamination of the product between two batches. Some blenders have a high shear blade to facilitate a high shear element to the tumbling action. The blended powder is fed to the tablet presses for compaction and formulation of tablets.

22.1.4 Packaging

The final step in the liquid nutraceutical manufacturing process is packaging. Based on the type of filling operation, filled containers must be brought to room temperature before labels are affixed to them. The labels provide information about brand, ingredients, shelf life, and safe use of the product. Most labels are made of paper, although some are made from a plastic film. Cans are generally preprinted with product information before filling.

Once the labels are applied to the containers, the containers are dropped into cartons or trays. The cases are then transported via conveyors to the next phase of the process.

The cases are then brought to a palletizer where they are stacked and organized into a larger pallet. The palletizer can be hand operated or a highly complex robotic device. These palletizers can pack any number and size of boxes. Pallets are designed in such a way so that they can be easily lifted by a forklift or crane, which can then be loaded into a truck, rail car, or ship for transit.

FURTHER READING

Arrowhead Systems. n.d. http://www.arrowheadsystems.com.

Concepts for Soft Drink Manufacturing. 2009. *Krones Process Technology Manual*. Neutraubling, Germany: Krones AG.

Palletizer. n.d. http://www.columbiaokura.com.

Shachman, M. 2004. *The Soft Drinks Companion: A Technical Handbook for the Beverage Industry*. Boca Raton, FL: CRC Press, 57–65.

Soft Drink Canning Process. n.d. http://www.solarnavigator.net/solar_cola/soft_drink_canning_process.htm.

Steen, D. P., and P. R. Ashurst. 2006. *Carbonated Soft Drinks—Formulation and Manufacture*. Hoboken, NJ: Wiley-Blackwell, 54–57.

Wildman, R. E. C. 2001. *Handbook of Nutraceuticals and Functional Foods*. Boca Raton, FL: CRC Press, 1–12.

Index

A

B

C

D

E

F

G

H

N

T

X

Y

Z